Y0-BWR-154

CONTEMPORARY MATHEMATICS

Nonlinear Partial Differential Equations

AMERICAN MATHEMATICAL SOCIETY

VOLUME 17

CONTEMPORARY MATHEMATICS

Titles in this Series

VOLUME 1 **Markov random fields and their applications**
Ross Kindermann and J. Laurie Snell

VOLUME 2 **Proceedings of the conference on integration, topology, and geometry in linear spaces**
William H. Graves, Editor

VOLUME 3 **The closed graph and P-closed graph properties in general topology**
T. R. Hamlett and L. L. Herrington

VOLUME 4 **Problems of elastic stability and vibrations**
Vadim Komkov, Editor

VOLUME 5 **Rational constructions of modules for simple Lie algebras**
George B. Seligman

VOLUME 6 **Umbral calculus and Hopf algebras**
Robert Morris, Editor

VOLUME 7 **Complex contour integral representation of cardinal spline functions**
Walter Schempp

VOLUME 8 **Ordered fields and real algebraic geometry**
D. W. Dubois and T. Recio, Editors

VOLUME 9 **Papers in algebra, analysis and statistics**
R. Lidl, Editor

VOLUME 10 **Operator algebras and K-theory**
Ronald G. Douglas and Claude Schochet, Editors

VOLUME 11 **Plane ellipticity and related problems**
Robert P. Gilbert, Editor

VOLUME 12 **Symposium on algebraic topology in honor of José Adem**
Samuel Gitler, Editor

Titles in this series

VOLUME 13 **Algebraists' homage: Papers in ring theory and related topics** Edited by S. A. Amitsur, D. J. Saltman and G. B. Seligman

VOLUME 14 **Lectures on Nielsen fixed point theory**
Boju Jiang

VOLUME 15 **Advanced analytic number theory. Part I: Ramification theoretic methods**
Carlos J. Moreno

VOLUME 16 **Complex representations of GL(2, K) for finite fields K**
Ilya Piatetski-Shapiro

VOLUME 17 **Nonlinear partial differential equations**
Joel A. Smoller, Editor

CONTEMPORARY MATHEMATICS

Volume 17

Nonlinear Partial Differential Equations

Joel A. Smoller, Editor

AMERICAN MATHEMATICAL SOCIETY

Providence • Rhode Island

The proceedings of the Summer Research Conference were prepared by the American Mathematical Society with partial support from the National Science Foundation Grant MCS 7924296.

1980 *Mathematics Subject Classification.* Primary 35-02, 35Q20, 73D05, 76C05.

Library of Congress Cataloging in Publication Data
Main entry under title:
Nonlinear partial differential equations.
(Contemporary mathematics, ISSN 0271-4132; v. 17)
Proceedings of a conference which was held at the University of New Hampshire, June 20–26, 1982, and was sponsored by the American Mathematical Society.
Bibliography: p.
1. Differential equations, Partial—Congresses. 2. Differential equations, Nonlinear—Congresses. I. Smoller, Joel. II. American Mathematical Society. III. Series: Contemporary mathematics (American Mathematical Society); v. 17.
QA377.N667 1983 515.3'53 83-2844
ISBN 0-8218-5017-2

TABLE OF CONTENTS

Preface ix

Papers Submitted

Part I

Theoretical problems and numerical results for nonlinear conservation laws
J. GLIMM 1

Quasilinear hyperbolic partial differential equations and the mathematical theory of shock waves
T. P. LIU 9

Analytical solutions of the Helmholtz equations
C. BARDOS 21

Large time regularity of viscous surface waves
J. T. BEALE 31

Shock waves and the Boltzmann equation
R. CAFLISCH AND B. NICOLAENKO 35

Formation of singularities for nonlinear hyperbolic partial differential equations
K. S. CHENG 45

Conservation laws
R. DiPERNA 57

Recent results for hyperbolic conservation laws
J. GREENBERG 61

On the smoothing of initial discontinuities and the development of singularities in certain quasilinear hyperbolic equations
D. HOFF 67

Non-strictly hyperbolic systems of conservation laws: formation of singularities
B. KEYFITZ AND H. KRANZER 77

On the variable sign penalty approximation of the Navier-Stokes equations
H. KHESHGI AND M. LUSKIN 91

Initial boundary value problems for the equations of motion of compressible viscous fluids
A. MATSUMURA AND T. NISHIDA 109

Convergence of finite difference approximations to nonlinear hyperbolic systems
T. NISHIDA AND J. SMOLLER 117

Linearized stability of extreme shock profiles for systems of conservation laws with viscosity
R. PEGO 125

On some "viscous" perturbations of quasi-linear first order hyperbolic systems arising in biology
M. RASCLE 133

Systems of conservation laws with coinciding shock and rarefaction curves
B. TEMPLE 143

Computational synergetics and mathematical innovation: waves and vortices
N. J. ZABUSKY 153

Part II

Reaction and diffusion in carbon combustion
N. AMUNDSON 165

Bifurcation for colinear solutions to a reaction-diffusion system
D. BARROW AND P. BATES 179

Travelling wave solutions to reaction-diffusion systems modeling combustion
H. BERESTYCKI, B. NICOLAENKO AND B. SCHEURER 189

Stable equilibria with variable diffusion
M. CHIPOT AND J. HALE 209

Diffusion and the predator-prey interaction: steady-state with flux at the boundaries
E. CONWAY 215

Asymptotic flow theory with complex chemistry
P. FIFE AND B. NICOLAENKO 235

Strongly nonlinear detonations
R. GARDNER 257

Differential equations and convergence almost everywhere in strongly monotone semiflows
M. HIRSCH 267

Some ideas in the proof that the Fitz-Hugh-Nagumo pulse is stable
C. JONES 287

Diffusion induced chaos
J. KEENER 293

Monotone and oscillatory equilibrium solutions of a problem arising in population genetics
H. KURLAND 323

Some convection diffusion equations arising in population dynamics
M. MIMURA 343

Finite-dimensional attracting manifolds in reaction-diffusion equations
X. MORA 353

Travelling wave solutions of multi-stable reaction-diffusion equations
D. TERMAN 361

Part III

Creation and breaking of self-duality symmetry - a modern aspect of calculus of variations
M. BERGER 379

Multiple solutions for a problem in buoyancy induced flow
S. HASTINGS AND N. KAZARINOFF 395

Frequency plateaus in a chain of weakly coupled oscillators
N. KOPELL 401

A system of nonlinear pde's arising in the optimal control of stochastic systems with switching costs
S. BELBAS AND S. LENHART 405

L^1 stability of travelling waves with applications to convective porous media flow
S. OSHER AND J. RALSTON 409

A one variable map analysis of bursting in the Belousov-Zhabotinskii reaction
J. RINZEL AND W. TROY 411

Stable and unstable states of nonlinear wave equations
W. STRAUSS 429

Equilibrium states of an elastic conductor in a magnetic field: a paradigm of bifurcation theory
P. WOLFE 443

PREFACE

This is the volume of the proceedings of a conference on Nonlinear Partial Differential Equations, which took place from June 20 - 26, 1982, at the University of New Hampshire in Durham, New Hampshire. The conference was sponsored by the American Mathematical Society, and was funded by the National Science Foundation.

There were 67 participants, of which 16 people gave 1 hour talks, 17 people gave 1/2 hour talks and there were 11 informal 1/2 hour evening talks. In addition there was an informal session on computational and numerical aspects of shock waves.

The theme of the conference was on time-dependent nonlinear partial differential equations; in particular, the majority of the speakers lectured either on shock waves or reaction-diffusion equations and related areas. The first day speakers were asked to give an overview of their field: to describe the main results, and the open problems.

Perhaps the most interesting feature of this conference was the constant interplay between analysis, topology and computational methods.

I would like to thank the members of the organizing committee, consisting of C. Conley, P. Fife, T. P. Liu for their help and constant encouragement.

I am extremely grateful to Carole Kohanski for doing a superb job of making the arrangements, and for being so helpful (and cheerful!) throughout the week.

Joel Smoller

Contemporary Mathematics
Volume 17, 1983

THEORETICAL PROBLEMS AND NUMERICAL RESULTS FOR NONLINEAR CONSERVATION LAWS

James Glimm[1]

Adaptive methods of numerical computation are most important for singular problems, where ordinary methods give slow convergence and poor performance on reasonable grid sizes. The singularities and the adaptation can refer both to the geometrical space x, y, z, t of independent variables and to the state space of dependent variables (pressure, velocity, temperature, . . .). The simplest and most common singularities are jump discontinuities. Examples are contact or material discontinuities, shock waves, chemical reaction fronts, flame fronts and moving phase boundaries such as melting and boiling fronts. In the simplest approximation, the governing equations are often nonlinear conservation laws. These are systems of hyperbolic equations of the form

$$u_t + \Delta F(u) = 0 \tag{1}$$

and express conservation of the components u_i of the vector

$$u = \begin{pmatrix} u_1 \\ \cdots \\ u_n \end{pmatrix} \in R^n$$

(mass, momentum, etc.) Associated with such systems, there is usually a second order diffusion equation, for example

$$u_t + \nabla F(u) = \varepsilon \Delta u. \tag{2}$$

A jump discontinuity is locally one dimensional in the appropriate (normal and tangential) coordinates, and is analyzed through the study of a Riemann problem. A Riemann problem is a Cauchy problem for (1) in one space dimension with data

$$\begin{aligned} u(x,t=0) &= u_{left} = \text{const}, \quad x < 0 \\ u(x,t=0) &= u_{right} = \text{const}, \quad x > 0. \end{aligned} \tag{3}$$

consisting of a single arbitrary jump discontinuity.

[1]Supported in part by the National Science Foundation, PHY80-09179, the ARO, contract DAAG29-79-C-1079 and the DOE, contract DEA-CO2-76ER-03077.

0271-4132/82/0000-1010/$02.50

Because both the equation (1) and the data (2) are invariant under the scale transformations

$$u \to u^{(s)} = u(sx,st),$$

we anticipate that solutions of the Riemann problem will be functions of $\xi = x/t$ alone. However much more is true. It has been known since the fundamental paper of Lax [8] that the solution will consist of coherent waves, generally either shock or rarefaction waves. For an $n \times n$ system, there will generally be n waves, separated by wedges in which u takes on a constant value. Thus in figure 1, we have drawn two waves ($n = 2$), three wedges and one

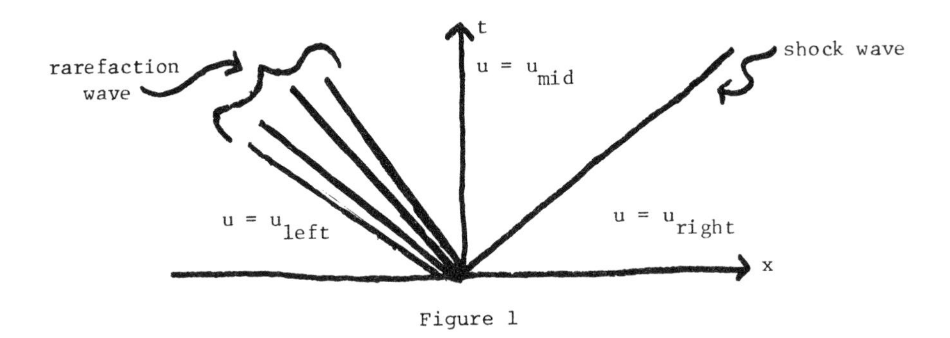

Figure 1

new constant state u_{mid}. The allowed elementary, or coherent waves, that is the shock, rarefaction and contact waves, are defined by solutions of *ordinary* differential equations or algebraic or functional equations in the state space R^n.

For a linear problem, that is $F(u) = Au$, with A an $n \times n$ matrix, the coherent waves result from the expansion of the jump discontinuity

$$u_{right} - u_{left} = \sum \alpha_i e_i$$

as a sum of right eigenvectors of the matrix u. In fact if the eigenvectors e_i are numbered in the order of increasing eigenvalues λ_i, then in the notation of figure 1,

$$u_{mid} - u_{left} = \alpha_1 e_1$$

$$u_{right} - u_{mid} = \alpha_2 e_2$$

Thus we see that the solution of the Riemann problem is equivalent to the expansion into normal modes, and that this expansion continues to have meaning in the nonlinear case. Having fixed the general ideas (see also [3,9,14] for more details), we now turn to specific problems.

<u>Riemann Problems in the large</u>. The picture sketched above is too simple, and we mention some of the complications which may arise. The research problem

is partly to study Riemann problems associated with specific equations (chemical reactors, magnetohydrodynamics, ...) and to determine which further complications arise and it is partly to find properties of the nonlinear flux function F which characterize the phenomena in the solution of the Riemann problem.

Nonconvexity of F occurs naturally in chemistry, elasticity and oil reservoir applications. It gives rise to composite waves, that is waves associated with a single nonlinear mode which are rarefaction waves with embedded shocks. The phenomena is fully understood in the large only for single equations. In fact the meaning of convexity for $F: R^n \to R^n$, $n > 1$ is not clear.

Degenerate wave speeds $\lambda_i = \lambda_{i+1}$ occur naturally (the gas dynamics vacuum is an example). The degeneracy may occur on an open subset of R^n (this is typical of surfactant based tertiary recovery petroleum applications) or on a set of codimension one, two, ... in the state space R^n. In the case of degenerate wave speeds, the waves in the Riemann problem can be discontinuous as a function of the data, (u_{left}, u_{right}), however the solution $u = u(x,t)$, or its time slice $u(x,t = t_0)$ remains continuous in some norm, as a function of the same data. The degeneracy set $\lambda_i = \lambda_{i+1}$ can be a boundary between hyperbolic behavior (all λ_i real) and elliptic behavior (some λ_i occur in complex conjugate pairs). The transition between elastic and plastic behavior is an example of this phenomena.

Existence of an entropy decreasing solution of the Riemann problem in the large should be expected for "reasonable" flux functions F, including all F which occur in equations describing physical phenomena. One expects the solution to be composed of distinct coherent waves. When is uniqueness also expected?

Riemann Problems with source terms. Source terms may be introduced into the right hand side of (1) due to curvature of a wavefront in two dimensions, where x represents the normal direction to the wavefront. Also curvature of a boundary (flow in a duct of variable cross section) or coordinate system gives rise to source terms, when the flow is represented as a one dimensional flow. The source terms in general produce extra waves, as well as a curvature of the wave path. In order to retain scale invariance, we consider a delta function source in (1) concentrated at $x = 0$. (We thank G. Marshall and D. Marchesin for calling this suggestion to our attention.) For the case of scalar equations, or isothermal or polytropic gas, the resulting Riemann problem has been solved [10, 11]. The phenomena includes nonunique solutions and solutions discontinuous in the data. Here the discontinuity of the solution is with respect to any usual norm for $u(x,t)$, as a function of x, t.

Riemann Problems with second order data. See [5] for a discussion of this problem.

Riemann Problems in two and three dimensions. In two and three space dimensions, discontinuities move and bend, remaining locally one dimensional. However they also cross and form cusps, and become intrinsically two or three dimensional in their local behavior. A coherent wave or diffraction pattern is collection of one dimensional waves, meeting at a point $(d = 2)$ or on a line $(d = 3)$. To idealize this problem, we suppose the waves are planar or centered rarefaction waves and the intermediate states are constant. However to be called coherent, such a configuration should be dynamically stable. In general such a configuration is not dynamically stable, but bifurcates into two or more patterns which are stable. (These patterns could include rarefaction waves as well as shocks.) The higher dimensional Riemann problem is to describe the coherent, or dynamically stable diffraction patterns and to describe the nature of the bifurcation process whereby an arbitrary such pattern evolves into several stable diffraction wave patterns. For a single equation, work on this problem has been given by Wagner [17], based on different motivations.

The numerical results I will discuss were obtained in collaboration with O. McBryan and others. The general method has been a tracking method, to follow discontinuity surfaces in two dimensions explicitly, and to propagate them dynamically in time using the wave speeds obtained from solution of Riemann problems. See [3, 4] for a discussion of these methods. We have applied these methods to three areas: gas dynamics, oil reservoirs and the Rayleigh-Taylor problem. The latter is a gravity driven fingering instability of the interface between two fluids of different densities (say air and water). For gas dynamics, we present sample runs of three types. I-Liang Chern, Brad Plohr, and O. McBryan participated in this work.

Circular waves. This is a required test problem because the answer can be obtained by an elementary and accurate one dimensional calculation in polar coordinates. In Figure 2, we show the result of an expanding circular wave, on a 15×15 grid. The pressure ratio between the inside and outside is 5:1. Both the shock and the contact wave are tracked.

Diffraction by a wedge. Good experimental data [1] makes this a good test problem. An incident plane wave collides with a wedge or ramp, producing a reflected wave. See Figure 3.

Results of a 10×10 grid calculation. The results agree with a 40×40 grid calculation.

Vortex rollup (Helmholtz instability). We present results from early and later stages of the calculation in Figure 4 and Figure 5. The grid is 20×20.

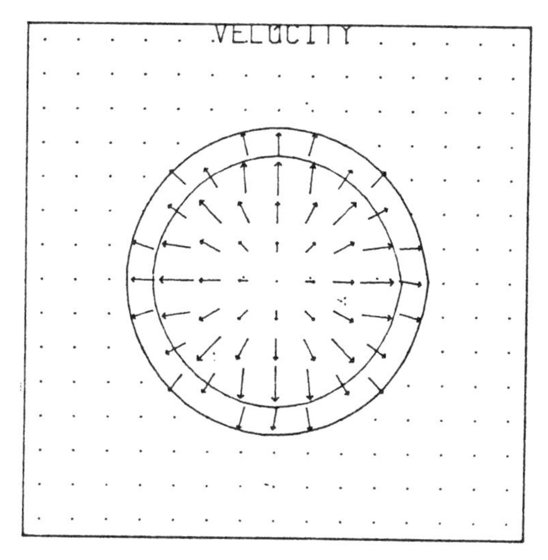

Figure 2

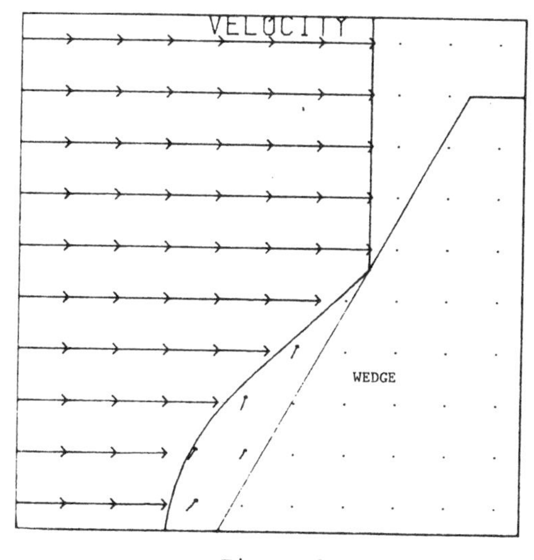

Figure 3

A statistical analysis of fingers in the Taylor-Saffman instability. For miscible displacement, we have studied the competing effects of an expanding circular geometry (i.e. the neighborhood of a single injection well) which gives stability and an adverse mobility ratio $M > 1$, which gives instability. The result of this competition is that the fingers stabilize at a finite size which is proportional to the distance from the source at the center, and thus have a fixed length on a logarithmic scale. This can be seen in Figure 6 where a number of runs with very distinct initial conditions give rise to the same (logarithmic) finger length. This work was in collaboration with E. Isaacson and O. McBryan.

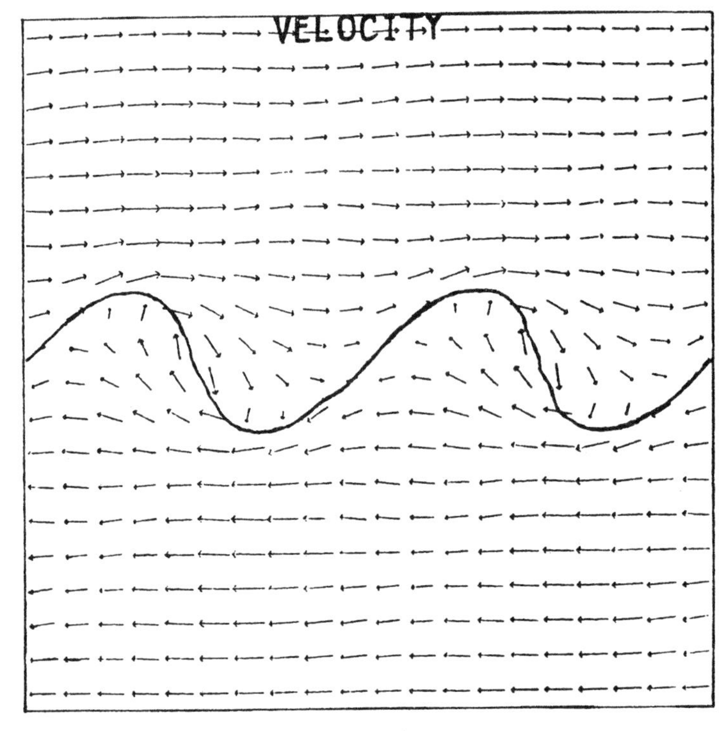

Figure 4

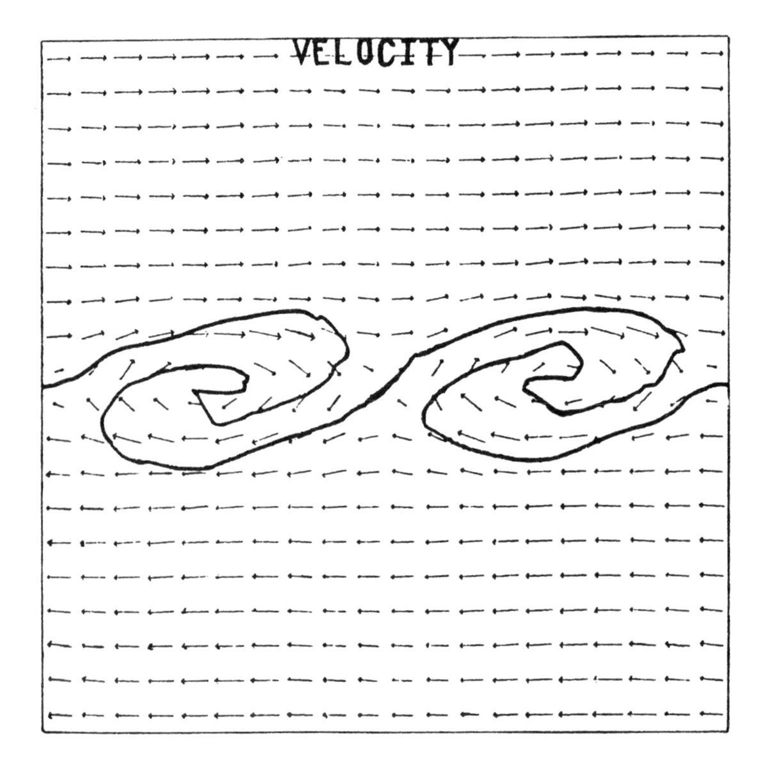

Figure 5

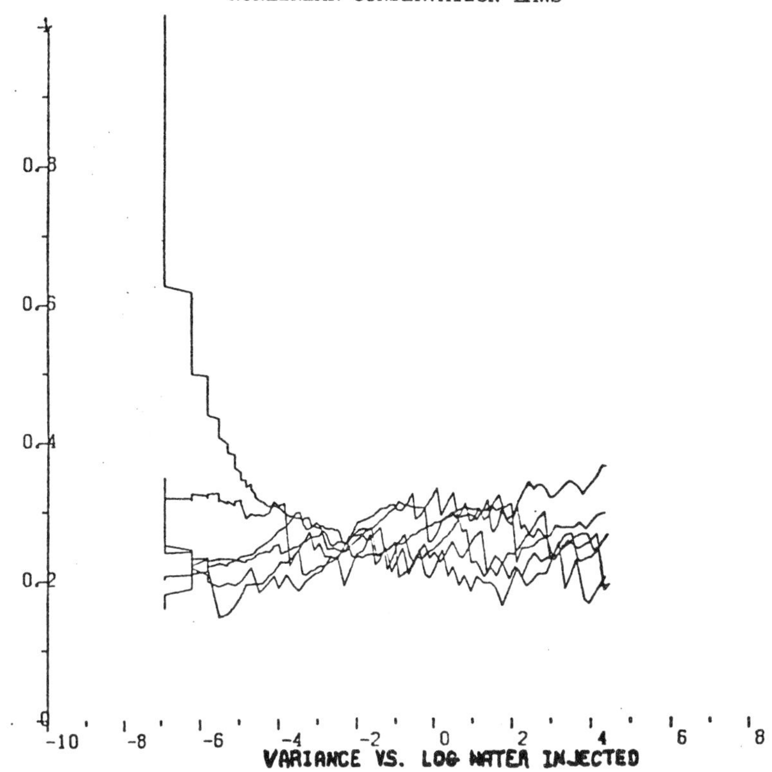

Figure 6

BIBLIOGRAPHY

1. G. Ben-Dor, "Regions and transitions of nonstationary oblique shock-wave diffractions in perfect and imperfect gases". University of Toronto, Institute of Aerospace Studies. Report No. 232.

2. C. Dafermos, "Structure of solutions of the Riemann problem for hyperbolic systems of conservation laws". Arch. Rat. Mech. Anal., 53, 203-217.

3. J. Glimm, "Tracking of interfaces for fluid flow: Accurate methods for piecewise smooth problems. In: Transonic, shock and multi-dimensional flows: Advances in scientific computing. Ed. R. Meyer. Academic Press, New York, 1982.

4. J. Glimm, E. Isaacson, D. Marchesin and O. McBryan, "Front tracking for hyperbolic systems". Adv. Appl. Math., 2, (1981), 91-119.

5. J. Glimm, G. Marshall and B. Plohr, "A generalized Riemann problem for quasi one-dimensional gas flows". Adv. Appl. Math. To appear.

6. E. Isaacson, "Global solution of a Riemann problem for a nonstrictly hyperbolic system of conservation laws arising in enhanced oil recovery". J. Comp. Phys. To appear.

7. B. Keyfitz and H. Kranzer, "A system of nonstrictly hyperbolic conservation laws arising in elasticity theory". Arch. Rat. Mech. Anal., 72, (1980), 219-241.

8. P. Lax, "Hyperbolic systems of conservation laws, II". Comm. Pure Appl. Math., 10, (1957), 537-566.

9. T. P. Liu, Lecture in this volume.

10. D. Marchesin, "Certain scalar conservation laws with source terms". In preparation.

11. D. Marchesin, "The Riemann problem for gas flow in ducts with variable area". In preparation.

12. D. Marchesin and P. Paes-Leme, "Shocks in gas pipelines". SIAM J. Sci. and Stat. Computing.

13. R. Smith, "The Riemann problem in gas dynamics". Trans. AMS, 249 (1979), 1-50.

14. J. Smoller, Shock Waves and Reaction Diffusion Equations, Springer-Verlag, New York, 1983.

15. B. Temple, "Global solution of the Cauchy problem for a class of 2 × 2 non-strictly hyperbolic conservations laws". Advances in Appl. Math. To appear.

16. B. Temple, Systems of conservation laws with coinciding shock and rarefaction curves. Rockefeller University preprint.

17. D. Wagner, "The Riemann problem in two space dimensions for a single conservation law". SIAM J. Math. Anal. To appear.

18. B. Wendroff, "The Riemann problem for materials with a nonconvex equation of state". J. Math. Anal. Appl., 38, 454-466.

19. G. Ben-Dor and I. D. Glass, "Domains and boundaries of non-stationary oblique shock wave reflexions." I. Diatomic gas. II. Monatomic gas. J. Fluid Mech. 92, (1979), 459-496 and 96, (1980), 735-756.

COURANT INSTITUTE
NEW YORK UNIVERSITY
251 MERCER STREET
NEW YORK, NEW YORK 10012

Contemporary Mathematics
Volume 17, 1983

QUASILINEAR HYPERBOLIC PARTIAL DIFFERENTIAL EQUATIONS AND THE MATHEMATICAL THEORY OF SHOCK WAVES

Tai-Ping Liu

ABSTRACT. This paper surveys the present status of the mathematical theory of shock waves. Basic features of the theory are illustrated. One of the main purposes is to raise open problems and possible future directions of research.

1. INTRODUCTION. Starting with the study of the compressible Euler equations, shock wave theory now encompasses the study of general systems of quasilinear hyperbolic equations in conservation form. Substantial progress has been made; however, much more remains to be done. Indeed, there still exist basic, open problems which can be stated simply. This is more than one can say about many other fields of mathematical research. The purpose of this paper is to survey the present state of the theory and raise open problems and possible future directions of research.

Let $u \in \mathbb{R}^n$ be the density vector and $f(u) \in \mathbb{R}^n$ the flux vector, then, in the absence of sinks and sources, we have <u>conservation laws</u>:

$$(*) \qquad \frac{d}{dt}\iint_{\Omega} u\,dx = \int_{\partial\Omega} f(u)\,d\sigma$$

for any domain Ω in the x-space; $\partial\Omega$ its boundary. Here (t,x) are the time-space variables. When $u = u(t,x)$ is <u>smooth</u>, the integral relation is equivalent to

$$(**) \qquad \frac{\partial u}{\partial t} + \nabla_x \cdot f(u) = 0.$$

In one space variable, $x \in \mathbb{R}^1$, it becomes

$$(1.1) \qquad \frac{\partial u}{\partial t} + \frac{\partial f(u)}{\partial x} = 0.$$

With source terms, the system becomes

$$(1.2) \qquad \frac{\partial u}{\partial t} + \frac{\partial f(u)}{\partial x} = g(t,x,u).$$

1980 Mathematics Subject Classification. 76L05, 35L65.

[1]John Simon Guggenheim Fellow, research supported in part by the National Science Foundation.

0271-4132/83/0000-1393/$04.00

An important example is the system of conservation laws for gas dynamics:

$$\frac{\partial\rho}{\partial t} + \frac{\partial(\rho u)}{\partial x} = 0 \qquad \text{(conservation of mass)}$$

$$(1.3) \qquad \frac{\partial(\rho u)}{\partial t} + \frac{\partial(\rho u^2 + p)}{\partial x} = 0 \qquad \text{(conservation of momentum)}$$

$$\frac{\partial(\rho E)}{\partial t} + \frac{\partial(\rho E u + pu)}{\partial x} = 0 \qquad \text{(conservation of energy),}$$

where ρ, u, p, e are, respectively, the density, velocity, pressure, internal energy of the gas, and $E = e + \frac{u^2}{2}$ the total energy. The system (1.3) has been studied since the beginning of the nineteenth century by Poisson, Riemann, Rankine, Hugoniot, Stokes, Taylor, von Neumann, Weyl, Bethe and many others. For the progress before 1950, see the excellent book of Courant and Friedrichs [3]. It becomes clear through these studies that (1.3) has to be supplemented by the second law of thermodynamics and that because of the nonlinearity of the system, singularities tend to develop in the solution which cause analytical and numerical difficulties.

We begin with the illustration of some of the basic features of the theory for the scalar conservation law, $u \in \mathbf{R}^1$. A smooth solution $u(t,x)$ is constant along the characteristic curve:

$$(1.4) \qquad \frac{dx}{dt} = \lambda(u(t,x)), \qquad \lambda(u) \equiv f'(u).$$

As a result, the characteristic speed $\lambda(u)$ is also unchanged along (1.4) and so characteristic curves are straight lines. This makes it very easy to find a local solution when a smooth initial profile $u(0,x)$ is given, Figure 1.

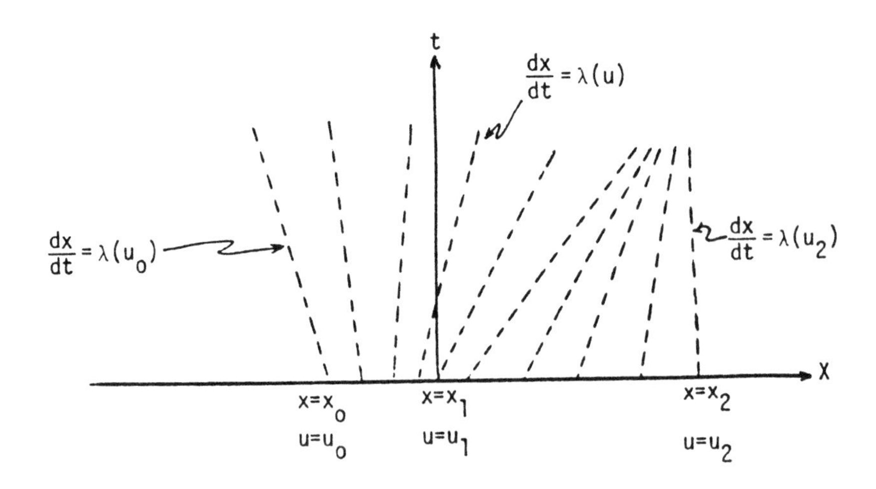

Figure 1

When $\lambda(u)$ depends on u, (1.2) is quasilinear and the characteristic lines have variable speed. In Figure 1, we have depicted a case where the characteristic lines are diverging between those issuing from $x = x_0$ and $x = x_1$ and converging between those issuing from $x = x_1$ and $x = x_2$. The wave connecting u_0 and u_1 is called a rarefaction wave, and u_1 and u_2 are connected by a compression wave, $u_i \equiv u(0,x_i)$, $i = 0,1,2$. It is clear that the focusing of the characteristic lines will cause the smooth solution to cease to exist in finite time. In this case, (1.1) has to be interpreted in terms of the original integral formulation (*), or equivelently, in the sense of distributions: A bounded measurable function $u(t,x)$ is a weak solution of (1.1) in the distributional sense if

$$\iint_{t\geq 0} [u \frac{\partial\varphi}{\partial t} + f(u) \frac{\partial\varphi}{\partial x}]dxdt + \int u(0,x)\varphi(0,x)dx = 0$$

for any smooth test function $\varphi(t,x)$ with compact support in $t \geq 0$. In either of these two equivalent integral formulations, a discontinuity along $x = x(t)$ satisfies the following jump (Rankine-Hugoniot) condition, Figure 2.

$$\text{(R-H)} \qquad (u_+ - u_-)x'(t) = f(u_+) - f(u_-)$$
$$u_\pm \equiv u(t, x(t) \pm 0)$$

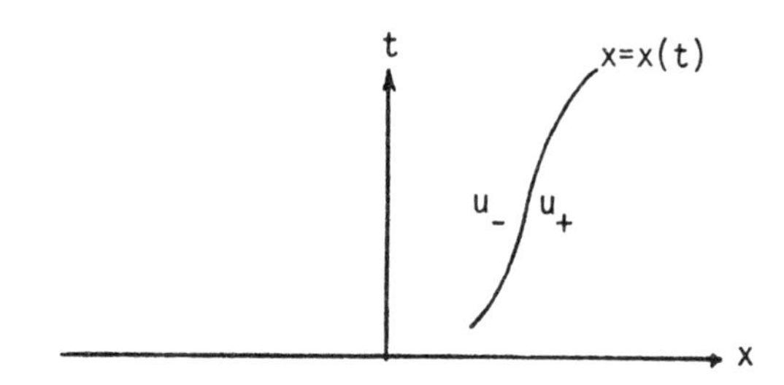

Figure 2

A fascinating feature of shock wave theory is that weak solutions to the initial value problem are not unique, and an admissibility criterion is needed to select the physically relevant solution. The simplest problem which exhibits this property is the Riemann problem, which is an initial value problem for (1.2) with

$$(1.5) \qquad u(x,0) = \begin{cases} u_\ell & \text{for } x < 0 \\ u_r & \text{for } x > 0 \end{cases}$$

for given constant states u_ℓ and u_r. The problem can be solved easily when

(1.2) is strongly nonlinear, that is, $\lambda(u)$ depends strictly on u:

$$\lambda'(u) \equiv f''(u) \neq 0. \tag{1.6}$$

There are two cases: First, assume that $\lambda(u_\ell) \leq \lambda(u_r)$. The characteristic lines $\frac{dx}{dt} = \lambda(u_\ell)$ and $\frac{dx}{dt} = \lambda(u_r)$ diverge and the solution is a centered rarefaction wave, Figure 3:

$$u(x,t) = v(\xi) = \begin{cases} u_\ell \\ \lambda^{-1}(\xi) & \text{for } \lambda(u_\ell) \leq \xi \leq \lambda(u_r), \\ u_r & \text{for } \xi \leq \lambda(u_r), \end{cases} \tag{1.7}$$

$$\xi \equiv x/t.$$

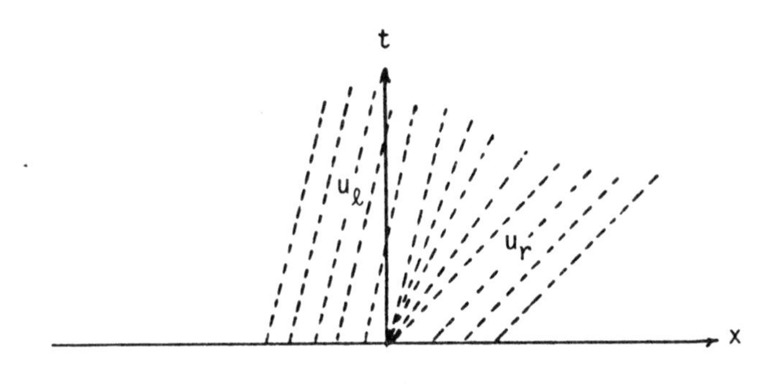

Figure 3

When $\lambda(u_\ell) > \lambda(u_r)$, characteristic lines $\frac{dx}{dt} = \lambda(u_\ell)$ and $\frac{dx}{dt} = \lambda(u_r)$ intersect, smooth solution does not exist and the solution is a <u>shock wave</u>, Figure 4:

$$u(x,t) = \begin{cases} u_\ell & \text{for } x < \sigma t \\ u_r & \text{for } x > \sigma t \end{cases} \tag{1.8}$$

where $\sigma = \sigma(u_\ell, u_r)$ is the shock speed determined by the jump condition:

$$\sigma(u_\ell, u_r) = \frac{f(u_\ell) - f(u_r)}{u_\ell - u_r}. \tag{1.9}$$

Notice that in the first case where $\lambda(u_\ell) \leq \lambda(u_r)$, besides the rarefaction wave just constructed, we may also use (1.8) and (1.9) to find a <u>rarefaction shock wave</u>, Figure 5. In other words, weak solutions may not be unique!

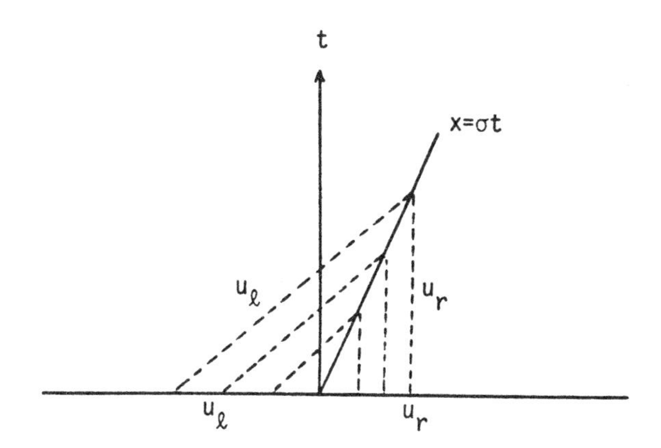

Figure 4

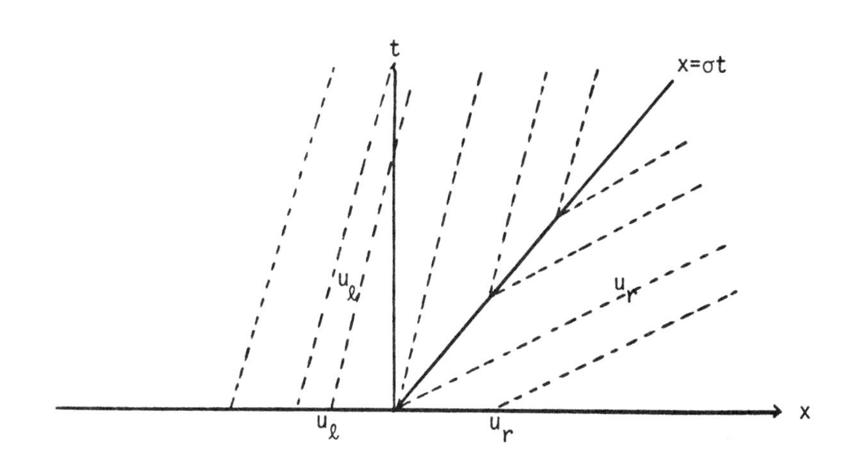

Figure 5

Thus we need to impose an additional <u>admissibility criterion</u>, which rules out the rarefaction shock wave,

$$\text{(E)} \qquad \lambda(u_-) > \sigma(u_-,u_+) > \lambda(u_+)$$

for any discontinuity (u_-,u_+) in the solution. Condition (E) is often called an <u>entropy condition</u> as it is an analogue of the second law of thermodynamics for the gas dynamics equations (1.3). Condition (E) insures the well-posedness of the initial value problem for convex scalar conservation laws,

(1.6). For nonconvex flux functions, a more sophisticated admissibility criterion is needed.

The most interesting feature of shock wave theory is the striking consequences of the nonlinearity of the system and the admissibility criterion on the qualitative behavior of the solutions. Consider a convex scalar conservation law. It is clear from condition (E) that two neighboring shock waves combine to form a single shock wave, Figure 6.

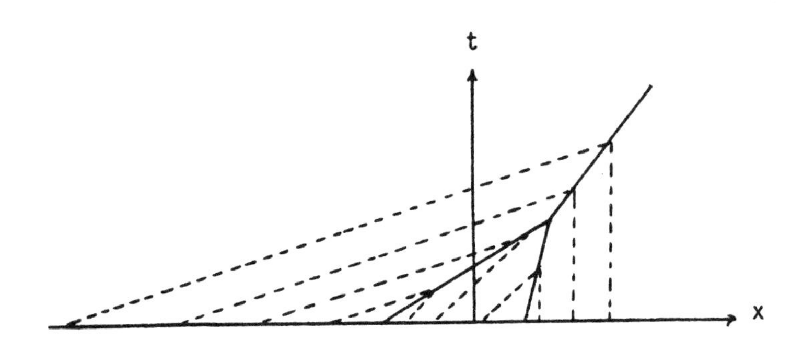

Figure 6

When a shock wave (u_1,u_2) is next to a rarefaction wave (u_2,u_3) they approach and cancel each other. As a result, both waves are weakened and the shock wave bends toward the rarefaction wave, Figure 7. In general, such combining and

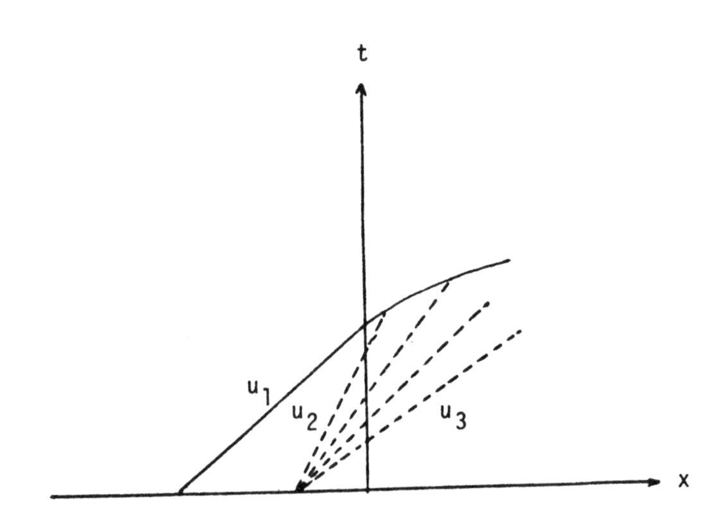

Figure 7

canceling of waves tend to regularize the solution. Indeed, though smooth initial data tend to develope singularities, a bounded measurable initial profile $u(0,x)$ yields a solution $u(t,x)$, $t > 0$, which has bounded local variation in x and becomes almost a piecewise smooth function. The asymptotic state of a general flow is also particularly simple. This is in stark contrast to linear or semilinear systems. The theory for a single conservation law is in good shape due to the efforts of Hopf, Lax, Oleinik and many others, [9], though the propagation of singularities for several space variables is yet to be understood, [31].

In the following sections we briefly describe various aspects of the much more difficult theory for systems. For brevity, we will only list the references which summarize the previous results.

2. RIEMANN PROBLEM FOR SYSTEMS. Suppose that (1.1) is strictly hyperbolic, that is, $\partial f/\partial u$ has real and distinct eigenvalues $\lambda_1(u) < \lambda_2(u) < \cdots < \lambda_n(u)$ with corresponding right eigenvectors $r_i(u)$, $i = 1,2,\ldots,n$. As in the scalar case, certain convexity conditions (c.f. (1.6)) makes wave propagation easier to analyze. For gas dynamics, the convexity condition is that pressure is a convex function of the specific volume for fixed entropy as discovered by Bethe. For general conservation laws, it is the genuinely nonlinear condition of Lax [15]:

$$\nabla\lambda_i \cdot r_i \neq 0.$$

The corresponding entropy condition then becomes

$$\lambda_i(u_+) < \sigma(u_-,u_+) < \lambda_i(u_-)$$

for some $i \in \{1,2,\ldots,n\}$. The Riemann problem is also solved in [15]. For the nonconvex case, the problem is solved by Wendroff [32] for gas dynamics and for general systems by Liu [17]. In general, the Riemann problem is solved only when u_ℓ is close to u_r. For instance, the Rieman problem for the important magneto-hydrodynamics equations has not been settled satisfactorily.

Several physical models are either nonstrictly hyperbolic or of mixed type. It would be interesting to investigate such a system both mathematically and physically.

The study of viscosity equations is important in settling the admissibility criterion, [2]. For equations of mixed type the viscosity approach does not seem to settle the question. Besides the pointwise admissibility criteria just mentioned, [15], [17], there is an integral admissibility criterion of Lax [16]. The condition is useful in checking the admissibility of numerical schemes and obtaining a priori estimates. For the limitation of the criterion and other variations see [4], [14].

3. EXISTENCE. Existence theory for general systems of conservation laws (1.1) was established in a celebrated paper [10] by Glimm. A random choice method was introduced to approximate a general solution using solutions of Riemann problem in Lax [15] as building blocks. Besides establishing existence of solutions with initial data having small total variation, the paper [10] contains several novel, and entirely new, features which are also useful in studying the qualitative behavior of the solutions. A deterministic version of the scheme was obtained in Liu [19] through a wave tracing technique. The technique yields new and more precise information about the solutions.

In the difficult and important paper of Glimm and Lax [11], a decay theory was developed for genuinely nonlinear systems of two conservation laws. It is shown that initial data $u(0,x)$ with small total oscillation yields a global solution $u(t,x)$ which has bounded local variation in x for any $t > 0$.

Global solutions with large initial data have been constructed by Nishida [25] for an ideal model of isentropic gas dynamics. For subsequent generalizations by Nishida-Smoller and others, see [29] and references therein. Global solutions with special data, not necessarily small, are constructed by Chang-Guo, Johnson-Smoller, Greenberg and others, see [28] and references therein.

The existence result of Glimm [10] has been generalized by Liu [22] to general systems not necessarily genuinely nonlinear.

An existence theorem for a system of nonstrictly hyperbolic and nondiagonizable system modeling flows through porous media is established by Temple [30].

Recently, DiPerna [8] has applied the theory of compensated compactness and Lax's construction of entropy—entropy flux, [16], to establish existence theorems for certain systems of two conservation laws.

In general, the existence of global solutions with large initial data to systems of conservation laws (1.1) remains an open problem.

4. UNIQUENESS. The uniqueness of solutions for the initial value problem for hyperbolic conservation laws (1.1) remains an open problem despite much effort. The problem is difficult because solutions in general contain discontinuities and the admissibility criterion has to be incorporated.

It has been shown that centered solutions for the Riemann problem are unique, [15], [17]. Oleinik [27] and Liu [18] use duality and Haar's method, respectively, to show that piecewise smooth solutions are unique. In an interesting paper [6], DiPerna has studied the uniqueness of solutions which are admissible in the sense of Lax [16]. He shows that two solutions with the same given initial data, one of them piecewise smooth and the other of bounded variation, must be the same. Except for [18], the above results apply only for genuinely nonlinear systems of two conservation laws.

5. REGULARITY. There exists a fairly satisfactory regularity theory for hyperbolic conservation laws (1.1). Although solutions are in general not smooth and may contain a dense set of discontinuities, the engineer's belief that solutions are piecewise smooth is more or less accurate.

Useful concepts such as generalized characteristics and conservation of waves were introduced in Glimm and Lax [11]. For systems of two conservation laws which are genuinely nonlinear and satisfy a shock interaction property, they show that initial data having small oscillation are immediately smoothed out to have finite local variation. The theory was subsequently applied by DiPerna [7] to show that the solution is continuous except for a countable Lipschitz curve of jump discontinuities.

For general systems, not necessarily genuinely nonlinear, complicated composite wave patterns may arise. A small perturbation of a discontinuity may give rise to several discontinuities. Nevertheless, a similar regularity result still holds, Liu [22]. This is done by employing a generalized version of the wave tracing technique of [19] and deriving a stability result for wave patterns. [22] also shows that random choice method approximates continuity points locally and discontinuities sharply.

It would be interesting to study the regularity problem for nonstrictly hyperbolic systems for which more intense nonlinear interactions may take place.

6. ASYMPTOTIC BEHAVIOR. As we have seen in the Introduction, nonlinearity may have a regularizing effect. For similar reasons, the mechanism of canceling and combining of waves leads to striking large-time behavior of the solution. Roughly speaking, solutions tend to a wave pattern which is noninteracting.

For general systems, nonlinear coupling of waves tend to produce waves of other families. This can be seen for very elementary interactions, see [13] and references therein. The effects of nonlinear coupling for two conservation laws are weaker than those for a system of more than two equations.

Glimm and Lax [11] obtain the decay result for genuinely nonlinear system of two conservation laws. They show that solutions with compact support decay at the rate $t^{-1/2}$, and periodic solutions decay at the rate t^{-1}. The result is applied by DiPerna [5] to show L_1-convergence to N-waves for solutions with compact support.

When initial data $u(0,x)$ have different limits at $x = \infty$ and $x = -\infty$, it has been shown by Liu [20] for a genuinely nonlinear system and Liu [22] for a general system that the solution tends to the elementary wave pattern of the solution of the Riemann problem (u_ℓ,u_r), $u_\ell \equiv u(0,-\infty)$, $u_r = (0,+\infty)$. The complicated coupling of nonlinear waves for general systems, not necessarily genuinely nonlinear, was analyzed in [22] by an elaborate wave tracing technique.

For nonstrictly hyperbolic systems, more persistent nonlinear interactions may take place which in turn affects the asymptotic behavior of the solution, see [23] for such a study for an elastic model proposed earlier by Keyfitz and Kranzer. It would be interesting to study more general nonstrictly hyperbolic, even nondiagonizable, systems, cf. [30] and references therein.

7. CONSERVATION LAWS WITH SOURCES. Conservation laws with physical or geometrical effects may take the form (1.2). When the source represents damping, the solution tends to smooth out. Solutions with compact support decay and remain smooth if the initial data is small and smooth, [26]. When the damping is uniform, all solutions decay. For physical, nonuniform, damping such as that dealt with in [26], nonlinear waves exist through constant supply of energy. It would be interesting to study the stability of such waves.

Conservation laws with a moving source, $g(x,t,u) = h(x-ct,u)$ in (1.2), exhibit very interesting wave phenomena when nonlinear resonance takes place, i.e., $c - \lambda_i$ may change sign for some i. It has been shown by Liu that waves may change type and become unstable, see [21] and references therein. In particular, it is shown that transonic flow along a converging duct is unstable; the first nonlinear instability result ever exhibited for the equations of gas dynamics.

Many physical situations, such as multiphase flows and flows with internal structures, exhibit rich nonlinear wave phenomena which deserve attention in the near future.

8. CONCLUDING REMARKS. We have surveyed shock wave theory for one space dimension. Of course, it is interesting to study the problem for several space dimensions. Most books on gas dynamics contain special solutions for several space dimensions, the book of Courant-Friedrichs [3] is recommended. Another important area of the theory is numerical calculations of shock waves for which intense investigations are currently under way. One is interested in numerical schemes which are stable, yield sharp discontinuities, calculate smooth flows accurately and are consistent with the entropy condition. Much is being gained by refining and generalizing the work of Godunov [12].

We conclude by remarking that although much that has been done on shock wave theory is sophisticated and by no means easy, there are many interesting and basic problems that remain to be settled even for one-space dimension. One expects that substantial progress will continue to be made in this mathematically interesting and physically important field of nonlinear partial differential equations.

REFERENCES

1. Conley, C. and Smoller, J.,"Topological methods in the theory of shock waves," in Partial Differential Equations, pp. 293-302, Proceedings of Symposia in Pure Mathematics, Vol. 23, Amer. Math. Soc., Providence, R.I., 1973.
2. Conlon, J. and Liu, P.-P., Admissibility criteria for hyperbolic conservation laws, Indiana U. Math. J., 30 (1981), 641-652.
3. Courant, R. and Friedrichs, K. O., Supersonic Flow and Shock Waves, Interscience, N.Y., 1948.
4. Dafermos, C. M., The entropy rate admissibility criterion for solutions of hyperbolic conservation laws, J. Differential Equations, 14 (1973), 202-212.
5. DiPerna, R., Decay and asymptotic behavior of solutions to nonlinear hyperbolic systems of conservation laws, Indiana U. Math. J., 24 (1975), 1047-1071.
6. ______, Uniqueness of solutions to hyperbolic conservation laws, Indiana U. Math J., 28 (1979), 137-188.
7. ______, Singularities of solutions of nonlinear hyperbolic systems of conservation laws, Arch. Rat. Mech. Anal., 60 (1975), 75-100.
8. ______, Convergence of approximate solutions of conservation laws.
9. Douglis, A., Layering methods for nonlinear partial differential equations of first order, Ann. Inst. Fourier (Grenoble) 22 (1972), 141-227.
10. Glimm, J., Solutions in the large for nonlinear hyperbolic systems of conservation laws, Comm. Pure Appl. Math. 18 (1065), 697-715.
11. Glimm, J. and Lax, P. D., Decay of solutions of nonlinear hyperbolic conservation laws, Memoirs Amer. Math. Soc., 101 (1970).
12. Godunov, S. K., Finite difference methods for numerical computation of discontinuous solutions of equations of fluid dynamics, Mat. Sb. 47 (1959), 271-295 (In Russian).
13. Greenberg, J., On the interaction of shocks and simple waves of the same family (II), Arch. Rat. Mech. Anal., 5 (1973), 209-277.
14. Hsiao, L., The entropy rate admissibility criterion in gas dynamics, J. Differential Equations, 38 (1980), 226-238.
15. Lax, P. D., Hyperbolic systems of conservation laws, II, Comm. Pure Appl. Math., 10 (1957), 537-566.
16. ______, "Shock waves and entropy," in Contributions to Nonlinear Functional Analysis, (E.A. Zarantonello, Ed.) pp. 603-634, Academic Press, N.Y., 1971.
17. Liu, T.-P., The Riemann problem for general systems of conservation laws, J. Differential Equations, 18 (1975), 218-234.
18. ______, Uniqueness of weak solutions of the Cauchy problem for general 2×2 conservation laws, J. Differential Equations, 20 (1976), 369-388.
19. ______, The deterministic version of the Glimm scheme, Comm. Math. Phys. 57 (1977), 135-148.
20. ______, Linear and nonlinear large-time behavior of solutions of general systems of hyperbolic conservation laws, Comm. Pure Appl. Math. 30 (1977), 769-796.
21. ______, Resonance for quasilinear hyperbolic equation, Bulletin Amer. Math. Soc., 6 (1982), 463-465.
22. ______, Admissible solutions of hyperbolic conservation laws, Memoirs, Amer. Math. Soc., 240 (1981).

23. Liu, T.-P. and Wang, C., On a nonstrictly hyperbolic system of conservation laws, J. Differential Equations, (to appear).

24. Majda, A., The existence of multi-dimensional shock fronts, Comm. Pure Appl. Math. (to appear).

25. Nishida, T., Global solutions for an initial boundary value problem of a quasilinear hyperbolic system, Proc. Japan, Acad., 44 (1968), 642-646.

26. Unpublished manuscript.

27. Oleinik, O. A., On the uniqueness of the generalized solution of the Cauchy problem for a nonlinear system of equations occurring in mechanics, Uspehi Mat. Nauk. 73 (1975), 169-176 (in Russian).

28. Smoller, J. A. and Johnson, J., Global solutions for an extended class of hyperbolic systems of conservation laws, Arch. Rat. Mech. Anal., 32 (1969), 169-189.

29. ______, Solutions in the large for some nonlinear hyperbolic conservation laws of gas dynamics, J. Diff. Equs., 41 (1981), 96-161.

30. ______, Global existence for a class of 2x2 nonlinear conservation laws with arbitrary Cauchy data, Adv. Appl. Math., to appear.

31. ______, The Riemann problem in two space dimensions for a simple conservation law (Michigan Ph.D. thesis, 1980).

32. Wenfroff, B., The Riemann problem for materials with nonconvex equations of state II: general flow, J. Math. Anal. 30 (1972), 640-658.

DEPARTMENT OF MATHEMATICS
UNIVERSITY OF MARYLAND
COLLEGE PARK, MARYLAND 20742

Contemporary Mathematics
Volume 17, 1983

ANALYTICAL SOLUTIONS OF THE KELVIN HELMHOLTZ EQUATIONS

Claude Bardos

ABSTRACT. This paper is devoted to the study of the Kelvin Holmholtz problem for an inviscid incompressible fluid in two space variables.
First we prove the existence of a weak solution for all time, then we discuss the instability of the separation curve and we prove the existence of a smooth solution for a small time, provided we assume that the initial dataare analytic.
The proofs rely on the conservation of the vorticity and on some abstract version of the Cauchy Kowalewsky theorem.

INTRODUCTION. In this talk we study the behaviour of an inviscid imcompressible fluid with singular initial data :

The initial datais singular in the following sense :

For $t = 0$ the vorticity is located on a smooth curve in two dimension and on a smooth surface in three dimension. This phenomena appeart for instance when two layers of the same fluids have exhibited for $t = 0$ a discontinuity of the velocity and an equation of the same type would be obtained for the free surface problem if the gravity is neglected. In section 1 we prove the existence of a weak solution in the large in two dimension. This uses mainley the conservation of the L^1 norm of the vorticity. In section 2 we derive from the Euler equation in conservative form the equation of the curve and we show in section 3 that the corresponding problem is unstable at least in the class of C^∞ functions, but that it is well posed in the class of analytic functions. Section 4 contains (following Baker, Meiron and Orszaz [5]) a discussion of the eventual appearance of singularities. Finally section 5 contains severals remarks concerning the extension of the result to the three dimensional case and the comparaison with some related problems.

I. WEAK SOLUTION OF THE EULER EQUATION IN TWO DIMENSION.

We consider the classical Euler equation of fluid mecanic in $\mathbb{R}^2$.

$$(1)\quad \frac{\partial u}{\partial t} + u.\nabla u = -\nabla p, \quad \nabla u = 0 \ , \ u(x,0) = u_0(x) \ .$$

For any smooth solution of (1) the following a priori estimates are available

0271-4132/83/0000-1376/$03.50

$$(2) \qquad \int_{\mathbb{R}^2} |u(x,t)|^2 dx = \int_{\mathbb{R}^2} |u(x,0)|^2\, dx$$

$$(3) \qquad \int_{\mathbb{R}^2} |(\nabla \wedge u)(x,t)|^p\, dx = \int_{\mathbb{R}^2} |(\nabla \wedge u)(x,0)|^p\, dx$$

$$(4) \qquad \sup_{x \in \mathbb{R}^2} |(\nabla \wedge u)(x,t)| = \sup_{x \in \mathbb{R}^2} |(\nabla \wedge u)(x,0)|$$

The estimation (2) is called the energy estimate and it turns out to be valid in any dimension. While the estimates (2) and (3) are only valid in two dimension. They are related to the following fact.

In two dimension the curl $\frac{\partial u_1}{\partial x_2} - \frac{\partial u_2}{\partial x_1} = \nabla \wedge u$ is a scalar which can be viewed as a vector orthogonal to the plan x_1, x_2 and therefore it is conserved along the characteristic curves :

$$\dot{x}(t) = u(x(t),t) \ .$$

Let u^ε be any convenient approximation of the solution then one can use the estimates (2) and (3) to show that the limit u satisfies the relation :

$$\lim_{\varepsilon \to 0} \frac{\partial}{\partial x_i} (u_i^\varepsilon u^\varepsilon) = \frac{\partial}{\partial x_i} (u_i\, u)$$

This leads to the following.

THEOREM 1. (Youdovitch [12], Bardos [1]). For any initial data u_0 belonging to the space $(L^2(\mathbb{R}^2))^2$ which satisfies the relation $\nabla \wedge u_0 \in L^p(\mathbb{R}^2)$ $(1 < p \leq \infty)$ there exist a weak solution $u \in L^\infty(\mathbb{R}_t ; (L^2(\mathbb{R}^2))^2)$ which satisfies the estimate : $\nabla \wedge u \in L^\infty(\mathbb{R}_t ; L^p(\mathbb{R}^2))$.

REMARK 1. For $p = \infty$ it is possible to obtain the uniqueness of the solution.

REMARK 2. For $p = 1$. The theorem 1 remains valid with some modifications : This is due to the non reflexivity of the space L^1. One has to introduce the closure of the space of functions which satisfy the estimates :

$$(5) \quad u \in L^2(\mathbb{R}^2) \ , \quad \int |(\nabla \wedge u)(x)| dx < +\infty \ , \ \nabla \cdot u = 0$$

This space is very similar to the space of functions with bounded variation. It has been studied by Strang and Teman [9], and it is called the space of functions of bounded deformation. Therefore with the estimate :

$$u_0 \in (L^2(\mathbb{R}^2))^2, \quad \nabla \wedge u_0 \in L^1(\mathbb{R}^2)$$

it is possible to construct a solution of the problem (1) which belongs to the space of functions of bounded variation.

Following this lines one can prove the :

THEOREM 2. Assume that the initial data u_0 belong to $L^2(\mathbb{R}^2)$ is divergence free with a vorticity located on a smooth curve Γ, satisfying the estimates

$$(6) \qquad \nabla \wedge u_0 = \omega_0(\sigma) \times \delta_{\Gamma_\sigma} \, , \quad \int_\Gamma |\omega_0(\sigma)| d\sigma < \infty \, ,$$

then for the problem (1) there exist at least on weak solution which belongs to space of function with bounded deformation.

The proof of the Theorem 2 can be done with the use of positive mollifier function $\rho_\varepsilon \in \mathcal{D}(\mathbb{R}^2)$ of total mass 1 :

$$\int \rho_\varepsilon(x)dx = 1$$

Then one solve (1) with the initial data $u_0^\varepsilon = \rho_\varepsilon * u_0$ and uses (2) and (3) (with $p = 1$) to pass to the limit for ε going to zero. Indeed one has :

$$(7) \qquad \int_{\mathbb{R}^2} |\nabla \wedge u_\varepsilon(x,t)| dx = \int_{\mathbb{R}^2} |\nabla \wedge (\rho_\varepsilon * u_0)| dx$$

$$= \int_{\mathbb{R}^2} \left| \int_\Gamma \rho_\varepsilon(x-\sigma) \, \omega(\sigma) d\sigma \right| dx \leq \int_\Gamma |\omega(\sigma)| d\sigma$$

And this is enough to pass to the limit in the non linear term.

II. THE KELVIN HELMHOLTZ EQUATIONS ON ACTIVE. The section is devoted to the behaviour of the curve Γ which separated two layers of the fluid.

First we derive an equation for this curve and for the corresponding density of vorticity.

PROPOSITION 1. [8]. Let $y = y(x,t)$ a family of curves of the plane $(x,y) \in \mathbb{R}^2$ and let u be a weak solution of the Euler equation. We assume that $\nabla \; u(\cdot,t)$ is zero outside Γ, and that u admits two limits u^+ and u^- above and below Γ.

We use the following notations :

$$v = (u^+ + u^-)/2 \, , \qquad w = u_\tau^+ - u_\tau^- \, , \qquad \Omega \quad \omega \sqrt{1 + y_x'^2}$$

Then for Γ, v and Ω we have the equations :

$$(8) \quad v_1 = \frac{-1}{2\pi} \int \frac{y(x,t) - y(x',t)}{(x-x')^2 + (y(x,t)-y(x',t))^2} \, \Omega(x',t) dx' \, .$$

$$(9) \quad v_2 = \frac{1}{2\pi} \int \frac{x - x'}{(x-x')^2 + (y(x,t) - y(x',t))^2} \, \Omega(x',t) dx' \, .$$

$$(10) \qquad \frac{\partial \Omega}{\partial t} + \frac{\partial}{\partial x} (\Omega\, v_1) = 0$$

$$(11) \qquad \frac{\partial y}{\partial t} + v_1 \frac{\partial y}{\partial x} = v_2 .$$

Proof. We denote by τ and η the unitary tangent and normal vector to Γ.

From the relation $\nabla . u = 0$ (in distribution sense) we deduce the formulas :

$$(12) \qquad u_h^+ = u_h^- , \quad \nabla \wedge u = (u_\tau^+ - u_\tau^-) \times \delta_\Gamma = \omega \times \delta_\Gamma$$

and the relation :

$$(13) \qquad \nabla . u = 0 , \quad \nabla \wedge u = \omega$$

gives the equations (8) and (9).

Since u is a weak solution, for any $\phi \in \mathcal{D}(\mathbb{R}^2)$ on has :

$$(14) \qquad \langle u, \frac{\partial}{\partial t} (\nabla \wedge \phi)\rangle + \sum_{i=1}^{2} \langle u_i u, \frac{\partial}{\partial x_i} (\nabla \wedge \phi)\rangle = 0 .$$

We denote by n the normal in the x,y plane to the curve $\Gamma = \{(x,y(x,t))\}$ and with several integrations by parts, we obtain (using the fact that the vorticity is zero outside Γ and the divergence zero every-where) the formula :

$$(15) \qquad \int_{-\infty}^{\infty} dt \int_\Gamma ([u] \wedge n) \frac{\partial \phi}{\partial t}\, d\sigma$$
$$+ \int_{\infty}^{\infty} dt \int_\Gamma ([u_i\; u] \wedge n) \frac{\partial \phi}{\partial x_i} - \sum_{i=1}^{n} n_i\, (\nabla \wedge (u_i\; u)\phi) d\sigma = 0.$$

Now we introduce the function $\psi(x,t) = \phi(x,y(x,t),t)$ and from (15) we deduce an identity in the sense of distribution in the (x,t) plane. With the formulas :

$$\frac{\partial \psi}{\partial x} = \frac{\partial \phi}{\partial x} + y'_x \frac{\partial \phi}{\partial y} , \qquad \frac{\partial \psi}{\partial t} = \frac{\partial \psi}{\partial t} + y'_t \frac{\partial \phi}{\partial y}$$

this leads to the relation :

$$(16) \qquad \int_{-\infty}^{\infty} dt \int_{-\infty}^{\infty} (\frac{\partial \Omega}{\partial t} + \frac{\partial}{\partial x} (\Omega\, v_1))\; \phi(x,y(x,t))dx$$
$$+ \int_{\infty}^{\infty} dt \int_{-\infty}^{\infty} \{\frac{\partial y}{\partial t} + y'_x\, v_1 - v_2\} \frac{\partial \phi}{\partial y} (x,y(x,t))dx = 0$$

Which is equivalent to the two equations (10) and (11).

REMARK 3. The equations (10) and (11) have been derived directly (for smooth solutions) by fluid-mechanic arguments (cf. Birkhoff [2] and the corresponding references).

(11) mean that in the space $\mathbb{R}^2_{(x,y)} \times \mathbb{R}_t$ the vector field $(v_1,v_2,1)$ is tangent to the surface $\Sigma = \{(x,y,t) | y=y(x,t)\}$. Therefore on this surface one can introduce the lagrangian coordinates defined by :

$$\dot{m}(\alpha,t) = v(m(\alpha,t),t), \quad m_1(\alpha,0) = \quad , \quad m_2(\alpha,0) = y(\alpha,0). \tag{17}$$

From (10) we deduce with the change of variable $x = m_1(\alpha,t)$ the relation :

$$\begin{aligned} &\frac{d}{dt}(\Omega(m_1(\alpha,t),t)\frac{\partial m_1}{\partial \alpha}(\alpha,t) = \\ &\frac{\partial m_1}{\partial \alpha}\left(\frac{\partial \Omega}{\partial t}+\frac{\partial \Omega}{\partial x}\frac{\partial m_1}{\partial t}\right) + \Omega\frac{\partial}{\partial t}\frac{\partial m_1}{\partial \alpha} = \\ &\frac{\partial m_1}{\partial \alpha}\left(\frac{\partial \Omega}{\partial t} + v_1\frac{\partial \Omega}{\partial x}\right) + \Omega\frac{\partial v}{\partial x}\frac{\partial m_1}{\partial \alpha} = 0 \end{aligned} \tag{18}$$

which implies that the quantity $\Omega(\alpha,t)\frac{\partial m_1}{\partial t}(\alpha,t)$ is independant of t .

Therefore a change of variable in the equations (8) and (9) will transform the problem in the following :

$$v_1 = -\frac{1}{2\pi}\int \frac{y(\alpha,t) - y(\alpha',t)}{(\alpha-\alpha')^2+(y(\alpha,t)-y(\alpha',t))^2}\,\Omega(\alpha',0)d\alpha' \tag{19}$$

$$v_2 = \frac{1}{2\pi}\int \frac{x(\alpha,t) - x(\alpha',t)}{(\alpha-\alpha')^2 + (y(\alpha,t) - y(\alpha',t))^2}\,\Omega(\alpha',0)d\alpha' \tag{20}$$

With the introduction of the complex variable $Z = m_1 + im_2$ the systems (19), (20) turns out to be the equivalent system

$$\frac{\partial \overline{Z}}{\partial t} = \frac{1}{2i\pi}\int \frac{\Omega(\alpha,0)\, d\alpha'}{Z(\alpha,t) - Z(\alpha',t)} \tag{21}$$

III. THE INSTABILITY OF THE LINEARISED SOLUTION AND THE SOLVABILITY OF THE PROBLEM IN THE CLASS OF ANALYTIC FUNCTIONS. It will be convenient to denote by U the set of variable (Ω,y), since v is given in term of Ω and y by a time independant equation, the problem (8), (11) can written formally

$$\frac{dU}{dt} = F(U) , \quad u(x,0) = U_0(x) . \tag{22}$$

The stationnary solution are the solution of the equation $F(U_0) = 0$. Now if the problem is well posed in the class of C^∞ function, for any small variation δU near U_0, the linearised equation

$$\frac{d\delta U}{dt} = \nabla F(U_0) \cdot \nabla\delta U \tag{23}$$

should be a well posed problem in the salue class.

A simple example is given by the interface beetween two fluid of opposite direction with :

$$U_0(x,0) = (\Omega(x,0),\ y(x,0)) = (2A,\ 0)\ .$$

For this example the linearised operator can be easily computed in the Fourier space. This gives :

$$\frac{\partial}{\partial t}\,\delta\hat{\Omega}(\zeta) + 2Ai\ \xi\ \delta\hat{v}_1(\xi) = 0$$

$$\frac{\partial}{\partial L}\,\delta\hat{y}(\xi) - \delta\hat{v}_2(\xi) = 0$$

$$\delta\hat{v}_1 = \frac{A}{\pi}\ i\xi\ \ \delta\,\hat{y}(\xi)\ \mathrm{sgn}\ \xi$$

$$\delta\hat{v}_2 = \frac{1}{2\pi}\ \mathrm{sgn}\ \xi\ \delta\ \Omega(\xi)$$

or equivalently .

$$(24)\qquad \frac{\partial}{\partial t}\begin{pmatrix}\delta\ \hat{\Omega}\\ \delta\ \hat{y}\end{pmatrix} - \frac{1}{\pi}\ \mathrm{sgn}\ \xi\begin{pmatrix}0, & 4A^2\ \xi^2\\ 1\ , & 0\end{pmatrix}\begin{pmatrix}\delta\ \Omega\\ \delta\ y\end{pmatrix} = 0$$

Now the matrix $M = \frac{1}{\pi}\ \mathrm{sign}\ \xi\begin{pmatrix}0, & 4A^2\ \xi^2\\ 1\ , & 0\end{pmatrix}$ for eigen values $\pm\ 2A\,\frac{\xi}{\pi}$ and has the corresponding evolution operator is expt M Which will not be a multiplicative operator in the space $\mathfrak{F}(\mathfrak{I})$. However it definies for t small enough a multiplicative operator in the space of function U such that $\hat{U}(\xi)$ decayes exponentially for $|\xi| \to \infty$. This space coïncide with the space of function $U(x,0)$ which have an analytic extension in a strip around the real axis.

Therefore we introduce a scale of spaces H^s. H^s will denote the space of real functions $f(x)$ which can be extended analitically in a strip $B_s = \{(x + i\eta)|\ |\eta| < s\}$ with the norm :

$$(25)\qquad \|f\|_s = \sup_{z\varepsilon B_s}\{|f(z)| + \sup_{z\varepsilon B_s, z\varepsilon B_{s'}}\frac{|f(z) - f(z')|}{|z - z'|^{\alpha}}\}$$

For this extension the function

$$(26)\qquad v(z,z') = (z-z')^2 + (y(z) - y(z'))^2$$

Which appears in the right hand side of (19), (20) may vanish for $z \neq z'$: we have

(27) $$|v(z,z)| > \mathrm{Re}\, v(z,z') > (x-x')^2 \left(1 - \|\tfrac{\partial y}{\partial z}\|_s\right) - (\eta-\eta')^2$$

Therefore it is necessary to have a uniform estimate for the derivative $\frac{\partial y}{\partial z}$. Therefore instead of an evolution equation for (Ω, y) we will derive an evolution equation for Ω, y, y_x and y_{xx}.

$$\frac{\partial\Omega}{\partial t} + \frac{\partial}{\partial x}(v_1\,\Omega) = 0$$

$$\frac{\partial y}{\partial t} + y_x\, v_1 - v_2 = 0$$

$$\frac{\partial}{\partial t} y_n + \frac{\partial}{\partial x}(y_x\, v_1 - v_2) = 0$$

(28) $$\frac{\partial}{\partial t} y_{xx} + \frac{\partial}{\partial x}\left(y_{xx}\, v_1 + y_x \frac{\partial v_1}{\partial x} - \frac{\partial v_2}{\partial x}\right) = 0$$

$$v_1 = -\frac{1}{2\pi}\int \frac{y(x,t) - y(x',t)}{(x-x')^2 + (y(x,t) - y(x',t))^2}\,\Omega(x',t)dx'$$

$$v_2 = \frac{1}{2\pi}\int \frac{x - x'}{(x-x')^2 + (y(x,t) - y(x',t))^2}\,\Omega(x',t)dx'$$

Then one can treat the system (28) as a non linear evolution equation for the unkown $\Phi = (\Omega, y, y_x, y_{xx})$, of the form $\frac{d\Phi}{dt} = G(\Phi)$.

For the operator G it is possible to use the result concerning the abstract Cauchy Kowalewsky theorem that were derived by Obciannikov [8], Nirenberg [6], Nishida [7] and Baouendi - Goulaouic [3]. The results of Baouendi and Goulaouic [3] turn out to be the most effecient for this blem giving the :

THEOREM 3. Assume that the initial data y and Ω belong to the space H^s and that in this space the norm of $\frac{dy}{dz}$ is less than 1/2 then there exists a strictly positive number T such that for $|t| < T$ the problem (28) has, for the corresponding initial data an analytic solution.

IV. NUMERICAL RESULTS CONCERNING THE SHAPE OF THE SINGULARITY. One of the drawback of the method described in the previous section is the necessity of having a resolution of the curve $\Gamma = \{(x, y(x,t))\}$ in term of x ; this may prevent the description of rolls. However this may be consistent if the first singularity appearing is not a roll.

This is "numerically" proven by Meiran Baker and Orzag [5]. Instead of the problem on the line they have considered the Kelvin Helmoltz equation in a periodic domain. In this case the equation (21) becomes the equation

(29) $$\frac{\partial \bar{z}}{\partial t}(x,t) = \frac{1}{4\pi i}\int \mathrm{cotg}\left(\frac{z(x,t)-z(x',t)}{z}\right)\,\Omega(x',t)dx'$$

with the initial data.

(30) $$z(x,0) = x\ ,\quad \Omega(x,0) = 1 + a\cos x\ .$$

The function $z(x,t)$ has a Taylor serie expansion

(31) $$z(x,t) = x + \sum_{n=1}^{\infty} t^n z_n(x)$$

In fact $z_n(x)$ can be expanded as a Fourier serie, and it is shown that this serie has only a finite number of modes

(32) $$z_n(x) = \sum_{m=0}^{n} z_{nm} \sin m x\ .$$

The z_{nm} modes can then be computed recursively.
Similarly the expression :

(33) $$\Omega^{(p)}(t) = \int_0^{2\pi} \left|\frac{\partial^p z}{\partial x^p}\right|^2 dx$$

can be expanded as a Taylor serie

(34) $$\Omega^{(p)}(t) = \sum_{n=0}^{\infty} \Omega_n^{(p)} t^{2n}$$

Because of time reversal invariance the serie (34) involves only even power of t and the computation of the coefficient (cf. [5]) shows that they are all positive. The first singularity is therefore real. In [5] the Domb-Sykes diagram are constructed for the ratios $\Omega_n^{(p)}/\Omega_{n-1}^{(p)}$, $p = 2,3,4,5$, for $a = 1$ giving a first singularity for $t = 1.008$.

Finally with the use of Padé approximants Meiron, Baker and Orszag [5] have estimated the shape of the curve $z(x,t)$ for t near t_c and have shown the appearance of a cusp for $x = 0$ instead of any kind of roll.

V. SOME FINAL REMARKS.

REMARK 3. The conservation of the vorticity in crucial for the existence of the weak solution in section 1. It does not plays any role (except for the simplification of the computation) in the analytic treatment of the solution in section 3. Therefore the result of section 3 can be extended to the description of the evolution of a surface interface in $\mathbb{R}^3$ instead of curve interface in $\mathbb{R}^2$. This is done in C. Sulem and als [10] , where the following problem is solved in the analytic frame :

(35) $$\frac{\partial z}{\partial t} + \frac{\partial z}{\partial x} v_1 + \frac{\partial z}{\partial y} v_2 = v_3$$

(36) $$\frac{\partial\Omega}{\partial t} + \frac{\partial}{\partial x}(\Omega\, v_1) + \frac{\partial}{\partial y}(\Omega\, v_2) = \Omega_1 \frac{\partial v}{\partial x} + \Omega_2 \frac{\partial v}{\partial y}$$

(37) $$v = -\frac{1}{4\pi}\int \frac{r(x,y,t) - r(x',y',t')}{|r(x,y,t) - r(x',y',t')|^3}\ \Omega(x',y',t)dx'dy'$$

with analytic initial data for Ω and z .

REMARK 4. The method of section 3 do not use any conformal mapping as it is done in Kano and Nishida [4], however with slight modification they could be applied to the water wave problem, with or without surface tension.

In the absence of surface tension the problem is the following : find u and $y(x,t)$ subject to the equations :

(38) $$\begin{aligned} &\frac{\partial u}{\partial t} + u\,\nabla u = -\nabla p - g\vec{j} \quad \text{in } \Omega_y(t) \\ &\nabla . u(x,0) = 0 \end{aligned}$$

$\vec{j}$ is the unit vector in the x_2 direction and $\Omega_y(t)$ is the open set.

(39) $$\Omega_y(t) = \{(x_1,x_2) \mid 0 < x_2 < y(x_1,t)\}$$

The derivation made in section II will lead to the following equations (which are equivalent to the classical equations studied in [4]) :

(40) $$\frac{\partial\Omega}{\partial t} + \frac{\partial}{\partial x}(\Omega\, v_1) = +g\, y'_x$$

(41) $$\frac{\partial y}{\partial t} + v_1 \frac{\partial y}{\partial x} = v_2$$

(42) $$\vec{v} = \int G(x,x',y(x,t),\ y(x',t))\ \Omega(x',t)dx$$

In (42) G is the Kernel of the operator obtained by the solution of the equations

(43) $$\Delta\,\phi(x_1,x_2) = \omega \times \delta_\Gamma \quad \text{in } \mathbb{R}_{x_1} \times \mathbb{R}^+_{x_2}$$

(44) $$\phi(x_1,0) = 0$$

(45) $$\vec{v} = \lim_{(x,y)\to\Gamma} \frac{1}{2}(\nabla \wedge \phi_-) \qquad (^1)$$

The same method applies also to three space variable therefore we obtain a generalisation of the result of Kano and Nishida [4], avoiding the use of any conformal mapping. However I don't know if the limiting process leading to

(1) ϕ_- is the value of ϕ below the curve $y(x,t)$, above it is taken equal to zero.

the shallow water equation would be possible in this situation.

REMARK 5. The existence of the term gj in the right hand side of (38) will give a contribution to the curl therefore the conservation of the vorticity is no more true and in this case it is not clear that the construction of the weak solution should still be valid.

REMARK 6. The section 1, 2 and 3 give the example of a well posed problem with a discontinuity curve and without any entropy condition. This is related to the fact that this is a contact discontinuity without any loss of energy. On thee other hand taking in account the surface forces may lead to a erniform decay of energy which will allow the existence of a smooth solution for all times as it is announced by Yasihara [11].

REFERENCES

[1] C. Bardos. J. Math. Anal. and Appl, 40, 769-790 (1972).

[2] G. Birkhoff. In hydrodynamic instability, p. 55-76. Proc. Symp. Appl. Math. 23, Ann. Math. Soc. (1962).

[3] S. Baouendi and C. Goulaouic. Comm. Part. Diff. Eq, 2. 1151-1162 (1977).

[4] T. Kano and T. Nishida. J. Math. Kyoto University, 19 (1979) 335-370.

[5] M.I. Meiron, G. Baker, S. Orszag. Analytic structure of vortex sheet dynamics, J. of Comp. Mech.

[6] L. Nirenberg J. Diff. Geometry, 6. 561-576 (1972).

[7] T. Nishida. J. Diff. Geometry, 12. 629-633 (1977).

[8] L. Ovsjannikov. Dokl Akad. Nank. USSR 200. Sov. Math. Dokl, 12. 1497-1502 (1971).

[9] G. Strang and R. Teman. J. Meca. 19 (1980) 493-527.

[10] C. Sulem, P.L. Sulem, C. Bardos and U. Frisch. Com. Math. Phys. 80. 485-516 (1981).

[11] H. Yasihara. Gravity waves on the free surface of an incompressible perfect fluid of finite depth preprint.

[12] V.I. Youdovich. Zh Vycid Math. Fiz. 3. 1032-1066 (1963).

Department of Mathematics
C.S.P. Université de Paris Nord
Avenue J.B. Clément,
93430 VILLETANEUSE
(France)

Contemporary Mathematics
Volume 17, 1983

LARGE-TIME REGULARITY OF VISCOUS SURFACE WAVES

J. Thomas Beale[1]

The work described here is concerned with existence and regularity questions for time-dependent motions of a viscous, incompressible fluid in a domain like an infinite ocean, under the influence of gravity and surface tension. The fluid is bounded above by an atmosphere of constant pressure and below by a rigid bottom. We show that motions which are initially fairly regular and close to equilibrium remain so for all later time. Thus the behavior is adequately represented by solutions of the usual equations of motion, in this case the Navier-Stokes equations with appropriate boundary conditions.

To summarize the principal results, let us suppose that the initial surface and velocity field are prescribed with mild regularity and compatibility conditions. Assuming this initial state is near equilibrium, we show that a unique solution of the equations of motion (described below) exists for all time in a space defined by Sobolev norms. In particular, the height of the free surface is always continuously differentiable as a function of the horizontal coordinates and time. Slightly after the initial time the solution becomes more smooth and satisfies the equations in the classical sense. (The regularity at time zero is limited by the compatibility conditions on the initial data.) The detailed results are given in [1].

We take the domain of the fluid to be $\{(x,y) : -b(x) < y < \eta(x,t)\}$. Here $x \in \mathbf{R}^2$, $b(x)$ is smooth, and $b(x) \to b_0 > 0$ as $x \to \infty$. The height $\eta(x,t)$ of the free surface above the equilibrium position $y = 0$ changes with the motion of the fluid. After nondimensionalizing and subtracting off

1980 Mathematics Subject Classifications: 76D05, 35Q10, 35R35.

[1]Sponsored by the United States Army under Contract No. DAAG29-80-C-0041, and partially supported by N.S.F. Grant MCS-81-01639.

the hydrostatic pressure, we write the Navier-Stokes equations as

$$u_t + (u \cdot \nabla)u - \nu\Delta u + \nabla p = 0 , \tag{1}$$

$$\nabla \cdot u = 0 . \tag{2}$$

On the bottom we have the usual no-slip condition

$$u = 0 \qquad (y = -b) . \tag{3}$$

At the free surface the tangential stress is zero and the normal stress is proportional to the mean curvature of the surface. Because of our previous adjustment of the pressure term, gravity appears explicitly in this condition. With n_i the normal to the surface and β the nondimensional coefficient of surface tension, we have

$$pn_i - \nu \sum_j (u_{i,j} + u_{j,i})n_j = \{g\eta - \beta\nabla \cdot [(1 + |\nabla\eta|^2)^{-1/2} \nabla\eta]\}n_i \qquad (y = \eta) . \tag{4}$$

Finally we have in addition on the free surface the kinematic condition that fluid particles do not cross the surface:

$$\eta_t = u_3 - (\partial_1\eta)u_1 - (\partial_2\eta)u_2 \qquad (y = \eta) . \tag{5}$$

These equations form a complete description of the motion of the fluid. Because time derivatives of u and η appear in (1) and (5), it is natural to regard the set of equations as a coupled system for these two unknowns.

The approach used in [1] is to construct the solution by an iteration based on the linearization about equilibrium. We first transform the entire problem to the equilibrium domain by a change of variables depending on the unknown free surface. Extra terms are introduced, and the nonlinearity becomes fairly general. The problem now has the form $Lz = F(z)$, where L is the linear part and F contains higher order terms. The most crucial part of the argument is in obtaining estimates for the operator L; once we know that the mapping $z \to L^{-1}F(z)$ is smooth in an appropriate sense, we can find the solution as a fixed point. It is particularly important how the estimates behave at large time; this argument could not apply to all semilinear parabolic equations, since in certain cases solutions do not exist for all time even with small data.

The linear problem associated with the operator L above can be written as

$$\eta_t = Ru$$
(6)
$$u_t + Au + R^*(g - \beta\Delta_x)\eta = f\ .$$

Here Ru is the restriction of u_3 to $y = 0$; R^* is the formal adjoint of R; and A is an operator containing the effect of viscosity. If A were not present, the system would be energy-preserving; on the other hand, the equation $u_t + Au = 0$ would behave like a parabolic equation. Good estimates for the solutions of (6) must take into account both types of behavior. It has seemed most promising to estimate the solution of the problem Laplace-transformed in time. After transforming (6) and eliminating $\hat{\eta}$ we have (assuming zero initial value for simplicity)

$$\lambda\hat{u} + A\hat{u} + \lambda^{-1}B\hat{u} = \hat{f}\ ,$$

where $B = R^*(g - \beta\Delta_x)R$. We find estimates of elliptic type for $\hat{u}$ and for $\hat{\eta} = \lambda^{-1}R\hat{u}$ with $\text{Re}\,\lambda > 0$; the dependence of the estimates on λ gives information about the time derivatives. By letting $\text{Re}\,\lambda \to 0$ and transforming back, we are led to a boundedness property for L^{-1} in certain space-time norms. We then estimate the nonlinear terms crudely and solve the full problem as indicated.

In earlier work [2] an existence result was given for the same problem for a short time interval, without assuming the initial data near equilibrium, and without including surface tension. A similar short-time result, but with more smoothness of the solution near time zero, was given by Solonnikov [3] for a fluid in a domain bounded entirely by a free surface.

BIBLIOGRAPHY

1. J. T. Beale, "Large-time regularity of viscous surface waves", submitted to Arch. Rational Mech. Anal.

2. J. T. Beale, "The initial value problem for the Navier-Stokes equations with a free surface", Comm. Pure Appl. Math. 34 (1980), 359-392.

3. V. A. Solonnikov, "Solvability of a problem on the motion of a viscous incompressible fluid bounded by a free surface", Math. U.S.S.R. Izvestiya 11 (1977), 1323-1358.

MATHEMATICS DEPARTMENT
TULANE UNIVERSITY
NEW ORLEANS, LA 70118

Contemporary Mathematics
Volume 17, 1983

SHOCK WAVES AND THE BOLTZMANN EQUATION

Russel Caflisch

and

Basil Nicolaenko

ABSTRACT. After a brief description of the Boltzmann equation, the Boltzmann theory of a steady shock wave is developed. We then prove the existence of shock profile solutions of the Boltzmann equation for shocks which are weak. This solution completely agrees with the Navier-Stokes profile to leading order. At higher order it may have different behaviour in its decay to a constant at $x = \pm\infty$. Finally, we discuss several unsolved problems.

1. INTRODUCTION

A steady shock wave in a gas can be described through several different theories. In each of them the shock wave connects two constant states at $x = \pm\infty$ and moves steadily at speed s. From the Euler equations the shock is described as a jump discontinuity in density, velocity, and temperature from $(\rho-,\underline{u}-,T-)$ on the left to $(\rho+,\underline{u}+,T+)$ on the right. From the Navier-Stokes equations it is described as a smooth profile connecting these two states at $x = -\infty$ and $x = +\infty$ respectively. Through the Boltzmann equation the gas is described by a velocity distribution function $F(x,\underline{\xi},t)$, and the shock wave is found as a smooth profile connecting two uniform Maxwellian states $F_+(\underline{\xi})$ at $x = \infty$ and $F_-(\underline{\xi})$ at $x = -\infty$ given by

$$F_\pm(\underline{\xi}) = \rho_\pm(2\pi T\pm)^{3/2} \exp\{-|\underline{\xi} - \underline{u}_\pm|^2/2T_\pm \quad . \tag{1.1}$$

A detailed comparison of these theories with experimental results has been made by Fiszdon, Herczynski, and Walenta [6]. For weak shocks the Navier-Stokes profiles are very close to the observed shock waves, but for stronger shocks they do not agree. For a wide range of shock strengths the Boltzmann

0271-4132/82/0000-1046/$03.75

shock profiles agree with experimental results. The agreement seems to be good for the distribution as a function of ξ, not just for its moments.

After a brief discussion of the Boltzmann equation in section 2, we develop the Boltzmann description of a shock in section 3. Our main result is the existence of shock profile solutions of the Boltzmann equation for weak shocks and the agreement of these solutions with the Navier-Stokes profiles for such shocks. In section 4 this theorem is stated and its proof is sketched. Finally, we discuss some unsolved problems in section 5.

The weak shock problem for the Boltzmann equation with hard sphere interactions was solved earlier by Nicolaenko and Thurber [12,13]. The method used here is a modification of their method. For a general reference on the Boltzmann equation, see Cercignani [4]; for shocks, see Courant and Friedericks [5].

2. THE BOLTZMANN EQUATION

In kinetic theory, a gas is described by a molecular distribution function $F(t,\underline{x},\underline{\xi}) > 0$ for particles of velocity $\underline{\xi}$ at point $\underline{x}$ and time t. The distribution evolves through time according to the nonlinear Boltzmann equation.

$$(2.1)\qquad \left(\frac{\partial}{\partial t} + \underline{\xi} \cdot \frac{\partial}{\partial \underline{x}} \right)F = Q(F,F)$$

in which

$$(2.2)\qquad Q(F,F)(\underline{\xi}) = \int_{R^3} \int_0^{2\pi} \int_0^{\pi/2} \{F(\underline{\xi}_1')F(\underline{\xi}') - F(\underline{\xi}_1)F(\underline{\xi})\}B(\theta,\underline{v})d\theta d\rho d\underline{\xi}_1 \ .$$

The variables x and t of F are merely parameters in (2.2) and

$$(2.3)\qquad \begin{aligned} v &= \underline{\xi}_1 - \underline{\xi} \\ \underline{\xi}_1 &= \underline{\xi} + \underline{\alpha}(\underline{\alpha}\cdot\underline{v}) \\ \underline{\xi}_1' &= \underline{\xi}_1 - \underline{\alpha}(\underline{\alpha}\cdot\underline{v}) \\ \underline{\alpha} &= (\cos\theta, \sin\theta\cos\phi, \sin\theta\cos\phi) \end{aligned}$$

in which $\underline{\alpha}$ is the apse line and θ is the scattering angle.

The left side of (2.1) describes the motion of particles according to their velocity; the right side gives the changes in F due to collisions. The integrand in the collision operator (2.2) consists of a difference of two terms. The first counts the number of collisions which produce a particle of velocity ξ; the second counts those in which a particle of velocity ξ is lost. It follows from (2.3) that

2.4)
$$\underline{\xi} + \underline{\xi}_1 = \underline{\xi}' + \underline{\xi}_1'$$
$$\underline{\xi}^2 + \underline{\xi}_1^2 = \underline{\xi}'^2 + \underline{\xi}_1'^2$$

so that particles of velocities $\underline{\xi}$ and $\underline{\xi}_1$ may collide and result in particles of velocities $\underline{\xi}'$ and $\underline{\xi}_1'$ or vice versa.

The collision kernel $B(\theta,\underline{v})$ depends on the intermolecular force law and is discussed in detail in [4]. The results presented in section 4 are valid for forces derived from hard cutoff potentials in the sense of Grad [7].

The local equilibrium distributions for the scattering process are distributions F with $Q(F,F) = 0$. From the symmetry of Q and the positivity of B, it can be shown that such an F satisfies

2.5)
$$F(\underline{\xi}_1')F(\underline{\xi}') = F(\underline{\xi}_1)F(\underline{\xi}),$$

which is called detailed balance. It follows from (2.4) that

2.6)
$$F(\underline{\xi}) = \rho(2\pi T)^{-3/2}\exp\{-(\underline{\xi} - \underline{u})^2/2T\}$$

in which $\rho,\underline{u},T$ are independent of $\underline{\xi}$. They may depend arbitrarily on x and t however, since Q does not act on the x and t dependence.

Corresponding to the conservation of mass, momentum and energy in individual collisions, there are the following conservation properties of Q:

2.7)
$$\langle 1,Q(F,G)\rangle = 0$$
$$\langle \xi_i,Q(F,G)\rangle = 0$$
$$\langle \xi^2,Q(F,G)\rangle = 0$$

in which $\langle f,g\rangle = \int_{R^3} f(\underline{\xi})g(\underline{\xi})d\underline{\xi}$ and $Q(F,G)$ is defined to be the symmetric extension of $Q(F,F)$. In addition, the symmetry and positivity properties of Q imply that

2.8)
$$\int \log F(\underline{\xi})Q(F,F)(\underline{\xi})d\underline{\xi} < 0$$

for any $F > 0$.

Finally, we define the fluid dynamic moments of the distribution F by

$$\rho_F = \langle 1, F \rangle$$

(2.9) $$\rho_F \underline{u}_F = \langle \underline{\xi}, F \rangle$$

$$\rho_F T_F = 1/3 \langle (\xi - \underline{u}_F)^2, F \rangle$$

These quantities are dual to the conservation laws (2.7). For a Maxwellian defined by (2.6), $\rho = \rho_F$, $\underline{u} = \underline{u}_F$, $T = T_F$.

3. THE BOLTZMANN SHOCK WAVE

A plane steady shock profile is a continuous solution $F(\underline{\xi},x,t) = F(\underline{\xi},x-st)$ which depends on only one space variable $x = x_1$ and translates at uniform speed s. Its values at $x = \pm\infty$ are Maxwellian $F_\pm$ given by (1.1) with $\rho_\pm, u_\pm, T$ each constant and $\underline{u}_\pm = (u_\pm, 0, 0)$. Therefore the Boltzmann equation (2.1) becomes

(3.1) $$(\xi_1 - s)\frac{\partial}{\partial x} F = Q(F,F)$$

in which ξ, is the first component of ξ, and there are asymptotic conditions

(3.2) $$F(\underline{\xi}, x = \pm\infty) = F_\pm(\underline{\xi}) \equiv \rho_\pm (2\pi T_\pm)^{-3/2} \exp\{-((\xi_1 - u_\pm)^2 + \xi_2^2 + \xi_3^2)/2T_\pm\}$$

The conservation properties (2.7) and the bound (2.8) result in conditions on F_+ and F_-. From (2.7) we obtain

$$\frac{\partial}{\partial x} \langle (\xi_1 - s), F \rangle = 0$$

(3.3) $$\frac{\partial}{\partial x} \langle \xi_1(\xi_1 - s), F \rangle = 0$$

$$\frac{\partial}{\partial x} \langle \xi^2(\xi_1 - s), F \rangle = 0$$

so that the quantities in brackets are constant in x. Equating their values at $x = \pm\infty$ and using the form of $F(x = \pm\infty)$ results in the <u>Rankine-Hugoniot conditions:</u>

$$(u_- - s)\rho_- = (u_+ - s)\;_+$$

3.4)
$$(u_- - s)\rho_- u_- + \rho_- T_- = (u_+ - s)\rho_+ u_+ + \rho_+ T_+$$

$$(u_- - s)\rho_-(\tfrac{3}{2}\, T_- + \tfrac{1}{2}\, u_-^2) + \rho_- u_- T_- = (u_+ - s)\rho_+(\tfrac{3}{2}\, T_+ + \tfrac{1}{2}\, u_+^2) + \rho_+ u_+ T_+$$

It follows from (2.8) and (3.1) that

3.5)
$$\frac{\partial}{\partial x}\int (\xi_i - s)\; F \log F \, d\xi < 0.$$

This is the analogue of the Boltzmann H-theorem for the shock problem. If we compare the values of this integral at $x = \pm\infty$ as before we obtain the <u>entropy inequality</u>:

3.6)
$$\operatorname{sgn}(u_- - s) S_- < \operatorname{sgn}(u_+ - s) S_+$$

in which $S = \log(\rho^{-2/3} T)$ is the fluid dynamic entropy. Note that $\operatorname{sgn}(u_- - s) = \operatorname{sgn}(u_+ - s)$.

The Rankine-Hugoniot conditions (3.4) and the entropy inequality (3.6) are exactly the description from the Euler equations of a steady planar shock. This shows the agreement between the Boltzmann theory and the Euler theory. Note that although (3.4) and (3.6) are relations only for the asymptotic values of F, the corresponding relations (3.3) and (3.5) apply for all x.

Before stating our main theorem, we discuss the Navier-Stokes equations for a shock. The equations are

$$\frac{\partial}{\partial x}\; \rho(u - s) = 0$$

3.6)
$$\frac{\partial}{\partial x}\,\rho u(u - s) + \frac{\partial}{\partial x}\,\rho T = \frac{4}{3}\,\eta\, \frac{\partial^2}{\partial x^2}\; u$$

$$\frac{\partial}{\partial x}\,\rho(u - s)(\tfrac{3}{2}\, T + \tfrac{1}{2}\, u^2) + \frac{\partial}{\partial x}\; uT = \frac{\partial}{\partial x}\,\lambda\frac{\partial}{\partial x}\, T + \frac{4}{3}\,\eta\frac{\partial}{\partial x}\,(u\, \frac{\partial}{\partial x}\, u)\; .$$

with asymptotic values

3.7)
$$\rho(x = \pm\infty) = \rho_\pm$$
$$u(x = \pm\infty) = u_\pm$$
$$T(x = \pm\infty) = T_\pm \quad .$$

Denote their solution by $\rho_{NS}, u_{NS}, T_{NS}$. The viscosity and heat conduction coefficiencts η and λ are to be those determined by the Chapman-Enskog expansion [4]. The end states (ρ_+, u_+, T_+) and (ρ_-, u_-, T_-) must satisfy the jump conditions (3.4) and (3.6).

To be definite we shall assume that the shock is moving to the right relative to the fluid, i.e.

$$u_- < s . \tag{3.8}$$

Then the Rankine-Hugoniot and entropy conditions (3.4) and (3.6) have a nontrivial solution [5] if and only if

$$0 < s - u_- < c_- \tag{3.9}$$

in which

$$c_- = (\tfrac{5}{3} T_-)^{1/2} \tag{3.10}$$

is the sound speed of the gas at $x = -\infty$.

Consider $F_- = F(x = -\infty)$ to be given. If $s - u_- \geq c_-$, then the only solution of the Boltzmann equation (3.1) is the constant solution

$$F(\underline{\xi}, x) = F_-(\underline{\xi}) . \tag{3.11}$$

We shall study only weak shocks with $\varepsilon = c_- - (s-u-) > 0$ small. This implies that

$$\begin{aligned} (\rho_-, u_-, T_-) - (\rho_+, u_+, T_+) &= \mathcal{O}(\varepsilon) \\ F_- - F_+ &= \mathcal{O}(\varepsilon) \end{aligned} \tag{3.12}$$

Also $\rho_{NS}, u_{NS}, T_{NS}$ are given approximately by the hyperbolic tangent shock profile of Taylor [15], e.g.

$$u_{NS} = \frac{u_- + u_+}{2} + \frac{u_+ - u_-}{2} \tanh(\tau \varepsilon x) + \mathcal{O}(\varepsilon^2) \tag{3.13}$$

with $\tau > 0$. There are similar formulae for ρ_{NS} and T_{NS}.

4. THE PROFILE OF A WEAK SHOCK

Define weighted sup norms on $\underline{\xi}$ by

$$\|f\|_{\alpha,r} = \sup_{\underline{\xi}} (1+\xi)^r e^{\alpha\xi^2} |f(\underline{\xi})|$$

4.1)

$$\|f\|_r = \|f\|_{o,r} .$$

Decay in x will be measured by the function

4.2) $$A(x) = e^{-\varepsilon\tau|x|} + \varepsilon e^{-\mu|x|^\beta}.$$

in which $0 < \beta \leq 1$ and μ are constants and τ is the factor in (3.13). Define the Maxwellian distribution corresponding to the solution $(\rho_{NS}, u_{NS}, T_{NS})$ of the Navier-Stokes equation (3.6), (3.7) as

4.3) $$F_{NS}(\underline{\xi},x) = \rho_{NS}(2\pi T_{NS})^{-3/2}\exp\{-((\xi_1 - u_{NS})^2 + \xi_2^2 + \xi_3^2)/2T_{NS}\}$$

Our main result, presented in detail in [3], is the following theorem:

Theorem. Let (ρ_-, u_-, T_-), (ρ_+, u_+, T_+) and s satisfy the Rankine-Hugoniot conditions (3.4) and the entropy conditons (3.6). Let $\varepsilon = c_- - (s - u_-) > 0$ be sufficiently small. Then there is a shock profile solution F of the Boltzmann equation (3.1), (3.2). It satisfies

4.4) $$\|F(x) - F_{NS}(x)\|_r \leq c\varepsilon^2 A(x)$$

4.5) $$\|F(x) - F_{NS}(x)\|_{\alpha,r} \leq c\varepsilon^2$$

in which $0 \leq \alpha < (2T_-)^{-1}$ and c depends only on α and r. Moreover, F is unique, up to translation in x, among those solutions satisfying (4.4), (4.5).

Remarks. 1) Since $(\rho_-, u_-, T_-) - (\rho_+, u_+, T_+) = \mathcal{O}(\varepsilon)$, the spatial variation of $(\rho_{NS}, u_{NS}, T_{NS})$ and F_{NS} are size ε. So the errors $\mathcal{O}(\varepsilon^2)$ in (4.4),(4.5) show nontrivial agreement of the Boltzmann and Navier-Stokes shock profiles.

2) Since $(\rho_{NS}, u_{NS}, T_{NS})$ are approximately hyperbolic tangents, F_{NS} approaches its limits $F_\pm$ at $x = \pm\infty$ at the rate $e^{-\varepsilon\tau|x|}$ for same τ. The exponent β in A (cf. (4.2)) is given by

4.6) $$\beta = \frac{s - 3}{s - 2}$$

if the intermolecular force law in $F(r) = r^{-s}$ with $s \geq 5$. For the hard sphere model with $s = \infty$, $\beta = 1$ and the decay rate of F is still exponential. This is the case solved by Nicolaenko and Thurber [12,13]. If $\infty > s \geq 5$, then $\beta < 1$ and the decay rate of A is less than exponential. Note that non-exponential part is only size ε. We believe that this decay as exponential of a power is optimal, but we have been unable to prove that.

Outline of a Proof of the Theorem

Think of the shock profile solution as a bifurcation from the constant solution $F \equiv F_-$ as the parameter ε becomes positive. The proof is by a projection method similar to the Lyapunov-Schmidt method.

Since $F_+ = F_- + \mathcal{O}(\varepsilon)$ and both satisfy $Q(F_1 F) = 0$, we expect that $(\xi_1 - s)\frac{\partial}{\partial y} F = Q(F,F) = \mathcal{O}(\varepsilon)$. So the spatial variation of F is slow. We look for F to have the form

4.7) $$F(x,\underline{\xi}) = F_-(\underline{\xi}) + \varepsilon f(y = \varepsilon x, \underline{\xi})$$

in which f satisfies

4.8) $$(\xi_1 - s)\frac{\partial}{\partial y} f = \frac{1}{\varepsilon}\mathcal{L} f + Q(f,f)$$

with

4.9) $$\mathcal{L} f = 2Q(M_-, f)$$

The dominant term in f is found as follows. Keep F_- fixed and let F_+ vary with ε in such a way that (3.4) and (3.6) remain true. Define

4.10) $$\Phi_o(\underline{\xi}) = \frac{d}{d\varepsilon} M_+(\underline{\xi}).$$

By adding a correction Φ_1 to Φ_o we obtain

4.11) $$\Phi_\varepsilon = \Phi_o + \varepsilon \Phi_1$$

which satisfies

4.12) $$\mathcal{L}\Phi_\varepsilon = \varepsilon\tau(\xi_1 - s)\Phi_\varepsilon + \mathcal{O}(\varepsilon^2)$$

for τ a constant.

As in the Lyapunov-Schmidt method, decompose f as

4.13) $$f(y,\underline{\xi}) = z_\varepsilon(y)\Phi_\varepsilon(\underline{\xi}) + \varepsilon w_\varepsilon(y,\underline{\xi}).$$

By taking projections of the equation (4.8) for f, we find that

4.14) $$\frac{\partial}{\partial y} z_\varepsilon = -\tau z_\varepsilon + \gamma(z_\varepsilon)^2 + \mathcal{O}(\varepsilon^2)$$

4.15) $$(\xi_1 - s)\frac{\partial}{\partial y} w_\varepsilon = -\frac{1}{\varepsilon} M w_\varepsilon + \frac{1}{\varepsilon} b$$

in which τ and γ are constants, M is a projection of $\mathscr{L}$, and b consists of terms which depend z_ε and small terms. These equations are solved by iteration.

The solution of (4.14) is

4.16) $$z_\varepsilon(y) = \frac{1}{2}(\tau/\gamma)\{\tanh(-\frac{1}{2}\tau y) + 1\} + \mathcal{O}(\varepsilon),$$

which can be shown to agree with the tanh profile for the Navier-Stokes solution. The analysis of (4.15) is difficult because the continuous spectrum of M reaches to the origin. It is these small spectral values which cause the non-exponential decay in (4.4).

5. FURTHER PROBLEMS

There are a number of interesting unanswered questions.

1. The optimal decay rate. We have been unable to prove that the optimal decay rate of F at $x = \pm\infty$ is at the rate $e^{-\mu|x|^\beta}$. This was suggested earlier by Lyubarski [9], Narasimha [11].

It seems that our estimation of the ξ decay rate is too crude - we use decay $e^{-\alpha\xi^2}$, $\alpha < (2T_-)^{-1}$. It should be exactly $e^{-\xi^2/2T_-}$.

2. Strong shocks. There is an approximate method by Mott-Smith [10] and Tamm [14] for calculating shock profiles at all strengths. It is seen to be quite accurate [6], but there is no mathematical derivation of it. Such a derivation could estimate its accuracy and suggest possible improvements. Grad [8] has suggested that the method is asymptotically correct for infinitely strong shocks.

3. Time-dependent shocks. An analysis of the Hilbert expansion away from shocks has been given in [1]. To handle time dependent weak shocks, the solution developed here should act as the first term in an inner expansion around the shock and must be matched with the Hilbert expansion. This procedure has been described formally for the Broadwell model in [2].

4. Shock-stability. We expect that the steady planar shock wave is dynamically stable. Besides its intrinsic interest, such a result may be needed in the previous problem.

REFERENCES

1. Caflisch, R., The fluid dynamic limit of the nonlinear Boltzmann equation, Comm. Pure Appl. Math. 33 (1980), 651-666.

2. Caflisch, R., The Broadwell equation with a time-dependent shock, INRIA report, 1981.

3. Caflisch, R. and Nicolaenko, B., Shock profile solutions of the Boltzmann equation, Comm. Math. Physics, to appear.

4. Cercignani, C., Theory and Application of the Boltzmann Equation, Elsevier, 1975.

5. Courant, R. and Friedricks, K.O., Supersonic Flow and Shock Waves, Wiley, 1948.

6. Fiszdon, W., Herczynski, R. and Walenta, Z., The structure of a plane shock wave of a monatomic gas: Theory and experiment, Ninth Int. Symp. Rarefied Gas Dyn., 1974, Appendix B.23, 1-57.

7. Grad, H., Asymptotic theory of the Boltzmann equation II, Third Int. Symp. Rarefied Gas Dyn., 1952, 25-59.

8. Grad, H., Singular and nonuniform limits of solutions of the Boltzmann equation, SIAM-AMS Proc. I, Transport Theory, 1967, 296-308.

9. Lyubarski, G., On the kinetic theory of shock waves, JETP, 13 (1961), 740-745.

10. Mott-Smith, H.M., The solution of the Boltzmann equation for a shock wave, Phys. Rev. 82 (1951) 885-892.

11. Narasimha, R., Asymptotic solutions for the distribution function in non-equilibrium flours, Part I: The shock wave, J. Fluid. Mech., 34 (1968), 1-24.

12. Nicolaenko, B., A general class of nonlinear bifurcation problems from a point in the essential spectrum, application to shock wave solutions of kinetic equations, in Applications of Bifurcation Theory, Academic Press, 1977, 333-357.

13. Nicolaenko, B. and Thurber, J.K., Weak shocks and bifurcating solutions of the non-linear Boltzmann equation. J. de Mech. 14 (1975), 305-338.

14. Tamm, I.E., Width of High-Intensity Shock Waves, Proc. (Trudy) Lebedev Phys. Inst. 29 (1965), 231-241.

15. Taylor, G.I., The conditions necessary for discontinuous motion in gases, Proc. Roy. Soc., London, A 84 (1910), 371-377.

Russel Caflisch
DEPARTMENT OF MATHEMATICS
STANFORD UNIVERSITY
STANFORD, CA. 94305

Basil Nicolaenko
LOS ALAMOS SCIENTIFIC LABORATORY
LOS ALAMOS, N.M. 87545

Contemporary Mathematics
Volume 17, 1983

FORMATION OF SINGULARITIES FOR NONLINEAR HYPERBOLIC PARTIAL DIFFERENTIAL EQUATIONS*

Kuo-Shung Cheng

ABSTRACT. Without genuinely nonlinear assumptions, we prove that smooth solutions of a system of two quasilinear partial differential equations with compact support initial data or with small-amplitude periodic initial data in general develop singularities in finite time.

1. INTRODUCTION

In this paper we study the formation of singularities of solutions for a first-order strictly hyperbolic system of two partial differential equations

$$\begin{aligned} u_t + au_x + bv_x &= 0, \\ v_t + cu_x + dv_x &= 0 \end{aligned} \tag{1.1}$$

with nonconstant smooth initial data of compact support or with nonconstant smooth periodic initial data, where a,b,c and d are smooth functions of u and v.

If system (1.1) is "genuinely nonlinear" in the sense to be defined below (see also Lax [4]), then it is well known that solutions of (1.1) will develop singularities in finite time for any nonconstant smooth initial data of compact support or any nonconstant periodic initial data (see Lax [5], [6], Glimm and Lax [1], John [2] and Klainerman and Majda [3]). If system (1.1) is not "genuinely nonlinear" only partial results are known (see Liu [7], Klainerman and Majda [3]). In [3], Klainerman and Majda consider the nonlinear wave equations

$$w_{tt} = (K(w_x))_x, \quad K' > 0, \tag{1.2}$$

which can be written as

$$\begin{aligned} u_t - v_x &= 0, \\ v_t - K(u)_x &= 0, \end{aligned} \tag{1.3}$$

where $u = w_x$ and $v = w_t$. They show that if there exists a positive integer p such that $K^{(p+1)}(0) \neq 0$, then for sufficiently small nonconstant initial

---

*The work presented in this paper is supported by the National Science Council of the Republic of China.

0271-4132/82/0000-1012/$03.75

data, of compact support or periodic, the classical C^2 solution of (1.2) will develop a singularity in finite time.

It is the purpose of this paper to give an almost complete treatment of (1.1) when (1.1) is not "genuinely nonlinear". Our results generalize that of Klainerman and Majda's. But the basic techniques are the same as theirs.

It is well known that there exist functions $r(u,v)$ and $s(u,v)$, called Riemann invariants for system (1.1), such that system (1.1) is equivalent to

$$\begin{aligned} r_t - \lambda r_x &= 0, \\ s_t + \mu s_x &= 0 \end{aligned} \tag{1.4}$$

for classical solutions, where $(-\lambda)$ and μ are the real and distinct eigenvalues of the matrix

$$\begin{pmatrix} a & b \\ c & d \end{pmatrix}.$$

Here for convenience we choose a proper reference frame to ensure that λ and μ are positive for all relevant values of r and s. From now on, we will consider system (1.4) instead of system (1.1). The system (1.4) is called genuinely nonlinear if λ_r and μ_s are never zero for all relevant values of r and s. We will consider (1.4) with the initial data

$$\begin{aligned} r(x,0) &\equiv r_0(x) = F(x), \ |F(x)| \le \delta, \\ s(x,0) &\equiv s_0(x) = G(x), \ |G(x)| \le \delta, \end{aligned} \tag{1.5}$$

where $F(\cdot)$ and $G(\cdot)$ are nonconstant C^1 functions with compact support. We will also consider (1.4) with the initial data

$$\begin{aligned} r(x,0) &= r_0(x) = \varepsilon f(x), \ |f(x)| \le \delta, \\ s(x,0) &= s_0(x) = \varepsilon g(x), \ |g(x)| \le \delta, \end{aligned} \tag{1.6}$$

where $0 < \varepsilon \le 1$, $f(\cdot)$ and $g(\cdot)$ are nonconstant C^1 periodic functions with periods L_1 and L_2 respectively and δ is an appropriate small constant.

We assume that $\lambda_r(0,0) = \mu_s(0,0) = 0$. Hence system (1.4) is not genuinely nonlinear in any neighborhood of $(0,0)$. λ and μ may satisfy

$$\begin{aligned} \lambda_r(0,0) &= 0 \text{ but } \lambda_r(r,s) \ne 0 \text{ for all } (r,s) \ne (0,0), \\ \mu_s(0,0) &= 0 \text{ but } \mu_s(r,s) \ne 0 \text{ for all } (r,s) \ne (0,0). \end{aligned} \tag{1.7}$$

We will see that the cases when λ and μ satisfy (1.7) and (1.8) can be treated almost as in the genuinely nonlinear case. Now assume that λ_r and μ_s take both positive and negative values in the vicinity of the origin $(0,0)$. We will assume that the set $\{(r,s) \mid \lambda_r(r,s) = 0\}$ and $\{(r,s) \mid \mu_s(r,s) = 0\}$ are union of smooth one-dimensional manifolds. Hence we assume that

$$\lambda(r,s) = \lambda_0 + h_1(s) + \int_0^r (a_1 r'+b_1 s)^{p_1} \cdots (a_N r'+b_N s)^{p_N} dr' + O((|r|+|s|)^{p+1}), \quad (1.9)$$

$$\mu(r,s) = \mu_0 + h_2(r) + \int_0^s (c_1 r+d_1 s')^{q_1} \cdots (c_M r+d_M s')^{q_M} ds' + O((|r|+|s|)^{q+1}), \quad (1.10)$$

where $\lambda_0 = \lambda(0,0)$ and $\mu_0 = \mu(0,0)$, $h_1(s)$ is a polynomial of order p in s, $h_2(r)$ is a polynomial of order q in r, a_i, b_i, c_i, d_i are constants which satisfy $a_i^2 + b_i^2 > 0$ and $c_i^2 + d_i^2 > 0$ for all relevant i, $p = (p_1+\dots+p_N + 1)$ and $q = (q_1+\dots+q_M + 1)$, p_i's and q_i's are positive integers and $M \geq 1$, $N \geq 1$. For notational convenience, we let

$$H_1(r,s) = (a_1 r + b_1 s)^{p_1} \cdots (a_N r + b_N s)^{p_N},$$

$$H_2(r,s) = (c_1 r + d_1 s)^{q_1} \cdots (c_M r + d_M s)^{q_M}.$$

If λ_r is zero in an open neighborhood of $(0,0)$, we say that r-field is "linear degenerate". We shall see that the linear degenerate case is easy to treat.

Our main results are

THEOREM 1. Consider system (1.4) with the initial data (1.5). Let λ and μ satisfy one of the following:

(i) λ satisfies (1.7) and μ satisfies (1.8),

(ii) λ satisfies (1.7), μ satisfies (1.10) and $F(\cdot)$ is nonconstant,

(iii) λ satisfies (1.9), μ satisfies (1.8) and $G(\cdot)$ is nonconstant,

(iv) λ satisfies (1.9), μ satisfies (1.10), $a_i \neq 0$ for all i and $F(\cdot)$ is nonconstant,

(v) λ satisfies (1.9), μ satisfies (1.10), $d_i \neq 0$ for all i and $G(\cdot)$ is nonconstant.

Then for small $\delta > 0$, solution of (1.4) will develop a singularity in finite time.

THEOREM 2. Consider system (1.4) with the initial data (1.6). If λ satisfies (1.7) or (1.9) and μ satisfies (1.8) or (1.10), then there exists ε_0 such that for all $0 < \varepsilon \leq \varepsilon_0$, solution of (1.4) will develop a singularity in finite time.

We should like to note that, when λ satisfies (1.9) with some $a_i = 0$ and μ satisfies (1.10) with some $d_i = 0$, solution of (1.4) can be smooth for all time even for nonconstant $F(\cdot)$ and $G(\cdot)$. We shall explain this peculiar situation in Section 3.

2. PRELIMINARIES

We introduce the characteristic curves $X_1(\alpha,t)$ and $X_2(\alpha,t)$ in the x,t-plane for a given solution (r,s). They are defined by

$$\frac{\partial X_1(\alpha,t)}{\partial t} = -\lambda(r(x_1(\alpha,t),t), s(X_1(\alpha,t),t)),\ X_1(\alpha,0) = \alpha, \tag{2.1}$$

$$\frac{\partial X_2(\beta,t)}{\partial t} = \mu(r(X_2(\beta,t),t),\ s(X_2(\beta,t),t)),\ X_2(\beta,0) = \beta. \tag{2.2}$$

It is easy to see from (1.4) that

$$r(X_1(\alpha,t),t) = r_0(\alpha), \tag{2.3}$$

$$s(X_2(\beta,t),t) = s_0(\beta). \tag{2.4}$$

Let

$$Z_1(\alpha,t) = \frac{\partial X_1(\alpha,t)}{\partial \alpha} \tag{2.5}$$

$$Z_2(\beta,t) = \frac{\partial X_2(\beta,t)}{\partial \beta}\ . \tag{2.6}$$

Taking the derivative of (2.1) with respect to α we derive from (2.1), (2.3)

$$\frac{\partial}{\partial t} Z_1(\alpha,t) = -\lambda_r r_0'(\alpha) - \lambda_s s_x Z_1(\alpha,t) \tag{2.7}$$

On the other hand, by (1.4), we have along $(X_1(\alpha,t),t)$

$$\frac{d}{dt} s(X_1(\alpha,t),t) = s_t - \lambda s_x = -(\lambda + \mu)s_x.$$

Hence along $(X_1(\alpha,t),t)$

$$s_x = -\frac{1}{\lambda+\mu}\frac{d}{dt} s. \tag{2.8}$$

Let

$$\Lambda_1(r,s) = \int_0^s \frac{\lambda_s(r,s')}{\lambda(r,s')+\mu(r,s')}\, ds' \tag{2.9}$$

and

$$\rho_1(\alpha,t) = \exp\{\Lambda_1(r(X_1(\alpha,t),t), s(X_1(\alpha,t),t))\}. \tag{2.10}$$

We have from (2.8), (2.3), (2.9) and (2.10)

$$-\lambda_s s_x - \frac{\lambda_s}{\lambda+\mu}\frac{d}{dt} s = \frac{d}{dt}\log \rho_1(\alpha,t). \tag{2.11}$$

Integrating (2.7) we obtain

$$Z_1(\alpha,t) = \frac{\rho_1(\alpha,t)}{\rho_1(\alpha,0)}\left[1 - r_0'(\alpha)\int_0^t \lambda_r \frac{\rho_1(\alpha,0)}{\rho_1(\alpha,t')}\, dt'\right]. \tag{2.12}$$

Similarly we have along $(X_2(\beta,t),t)$

$$Z_2(\beta,t) = \frac{\rho_2(\beta,t)}{\rho_2(\beta,0)}\left[1 + s_0'(\beta)\int_0^t \mu_s \frac{\rho_2(\beta,0)}{\rho_2(\beta,t')}\, dt'\right], \tag{2.13}$$

where

$$\rho_2(\beta,t) = \exp\{\Lambda_2(r(X_2(\beta,t),t), s(X_2(\beta,t),t))\}, \tag{2.14}$$

and

$$\Lambda_2(r,s) = \int_0^r \frac{\mu_r(r',s)}{\lambda(r',s)+\mu(r',s)}\, dr'. \tag{2.15}$$

If for some α and T, $r_0'(\alpha) \neq 0$ and $Z_1(\alpha,T) = 0$. Then from (2.3)

$$r_x \cdot Z_1(\alpha,T) = r_0'(\alpha).$$

Hence $|r_x(X_1(\alpha,T),T)| = \infty$.

Thus to establish Theorem 1 and Theorem 2 we only have to prove that there exists either an α or a β with a corresponding time, T, such that

$$r_0'(\alpha) \neq 0 \quad \text{and} \quad Z_1(\alpha,T) = 0, \tag{2.16}$$

or

$$s_0'(\beta) \neq 0 \quad \text{and} \quad Z_2(\beta,T) = 0. \tag{2.17}$$

3. PROOF OF THEOREM 1

Let

$$\text{support of } (F(\cdot),G(\cdot)) \subset [\alpha_0,\beta_0].$$

Let T be the time which $X_1(\beta_0,t)$ and $X_2(\alpha_0,t)$ intersect each other. If $(r(x,t),s(x,t))$ is still smooth for $t \geq T$, then all waves are simple waves for $t \geq T$ (see Lax [4], John [2]). Hence

$$(r(X_1(\alpha,t),t),\ s(X_1(\alpha,t),t)) = (r_0(\alpha),0),$$

$$(r(X_2(\beta,t),t),\ s(X_2(\beta,t),t)) = (0,s_0(\beta))$$

for all $t \geq T$. For all cases in Theorem 1, we either have $\lambda_r(r,0) \neq 0$ for $0 < |r| \leq \delta$ and $F(\cdot)$ is nonconstant or $\mu_s(0,s) \neq 0$ for $0 < |s| \leq \delta$ and $G(\cdot)$ is nonconstant. Hence we have

$$\max_\alpha \frac{d}{d\alpha}\lambda(r_0(\alpha),0) > 0 \quad \text{or} \quad \min_\beta \frac{d}{d\beta}\mu(0,s_0(\beta)) < 0.$$

From (2.12) and (2.13) it is easy to see that we can find $T_1 \geq T$ or $T_2 \geq T$ such that $Z_1(\alpha^*,T_1) = 0$ or $Z_2(\beta^*,T_2) = 0$, where α^* and β^* are some maximum and minimum points for $\frac{d}{d\alpha}\lambda(r_0(\alpha),0)$ and $\frac{d}{d\beta}\mu(0,s_0(\beta))$ respectively. Q.E.D.

If λ satisfies (1.9) with some $a_i = 0$ and μ satisfies (1.10) with some $d_i = 0$, for example, $\lambda = \lambda_0 + rs + h_1(s)$ and $\mu = \mu_0 + rs + h_2(r)$, then it is possible that $(r(x,t),s(x,t))$ is still smooth up to time T (sufficiently small δ and smooth F and G). In this case there is not enough time for nonlinear waves to interact and develop singularities. We have globally smooth solutions.

4. PROOF OF THEOREM 2

First we need some simple lemmas. In the following $O(1)$ is a constant which depends on system (1.4), f and g only.

LEMMA 1.

$$|\Lambda_i(r,s)| = O(1)\varepsilon, \quad i = 1,2,$$
$$\rho_1(\alpha,t) = 1 + O(1)\varepsilon, \tag{4.1}$$
$$\rho_2(\beta,t) = 1 + O(1)\varepsilon.$$

Proof. It is obvious from (2.9), (2.10), (2.14), (2.15), (2.3), (2.4) and (1.6). Q.E.D.

LEMMA 2. Let

$$m_1 = \min_{\alpha,\beta} \lambda(r_0(\alpha),s_0(\beta)), \quad M_1 = \max_{\alpha,\beta} \lambda(r_0(\alpha),s_0(\beta)),$$
$$m_2 = \min_{\alpha,\beta} \mu(r_0(\alpha),s_0(\beta)), \quad M_2 = \max_{\alpha,\beta} \mu(r_0(\alpha),s_0(\beta)).$$

Then

$$m_1 = \lambda_0 + O(1)\varepsilon, \quad M_1 = \lambda_0 + O(1)\varepsilon, \tag{4.2}$$
$$m_2 = \mu_0 + O(1)\varepsilon, \quad M_2 = \mu_0 + O(1)\varepsilon.$$

Furthermore, for a given α and t, define $\beta(\alpha,t) \leq \alpha$ by

$$X_1(\alpha,t) = X_2(\beta(\alpha,t),t).$$

Similarly, for a given β and t, define $\alpha(\beta,t) \geq \beta$ by

$$X_2(\beta,t) = X_1(\alpha(\beta,t),t).$$

Then we have

$$(m_1+m_2)t \leq \alpha - \beta(\alpha,t) \leq (M_1+M_2)t, \tag{4.3}$$
$$(m_1+m_2)t \leq \alpha(\beta,t) - \beta \leq (M_1+M_2)t.$$

Proof. (4.2) are obvious from (1.7), (1.8), (1.9) and (1.10). To prove (4.3) we see that

$$\alpha - M_1 t \leq X_1(\alpha,t) \leq \alpha - m_1 t.$$

Hence

$$(\alpha-M_1 t) - M_2 t \leq \beta(\alpha,t) \leq (\alpha-m_1 t) - m_2 t.$$

This completes the proof of (4.3) for $\beta(\alpha,t)$. Similarly we can prove the other half of (4.3). Q.E.D.

LEMMA 3. For a given α, we define $t_1(\beta;\alpha)$ for every $\beta \leq \alpha$, such that

$$X_1(\alpha,t_1(\beta;\alpha)) = X_2(\beta,t_1(\beta;\alpha)). \tag{4.4}$$

For a given β, we define $t_2(\alpha;\beta)$ for every $\alpha \geq \beta$, such that

$$X_2(\beta,t_2(\alpha;\beta)) = X_1(\alpha,t_2(\alpha;\beta)). \tag{4.5}$$

Then

$$\frac{dt_1(\beta;\alpha)}{d\beta} = -\left.\frac{z_2}{\lambda+\mu}\right|_{(\beta,t_1(\beta;\alpha))} < 0, \tag{4.6}$$

and

$$\frac{dt_2(\alpha;\beta)}{d\alpha} = \left.\frac{z_1}{\lambda+\mu}\right|_{(\alpha,t_2(\alpha;\beta))} > 0. \tag{4.7}$$

Proof. Taking the derivatives of (4.4) and (4.5) with respect to β and α respectively and using (2.1) and (2.2), we obtain the respective results. Q.E.D.

LEMMA 4. If λ satisfies (1.9) and μ satisfies (1.10), then

(i) (a) $z_1(\alpha,t) = O(1)\ [1 + \varepsilon^p t]$, (4.8)

(b) $z_2(\beta,t) = O(1)\ [1 + \varepsilon^q t]$, (4.9)

(ii) (a) if $\beta_2 \le \beta_1 \le \alpha$, then

$$0 \le t_1(\beta_2;\alpha) - t_1(\beta_1;\alpha) \le O(1)[1+\varepsilon^q t](\beta_1-\beta_2), \tag{4.10}$$

(b) if $\beta \le \alpha_1 \le \alpha_2$, then

$$0 \le t_2(\alpha_2;\beta) - t_2(\alpha_1;\beta) \le O(1)[1+\varepsilon^p t](\alpha_2-\alpha_1), \tag{4.11}$$

(iii) (a) $$|z_1(\alpha,t_2)-z_1(\alpha,t_1)| = O(1)\varepsilon^p(t_2-t_1) + O(1)\varepsilon(1+\varepsilon^p t_1), \quad t_2 > t_1 \tag{4.12}$$

(b) $$|z_2(\beta,t_2)-z_2(\beta,t_1)| = O(1)\varepsilon^q(t_2-t_1) + O(1)\varepsilon(1+\varepsilon^q t_1), \quad t_2 > t_1. \tag{4.13}$$

Proof. (i) (a) and (i) (b) are easy consequences of (2.12), (2.13), Lemma 1, (1.9) and (1.10). (ii) (a) and (ii) (b) are easy consequences of (4.6), (4.7) and (4.8), (4.9). To prove (iii) (a) we see that from (2.12)

$$\begin{aligned} z_1(\alpha,t_2)-z_1(\alpha,t_1) &= \frac{\rho_1(\alpha,t_2)-\rho_1(\alpha,t_1)}{\rho_1(\alpha,0)}\left[1-r_0'(\alpha)\int_0^{t_1}\lambda_r\frac{\rho_1(\alpha,0)}{\rho_1(\alpha,t')}\,dt'\right] \\ &\quad -\rho_1(\alpha,t_2)r_0'(\alpha)\int_{t_1}^{t_2}\lambda_r\frac{1}{\rho_1(\alpha,t')}\,dt' \\ &= \frac{\rho_1(\alpha,t_2)-\rho_1(\alpha,t_1)}{\rho_1(\alpha,t_1)}\,z_1(\alpha,t_1) \\ &\quad -\rho_1(\alpha,t_2)r_0'(\alpha)\int_{t_1}^{t_2}\lambda_r\frac{1}{\rho_1(\alpha,t')}\,dt'. \end{aligned}$$

Using (4.1), we obtain (iii) (a). Similarly we can prove (iii) (b). This completes the proof. Q.E.D.

LEMMA 5. If λ satisfies (1.9) and μ satisfies (1.10), then

$$\text{(i)}\quad Z_1(\alpha,t) = 1-r_0'(\alpha)\int_0^t H_1(r_0(\alpha),s(X_1(\alpha,t'),t'))dt' + O(1)\varepsilon[1+\varepsilon^p t]$$

$$= 1 + \frac{r_0'(\alpha)}{\lambda_0+\mu_0}\int_\alpha^{\beta(\alpha,t)} H_1(r_0(\alpha),s_0(\beta))Z_2(\beta,t_1(\beta;\alpha))d\beta$$

$$+ O(1)\varepsilon[1+\varepsilon^p t + (\varepsilon^p t)(\varepsilon^q t)],$$

$$\text{(ii)}\quad Z_2(\beta,t) = 1+s_0'(\beta)\int_0^t H_2(r(X_2(\beta,t'),s_0(\beta))dt' + O(1)\varepsilon[1+\varepsilon^q t]$$

$$= 1 + \frac{s_0'(\beta)}{\lambda_0+\mu_0}\int_\beta^{\alpha(\beta,t)} H_2(r_0(\alpha)+s_0(\beta))Z_1(\alpha,t_2(\alpha;\beta))d\alpha$$

$$+ O(1)\varepsilon[1+\varepsilon^q t + (\varepsilon^p t)(\varepsilon^q t)].$$

Proof. From (2.12), (1.9) and (4.1) we have

$$Z_1(\alpha,t) = (1+O(\varepsilon))[1-r_0'(\alpha)\int_0^t [H_1(r_0(\alpha),s(X_1(\alpha,t'),t')) + O(\varepsilon^p)][1+O(\varepsilon)]dt'$$

$$= 1 - r_0'(\alpha)\int_0^t H_1(r_0(\alpha),s(X_1(\alpha,t'),t'))dt' + O(1)\varepsilon[1+\varepsilon^p t]$$

$$= 1 - r_0'(\alpha)\int_\alpha^{\beta(\alpha,t)} H_1(r_0(\alpha),s_0(\beta))\frac{dt_1(\beta;\alpha)}{d\beta}\,d\beta + O(1)\varepsilon[1+\varepsilon^p t]$$

$$= 1 + r_0'(\alpha)\int_\alpha^{\beta(\alpha,t)} H_1(r_0(\alpha),s_0(\beta))\frac{Z_2(\beta,t_1(\beta;\alpha))}{\lambda_0+\mu_0}(1+O(\varepsilon))d\beta+O(1)\varepsilon[1+\varepsilon^p t]$$

$$= 1 + \frac{r_0'(\alpha)}{\lambda_0+\mu_0}\int_\alpha^{\beta(\alpha,t)} H_1(r_0(\alpha),s_0(\beta))Z_2(\beta,t_1(\beta;\alpha))d\beta$$

$$+ O(1)\varepsilon[1+\varepsilon^p t + (\varepsilon^p t)(\varepsilon^q t)].$$

This proves (i). Similarly we can prove (ii). Q.E.D.

LEMMA 6. If λ satisfies (1.9), μ satisfies (1.10) and

$$\alpha - (N_2+1)L_2 < \beta(\alpha,t) \le \alpha-N_2L_2,$$

$$\beta + N_1L_1 \le \alpha(\beta,t) < \beta + (N_1+1)L_1,$$

Then

$$Z_1(\alpha,t) = 1-\frac{M_1'(\alpha)}{\lambda_0+\mu_0}N_2L_2\varepsilon^p+O(1)\varepsilon[1+\varepsilon^p t+\varepsilon^q t+(\varepsilon^p t)(\varepsilon^q t)+(\varepsilon^p t)^2(\varepsilon^q t)+(\varepsilon^p t)(\varepsilon^q t)^2],$$

$$Z_2(\beta,t) = 1+\frac{M_2'(\beta)}{\lambda_0+\mu_0}N_1L_1\varepsilon^q+O(1)\varepsilon[1+\varepsilon^q t+\varepsilon^p t+(\varepsilon^p t)(\varepsilon^q t)+(\varepsilon^q t)^2(\varepsilon^p t)+(\varepsilon^q t)(\varepsilon^p t)^2],$$

where

$$M_1(\alpha) = \frac{1}{L_2}\int_0^{L_2}\int_0^{f(\alpha)} \varepsilon^{-p+1} H_1(\varepsilon\eta,\varepsilon g(\beta))d\eta d\beta,$$

$$M_2(\beta) = \frac{1}{L_1}\int_0^{L_1}\int_0^{g(\beta)} \varepsilon^{-q+1} H_2(\varepsilon f(\alpha),\varepsilon\xi)d\xi d\alpha.$$

Proof. From the assumptions, we have

$$Z_1(\alpha,t) = Z_1(\alpha,t) - Z_1(\alpha,t_1(\alpha-N_2L_2;\alpha)) + Z_1(\alpha,t_1(\alpha-N_2L_2;\alpha)).$$

From (4.12) and (4.10) we obtain

$$\begin{aligned}
&|Z_1(\alpha,t) - Z_1(\alpha,t_1(\alpha-N_2L_2;\alpha))| \\
&\quad = O(1)\varepsilon^p(t-t_1(\alpha-N_2L_2;\alpha)) + O(1)\varepsilon(1+\varepsilon^p t_1(\alpha-N_2L_2;\alpha)) \\
&\quad = O(1)\varepsilon^p[1+\varepsilon^q t](\beta-\alpha+N_2L_2) + O(1)\varepsilon(1+\varepsilon^p t) \\
&\quad = O(1)\varepsilon^p[1+\varepsilon^q t]L_2 + O(1)\varepsilon(1+\varepsilon^p t) \\
&\quad = O(1)\varepsilon(1+\varepsilon^p t+\varepsilon^q t) \qquad (\text{since } p \geq 2).
\end{aligned} \tag{4.14}$$

Now from Lemma 5 (i) we have

$$\begin{aligned}
Z_1(\alpha,t_1(\alpha-N_2L_2;\alpha)) &= 1+\frac{r_0'(\alpha)}{\lambda_0+\mu_0}\int_\alpha^{\alpha-N_2L_2} H_1(r_0(\alpha),s_0(\beta))Z_2(\beta,t_1(\beta;\alpha))\,d\beta \\
&\quad + O(1)\varepsilon[1+\varepsilon^p t+(\varepsilon^p t)(\varepsilon^q t)] \\
&= 1 + \sum_{n=0}^{(N_2-1)} \frac{A_n}{\lambda_0+\mu_0} + O(1)\varepsilon[1+\varepsilon^p t+(\varepsilon^p t)(\varepsilon^q t)],
\end{aligned} \tag{4.15}$$

where

$$A_n = r_0'(\alpha)\int_{\alpha-nL_2}^{\alpha-(n+1)L_2} H_1(r_0(\alpha),s_0(\beta))Z_2(\beta,t_1(\beta;\alpha))\,d\beta \tag{4.16}$$

Using Lemma (5) (ii) we have

$$A_n = \int_{\alpha-nL_2}^{\alpha-(n+1)L_2} r_0'(\alpha)H_1(r_0(\alpha),s_0(\beta))\,d\beta+B_n+O(1)\varepsilon[1+\varepsilon^q t]\varepsilon^p L_2, \tag{4.17}$$

where

$$B_n = \int_{\alpha-nL_2}^{\alpha-(n+1)L_2} r_0'(\alpha)H_1(r_0(\alpha),s_0(\beta))s_0'(\beta)\int_0^{t_1(\beta;\alpha)} H_2(r(X_2(\beta,t'),t'),s_0(\beta))\,dt'\,d\beta.$$

A change of integration variable from t' to α' we get

$$\begin{aligned}
B_n &= \int_{\alpha-nL_2}^{\alpha-(n+1)L_2} r_0'(\alpha)H_1(r_0(\alpha),s_0(\beta))s_0'(\beta)\int_\beta^\alpha H_2(r_0(\alpha'),s_0(\beta))\frac{Z_1(\alpha',t_2(\alpha';\beta))}{\lambda+\mu}\,d\alpha'\,d\beta \\
&= \frac{1}{\lambda_0+\mu_0}C_n + O(1)\varepsilon[1+\varepsilon^p t]\varepsilon^q(n+1)L_2\varepsilon^p L_2,
\end{aligned} \tag{4.18}$$

where

$$C_n = \int_{\alpha-nL_2}^{\alpha-(n+1)L_2} r_0'(\alpha)H_1(r_0(\alpha),s_0(\beta))s_0'(\beta)\int_\beta^\alpha H_2(r_0(\alpha'),s_0(\beta))Z_1(\alpha',t_2(\alpha';\beta))\,d\alpha'\,d\beta.$$

We write C_n as

$$C_n = D_n + E_n + F_n,$$

where

$$D_n = \int_{\alpha-nL_2}^{\alpha-(n+1)L_2} r_0'(\alpha)H_1(r_0(\alpha),s_0(\beta))s_0'(\beta)\int_{\beta}^{\alpha-nL_2} H_2(r_0(\alpha'),s_0(\beta))\cdot$$
$$\cdot Z_1(\alpha',t_2(\alpha';\beta))\,d\alpha' d\beta$$

$$E_n = \int_{\alpha-nL_2}^{\alpha-(n+1)L_2} r_0'(\alpha)H_1(r_0(\alpha),s_0(\beta))s_0'(\beta)\int_{\alpha-nL_2}^{\alpha} H_2(r_0(\alpha'),s_0(\beta))\cdot$$
$$\cdot Z_1(\alpha',t_2(\alpha';\alpha-nL_2))\,d\alpha' d\beta$$

$$F_n = \int_{\alpha-nL_2}^{\alpha-(n+1)L_2} r_0'(\alpha)H_1(r_0(\alpha),s_0(\beta))s_0'(\beta)\int_{\alpha-nL_2}^{\alpha} H_2(r_0(\alpha'),s_0(\beta))\cdot$$
$$\cdot[Z_1(\alpha',t_2(\alpha';\beta)) - Z_1(\alpha,t_2(\alpha';\alpha-nL_2))]\,d\alpha' d\beta.$$

But we have

$$D_n = O(1)[1+\varepsilon^p t](\varepsilon^q L_2)(\varepsilon^p L_2), \tag{4.19}$$

$$E_n = \int_{\alpha-nL_2}^{\alpha} d\alpha' r_0'(\alpha)Z_1(\alpha' t_2(\alpha';\alpha-nL_2))\int_{\alpha-nL_2}^{\alpha-(n+1)L_2} d\beta\,\frac{\partial}{\partial\beta}K = 0, \tag{4.20}$$

$$F_n = \int_{\alpha-nL_2}^{\alpha-(n+1)L_2} r_0'(\alpha)H_1(r_0(\alpha),s_0(\beta))s_0'(\beta)\int_{\alpha-nL_2}^{\alpha} H_2(r_0(\alpha'),s_0(\beta))$$
$$\times\ [Z_1(\alpha',t_1(\beta;\alpha'))-Z_1(\alpha',t_1(\alpha-nL_2;\alpha'))]\,d\alpha' d\beta,$$

where K is some polynominal in $s_0(\beta)$. From (4.12) and (4.10) we have

$$\begin{aligned}
&|Z_1(\alpha',t_1(\beta;\alpha')) - Z_1(\alpha',t_1(\alpha-nL_2;\alpha'))| \\
&\quad = O(1)\varepsilon^p(t_1(\beta;\alpha') - t_1(\alpha-nL_2;\alpha')) + O(1)\varepsilon(1+\varepsilon^p t) \\
&\quad = O(1)\varepsilon^p[1+\varepsilon^q t]L_2 + O(1)\varepsilon(1+\varepsilon^p t).
\end{aligned}$$

Hence

$$\begin{aligned}
F_n &= O(1)\varepsilon^p[1+\varepsilon^q t](\varepsilon^q nL_2)(\varepsilon^p L_2) \\
&\quad + O(1)\varepsilon(1+\varepsilon^p t)(\varepsilon^q nL_2)(\varepsilon^p L_2) \\
&= O(1)\varepsilon(1+\varepsilon^q t+\varepsilon^p t)(\varepsilon^q nL_2)(\varepsilon^p L_2).
\end{aligned} \tag{4.21}$$

Combining (4.14), (4.15), (4.16), (4.17), (4.18), (4.19), (4.20), (4.21) and using the fact that $N_2L_2 = O(1)t$ we finally have

$$\begin{aligned}
Z_1(\alpha,t) = 1 + \sum_{n=0}^{(N_2-1)} \frac{G_n}{\lambda_0+\mu_0} &+ O(1)\varepsilon[1+\varepsilon^p t+\varepsilon^q t+(\varepsilon^p t)(\varepsilon^q t) \\
&+ (\varepsilon^p t)^2(\varepsilon^q t)+(\varepsilon^p t)(\varepsilon^q t)^2],
\end{aligned}$$

where

$$\begin{aligned} G_n &= \int_{\alpha-nL_2}^{\alpha-(n+1)L_2} r_0'(\alpha)H_1(r_0(\alpha),s_0(\beta))d\beta \\ &= -\frac{d}{d\alpha}\int_0^{L_2} d\beta \int_0^{r_0(\alpha)} dr' H_1(r',s_0(\beta)) \\ &= -\varepsilon^p \frac{d}{d\alpha}\int_0^{L_2} d\beta \int_0^{f(\alpha)} \varepsilon^{-p+1} H_1(\varepsilon\eta,\varepsilon g(\beta))d\eta \\ &= -L_2\varepsilon^p \frac{d}{d\alpha} M_1(\alpha). \end{aligned}$$

This completes the proof for half of the lemma. The other half can be similarly proved. Q.E.D.

Now we are in a position to prove Theorem 2.

Proof of Theorem 2. If λ satisfies (1.7) and r_0 is not a constant function, then from (2.12)

$$Z_1(\alpha,t) \le \frac{\rho_1(\alpha,t)}{\rho_1(\alpha,0)}[1 - C\eta t],$$

where C is a positive lower bound of $\rho_1(\alpha,t)/\rho_1(\alpha,0)$ and

$$\eta = \max_{0\le\alpha\le L_1} \{r_0'(\alpha)[\min_{0\le\beta\le L_2} \lambda_r(r_0(\alpha),s_0(\beta))]\} \text{ if } \lambda_r \ge 0,$$

$$\eta = \max_{0\le\alpha\le L_1} \{r_0'(\alpha)[\max_{0\le\beta\le L_2} \lambda_r(r_0(\alpha),s_0(\beta))]\} \text{ if } \lambda_r \le 0.$$

Since $r_0(\cdot)$ is not a constant function and $r_0(\cdot)$ and $s_0(\cdot)$ are periodic, we have $\eta > 0$. Hence there exist an α and a T such that $Z_1(\alpha,T) = 0$. This proves the theorem when λ satisfies (1.7). Now assume that λ satisfies (1.9) and μ satisfies (1.10). We have

$$\max_{0\le\alpha\le L_1} M_1'(\alpha) = M_1'(\alpha^*) > 0$$

and

$$-\min_{0\le\beta\le L_2} M_2'(\beta) = -M_2'(\beta^*) > 0.$$

From Lemma 6 and using the facts that

$$N_2L_2 = (\lambda_0+\mu_0)t + O(1)\varepsilon t,$$

$$N_1L_1 = (\lambda_0+\mu_0)t + O(1)\varepsilon t,$$

we have

$$Z_1(\alpha^*,t) = 1 - M_1'(\alpha^*)t\varepsilon^p + O(1)\varepsilon Q(\varepsilon^p t,\varepsilon^q t), \tag{4.22}$$

$$Z_2(\beta^*,t) = 1 + M_2'(\beta^*)t\varepsilon^q + O(1)\varepsilon Q(\varepsilon^p t,\varepsilon^q t), \tag{4.23}$$

where

$$Q(\varepsilon^p t,\varepsilon^q t) = 1+\varepsilon^p t+\varepsilon^q t+(\varepsilon^p t)(\varepsilon^q t)+(\varepsilon^p t)^2(\varepsilon^q t)+(\varepsilon^p t)(\varepsilon^q t)^2.$$

Choose $C_1 > 0$ and $C_2 > 0$ such that

$$1 - M_1'(\alpha^*)C_1 < 0 \quad\text{and}\quad 1 + M_2'(\beta^*)C_2 < 0.$$

Let

$$T(\varepsilon) = \min\{C_1\varepsilon^{-p}, C_2\varepsilon^{-q}\}.$$

Then

$$\varepsilon^p T(\varepsilon) \leq C_1, \ \varepsilon^q T(\varepsilon) \leq C_2$$

and

$$Q(\varepsilon^p T(\varepsilon), \varepsilon^q T(\varepsilon)) \quad \text{is bounded.}$$

Therefore there exists a ε_0, such that for all $0 < \varepsilon \leq \varepsilon_0$

$$Z_1(\alpha^*, T(\varepsilon)) = 1 - M_1'(\alpha^*)\varepsilon^p T + O(1)\varepsilon Q(\varepsilon^p T(\varepsilon), \varepsilon^q T(\varepsilon)) \leq 0$$

or

$$Z_2(\beta^*, T(\varepsilon)) = 1 + M_2'(\beta^*)\varepsilon^q T + O(1)\varepsilon Q(\varepsilon^p T(\varepsilon), \varepsilon^q T(\varepsilon)) \leq 0.$$

This completes the proof of Theorem 2. Q.E.D.

BIBLIOGRAPHY

[1] Glimm, James, and Lax, Peter D., "Decay of solutions of systems of nonlinear hyperbolic conservation laws", Memoirs of the Amer. Math. Soc., 101, 1970.

[2] John, Fritz, "Formation of singularities in one-dimensional nonlinear wave propagation," Comm. Pure Appl. Math., 27, 1974, 377-405.

[3] Klainerman, Sergiu, and Majda, Andrew, "Formation of singularities for wave equations including the nonlinear vibrating string," Comm. Pure Appl. Math., 33, 1980, 241-263.

[4] Lax, Peter D., "Hyperbolic systems of conservation laws, II", Comm. Pure Appl. Math., 10, 1957, 227-241.

[5] Lax, Peter D., "The formation and decay of shock waves," Amer. Math. Monthly, 79, 1972, 227-241.

[6] Lax, Peter D., "Development of singularities of solutions of nonlinear hyperbolic partial differential equations," Journal of Math. Physics, 5, 1964, 611-613.

[7] Liu, T. P., "Development of singularities in the nonlinear waves for quasi-linear hyperbolic partial differential equations," J. Diff. Eqns., 33, 1979, 92-111.

DEPARTMENT OF APPLIED MATHEMATICS
NATIONAL CHIAO TUNG UNIVERSITY
HSINCHU, TAIWAN 300
REPUBLIC OF CHINA

Contemporary Mathematics
Volume 17, 1983

CONSERVATION LAWS

Ronald J. DiPerna

We shall discuss some recent results in the theory of conservation laws concerning the convergence of approximate solutions. The setting is provided by a strictly hyperbolic system of conservation laws in one space dimension

$$u_t + f(u)_x = 0 \ . \tag{1}$$

Here $u = u(x,t) \in R^n$ and f is a smooth nonlinear map on R^n whose Jacobian ∇f has n real and distinct eigenvalues:

$$\lambda_1(u) < \lambda_2(u) < \ldots < \lambda_n(u) \ .$$

We are interested in the approximation of the solution to the Cauchy problem by a diffusion process

$$u_t + f(u)_x = \varepsilon D u_{xx}; \quad u = u_\varepsilon(x,t) \tag{2}$$

and by discrete solutions generated by finite difference schemes which are conservative in the sense of Lax and Wendroff [5].

$$\partial_t u + \partial_x f(u) = 0; \quad u = u(x,t;\Delta x) \ .$$

In the setting of hyperbolic conservation laws, the L^∞ norm and the total variation norm provide a useful pair of metrics in which to investigate stability: the L^∞ norm measures the solution amplitude and the total variation norm measures the solution gradient. The role of the function spaces is indicated by a fundamental theorem of Glimm [3] concerning the stability and convergence of his random choice approximations.

THEOREM 1. If the total variation of the initial data u_0 is sufficiently small then the random choice approximations converge pointwise almost everywhere to a globally defined distribution solution and maintain uniform control on the amplitude and spatial variation, i.e.

$$|u(\cdot,t;\Delta x)|_\infty \leq \text{const}\, |u_0(\cdot)|_\infty$$

$$T\,V\, u(\cdot,t;\Delta x) \leq \text{const}\ T\,V\, u_0,$$

where the constants are independent of the mesh length Δx.

0271-4132/83/0000-1382/$01.75

Here the convergence of a sequence of random choice approximations follows from a compactness argument in the strong topology based upon uniform control of the function and its derivatives. The proof of Theorem 1 is based upon a general study of elementary wave interactions in the exact solution and in the random choice approximations. In this note we shall mention some recent work providing convergence of parabolic regularizations and conservative difference approximations. The analysis is based on a viewpoint which appeals to averaged quantities and the weak topology. The proof employs the theory of compensated compactness which originates in the work of Tartar [7] and Murat [6]. The main step gives a proof of a conjecture of Tartar [7]. Roughly, the statement is that, for a class of approximation methods, L^∞ stability together with correct entropy production implies convergence.

We shall state a representative result in the setting of a diffusion process. Consider a system of two conservation laws which is genuinely nonlinear in Lax's sense [4] that wave speeds $\lambda_j(u)$ change monotonically with wave amplitude, i.e.

$$r_j \cdot \nabla \lambda_j \neq 0$$

where r_j denotes the right eigenvector of ∇f associated with λ_j. Assume that the diffusion matrix D in system (2) induces correct entropy production, cf [2,8].

THEOREM. If u_ε is a sequence of smooth solutions of (2) satisfying

$$|u_\varepsilon|_\infty \leq M ,$$

for some constant M independent of ε, there exists a subsequence which converges pointwise almost everywhere to a solution u of the hyperbolic system (1).

Hence the solution operators S_ε of (2) form a family of mappings which is compact from L^∞ to L^1_{loc} uniformly with respect to ε. The source of the compactness lies in the nonlinear structure of the eigenvalues and in the loss of information induced by the entropy structure of process.

We refer the reader to [6] for results on the scalar conservation law and to [2] for results on systems of two equations and additional references. Regarding the general theory of compensated compactness we mention the lecture notes of Dacorogna [13] and a forthcoming article by Tartar in the NATO/LMS Advanced Study Institute on Systems of Nonlinear Partial Differential Equations held in Oxford, July 25 - August 7, 1982 and organized by J. M. Ball et al. With regard to the elliptic conservation laws and the weak topology we refer the reader to Ball [1].

References

1. Ball, J. M., Convexity conditions and existence theorems in nonlinear elasticity, Arch. Rational Mech. Anal.,63 (1977), 337-403.

2. DiPerna, R. J., Convergence of approximate solutions to conservation laws, Arch. Rational Mech. Anal., to appear (1983).

3. Glimm, J., Solutions in the large for nonlinear hyperbolic systems of equations, Comm. Pure Appl. Math., 18 (1965), 697-715.

4. Lax, P. D., Hyperbolic systems of conservation laws II, Comm. Pure Appl. Math., 10 (1957), 537-566.

5. Lax, P. D. and B. Wendroff, Systems of conservation laws, Comm. Pure Appl. Math., 13 (1960), 217-237.

6. Murat, F., Compacité par compensation, Ann. Scuola Norm. Sup. Pisa, 5 (1978), 489-507.

7. Tartar, L., Compensated compactness and applications to partial differential equations, in Research Notes in Mathematics, Nonlinear Analysis and Mechanics: Heriot-Watt Symposium, Vol. 4, ed. R. J. Knops, Pitman Press, 1979.

8. Pego, R., Viscosity matrices for systems of conservation laws, preprint, Center for Pure and Applied Mathematics, University of California, Berkeley.

9. Dacorogna, B., Weak continuity and weak lower semicontinuity of nonlinear functionals, Lefschitz Center for Dynamical Systems Lecture Notes #81-77, Brown University (1981).

DEPARTMENT OF MATHEMATICS
DUKE UNIVERSITY
DURHAM, NORTH CAROLINA 27706

Contemporary Mathematics
Volume 17, 1983

RECENT RESULTS FOR HYPERBOLIC CONSERVATION LAWS

J. M. Greenberg

This note shall review some recent results for non-linear hyperbolic partial differential equations. No proofs will be presented since the details will appear in [1] and [2].

The first result deals with solutions of the quasilinear wave equation

$$\frac{\partial u}{\partial t} - \frac{\partial v}{\partial x} = 0 \quad \text{and} \quad \frac{\partial v}{\partial t} - \frac{\partial \sigma(u)}{\partial x} = 0 . \tag{1}$$

The function σ satisfies $\sigma'(u) > 0$ for all u and the nonessential normalization conditions $\sigma(0) = 0$ and $\sigma'(0) = 1$. Solutions are sought on $0 < x < L$ and $0 < t$ which satisfy the boundary conditions

$$\sigma(u(0,t)) = rv(0,t), \quad 0 < r < \infty, \text{ and } \quad v(L,t) = 0 . \tag{2}$$

If one thinks of (1) as modelling the longitudinal motions of an elastic bar, then the condition $v(L,t) = 0$ implies that the end $x = L$ is fixed while the condition $\sigma(u(0,t)) = rv(0,t)$ implies that end $x = 0$ is connected to a viscous dashpot. Solutions of (1) and (2) are dissipative; that is the following energy identify holds:

$$\frac{d}{dt}\int_0^L \left(\frac{v^2}{2} + \int_0^u \sigma(\eta)d\eta\right)(x,t)dx = -rv^2(0,t) . \tag{3}$$

The principal result of [1] deals with solutions of (1) and (2) which satisfy the initial condition

$$\lim_{t\to 0^+}(u,v)(x,t) = (u_o,v_o)(x), \quad 0 \le x \le L , \tag{4}$$

and is summarized in Theorem 1.

THEOREM 1. If the data (u_o,v_o) are small in the C^1 norm and satisfy the compatibility conditions

$$\sigma(u_o(0)) = rv_o(0), \quad v_o'(0) = ru_o'(0), \quad v_o(L) = 0, \quad \text{and} \quad u_o'(L) = 0, \tag{5}$$

then the problem (1), (2), and (4) has a unique smooth solution which decays

0271-4132/83/0000-1385/$02.25

to zero exponentially (in time) in the C^1 norm. Moreover, if $r = 1$, the solutions decay at a superexponential rate. □

The proof of Theorem 1 relies on an analysis of the Riemann Invariants

$$\alpha = \frac{1}{2}\left(v - \int_0^u \sqrt{\sigma'(\eta)}\,d\eta\right) \quad \text{and} \quad \beta = \frac{1}{2}\left(v + \int_0^u \sqrt{\sigma'(\eta)}\,d\eta\right) \tag{6}$$

and their x derivatives. The decay rates are essentially the same as those obtained for the linearized version of (1) and (2), namely the equations

$$u_t - v_x = 0 \quad \text{and} \quad v_t - u_x = 0, \quad 0 < x < L \quad \text{and} \quad 0 < t, \qquad (1)_{Lin}$$

and

$$u(0,t) = rv(0,t), \quad 0 < r < \infty, \quad \text{and} \quad v(L,t) = 0. \qquad (2)_{Lin}$$

When $r \neq 1$, solutions of $(1)_{Lin}$ and $(2)_{Lin}$ decay as $e^{-\lambda t}$ where $\lambda = \frac{1}{2L}\log\left|\frac{1+r}{1-r}\right|$. When $r = 1$, solutions of $(1)_{Lin}$ and $(2)_{Lin}$ are identically zero for all $t > 2L$.

The second set of results deal with solutions of the system

$$u_t + \sigma_x = 0 \quad \text{and} \quad (\sigma - f(u))_t + (\sigma - \mu f(u)) = 0. \tag{7}$$

This system serves as a model for the equations of motion of a viscoelastic solid. Equation $(7)_1$ is the balance law and $(7)_2$ represents the constitutive equation for the material. The relation $\sigma = f(u)$ defines the instantaneous response of the material and $\sigma = \mu f(u)$ gives the equilibrium response of the material. Typical assumptions on f on $u \geq 0$ are $f(0) = 0$, $0 < f'(u)$, and $0 < f''(u)$. In order to guarantee that stress relaxation and creep occur the paramater μ is constrained by the inequality $0 < \mu < 1$.

For any positive constant u there are traveling wave solutions of (7) which satisfy $\lim_{x \to -\infty} (u,\sigma) = (u_-, \mu f(u_-))$ and $\lim_{x \to +\infty} (u,\sigma) = (0,0)$. These solutions can be regarded as functions of $\tau = t - \frac{x}{c}$ where $c = \frac{\mu f(u_-)}{u_-}$. The u component of these traveling waves satisfies

$$\frac{d}{d\tau}(f(\tilde{u}) - c\tilde{u}) = (c\tilde{u} - \mu f(\tilde{u})) \tag{8}$$

and the boundary conditions

$$\lim_{\tau \to -\infty} \tilde{u}(\tau) = 0 \quad \text{and} \quad \lim_{\tau \to +\infty} \tilde{u}(\tau) = u_- > 0. \tag{9}$$

The σ component of the traveling wave is given by $\sigma = c\tilde{u}(\tau)$.

When $c < f'(0)$ the traveling wave solutions to (8) are smooth, strictly increasing, satisfy (9), and are unique to within a translation. When $c > f'(0)$ the traveling wave solutions are nondecreasing and unique to within a translation. One particular solution has the following properties:

(1) $\tilde{u}(\tau) = 0, \quad -\infty < \tau < 0,$

(2) $\lim_{\tau \to 0^+} \tilde{u}(\tau) = u_*$ where $u_* > 0$ satisfies the instantaneous Rankine-Hugoniot equation $f(u_*) = cu_*$, and

(3) $\tilde{u}(\tau)$ satisfies (8) on $(0,\infty)$ and meets the boundary condition (9) at plus infinity.

The goal of [2] was to establish a connection between the traveling wave solutions of (7) and solutions of the impact loading problem:

$$u_t + \sigma_x = 0 \quad \text{and} \quad (\sigma - f(u))_t + (\sigma - \mu f(u)) = 0, \quad x > 0 \quad \text{and} \quad t > 0, \tag{7}$$

$$\lim_{t \to 0^+} (u,\sigma)(x,t) = (0,0), \quad x > 0, \quad \text{and} \tag{10}$$

$$\lim_{x \to 0^+} \sigma(x,t) = \mu f(u_-), \quad t > 0. \tag{11}$$

One set of results dealt with the case where $c = \dfrac{\mu f(u_-)}{u_-} < f'(0)$ and $0 < u_- \ll 1$. In this case the following theorem was established.

<u>THEOREM 2</u>.

(1) There is a $C^2[0,\infty)$ curve $x = s(t)$ satisfying $s(0)$ and $\frac{ds}{dt} > 0$ such that $(u,\sigma) \equiv (0,0)$, $x > s(t)$ and $t \geq 0$,

(2) (u,σ) is C^1 and satisfies (7) and (11) in $0 < x < s(t)$ and $t > 0$. Moreover, in this region u and σ satisfy the following inequalities: $0 < u < u_-$, $0 < \sigma < \mu f(u_-)$, $0 < u_t$, $0 < \sigma_t$, $u_x < 0$, and $\sigma_x < 0$.

(3) The curve $x = s(t)$ is an admissible shock wave and satisfies the Rankine-Hugoniot relations for (7); that is $0 < u_-(s(t),t) = \lim_{\substack{x \to s(t) \\ x < s(t)}} u(x,t)$, $\sigma_-(s(t),t) = \lim_{\substack{x \to s(t) \\ x < s(t)}} \sigma(x,t) = f(u_-(s(t),t))$, $\dfrac{ds}{dt} = \dfrac{f(u_-(s(t),t))}{u_-(s(t),t)}$ and $s(0) = 0$. Moreover, $u_-(s(t),t)$ satisfies $\dfrac{du_-(s(t),t)}{dt} < 0$ and $\lim_{t \to \infty} u_-(s(t),t) = 0$. □

When $c = \dfrac{\mu f(u_-)}{u_-} < f'(0)$ and $0 < u_- \ll 1$ there is a relation between solutions of the impact loading problem and the smooth traveling wave solutions of (7). If one lets $x = x(\alpha,t)$, $t \geq \tau(\alpha)$, be a level line of u; i.e. $u(x(\alpha,t),t) = \alpha$, $0 < \alpha < u_-$ and $t \geq \tau(\alpha)$ where $\tau(\alpha) \geq 0$ is the first $\tau \geq 0$ such that either $u(0^+,\tau) = \alpha$ or $u_-(s(\tau),\tau) = \alpha$, and defines

$$c_{AV}(t) = \frac{1}{u(0^+,t) - u_-(s(t),t)} \int_{u_-(s(t),t)}^{u(0^+,t)} \frac{\partial x}{\partial t}(\alpha,t)d\alpha ,$$

then the arguments developed in [2] yield

$$\lim_{t\to\infty} c_{AV}(t) = c = \frac{\mu f(u_-)}{u_-} . \tag{12}$$

Equation (12) is a weak statement about the convergence of solutions to the impact loading problem to a traveling wave; it states that the average speed of propagation of the level lines of u converge to the speed at which the traveling waves propagate.

When $c = \frac{\mu f(u_-)}{u_-} \gg f'(0)$ the results of Theorem 2 are again true. In this case one is also able to show that

$$\tilde{u}(t - \frac{x}{c}) < u < u_- \quad \text{and} \quad c\tilde{u}(t - \frac{x}{c}) < \sigma(x,t) < \mu f(u_-) \tag{13}$$

in the region $0 \leq x \leq ct$ and $0 \leq t$ and that the shock wave $x = s(t)$ satisfies

$$ct < s(t) < ct + \frac{1}{c(c - f'(0))} \int_{u_*}^{u_-} \frac{(u_- - u)(f'(u) - c)}{u(c - \frac{\mu f(u)}{u})} du . \tag{14}$$

The function $\tilde{u}(\cdot)$ is the u component of the traveling wave with a jump discontinuity at $\tau = 0$ and the number $u_* > 0$ satisfies $cu_* = f(u_*)$. The bounds (13) and (14) imply that

$$\lim_{t\to\infty} (u,\sigma)(\lambda t,t) = \begin{cases} (u_-,\mu f(u_-)) , & \lambda < c = \frac{\mu f(u_-)}{u_-} \\ (0,0) , & \lambda > c = \frac{\mu f(u_-)}{u_-} \end{cases} . \tag{15}$$

The limit relations (12) and (15) point out the distinguished role played by the traveling waves. It is desired to obtain stranger convergence results. In the case where the traveling waves are smooth we conjecture that $\lim_{t\to\infty} x_t(\alpha,t) = \frac{\mu f(u_-)}{u_-}$ for any $0 < \alpha < u_-$ whereas in the case of the discontinuous traveling waves we would like to know about $\lim_{t\to\infty}(u,\sigma)(\varphi+ct,t)$ where $c = \frac{\mu f(u_-)}{u_-}$. We would also like to know that $\lim_{t\to\infty}\dot{s}(t) = c$. This last relation does not follow from the bounds (14) and to date we have not been able to establish this result.

REFERENCES

[1] J. M. Greenberg and Li Ta Tsien, The effect of boundary damping for the quasilinear wave equation. To appear in Jour. Diff. Equations.

[2] J. M. Greenberg and Ling Hsiao, The Riemann Problem for the system $u_t + \sigma_x = 0$ and $(\sigma - f(u))_t + (\sigma - \mu f(u)) = 0$. MRC Technical Summary Report 2281, University of Wisconsin, September 1981. To appear in Arch. Rat. Mech. Anal.

DEPARTMENT OF MATHEMATICS
THE OHIO STATE UNIVERSITY
COLUMBUS, OHIO 43210

Contemporary Mathematics
Volume 17, 1983

ON THE SMOOTHING OF INITIAL DISCONTINUITIES AND THE DEVELOPMENT OF SINGULARITIES IN SOLUTIONS OF CERTAIN QUASILINEAR HYPERBOLIC EQUATIONS

by David Hoff

ABSTRACT. In this paper we discuss certain questions relating to the smoothing of initial discontinuities and to the development of singularities in solutions of hyperbolic conservation laws. First, we describe necessary and sufficient conditions on the L^∞ Cauchy data for a single conservation law in several space variables under which the Volpert-Kruzkov solution will be locally Lipschitz continuous up to time T. When T is positive, our condition requires that the initial data have locally bounded variation. Next, we study systems of conservation laws in one space variable for which there exist a coordinate set of Riemann invariants. We describe conditions which guarantee the global existence of Lipschitz solutions, and we prove a partial converse. These results are achieved by a device which effectively uncouples the system of equations for $\partial u/\partial x$.

§1. Introduction

In this paper we discuss certain questions relating to the smoothing of initial discontinuities and to the development of singularities in solutions of certain hyperbolic conservation laws. In §2 we give a complete analysis of the situation for a scalar equation in several space variables by formulating necessary and sufficient conditions on the L^∞ Cauchy data under which the Volpert-Kruzkov solution will be locally Lipschitz up to time T. When T is positive, our condition requires that the initial data has locally bounded variation. Therefore the solution operator for this problem can never be smoothing on the class $L^\infty - LBV$. In §3 we extend some of these results to systems of conservation laws in one space variable for which there exist coordinate sets of Riemann invariants. The key feature in this analysis is a device which effectively uncouples the system of equations for $\frac{\partial u}{\partial x}$.

§2. The Scalar Conservation Law

In this section we formulate conditions on F and on $u_0 \in L^\infty(\mathbb{R}^n)$ under which the solution of the initial value problem

$$u_t + \operatorname{div} F(u) = 0 \tag{2.1}$$

$$u(x,0) = u_0(x) \tag{2.2}$$

0271-4132/83/0000-1387/$03.75

will be locally Lipschitz continuous in a strip $0 < t < T$. Here $F : \mathbb{R} \to \mathbb{R}^n$ and the div is with respect to x. We remark that Volpert in [10] and Kruzkov in [6] have established the global existence of weak solutions of (2.1)-(2.2) subject to a certain entropy condition. By using the chain rule for Sobolev functions, one can easily show that locally Lipschitz solutions of (2.1) always satisfy this entropy condition. The solutions which we shall study will always be locally Lipschitz in $\{t > 0\}$; they therefore will agree with the unique solutions in the sense of Volpert-Kruzkov.

It is well known that the solution of (2.1)-(2.2) need not be C^1 for all $t > 0$ no matter how smooth u_0 is. On the other hand, it is also possible that discontinuities in u_0 become smoothed out in positive time. Our task is therefore to formulate the precise conditions on u_0 and F which both guarantee the preservation of smoothness as well as provide for the smooth resolution of initial discontinuities.

We begin by making some simple observations about the case $n = 1$. We write (2.1) as

$$u_t + f(u)_x = 0 . \tag{2.3}$$

This equation requires that a C^1 solution be constant on the characteristic curves

$$x = x_0 + tf'(u_0(x_0)) ,$$

which are therefore lines in $x - t$ space. Such a solution ceases to be continuous as soon as two characteristic lines cross. This occurs at the smallest T for which

$$\frac{1}{T} = -\frac{f'(u_0(x_2)) - f'(u_0(x_1))}{x_2 - x_1}$$

for some x_1 and x_2. In this case all the characteristics emanating from $[x_1, x_2]$ meet at a point, and T may be characterized by

$$\frac{1}{T} = \max_x \{ \frac{d}{dx} f'(u_0(x)) \} .$$

In particular, when $f'(u_0(x))$ is increasing, the solution $u(x,t)$ will be smooth in all of $\{t > 0\}$.

On the other hand, when u_0 itself is not smooth, extra conditions must be imposed to insure that initial discontinuities are smoothed out in positive time. Consider, for example, the initial value problem for (2.3) with "Riemann" data

$$(2.4) \qquad u_0(x) = \begin{cases} u_\ell \ , & x < 0 \\ u_r \ , & x > 0 \ . \end{cases}$$

Assuming that $f'(u_\ell) < f'(u_r)$ (so that characteristics do not meet at positive times), we can display a smooth solution of (2.3)-(2.4) provided that f' has a C^1 inverse function a defined on $\text{conh}\{u_\ell , u_r\}$. This solution is the "rarefaction wave"

$$u(x,t) = \begin{cases} u_\ell \ , & x/t \le f'(u_\ell) \\ a(x/t) \ , & f'(u_\ell) \le x/t \le f'(u_r) \\ u_r \ , & f'(u_r) \le x/t \ . \end{cases}$$

Observe that $u(x,t)$ is locally Lipschitz in $\{t > 0\}$ but is not C^1. The smoothing in this example appears to depend crucially on the strict monotonicity of f' across the jump in u_0.

Heuristically, then, it appears that the solution of (2.3) will be smooth up to time T provided that $\frac{1}{T} \ge -\frac{d}{dx} f'(u_0(x))$ and that f'' should be nonzero across jumps in u_0. The precise statement is as follows:

<u>Theorem 1</u>: Let f be C^2 and $u_0 \in L^\infty(\mathbb{R})$. Then the Volpert-Kruzkov solution of (2.3) with initial data u_0 will be locally Lipschitz in the strip $0 < t < T$ if and only if

a) $f'(u_0(x)) + \frac{x}{T}$ is increasing in x ;

b) $\sigma f'' \ge 0$ on $\text{conh}\{u_0(x)\}$, where σ is either +1 or -1;

c) for each x, either u_0 is Lipschitz continuous in a neighborhood of x, or u_0 has left and right limits at x and $\sigma f'' > 0$ on the interval $[u_0(x\mp), u_0(x\pm)]$.

One sees immediately that, when a)-b)-c) hold, u_0 is a function of locally bounded variation. Theorem 1 is a subcase of Theorem 2 below.

We turn now to the case of several space variables. The characteristic curves for the equation

$$u_t + \text{div}\, F(u) = 0$$

are again the lines

$$(2.5) \qquad x = x_0 + tF'(u_0(x_0)) \ .$$

Conway shows in [1] that these characteristic lines remain disjoint for all time provided that

$$\text{(2.6)} \qquad \operatorname{div} F'(u_0(x)) \geq 0 \quad \text{for all} \quad x .$$

Condition (2.6) thus guarantees the global existence of C^1 solutions when u_0 itself is C^1. In this case we can give (2.6) the following simple geometrical interpretation: Let $\Omega_0 \subseteq \mathbb{R}^n$ and let Ω_t be the set obtained by propogating Ω_0 along the characteristics up to time t. Then, using the mapping (2.5),

$$\begin{aligned} m\Omega_t &= \int_{\Omega_t} dx \\ &= \int_{\Omega_0} \det[I + tF''(u_0(x)) \cdot \nabla u_0(x)]dx \\ &= \int_{\Omega_0} [1 + t \operatorname{div} F'(u_0(x))]dx \\ &= m\Omega_0 + t \int_{\Omega_0} \operatorname{div} F'(u_0(x))dx . \end{aligned}$$

Thus the characteristics are seen to diverge when (2.6) holds. More precisely, we shall obtain as a consequence of Theorem 2 below that, when u_0 itself is locally Lipschitz continuous, the solution $u(x,t)$ will also be locally Lipschitz continuous up to time T if and only if

$$T^{-1} + \operatorname{div} F'(u_0) \geq 0 \quad \text{a.e.}$$

Next, we need to investigate the conditions under which discontinuities in u_0 are smoothly resolved in positive time. In the case of one space dimension, we saw that this smoothing occurred when f was strictly convex across jumps in u_0. When $n > 1$ the situation is somewhat more complicated, as is shown by the following example.

Let (2.1) take the form

$$u_t + \left(\frac{u^2}{2}\right)_x + \left(\frac{u^2}{2}\right)_y = 0$$

so that the components of F are strictly convex. If g is any smooth function of a scalar, then

$$u(x,y,t) = g(x-y)$$

is a solution which incidentally satisfies (2.6). By approximating a discontinuous g by smooth g's, we thus obtain a Volpert-Kruzkov solution satisfying (2.6) in the sense of distributions, but which fails to be continuous despite the fact that the components of F are strictly convex.

This is evidently the wrong measure of convexity for this problem. The correct condition is formulated below as "property P_T." It is the necessary and sufficient condition that u be locally Lipschitz for $0 < t < T$.

Definition: We say that $u_0 \in L^\infty(\mathbb{R}^n)$ satisfies P_T $(0 < T \leq \infty)$ if there is a sequence $\{u_k\}$ of locally Lipschitz functions such that

a) $u_k \to u_0$ in $L^1_{loc}(\mathbb{R}^n)$ as $k \to \infty$;

b) $\|u_k\|_{L^\infty} \leq \|u_0\|_{L^\infty}$; and

c) for every $\varepsilon > 0$ and every compact set $K \subseteq \mathbb{R}^n$ there is a positive number $L = L(K,\varepsilon)$ such that

$$(2.7) \qquad |\nabla u_k(x)| \leq L[\operatorname{div} F'(u_k(x)) + \varepsilon + T^{-1}]$$

holds a.e. in K for all k sufficiently large.

The following contains some easy consequences of the definition:

Proposition: a) If u_0 satisfies P_T then u_0 is a function of locally bounded variation and $T^{-1} + \operatorname{div} F'(u_0)$ is a nonnegative measure.

b) If u_0 is locally Lipschitz, then u_0 satisfies P_T iff $T^{-1} + \operatorname{div} F'(u_0) \geq 0$ a.e.

Proof: (2.7) requires that $T^{-1} + \operatorname{div} F'(u_0)$ is a nonnegative distribution, hence a nonnegative measure. Taking the distribution limit as $k \to \infty$ in

$$L[\operatorname{div} F'(u_k) + \varepsilon + T^{-1}] \pm \frac{\partial u_k}{\partial x_i} \geq 0 ,$$

we thus obtain that each $\frac{\partial u_0}{\partial x_i}$ is a measure and hence that u_0 has locally bounded variation. For b), we observe that if u_0 is locally Lipschitz and $T^{-1} + \operatorname{div} F'(u_0) \geq 0$ a.e., then we may take $u_k = u_0$ for all k and

$$L(K,\varepsilon) = \varepsilon^{-1}\|\nabla u_0\|_{L^\infty(K)}$$

in the definition. //

The main result of this section is the following.

Theorem 2: Let u be the Volpert-Kruzkov solution of (2.1) with data $u_0 \in L^\infty(\mathbb{R}^n)$. Then u is locally Lipschitz continuous for $0 < t < T$ iff u_0 satisfies P_T.

The proof of Theorem 2 is given in [3]. A related discussion appears in [2].

By combining part b) of the proposition with Theorem 2, we obtain a sharpening of Conway's result:

Corollary: If u_0 itself is locally Lipschitz continuous, then the solution of (2.1) with data u_0 will also be locally Lipschitz for $0 < t < T$ iff $T^{-1} + \text{div } F'(u_0) \geq 0$ a.e.

And by combining part a) of the proposition with Theorem 2, we obtain the fact that the solution operator for (2.1) is never smoothing on the class $L^\infty - LBV$:

Corollary: If the Volpert-Kruzkov solution of (2.1)-(2.2) is locally Lipschitz in any time interval $0 < t < T$, then u_0 must have locally bounded variation.

The condition P_T may seem somewhat unsatisfactory because it requires the existence of an approximating sequence displaying certain properties uniformly. One may therefore ask whether it is sufficient that u_0 itself has those properties. More precisely, suppose that $u_0 \in L^\infty(\mathbb{R}^n)$ and that for each $\varepsilon > 0$ and each compact set $K \subseteq \mathbb{R}^n$ there is an $L(K,\varepsilon)$ for which

$$L[\text{div } F'(u_0) + \varepsilon + T^{-1}] \pm \frac{\partial u_0}{\partial x_i} \geq 0 \text{ , } i = 1,\ldots,n \text{ ,}$$

in the sense of distributions. Will the solution of (2.1) with data u_0 then necessarily be locally Lipschitz for $0 < t < T$?

The answer is no, and for a counterexample we can take $f(u) = u^3/3$ and

$$u_0(x) = \begin{cases} -1 \text{ , } x < 0 \\ 2 \text{ , } x > 0 \text{ .} \end{cases}$$

One easily computes that $\frac{du_0}{dx} = f'(u_0)_x = 3\delta$, where δ is the Dirac measure at 0. Thus for any T and ε,

$$f'(u_0)_x + \varepsilon + T^{-1} \pm \frac{du_0}{dx} \geq 0 \text{ .}$$

However, u_0 cannot satisfy P_T for any T. One way to see this is to observe that the Volpert-Kruzkov solution for the problem at hand contains a contact discontinuity which is present at all time levels. Alternatively, if there is a sequence $\{u_k\}$ as in the definition of P_T, then there must be points $y_k < x_k$ converging to 0 such that $u_k(y_k) = -\frac{1}{2}$ and $u_k(x_k) = 0$. Then

$$\frac{f'(u_k(y_k)) - f'(u_k(x_k))}{y_k - x_k} = \frac{\frac{1}{4}}{y_k - x_k} \to -\infty ,$$

whereas (2.7) requires that this difference quotient be bounded below by $-T^{-1}$.

§3. Systems of Equations in Diagonal Form

In this section we discuss a few extensions of the previous results to certain systems of hyperbolic equations. The situation is now complicated by the fact that there will be n families of characteristics. We therefore restrict attention to systems of homogeneous equations in one space variable which are in diagonal form:

$$(3.1) \qquad u_t + \Lambda(u)u_x = 0 .$$

Here $u \in \mathbb{R}^n$ and $\Lambda = \mathrm{diag}(\lambda_1,\ldots,\lambda_n)$ is C^1. Because these equations are only weakly coupled, our results will parallel those summarized in Theorem 1 for a scalar equation. Still, the discussion will apply to the system of conservation laws

$$u_t + f(u)_x = 0$$

$(u \in \mathbb{R}^n)$ when there exists a coordinate set of Riemann invariants. These are scalar functions $v_i(u)$, $i = 1,\ldots,n$, whose gradients form a linearly independent set of eigenvectors for $\frac{df}{du}$. The vector $v = (v_1,\ldots,v_n)$ will then satisfy a system of the form (3.1). Such Riemann invariants always exist locally when $n = 2$.

For simplicity we consider initial data u_0 with locally bounded variation, and we ask whether there is a globally smooth (locally Lipschitz) solution of (3.1) which approaches u_0 in L^1_{loc} as $t \downarrow 0$. (Lax discusses in [7] the local existence of such solutions when u_0 itself is smooth.) We also assume that there is a bounded rectangle S in u-space such that u_0 has values in S. It is clear from (3.1) that S will be positively invariant.

We begin by stating a necessary condition for the existence of a globally smooth solution. For this we index the eigenvalues so that $\lambda_1 < \lambda_2 < \cdots < \lambda_n$.

Theorem 3: Suppose that (3.1) is strictly hyperbolic in the sense that the closed sets $\lambda_i(S)$ are disjoint. Assume that u_0 has values in S and that the limits $u_0(\pm\infty)$ exist. If there is a globally smooth solution of (3.1) with initial data u_0, then for each i, the function $\lambda_i(w_i(x))$ must be increasing in x, where $w_i = (w_i^1,\ldots,w_i^n)$ is defined by

$$w_i^j(x) = \begin{cases} u_0^j(-\infty) & , j < i \\ u_0^i(x) & , j = i \\ u_0^j(+\infty) & , j > i . \end{cases}$$

Thus $\frac{\partial\lambda_i}{\partial u^i}$ is of one sign on the range of w_i, and if the zeroes of $\frac{\partial\lambda_i}{\partial u^i}$ in S are isolated, u_0^i must be monotone in x.

Proof: (3.1) shows that $u^j(x,t)$ is constant on the j-characteristic $\frac{dx}{dt} = \lambda_j(u(x,t))$. Combining this with the strict hyperbolicity, one sees that, as $t \to \infty$ along the i-characteristic through a point (x,ε), u approaches the value $\tilde{w}_i$ defined by

$$\tilde{w}_i^j(x) = \begin{cases} u_0^j(-\infty) & , j < i \\ u^i(x,\varepsilon) & , j = i \\ u_0^j(+\infty) & , j > i . \end{cases}$$

Evidently this i-characteristic approaches a line with speed $\frac{dx}{dt} = \lambda_i(\tilde{w}_i)$. Since these asymptotic characteristics cannot intersect, $\lambda_i(\tilde{w}_i(x))$ must be increasing in x. Letting $\varepsilon \to 0$, we find that $\lambda_i(w_i(x))$ is also increasing in x. The statements about $\frac{\partial\lambda_i}{\partial u^i}$ and the monotonicity of u_0^i then follow easily. //

The next task is therefore to formulate the conditions on u_0 and Λ which guarantee that initial discontinuities are smoothed out in positive time. In light of the above result, we shall assume that, for each i, there is a σ_i, which is either +1 or -1, such that $\sigma_i \frac{\partial\lambda_i}{\partial u^i} \geq 0$ in S and such that $\sigma_i u_0^i$ is increasing. These assumptions preclude the intersecting of different characteristics from the same family at large times.

Our technique is simply to mollify a discontinuous u_0, solve (3.1) globally with smooth data, then pass to the limit as the mollifier approaches a δ-function. We can justify this passage to the limit as soon as we establish a bound for $\|\frac{\partial u}{\partial x}(\cdot,t)\|_{L^\infty(\mathbb{R})}$ which is independent of $\frac{\partial u_0}{\partial x}$. We shall give a brief description of how such a bound can be achieved.

Letting $v^i = \frac{\partial u^i}{\partial x}$, we have from (3.1) that, along an i-characteristic,

$$\frac{dv^i}{dt} = -\left(\sum_{j=1}^n \frac{\partial \lambda_i}{\partial u^j}(u)v^j\right)v^i . \tag{3.2}$$

This immediately implies the preservation of monotonicity in x of u^i. More important, we have discovered a simple device which effectively uncouples the above equations in the sense that the $j \neq i$ terms on the right are rendered irrelevant. If we then impose the genuine nonlinearity assumption $\frac{\partial \lambda_i}{\partial u^i} \neq 0$ across jumps in the unmollified u_0^i, (3.1) will become a Ricatti-type equation from which we can easily deduce the required smoothing.

More specifically, we let

$$z^i = \sum_{j\neq i} (u^j + c^j)^{p_{ij}}$$

where c^j is chosen so that $u^j + c^j > 0$ for $u \in S$, and p_{ij} is a free parameter to be chosen later. Next, let

$$w^i = z^i v^i = z^i \frac{\partial u^i}{\partial x} .$$

Then one can compute that, along i-characteristics,

$$\frac{dw^i}{dt} = -\left(\frac{\partial \lambda_i}{\partial u^i}(u)/z^i\right)(w^i)^2$$

plus terms whose signs can be controlled by adjusting p_{ij} appropriately. The result is that $\sigma_i w^i(\cdot,t)$, and hence $\sigma_i \frac{\partial u^i}{\partial x}(\cdot,t)$, must lie in an interval $[0, K/t]$ for some constant which is independent of $\frac{\partial u_0}{\partial x}$. This is the required estimate. The precise statement of our result is the following:

Theorem 4: Assume that in S,

a) the λ_i are C^1;

b) there are numbers $\sigma_i \in \{-1,1\}$ such that $\sigma_i \frac{\partial \lambda_i}{\partial u^i} \geq 0$; and

c) $\lambda_i \neq \lambda_j$ for $i \neq j$.

Then there is a $\delta > 0$ such that if $u_0 : \mathbb{R} \to S$ satisfies

d) $\sigma_i u_0^i$ is increasing;

e) $|u_0(\infty) - u_0(-\infty)| \leq \delta$; and

f) for each i and x , either u_0^i is Lipschitz in a neighborhood of x , or $\frac{\partial \lambda_i}{\partial u^i}$ is nonvanishing on the set

$$\Pi_{j=1}^{n} [u_0^j(x\mp) , u_0^j(x\pm)] ,$$

then there is a globally smooth solution of (3.1) with initial data u_0 . When $n = 2$, hypothesis e) may be omitted.

A detailed proof of Theorem 4 may be found in [4].

We conclude by mentioning some related work concerning systems of equations in one space dimension. Lax in [8] estimates the blow-up time for a system of two conservation laws. John in [5] proves the blow-up of C^2 solutions of the (nondiagonal) hyperbolic system

$$u_t + a(u)u_x = 0$$

when u_0 is C^2 , has compact support, and is small in some sense. This result is extended by Liu in [9]. And finally, Yamaguti and Nishida prove in [11] the special case of Theorem 4 in which $n = 2$ and the λ_i are genuinely nonlinear throughout S .

References

1. E.D. Conway, The Formation and Decay of Shocks for a Conservation Law in Several Dimensions, Arch. Rat. Mech. Anal. 64, No. 1 (1977).

2. David Hoff, Locally Lipschitz Solutions of a Single Conservation Law in Several Space Variables, J. Diff. Eqns. 42, No. 2 (1981), 215-233.

3. David Hoff, A Characterization of the Blow-Up Time for the Solution of a Conservation Law in Several Space Variables, Comm. in Partial Diff. Eqns. 7(2), 141-151 (1982).

4. David Hoff, Globally Smooth Solutions of Quasilinear Hyperbolic Systems in Diagonal Form, J. Math. Ana. and Appl. 86, No. 1 (1982), 221-236.

5. F. John, Formation of Singularities in One-Dinensional Nonlinear Wave Propogation, Comm. Pure and Appl. Math. 27(1974).

6. S.N. Kruzkov, First Order Quasilinear Equations in Several Independent Variables, Math. USSR Sbornik 10, No. 2 (1970).

7. P.D. Lax, Nonlinear Hyperbolic Equations, Comm. Pure and Appl. Math 6(1953).

8. P.D. Lax, Development of Singularities of Solutions of Nonlinear Hyperbolic Partial Differential Equations, J. Math. Phys. 5, No. 5 (1964).

9. Tai-Ping Liu, Development of Singularities in the Nonlinear Waves for Quasilinear Hyperbolic Partial Differential Equations, J. Diff. Eqns. 33, No. 1 (1979).

10. A.I. Volpert, The Spaces BV and Quasilinear Equations, Math. USSR Sbornik 2(1967), 225-267.

11. M. Yamaguti and T. Nishida, On Some Global Solutions of Quasilinear Hyperbolic Systems, Funkcial. Ekvac. 11(1968).

Contemporary Mathematics
Volume 17, 1983

NON-STRICTLY HYPERBOLIC SYSTEMS OF CONSERVATION LAWS: FORMATION OF SINGULARITIES

Barbara L. Keyfitz[1] and Herbert C. Kranzer

ABSTRACT. Two model non-strictly hyperbolic systems are considered. Each system has two real eigenvalues which become equal on a single curve Σ in the phase plane. One system is genuinely nonlinear, while the other has one nonlinear and one linearly degenerate eigenvalue. Although the Riemann problems for these systems exhibit behavior considerably different from that of their strictly hyperbolic counterparts, the formation of singularities from smooth Cauchy data occurs in precisely the same situations as in the strictly hyperbolic case. The proof uses explicit formulas for the Riemann invariants on the two sides of Σ.

1. INTRODUCTION. The Cauchy problem for a nonlinear hyperbolic system of conservation laws:

$$U_t + F(U)_x = 0 \ , \qquad U(x,0) = U_0(x) \ , \tag{1.1}$$

with smooth initial data of compact support, has the property that smooth solutions generally exist for only a finite time. After this time, only weak solutions (usually containing shocks or other discontinuities) can be defined.

In 1964, Lax [7] obtained an estimate for the "blow up" time for the case of two equations when the system is genuinely nonlinear, using Riemann invariants and deriving an equation of the form $z' = a(t)\, z^2$ along the characteristic curves of the system. Then the result from ordinary differential equations proves non-existence of a solution for large t.

Keller and Ting [3] obtained non-existence for two equations by integrating directly along the characteristics (rather than reducing to an ordinary differential equation).

F. John [2] extended this to a genuinely nonlinear system of n equations, again using characteristics and reducing to $z' = az^2$, though without employing Riemann invariants, which do not necessarily exist in this case.

1980 Mathematics Subject Classification. 35L65, 35L67.

[1]Supported in part by NSF Grant MCS 81-03441.

T.P. Liu [8] extended John's result by allowing some families to be linearly degenerate, but assuming some restrictions on the interaction between linear and nonlinear families. He used the same techniques, although he replaces F. John's continuous integration along characteristics by a discrete process.

Klainerman and Majda [6] studied the case $n = 2$ when both characteristic families are linearly degenerate along a curve in the phase plane. In fact, they looked at the wave equation $u_{tt} = (K(u_x))_x$ where $K(u_x) = u_x + O(u_x^3)$ near $u_x = 0$.

All these approaches assume that the characteristics are distinct (strict hyperbolicity). The reason for this is made explicit in John's and in Liu's work, where each family of characteristics carrying nontrivial initial data is shown, under the assumptions, to intersect all such members of other families before some time T, so that after time T one has essentially n uncoupled single conservation laws.

In this paper, we consider two model systems of conservation laws for which the characteristics are not always distinct, so that none of the above results apply. These systems are defined as follows. The first is

$$\begin{aligned} u_t + (\phi u)_x &= 0 \\ v_t + (\phi v)_x &= 0 \end{aligned} \tag{1.2}$$

where $\phi = \phi(u,v) = \phi(r,\theta)$, $r = \sqrt{u^2+v^2}$, $\theta = \arctan(v/u)$. Here the characteristic speeds are $\lambda_1 = \phi$ and $\lambda_2 = \phi + r\phi_r$ and become equal on the curve $\Sigma = \{(u,v)|\phi_r = 0\}$. The eigenvalue λ_1 is linearly degenerate, while λ_2 is genuinely nonlinear provided $(r\phi)_{rr} \neq 0$, which we assume.

As we showed in [4], where we solved the Riemann problem for system (1.2), there are two essentially different cases:

(1.2a) $\phi_\theta = 0$ on Σ also, and $A(U)$, the matrix $\partial F/\partial U$, is diagonalizable on Σ.

(1.2b) $\phi_\theta \neq 0$ on Σ, $A(U)$ is not diagonalizable and there is a parabolic degeneracy on Σ.

Physically, (1.2a) models the forward-propagating longitudinal (speed λ_2) and transverse (speed λ_1) waves on a nonlinearly elastic string moving in the plane. System (1.2b) appears as a model for secondary oil recovery problems in one space dimension; cf. Isaacson [1] and Temple [9].

Our second model system is

$$\begin{aligned} u_t - v_x + f(u)_x &= 0\ , \\ v_t + g(u)_x &= 0\ , \end{aligned} \tag{1.3}$$

where $f(u) \approx \gamma u^2$, $\gamma > 0$, and $g(u) \approx -u^3/3$ near $u = 0$. This system is genuinely nonlinear in both characteristic speeds, but strict hyperbolicity fails on the line $\Sigma : u = 0$, where both speeds vanish and a parabolic degeneracy occurs. A special system of this type (which we call the prototype system) has $f(u) = 0$ and $g(u) = -u^3/3$; it is equivalent to the nonlinear wave equation $u_{tt} = (c^2 u_x)_x$ with "sound speed" $c = u$. The Riemann problem for system (1.3) is treated in [5].

For both models, the problem is to prove that smooth initial data of compact support must lead to a singularity in finite time, or alternatively to characterize precisely those classes of exceptional initial data which do not lead to a singularity. By smooth initial data of compact support we mean a C_1 function $U_0(x) = \{u_0(x), v_0(x)\}$ which is equal to a constant W outside some interval $|x| \leq X$. We focus on initial data which intersect or come close to Σ, since solutions which are bounded away from Σ can be handled by the strictly hyperbolic methods previously cited.

Our major results are contained in the following theorems.

THEOREM 1. For smooth initial data $U_0(x)$ of compact support, system (1.2) has a C_1 solution $U(x,t) = \{u(x,t), v(x,t)\}$ for $0 \leq t < \infty$ if and only if $\phi(U_0(x))$ is constant.

THEOREM 2. A solution of system (1.3) with nonconstant smooth initial data of compact support must become singular (in particular, U_x must become infinite) at a finite value of t.

2. THE DIAGONALIZABLE CASE.

In this section, we prove Theorem 1 for the diagonalizable case (1.2a). We follow the general approach of Keller and Ting [3], integrating derivatives of solutions of (1.2) along the characteristic curves of this system until one of these derivatives becomes infinite. The estimates for derivatives which are obtained in this process will be used also in Section 3 to treat the nondiagonalizable case. The same procedure (with different estimates) will reappear in Section 4 where Theorem 2 is proved.

Since $(r\phi)_{rr} \neq 0$ by genuine nonlinearity, we may assume without loss of generality that $(r\phi)_{rr} > 0$. Now the curve Σ on which the characteristic speeds are equal is just the locus $\phi_r = 0$. (The origin $r = 0$ is a singular point of the system and must be excluded from consideration.) Hence ϕ has for fixed θ a minimum value at the point (if any) where the ray $\theta = \text{const}$ crosses Σ. Thus Σ divides the u,v-plane (excluding the origin) into two regions, the inside of Σ where $\phi_r < 0$ and $\lambda_1 > \lambda_2$ and the outside where $\phi_r > 0$ and $\lambda_1 < \lambda_2$.

The functions $\theta(u,v)$ and $\phi(u,v)$ are Riemann invariants for system (1.2). They form a locally invertible coordinate net in the (u,v)-plane on each side of Σ, but the coordinate transformation $(u,v) \to (\theta,\phi)$ becomes singular on Σ itself. The Riemann invariants satisfy the differential equations

$$\text{(2.1)} \qquad \begin{aligned} \frac{\partial}{\partial\lambda_1}\theta &\equiv \theta_t + \phi\theta_x = 0 \quad ; \\ \frac{\partial}{\partial\lambda_2}\phi &\equiv \phi_t + (r\phi)_r\phi_x = 0 \ . \end{aligned}$$

They are related as follows to the characteristic curves $x_1(\alpha,t)$ and $x_2(\beta,t)$ defined by

$$\text{(2.2)} \qquad \begin{aligned} C_1 : \ \frac{dx_1}{dt} &= \lambda_1(x_1,t) = \phi(x_1,t), \qquad x_1(\alpha,0) = \alpha \quad ; \\ C_2 : \ \frac{dx_2}{dt} &= \lambda_2(x_2,t) = (r\phi)_r(x_2,t), \qquad x_2(\beta,0) = \beta \ . \end{aligned}$$

On C_1, θ is constant, while on C_2, ϕ is constant. These characteristic curves themselves form an invertible coordinate net (α,β) in the (x,t)-plane as long as the solution $U(x,t)$ remains smooth.

Now consider a single characteristic curve C_2 with parameter β, and assume for the moment that $U(x,t)$ does not lie on Σ for any $(x,t) \in C_2$. Then along C_2 we may compute

$$\text{(2.3)} \qquad \begin{aligned} \frac{d}{dt}\left(\frac{\partial x_2}{\partial\beta}\right) &= \frac{\partial}{\partial\beta}\left(\frac{dx_2}{dt}\right) = \frac{\partial}{\partial\beta}(\lambda_2(x_2(\beta,t),t)) \\ &= \frac{\partial\lambda_2}{\partial\phi}\frac{\partial\phi}{\partial\beta} + \frac{\partial\lambda_2}{\partial\theta}\frac{\partial\theta}{\partial x}\frac{\partial x_2}{\partial\beta} \end{aligned}$$

(since all coordinate transformations are invertible). Now in the first term

$$\text{(2.4)} \qquad \frac{\partial\phi}{\partial\beta} = \phi_0'(\beta) \ ,$$

where $\phi_0(\beta)$ denotes the initial value $\phi(U_0(\beta))$, since ϕ is constant on each $C_2(\beta)$. For the second term, note that on C_2

$$\frac{d\theta}{dt} \equiv \frac{\partial\theta}{\partial t} + \lambda_2\frac{\partial\theta}{\partial x} = (\lambda_2-\lambda_1)\frac{\partial\theta}{\partial x}$$

since $\theta_t = -\lambda_1\theta_x$ by (2.1). Hence

$$\frac{\partial\theta}{\partial x} = \frac{1}{\lambda_2-\lambda_1}\frac{d\theta}{dt},$$

which, by our assumption, is well-defined on C_2. Therefore

$$\frac{\partial\lambda_2}{\partial\theta}\frac{\partial\theta}{\partial x} = \frac{1}{\lambda_2-\lambda_1}\frac{\partial\lambda_2}{\partial\theta}\frac{d\theta}{dt} = \frac{1}{\lambda_2-\lambda_1}\frac{d\lambda_2}{dt}$$

$$= \frac{d}{dt}\log[\lambda_2(\theta,\phi)-\lambda_1(\phi)]$$

since $\phi = \lambda_1$ is constant on C_2. Hence

$$(2.5)\qquad \frac{d}{dt}\left(\frac{\partial x_2}{\partial\beta}\right) = \frac{\partial\lambda_2}{\partial\phi}\phi_0'(\beta) + \frac{d}{dt}\log(\lambda_2-\lambda_1)\frac{\partial x_2}{\partial\beta} \quad \text{along } C_2,$$

which is a differential equation for $\frac{\partial x_2}{\partial\beta}$, together with the initial condition

$$(2.6)\qquad \frac{\partial x_2}{\partial\beta}(0) = 1 .$$

Now the solution of $y' = p(t)y + q(t)$, $y(0) = y_0$ is

$$(2.7)\qquad y(t) = \mu(t)\left[y_0 + \int_0^t \frac{q(s)}{\mu(s)}ds\right], \quad \text{where } \mu(t) = e^{\int_0^t p(s)ds} .$$

Hence

$$(2.8)\qquad \mu(t) = \exp\left(\int_0^t \frac{d}{ds}\log(\lambda_2-\lambda_1)ds\right) = \frac{(\lambda_2-\lambda_1)(t)}{(\lambda_2-\lambda_1)(0)}$$

on C_2.

For initial data of compact support, θ and ϕ are initially bounded, and therefore remain bounded for all time. By invertibility, u and v remain bounded, and thus so do λ_1 and λ_2. Hence there exists a constant A so that

$$(2.9)\qquad 0 < \mu(t) \leq A \quad \text{on } C_2$$

as long as $U(x,t)$ remains regular and does not intersect Σ.

Moreover,

$$(2.10)\qquad \int_0^t \frac{q(s)}{\mu(s)}ds = \int_0^t \frac{\partial\lambda_2}{\partial\phi}\phi_0'(\beta)\frac{1}{\mu(s)}ds = \phi_0'(\beta)\int_0^t \frac{1}{\mu(s)}\frac{\partial\lambda_2}{\partial\phi}ds$$

with $\frac{\partial\lambda_2}{\partial\phi} = \frac{\partial\lambda_2}{\partial r}\frac{\partial r}{\partial\phi} = (r\phi)_{rr}/\phi_r$. The numerator of this fraction is positive everywhere by genuine nonlinearity, while the denominator is positve outside Σ, negative inside, and zero on Σ. By compactness, there exists a positive constant M for which

$$\text{(2.11)}\qquad \begin{aligned} &\frac{\partial\lambda_2}{\partial\phi} \leq -M \quad \text{on } C_2 \text{ for } U(x,t) \text{ inside } \Sigma \ ; \\ &\frac{\partial\lambda_2}{\partial\phi} \geq M \quad \text{on } C_2 \text{ for } U(x,t) \text{ outside } \Sigma \ . \end{aligned}$$

Combining (2.7), (2.9), (2.10) and (2.11) yields the following inequalities:

$$\text{(2.12a)}\qquad \frac{\partial x_2}{\partial\beta} \leq \mu(t)[1-\phi_0'(\beta)\frac{M}{A}t]$$

on $C_2(\beta)$ if $\phi_0'(\beta) > 0$ and $U(x_2(\beta,t),t)$ is inside Σ, while

$$\text{(2.12b)}\qquad \frac{\partial x_2}{\partial\beta} \leq \mu(t)[1+\phi_0'(\beta)\frac{M}{A}t]$$

on $C_2(\beta)$ if $\phi_0'(\beta) < 0$ and $U(x_2(\beta,t),t)$ is outside Σ. But if either of these inequalities holds, $\frac{\partial x_2}{\partial\beta}$ becomes negative along C_2 for sufficiently large t, so that the C_2-characteristics eventually cross and the solution becomes singular. Thus we have proved

LEMMA 1. Suppose that the image $U(x_2(\beta,t),t)$ of a characteristic curve $C_2(\beta)$ of system (1.2), with initial data $U_0(x)$ of compact support, remains for all time on one side of Σ. Suppose further that

$$\text{(2.13a)}\qquad \phi_0'(\beta) > 0 \quad \text{with} \quad U_0(\beta) \text{ inside } \Sigma \ ,$$

or alternatively that

$$\text{(2.13b)}\qquad \phi_0'(\beta) < 0 \quad \text{with} \quad U_0(\beta) \text{ outside } \Sigma \ .$$

Then the solution $U(x,t)$ becomes singular along this characteristic at some finite time.

PROOF OF THEOREM 1 IN THE DIAGONALIZABLE CASE. First, suppose that $\phi(U_0) = \phi_0 =$ constant. Then $\phi(U) = \phi_0$ along every C_2-characteristic, so $\phi = \phi_0$ everywhere. Therefore θ evolves according to the linear differential equation $\theta_t + \phi_0\theta_x = 0$ and remains regular for all time. Hence $U(x,t)$ is regular for all $0 \leq t < \infty$.

If, on the other hand, $\phi(U_0)$ is not constant, $U_0(x)$ cannot lie entirely in Σ, because in the diagonalizable case $\phi_r = \phi_\theta = 0$ on Σ and so ϕ is constant (say $\phi = \phi_1$) on Σ. Moreover, the assumption of compact support implies that $\phi_0(-\infty) = \phi_0(+\infty)$, so that $\phi_0'(\beta)$ must have both positive and negative values. Since the absolute minimum ϕ_1 of ϕ occurs on Σ, values of $\phi_0'(\beta)$ having opposite sign must occur on the same side of Σ. Thus either (2.13a) or (2.13b) must hold for some β. But for this β we have $\phi_0(\beta) > \phi_1$, so that $\phi(U) = \phi_0(\beta) \neq \phi_1$ everywhere on $C_2(\beta)$ and so $U \notin \Sigma$. Therefore Lemma 1 applies, and the solution must become singular.

3. THE NON-DIAGONALIZABLE, LINEARLY DEGENERATE MODEL. We complete the proof of Theorem 1 in this section by considering the non-diagonalizable case (1.2b). Lemma 1 and its proof use only the general properties of the model equations (1.2) and so remain valid in this case. The argument when $\phi(U_0)$ is constant also retains its validity. However, ϕ is no longer constant on Σ, so that it is not certain that an initial point satisfying (2.13) must always exist when $\phi(U_0)$ is non-constant, nor is it clear that the C_2-characteristic through such a point cannot reach Σ.

We deal with the latter problem first. Suppose that the initial point $(\beta,0)$ satisfies (2.13). Then Lemma 1 asserts that a singularity must eventually develop along $C_2(\beta)$ unless $U(x_2(\beta,t),t)$ reaches Σ first. But if $U(x,t)|_{C_2}$ first reaches Σ at $t = t_1$, we have

(3.1) $$\frac{\partial\phi}{\partial\beta}\Big|_{C_2} = \phi_0'(\beta) \neq 0 \quad \text{for} \quad 0 \leq t < t_1 .$$

By continuity, this holds also at $t = t_1$, where we find

(3.2) $$\phi_x = \phi_x(x_2(\beta,t_1),t_1) = \frac{\partial\phi}{\partial\beta}\frac{\partial\beta}{\partial x} = \phi_0'(\beta) \cdot \beta_x$$

with $\beta_x = (\frac{\partial x_2}{\partial\beta})^{-1} > 0$. But on Σ we have $\phi_x = \phi_\theta\theta_x$, since $\phi_r = 0$, and therefore at $t = t_1$ on C_2

(3.3) $$\theta_x = \frac{1}{\phi_\theta}\phi_0'(\beta)\beta_x$$

with $\phi_\theta \neq 0$ in the non-diagonalizable case. Therefore

(3.4) $$\text{sgn}\ \theta_x = \text{sgn}\ \phi_\theta\ \text{sgn}\ \phi_0'(\beta)$$

in an entire neighborhood $t_1 - \varepsilon \leq t \leq t_1$ on $C_2(\beta)$. But now for $t < t_1$ (2.1) yields

(3.5) $$\frac{d\theta}{dt}\Big|_{C_2} = \theta_t + \lambda_2\theta_x = (\lambda_2-\lambda_1)\theta_x$$

with $(\lambda_2-\lambda_1)\phi_0'(\beta) < 0$ by (2.13). Therefore the sign of $\frac{d\theta}{dt}$ in $[t_1-\varepsilon,t_1)$ is opposite to the sign of ϕ_θ. However, ϕ is constant on C_2, so that

$$\frac{d\phi}{dt} = \phi_r \frac{dr}{dt} + \phi_\theta \frac{d\theta}{dt} = 0 \tag{3.6}$$

and therefore $\phi_r \frac{dr}{dt} > 0$ in $[t1-\varepsilon,t1)$. But ϕ_r is negative inside Σ and positive outside, and the same is therefore true for $\frac{dr}{dt}$ in the t-interval just preceding t_1. In other words, U is moving away from Σ just prior to t_1, and so can never reach Σ at any finite time. Hence if there is an initial point satisfying (2.13) in the non-diagonalizable case, a singularity must develop along its C_2-characteristic.

The only other possibility for a nonconstant $\phi(U_0)$ of compact support is that $\phi_0'(\beta) \leq 0$ everywhere inside Σ and $\phi_0'(\beta) \geq 0$ everywhere outside Σ. Then the maximum and minimum values of $\phi(U_0)$ must occur on Σ, since $\phi_0'(\beta) = 0$ and changes sign at both the minimum point $\beta = \beta_1$ and the maximum point $\beta = \beta_2$. Now for $j = 1,2$

$$\frac{\partial\phi}{\partial\beta}\Big|_{C_2(\beta_j)} = \phi_0'(\beta_j) = 0 \quad \text{for all } t \tag{3.7}$$

and therefore $\phi_x = 0$ on $C_2(\beta_j)$ and U_x is tangent to the curve $\phi = \phi_j \equiv \phi_0(\beta_j)$. Since this curve is a rarefaction curve of the λ_1 family (see [4] for details), U_x is an eigenvector of $A(U)$ with eigenvalue λ_1. Thus along $C_2(\beta_j)$

$$\frac{dU}{dt} = U_t + \lambda_2 U_x = (\lambda_2-A)U_x = (\lambda_2-\lambda_1)U_x . \tag{3.8}$$

But initially U starts out on Σ, where $\lambda_2 = \lambda_1$ and $\frac{dU}{dt} = 0$, so the uniqueness theory for ordinary differential equations guarantees that $U = \text{const} = U_0(\beta_j)$ on $C_2(\beta_j)$ for all time. Thus the propagation speed $\lambda_2(U)$ has the constant value $\lambda_1(U_0(\beta_j)) = \phi_j$ and the characteristic curve $C_2(\beta_j)$ becomes a straight line. Since $\phi_1 < \phi_2$, these two straight lines will intersect at a positive t when $\beta_1 > \beta_2$, while if $\beta_1 < \beta_2$ one of them must intersect one of the two constant states generated from the initial value $U(\pm\infty) = W$. In either case, a singularity will form in finite time, and Theorem 1 is fully established.

4. THE FULLY NONLINEAR MODEL. We now proceed to the proof of Theorem 2. The characteristic speeds for system (1.3) are

$$\lambda_1 = \lambda_1(u) = \frac{1}{2}\left(f' - \sqrt{(f')^2-4g'}\right) \leq 0 \; ; \tag{4.1}$$
$$\lambda_2 = \lambda_2(u) = \frac{1}{2}\left(f' + \sqrt{(f')^2-4g'}\right) \geq 0 \; .$$

Under the assumptions

$$f(u) = \gamma u^2 + o(u^2), \; 2\gamma = f''(0) \geq 0 \; ; \tag{4.2}$$
$$g(u) = -u^3/3 + o(u^3), \; g'(u) = -u^2 + o(u^2) \leq 0$$

the speeds λ_1 and λ_2 behave near $\Sigma : u = 0$ as follows. In $D_1 : u > 0$

$$\lambda_1 = c_- u + o(u) \; , \qquad c_\pm = \gamma \pm \sqrt{1+\gamma^2} \gtrless 0 \; , \tag{4.3}$$
$$\lambda_2 = c_+ u + o(u) \; ,$$

while in $D_2 : u < 0$ these formulas are reversed. In both regions, we have

$$\lambda_2 - \lambda_1 = (c_+ - c_-)\,|u| + o(u) > 0 \; . \tag{4.4}$$

The Riemann invariants are

$$\ell(u,v) = \frac{1}{2}\left(v - \int_0^u \lambda_1(w)\,dw\right) \quad \text{and} \tag{4.5}$$
$$r(u,v) = \frac{1}{2}\left(v - \int_0^u \lambda_2(w)\,dw\right) \; .$$

They satisfy

$$\frac{\partial \ell}{\partial t} + \lambda_1 \frac{\partial \ell}{\partial x} = 0 \tag{4.6}$$
$$\frac{\partial r}{\partial t} + \lambda_2 \frac{\partial r}{\partial x} = 0 \; .$$

The characteristic curves $C_1(\alpha)$ and $C_2(\beta)$ are defined by

$$C_1(\alpha) = \{(x_1(\alpha,t),t)\} : \frac{dx_1}{dt} = \lambda_1, \; x_1(\alpha,0) = \alpha \; ; \tag{4.7}$$
$$C_2(\beta) = \{(x_2(\beta,t),t)\} : \frac{dx_2}{dt} = \lambda_2, \; x_2(\beta,0) = \beta \; .$$

On C_1, ℓ is constant while r is constant on C_2. As in Section 2, the transformations $(u,v) \to (\ell,r)$ and $(\alpha,\beta) \to (x_1,x_2)$ are invertible in D_1 and D_2 as long as the solution $U(x,t)$ remains regular.

The analogue of Lemma 1 for this model is

LEMMA 2. Suppose that the image $U(x_i,t)$, $i = 1$ or 2, of a characteristic curve $C_1(\alpha)$ or $C_2(\beta)$ of system (1.3), with initial data $U_0(x)$ of compact support, remains for all time on one side of Σ. Suppose further that one of the following conditions holds at the initial point $(x_i,0)$ of C_i:

$$\ell_0'(\alpha) > 0 \text{ or } r_0'(\beta) > 0, \text{ with } U_0(x_i) \text{ in } D_1 , \tag{4.8a}$$

or alternatively

$$\ell_0'(\alpha) < 0 \text{ or } r_0'(\beta) < 0, \text{ with } U_0(x_i) \text{ in } D_2 . \tag{4.8b}$$

Then the solution $U(x,t)$ becomes singular along C_i at some finite time.

PROOF. For simplicity we consider only the case $i = 2$ of a C_2-curve; the calculations for a C_1-curve are exactly the same. Along C_2 we have

$$\frac{d}{dt}\left(\frac{\partial x_2}{\partial \beta}\right) = \frac{\partial}{\partial \beta}\left(\frac{dx_2}{dt}\right) = \frac{\partial}{\partial \beta}\lambda_2(x_2(\beta,t),t) \tag{4.9}$$

$$= \frac{\partial \lambda_2}{\partial r}\frac{\partial r}{\partial \beta} + \frac{\partial \lambda_2}{\partial \ell}\frac{\partial \ell}{\partial x}\frac{\partial x_2}{\partial \beta} .$$

But

$$\frac{\partial r}{\partial \beta} = r_0'(\beta) \tag{4.10}$$

and $\frac{\partial \ell}{\partial x}$ may be computed from

$$\frac{d\ell}{dt} = \frac{\partial \ell}{\partial t} + \lambda_2\frac{\partial \ell}{\partial x} = (\lambda_2-\lambda_1)\frac{\partial \ell}{\partial x} . \tag{4.11}$$

Furthermore, we see from (4.5) that

$$\frac{\partial u}{\partial \ell} = -\frac{\partial u}{\partial r} = \frac{2}{\lambda_2-\lambda_1} \tag{4.12}$$

which implies

$$\frac{\partial \lambda_2}{\partial \ell} = -\frac{\partial \lambda_2}{\partial r} = \frac{2\lambda_2'(u)}{\lambda_2-\lambda_1} . \tag{4.13}$$

Hence $y = \frac{\partial x_2}{\partial \beta}$ satisfies along $C_2(\beta)$ the differential equation $y' = p(t)y + q(t)$, $y(0) = 1$ with

$$q(t) = -\frac{2\lambda_2'(u)}{\lambda_2-\lambda_1}r_0'(\beta) \tag{4.14}$$

and

$$p(t) = \frac{\partial \lambda_2}{\partial \ell} \frac{\partial \ell}{\partial x} = \frac{2\lambda_2'(u)}{(\lambda_2-\lambda_1)^2} \frac{d\ell}{dt} .$$

But $\frac{d\ell}{dt} = \frac{d}{dt}(\ell - r) = \frac{1}{2}(\lambda_2-\lambda_1)\frac{du}{dt}$, so that

$$p(t) = \frac{1}{\lambda_2-\lambda_1} \frac{d\lambda_2}{dt} . \tag{4.15}$$

In the solution formula (2.7) for y, we then have

$$\mu(t) = \exp \int_0^t p(s)ds = \exp \int_0^t \frac{1}{(\lambda_2-\lambda_1)(s)} \frac{d\lambda_2}{ds} ds \tag{4.16}$$

$$= \exp \int_{u_0}^{u(t)} \frac{1}{(\lambda_2-\lambda_1)(u)} \lambda_2'(u)du .$$

But from (4.3) we may deduce that

$$0 < \lambda_2 < \lambda_2 - \lambda_1 < c\lambda_2 \tag{4.17}$$

for some global constant c. Therefore, with $k = 1$ or $1/c$ depending on the sign of $d\lambda_2$,

$$0 < \mu(t) \leq \exp[k \int_{u_0}^{u(t)} \frac{\lambda_2'(u)}{\lambda_2(u)} du] \tag{4.18}$$

$$= \exp [k(\log \lambda_2(t) - \log \lambda_2(0))]$$

$$= [\lambda_2(t)/\lambda_2(0)]^k \leq A < \infty$$

by compactness, since $U_0 \notin \Sigma$ and so $\lambda_2(0) \neq 0$. Furthermore,

$$\int_0^t \frac{q(s)}{\mu(s)} ds = - r_0'(\beta) \int_0^t \frac{2\lambda_2'(u(s))}{(\lambda_2-\lambda_1)(s)\ \mu(s)} ds \tag{4.19}$$

with $\lambda_2-\lambda_1$ positive and bounded above, while from (4.3) $\lambda_2'(u)$ is bounded away from zero and has the same sign as u : positive in D_1, negative in D_2. Thus an estimate of the type (2.11) holds for the integrand in (4.19), and therefore $y = \frac{\partial x_2}{\partial \beta}$ satisfies one of the inequalities (2.12) with $\phi_0'(\beta)$ replaced by $r_0'(\beta)$. Specifically, (2.12a) holds for U_0 in D_1 and (2.12b) for U_0 in D_2. Thus $\frac{\partial x_2}{\partial \beta}$ eventually becomes negative along $C_2(\beta)$, and a singularity must develop. This proves Lemma 2.

Proceeding with the proof of Theorem 2, we first assume that one of the inequalities (4.8) holds for some initial point $(x_i,0)$. We show that U cannot reach Σ along the corresponding characteristic $C_i(x_i)$ without first becoming singular. For suppose U does first reach Σ at time T. Consider for definiteness (4.8a) with $i = 2$; the arguments in the other three cases are isomorphic. We have $u(x_2,T) = 0$, so that by (4.5)

$$\frac{\partial r}{\partial \beta}(T) = \frac{1}{2}\frac{\partial v}{\partial \beta}(T) - \frac{1}{2}\lambda_2(0)\frac{\partial u}{\partial \beta}(T) = \frac{1}{2}\frac{\partial v}{\partial \beta}(T) , \tag{4.20}$$

since $\lambda_2(0) = 0$. But $\frac{\partial r}{\partial \beta}(T) = r_o'(\beta) > 0$ and therefore $\frac{\partial v}{\partial \beta}(T) > 0$; i.e., $\frac{\partial v}{\partial x}\frac{\partial x_2}{\partial \beta} > 0$. Now $\frac{\partial x_2}{\partial \beta} > 0$ in the absence of a singularity, and therefore $v_x(x_2,T) > 0$. Then the differential equation (1.3) would imply $u_t + f(u)_x > 0$ at (x_2,T). But $f(u)_x = f'(0)\, u_x = 0$ by (4.2) and so we would have

$$\frac{du}{dt} = u_t + \lambda_2(0)\, u_x = u_t > 0 , \qquad t = T . \tag{4.21}$$

Hence u can never reach zero from above along $C_2(\beta)$; in other words, U cannot reach Σ from D_1 if (4.8a) holds.

This result, combined with Lemma 2, demonstrates that a singularity must develop in finite time if any initial point satisfies any of the four inequalities (4.8).

There remains only the case in which $\ell_o'(x) \leq 0$ and $r_o'(x) \leq 0$ whenever $U_o(x) \in D_1$ and also $\ell_o'(x) \geq 0$ and $r_o'(x) \geq 0$ whenever $U_o(x) \in D_2$. Then, just as in Section 3, the maximum and minimum values of both $\ell_o(\alpha)$ and $r_o(\beta)$ must occur on Σ. Since $U_o(x)$ is nonconstant, at least one of these functions (say r_o) must have distinct maximum and minimum values. Let

$$r_1 = \min r_o = r_o(x_1), \qquad r_2 = \max r_o = r_o(x_2) . \tag{4.22}$$

We have $r_1 < r_2$ and therefore $x_1 \neq x_2$, and we also have

$$\begin{aligned} u_o(x_j) &\equiv u(x_j,0) = 0 , \\ v_o(x_j) &\equiv v(x_j,0) = 2r_j , \end{aligned} \qquad j = 1,2 , \tag{4.23}$$

and

$$r_o'(x_j) = \frac{1}{2} v_o'(x_j) = 0 . \tag{4.24}$$

Therefore $v_x(x_j,0) = 0$, and so the vector derivative

$$\frac{\partial U}{\partial t} = \binom{u}{v}_t = \binom{v-f(u)}{-g(u)}_x = \binom{v_x - f'(0)u_x}{-g'(0)u_x} \tag{4.25}$$

vanishes initially, hence identically, at $x = x_j$. Thus

$$u(x_j,t) = 0, \quad v(x_j,t) = 2r_j, \quad j = 1,2 , \quad 0 \leq t < \infty . \tag{4.26}$$

Up to this point, we have followed substantially the same method of proof as in Section 3. We must diverge now, however, because the characteristic speeds at (x_j,t) are all zero, so that the characteristic lines through $(x_1,0)$ and $(x_2,0)$ are parallel, and the intersection argument of Section 3 does not apply here. Instead, we compute

$$\frac{d}{dt}\int_{x_1}^{x_2} u(x,t)dx = \int_{x_1}^{x_2} u_t dx = \int_{x_1}^{x_2} [v-f(u)]_x dx \tag{4.27}$$

$$= [\, v-f(u)]_{x_1}^{x_2} = 2(r_2-r_1) \; ,$$

a positive constant. Hence

$$\int_{x_1}^{x_2} u \; dx > 0 \tag{4.28}$$

for sufficiently large t. Thus there exist points (x,t) with x between x_1 and x_2 at which

$$\operatorname{sgn} u = \operatorname{sgn}(x_2-x_1) = \operatorname{sgn} r_0'(\beta) = \operatorname{sgn} r_x(x,t) \; , \tag{4.29}$$

where β denotes the foot of the C_2-characteristic through (x,t). If we consider a new initial value problem beginning at time t, the initial curve for this new problem will satisfy (4.8). Therefore a singularity will eventually form in this case also.

The proof of Theorem 2 is now complete.

BIBLIOGRAPHY

1. E.L. Isaacson, "Global solution of a Riemann problem for a non-strictly hyperbolic system of conservation laws arising in enhanced oil recovery," preprint.

2. F. John, "Formation of singularities in one-dimensional nonlinear wave propagation," Comm. Pure Appl. Math. 27 (1974) 337-405.

3. J.B. Keller and L. Ting, "Periodic vibrations of systems governed by nonlinear partial differential equations," Comm. Pure Appl. Math. 19 (1966) 371-420.

4. B.L. Keyfitz and H.C. Kranzer, "A system of non-strictly hyperbolic conservation laws arising in elasticity theory," Arch. Rat. Mech. Anal. 72 (1980) 219-241.

5. B.L. Keyfitz and H.C. Kranzer, "The Riemann problem for a class of hyperbolic conservation laws exhibiting a parabolic degeneracy," Jour. Diff. Eqns., to appear November 1982.

6. S. Klainerman and A. Majda, "Formation of singularities for wave equations including the nonlinear vibrating string," Comm. Pure Appl. Math. 33 (1980) 241-253.

7. P.D. Lax, "Development of singularities of solutions of nonlinear hyperbolic partial differential equations," Jour. Math. Phys. 5 (1964) 611-613.

8. T.P. Liu, "Development of singularities in the nonlinear waves for quasi-linear hyperbolic partial differential equations," Jour. Diff. Eqns, 33 (1979) 92-111.

9. J.B. Temple, "Global existence for a class of 2 x 2 nonlinear conservation laws with arbitrary Cauchy data," preprint.

DEPARTMENT OF MATHEMATICS
ARIZONA STATE UNIVERSITY
TEMPE, ARIZONA 85287

DEPARTMENT OF MATHEMATICS AND COMPUTER SCIENCE
ADELPHI UNIVERSITY
GARDEN CITY, NEW YORK 11530

Contemporary Mathematics
Volume 17, 1983

ON THE VARIABLE SIGN PENALTY APPROXIMATION OF THE NAVIER-STOKES EQUATION

Haroon Kheshgi[1]
and
Mitchell Luskin[2]

ABSTRACT. The penalty approximation of the Navier-Stokes equation replaces the continuity equation $u_{i,i} = 0$ by the approximation $u_{i,i} = -\epsilon p$ where $\epsilon > 0$ is a small parameter. This procedure is used in numerical algorithms to eliminate the pressure variables and thus to reduce the size of the system of nonlinear equations as well as the bandwidth. We propose that the continuity equation be replaced by $u_{i,i} = \epsilon \varphi_h p$ where φ_h alternates sign on a grid of size h. We prove that the approximation error in L^2 of our variable sign penalty approximation is $O(\epsilon h + \epsilon^2)$ whereas the error of the traditional penalty approximation in L^2 is $O(\epsilon)$.

1. INTRODUCTION. We give an analysis in this paper of an approximation of the Navier-Stokes system of equations for viscous, incompressible flow. This approximation can be discretized to give an efficient numerical method for the computation of low Reynolds number flows.

The Navier-Stokes system for viscous, incompressible flow is given by

$$(1.1)\qquad \begin{aligned} R(\underline{u}\cdot\nabla)\underline{u} + \nabla p &= 2\nabla\cdot\underline{\underline{D}} + \underline{f}\,, && x\in\Omega\,, \\ \nabla\cdot\underline{u} &= 0\,, && x\in\Omega\,, \end{aligned}$$

where $\underline{u} = (u_1, u_2)$ is velocity, p is pressure, $\underline{f}$ is a given body force, R is the Reynolds number, Ω is a bounded domain in $\mathbb{R}^2$,

1980 Mathematics Subject Classification. 35A40, 65N15, 65N30.

[1] Supported by the NSF.

[2] Supported by the NSF, Grant MCS-810-1631.

0271-4132/83/0000-1390/$05.50

$$D_{ij}(\underline{u}) = \frac{1}{2}\left(\frac{\partial u_i}{\partial x_j} + \frac{\partial u_j}{\partial x_i}\right), \qquad i,j = 1,2,$$

$$(\nabla \cdot \underline{\underline{D}})_i = \sum_{j=1}^{2} \frac{\partial}{\partial x_j} D_{ij}, \qquad i = 1,2 \quad .$$

The traditional penalty approximation of (1.1) is to replace the continuity equation

$$\nabla \cdot \underline{u} = 0 , \qquad x \in \Omega ,$$

by the perturbed equation

$$(1.2) \qquad \nabla \cdot \underline{u} = -\epsilon p , \qquad x \in \Omega ,$$

where $\epsilon > 0$ is a small parameter (see the references in [3,8]). The purpose of introducing the approximation (1.2) is that it allows the pressure variable to be eliminated from (1.1) to give the system of equations

$$(1.3) \qquad R(\underline{u}\cdot\nabla)\underline{u} - \nabla\left(\frac{1}{\epsilon}\nabla\cdot\underline{u}\right) = 2\nabla\cdot\underline{\underline{D}} + \underline{f} , \qquad x \in \Omega \quad .$$

The penalty approximation is often applied to numerical approximations of (1.1) since it gives a reduction in the size of the system of nonlinear equations to be solved as well as a reduction in band-width [4].

It has been shown that the difference between the solution to (1.1) (with appropriate boundary conditions) and the solution to the traditional penalty approximation

$$(1.4) \qquad \begin{aligned} R(\underline{u}\cdot\nabla)\underline{u} + \nabla p &= 2\nabla\cdot\underline{\underline{D}} + \underline{f} , \qquad & x \in \Omega , \\ \nabla\cdot\underline{u} &= -\epsilon p , & x \in \Omega , \end{aligned}$$

is $O(\epsilon)$ in many cases [3,8]. Thus, ϵ must be sufficiently small to insure an accurate approximation. However, we must consider round-off error in addition to approximation error since the linear systems which result from the discretization in space of (1.3) must be solved in practice by digital computers in floating point arithmetic, and it is observed in practice that the round-off error for the solution of these linear systems by Gaussian elimination is $O(\frac{1}{\epsilon})$. This can be explained theoretically by noting that the "condition number" [2] of these linear systems is $O(\frac{1}{\epsilon h^2})$ where h is the mesh size. Hence, the penalty method is often sufficiently accurate for only a very restricted range of the parameter, ϵ (see figure 3 in section 6). Further, it may not be possible to achieve sufficient accuracy for any ϵ for some problems, especially if one is using a computer with a small word length.

In this paper, we propose and analyze a new penalty approximation which replaces the continuity equation by

$$\nabla \cdot \underline{u} = -\epsilon \varphi_h p , \qquad x \in \Omega , \tag{1.5}$$

where φ_h satisfies the properties

$$\|\varphi_h\|_{L^\infty(\Omega)} \leq 1 , \tag{1.6}$$

$$\int_\Omega \varphi_h \zeta \, dx \leq c_1 h \|\nabla \zeta\|_{L^1(\Omega)} . \tag{1.7}$$

For instance, if Ω is the disjoint union of squares of length, h, then φ_h can be taken to be the checkerboard function which takes the value 1 and -1 on alternate squares.

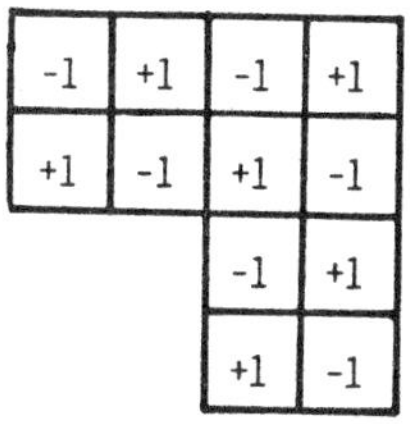

-1	+1	-1	+1
+1	-1	+1	-1
		-1	+1
		+1	-1

FIGURE 1

We call the approximation

$$\begin{aligned} R(\underline{u} \cdot \nabla)\underline{u} + \nabla p &= \nabla \cdot \underline{\underline{D}} + \underline{f} , \qquad x \in \Omega , \\ \nabla \cdot \underline{u} &= -\epsilon \varphi_h p , \qquad x \in \Omega , \end{aligned} \tag{1.8}$$

the variable sign penalty approximation. For certain finite element methods to be discussed in section 6, the variable sign penalty method is no more costly to implement than the traditional penalty method when the function, φ_h, is a piecewise constant function with respect to the finite element mesh (h denotes the size of the mesh).

We prove in this paper that the variable sign penalty approximation gives an $O(\epsilon h + \epsilon^2)$ L^2 error in the velocity and an $O(\epsilon)$ L^2 error in the pressure. If the approximate pressure is appropriately smoothed (post-processed), then we prove that an $O(\epsilon h + \epsilon^2)$ L^2 error is also achieved by the pressure on interior domains. Thus, the variable sign penalty approximation is seen to be asymptotically more accurate than the traditional penalty approximation. Numerical experiments have indicated that both penalty methods are equally sensitive to round-off error (see fig. 3 and [6]). Thus, the above estimates indicate that a given accuracy is achieved by the variable sign penalty method for a greater range of ϵ than for the traditional penalty method. This has also been confirmed in numerical experiments for practical values of the mesh length, h (See fig. 3 and [6]). Furthermore, for a given mesh there are levels of accuracy which can only be attained by the variable sign penalty method.

In the following section, we define notation that will be subsequently used and we give a more explicit definition of our procedure and results. In section 3, we prove the basic estimates. Section 4 is devoted to recalling some properties of smoothing operators, and section 5 gives some interior estimates for solutions to the linearized Navier-Stokes equations and applies these estimates to difference quotients of the error in our approximation procedure. Finally, in section 6 we discuss a finite element implementation of the variable sign penalty approximation.

2. DEFINITIONS AND MAIN RESULTS. We assume that Ω has a C^∞ boundary, $\partial\Omega$, which is the union of an interior closed curve, Γ_1, and an exterior closed curve, Γ_2. We also assume that $\underline{f} \in C^\infty(\overline{\Omega})^2$.

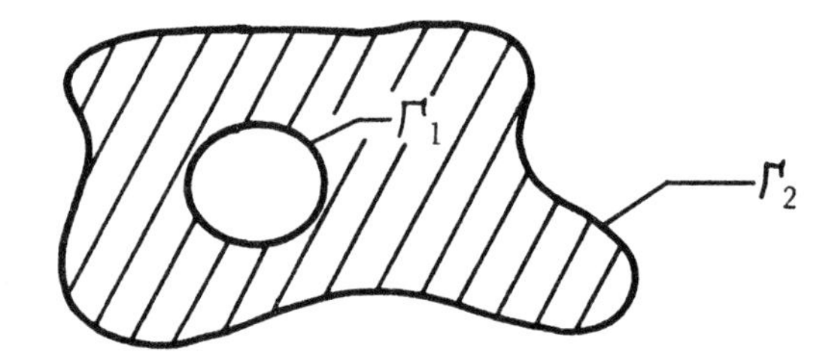

We will be interested in the boundary value problem

$$\text{(2.1)} \qquad \begin{aligned} &\underline{u} = 0 , && x \in \Gamma_1 , \\ &(2\underline{\underline{D}} - p\underline{\underline{I}})\underline{n} = 0 , && x \in \Gamma_2 , \end{aligned}$$

where $\underline{n}$ is the unit exterior normal to Γ_2. This choice of topology for Ω and this choice of boundary conditions guarantees that (1.1) for $R = 0$ is nonsingular and has smooth solutions for smooth data. Thus far, we have been unable to analyze the variable sign penalty approximation of (1.1) when values of the velocity are given on the entire boundary. However, the variable sign penalty approximation of this boundary value problem has been successfully used in numerical experiments (see section 6).

For scalar-valued functions $\xi, \eta \in L^2(\Omega)$, we define the inner product and norm

$$(\xi, \eta) = \int_\Omega \xi\eta \, dx , \quad \|\xi\|^2 = (\xi, \xi) .$$

We denote by $H^k(\Omega)$, k a nonnegative integer, the space of functions

$$H^k(\Omega) = \{\xi \in L^2(\Omega) | D^\alpha \xi \in L^2(\Omega) \quad \text{for} \quad |\alpha| \leq k\}$$

with norm

$$\|\xi\|_k^2 = \sum_{|\alpha| \leq k} \|D^\alpha \xi\|^2 \quad .$$

We further set for $k \geq 1$

$$H_0^k(\Omega) = \{\xi \in H^k(\Omega) | D^\alpha \xi \equiv 0 \quad \text{on} \quad \partial\Omega \qquad \text{for} \quad |\alpha| \leq k-1\} \quad .$$

We wish to also extend the above definitions in the usual fashion to vector-valued functions $\underline{v}: \Omega \to \mathbb{R}^2$ and we denote the resulting spaces by $L^2(\Omega)^2$, $H^k(\Omega)^2$, etc. For $\underline{v}: \Omega \to \mathbb{R}^2$, $\xi: \Omega \to \mathbb{R}$, and $\ell, m \geq 0$ integers we define the norm

$$\|\underline{v}, \xi\|_{\ell,m} = \|\underline{v}\|_\ell + \|\xi\|_m \quad , \qquad \underline{v} \in H^\ell(\Omega)^2 \; , \quad \xi \in H^m(\Omega) \quad .$$

Finally, it will be useful to define the space

$$V = \{\underline{v}, q | \underline{v} \in H^1(\Omega)^2 \; , \quad q \in L^2(\Omega) \; , \qquad \underline{v} \equiv 0 \quad \text{on} \quad \Gamma_1\} \; .$$

We next define the continuous bilinear form, $\tilde{B}: V \times V \to \mathbb{R}$, by

$$\tilde{B}(\underline{w}, r; \underline{v}, q) = 2(\underline{\underline{D}}(\underline{w}), \nabla \underline{v}) - (r, \nabla \cdot \underline{v}) - (\nabla \cdot \underline{w}, q) \; , \qquad \underline{w}, r; \; \underline{v}, q \in V \quad .$$

It is known that there exists a positive constant, c_2 , so that [3]

$$\sup_{\underline{v}, q \in V} \frac{\tilde{B}(\underline{w}, r; \underline{v}, q)}{\|\underline{v}, q\|_{1,0}} \geq c_2 \|\underline{w}, r\|_{1,0} \; , \qquad \underline{w}, r \in V \quad . \tag{2.2}$$

It will be useful to have the following consequence of the Sobolev embedding theorem. There exists c_3 so that

$$|((\underline{z} \cdot \nabla)\underline{w}, \; \underline{v})| \leq c_3 \|\underline{z}\|_1 \|\underline{w}\|_1 \|\underline{v}\|_1$$

where $\underline{z}, \underline{w}, \underline{v} \in H^1(\Omega)^2$.

We call $\underline{u}, p \in V$ a solution to (1.1), (2.1) if

$$\tilde{B}(\underline{u}, p; \underline{v}, q) + R((\underline{u} \cdot \nabla)\underline{u}, \underline{v}) = (\underline{f}, \underline{v}) \; , \quad \underline{v}, q \in V \quad . \tag{2.3}$$

It is known that there exists $R_0 > 0$ such that (2.3) has a solution $\underline{u}(R)$, $p(R)$ for $0 \leq R \leq R_0$ such that $\underline{u}, p: [0, R_0] \to V$ is continuously differentiable, $\underline{u}(R)$, $p(R) \in C^\infty(\overline{\Omega})^2 \times C^\infty(\overline{\Omega})$ for $0 \leq R \leq R_0$, and there exists $c_4 > 0$ such that

$$\sup_{\underline{v}, q \in V} \frac{\tilde{B}(\underline{w}, r; \underline{v}, q) + R((\underline{u} \cdot \nabla)\underline{w} + (\underline{w} \cdot \nabla)\underline{u}, \underline{v})}{\|\underline{v}, q\|_{1,0}} \geq c_4 \|\underline{w}, r\|_{1,0} \qquad \underline{w}, r \in V \quad . \tag{2.4}$$

This solution can be obtained by a continuation argument in the parameter, R.

The variable sign penalty solution to (1.8), (2.1) is the solution $\underline{u}_h, p_h \in V$ to

$$\tilde{B}(\underline{u}_h, p_h; \underline{v}, q) + R((\underline{u}_h \cdot \nabla)\underline{u}_h, \underline{v}) - \epsilon(\varphi_h p_h, q) = (\underline{f}, \underline{v}), \quad \underline{v}, q \in V. \tag{2.5}$$

It follows from (2.2) that if $\epsilon < c_2$ and $R = 0$, then (2.5) has a unique solution $\underline{u}_h, p_h \in V$. Furthermore, this solution can be continued for $0 \leq R \leq R_1$, where $R_1 > 0$. In section 3 we prove Theorem 1 which gives an estimate for the dependence of R_1 on ϵ.

THEOREM 1. *Let* $0 < R_1 < R_0$ *and* $\epsilon \geq 0$ *be such that*

$$\frac{16\, \epsilon R \|p(R)\| c_3}{c_4^2} \leq 1 \tag{2.6}$$

and

$$\epsilon + 2c_3 R d(\epsilon, R) \leq \frac{c_4}{2}$$

for $0 \leq R \leq R_1$ *where*

$$d(\epsilon, R) \equiv \frac{c_4}{2Rc_3}\left[1 - \sqrt{1 - 16\, \epsilon R \|p(R)\| c_3 c_4^{-2}}\right] = O(\epsilon).$$

Then the solution to (2.5), $\underline{u}_h(R)$, $p_h(R) \in V$, *can be continued for* $0 \leq R \leq R_1$ *such that* $\underline{u}_h, p_h : [0, R_1] \to V$ *is continuously differentiable.*

Furthermore, we will prove:

THEOREM 2. *The estimate*

$$\|\underline{u} - \underline{u}_h, p - p_h\|_{1,0} \leq d(\epsilon, R)$$

holds for $0 \leq R \leq R_1$ *where* R_1 *satisfies the conditions of Theorem 1 and* $\underline{u}_h(R)$, $p_h(R)$ *is the solution obtained by continuation.*

Both Theorem 1 and Theorem 2 only utilize the property

$$\|\varphi_h\|_{L^\infty(\Omega)} \leq 1$$

and are thus valid also for the traditional penalty method. The next results utilize property (1.7) and are thus valid only for the variable sign penalty method. They demonstrate the asymptotic superiority of the variable sign penalty method.

For $r \in L^2(\Omega)$ we define the norm

$$\|r\|_{-1} = \sup_{q \in H_0^1(\Omega)} \frac{(r,q)}{\|q\|_1} .$$

We then can obtain the following result:

THEOREM 3. There exists c_5 so that

$$\|\underline{u} - \underline{u}_h\| + \|p - p_h\|_{-1} \le c_5[\epsilon h + \epsilon^2]$$

for $0 \le R \le R_1$.

Now Theorem 3 gives the $O(\epsilon h + \epsilon^2)$ estimate for the error in the velocity. To achieve a similar estimate for the pressure we must assume that $\Omega_1 \subset\subset \Omega$ and that

$$(2.7) \qquad \partial_{i,h}\varphi_h = 0 , \quad i = 1,2 , \quad h < \mathrm{dist}(\Omega_1, \partial\Omega) ,$$

where

$$\partial_{i,h}\xi(x) = \frac{\xi(x + he_i) - \xi(x - he_i)}{2h} , \quad x \in \Omega_1 ,$$

and $e_1 = (1,0)$, $e_2 = (0,1)$. We note that the checkerboard function (fig. 1) satisfies (2.7). Now let

$$K(x) = K(x_1,x_2) = \begin{cases} (2h)^{-2} , & \text{if } |x_i| \le h , \ i = 1,2, \\ 0 , & \text{otherwise.} \end{cases}$$

Then we have the following result:

THEOREM 4. Assume that (2.7) holds and that $0 \le R \le R_1$. It then follows for $H < \mathrm{dist}(\Omega_1, \partial\Omega)$ where $H = \sqrt{2}\, h$ that

$$\|p - K*p_h\|_{0,\Omega_1} \le c_6(h^2 + \epsilon h + \epsilon^2)$$

where

$$\|\xi\|^2_{0,\Omega_1} = \int_{\Omega_1} \xi^2 \, dx ,$$

and

$$K*p_h(x) \equiv \int_\Omega K(x-y)p_h(y)dy , \qquad x \in \Omega_1 .$$

Higher order variable sign penalty methods can also be constructed and analyzed by the techniques presented here. For instance, the checkerboard function, φ_h , described in section 1 satisfies the higher order estimate

$$\int_\Omega \varphi_h \zeta \, dx \le ch^2 \sum_{|\alpha|=2} \|D^\alpha \zeta\|_{L^1(\Omega)} ,$$

and if the approximate velocity and pressure are appropriately smoothed, then higher order $O(h^2 + \epsilon h^2 + \epsilon^2)$ estimates can be proven for the error in L^2 in interior domains by our techniques.

We note that our estimates are for solutions obtained by the numerically practical method of continuation in the Reynold's number, R. Techniques similar to those presented here can be used with continuation in ϵ to also derive $O(\epsilon h + \epsilon^2)$ error estimates for the variable sign penalty approximation of any solution, u, of (1.1) such that the linearization of (1.1) about u is nonsingular. Finally, we note that the variable sign penalty approximation and its analysis generalizes to R^3 without modification.

3. BASIC ESTIMATES. We noted in section 2 that (2.5) has a solution $\underline{u}_h, p_h \in V$ for $R = 0$ if $\epsilon < c_2$. If $\epsilon < c_2$, then $\underline{u}_h(R), p_h(R) \in V$ can be continued by solving the ordinary differential equation

$$\frac{d}{dR}\underline{u}_h(R) = \underline{w}_h(R),$$

$$\frac{d}{dR}p_h(R) = r_h(R),$$

where $\underline{w}_h(R), r_h(R) \in V$ satisfy

$$(3.1)\quad \tilde{B}(\underline{w}_h, r_h; \underline{v}, q) + R((\underline{u}_h\cdot\nabla)\underline{w}_h + (\underline{w}_h\cdot\nabla)\underline{u}_h, \underline{v}) - \epsilon(\varphi_h r_h, q) = -((\underline{u}_h\cdot\nabla)\underline{u}_h, \underline{v}), \quad \underline{v}, q \in V.$$

In order to get a bound on how far we can continue $\underline{u}_h(R), p_h(R)$ we prove the following lemma.

LEMMA 1. *Suppose that for* $R_* \le R_1$ *the following hold:*

(3.2) (2.5) *has a solution* $\underline{u}_h, p_h \in V$ *for* $0 \le R \le R_*$ *such that* $\underline{u}_h, p_h\colon [0, R_*] \to V$ *is continuously differentiable*,

$$(3.3)\quad \epsilon \le \frac{c_4}{2} \quad \text{for } 0 \le R \le R_*, \text{ and}$$

$$(3.4)\quad \frac{16\,\epsilon R\|p(R)\|c_3}{c_4^2} \quad \text{for } 0 \le R \le R_*,$$

Then

$$(3.5)\quad \|\underline{u} - \underline{u}_h, p - p_h\|_{1,0} \le \frac{c_4}{2Rc_3}\left[1 - \sqrt{1 - 16\,\epsilon R\|p(R)\|c_3 c_4^{-2}}\right] \equiv d(\epsilon, R).$$

PROOF. We note that $\underline{e} = \underline{u} - \underline{u}_h$ and $E = p - p_h$ satisfy

$$B_\epsilon(\underline{e},E;\underline{v},q) = -\epsilon(\varphi_h p,q) + R((\underline{e}\cdot\nabla)\underline{e},\underline{v}), \qquad \underline{v},q \in V \quad , \tag{3.6}$$

where

$$B_\epsilon(\underline{w},r;\underline{v},q) = \tilde{B}(\underline{w},r;\underline{v},q) + R((\underline{u}\cdot\nabla)\underline{w} + (\underline{w}\cdot\nabla)\underline{u},\underline{v}) - \epsilon(\varphi_h r,q) \quad ,$$

$$\underline{w},r;\underline{v},q \in V \ .$$

It follows from (2.4) that

$$\sup_{\underline{v},q\in V} \frac{B_\epsilon(\underline{w},r;v,q)}{\|\underline{v},q\|_{1,0}} \geq [c_4 - \epsilon]\|w,r\|_{1,0} \geq \frac{c_4}{2}\|\underline{w},r\|_{1,0} \ .$$

Hence, it follows from (3.6) that

$$\frac{c_4}{2}\|\underline{e},E\|_{1,0} \leq \epsilon\|p\| + Rc_3\|\underline{e},E\|_{1,0}^2 \tag{3.7}$$

Lemma 1 now follows from (3.7) and the continuity of $\|\underline{e},E\|_{1,0}$ as a function of R ∎

PROOF OF THEOREM 1. We note that as long as $\underline{u}_h, p_h$ can be continued for $0 \leq R \leq R_1$ we have from Lemma 1 that

$$\begin{aligned}\sup_{\underline{v},q\in V} \frac{B_\epsilon(\underline{w}_h,r_h;\underline{v},q) + R((\underline{e}\cdot\nabla)\underline{w}_h + (\underline{w}_h\cdot\nabla)\underline{e},\underline{v})}{\|\underline{v}.q\|_{1,0}} &\geq [c_4 - \epsilon - 2c_3R\|\underline{e}(R)\|_1]\|\underline{w}_h,r_h\|_{1,0} \\ &\geq [c_4 - \epsilon - 2c_3Rd(\epsilon,R)]\|\underline{w}_h,r_h\|_{1,0} \\ &\geq \frac{c_4}{2}\|\underline{w}_h,r_h\|_{1,0} \ .\end{aligned} \tag{3.7}$$

Hence, we have from (3.1) that

$$\begin{aligned}\|\underline{w}_h,r_h\|_{1,0} &\leq \frac{2}{c_4} c_3\|\underline{u}_h\|_1^2 \\ &\leq \frac{2}{c_4} c_3(\|\underline{u}\|_1 + d)^2 \ .\end{aligned} \tag{3.8}$$

The bound (3.8) shows that $\underline{u}_h, p_h$ can be continued for $0 \leq R \leq R_1$ ∎

We note now that Theorem 2 follows immediately from Theorem 1 and Lemma 1. We next turn to the proof of Theorem 3. We will need the <u>a priori</u> inequality that if $\underline{v},q \in L^2(\Omega)^2 \times H_0^1(\Omega)$ and $\underline{z},m \in V$ satisfy

$$(3.9) \qquad B_0(\underline{w},r;\underline{z},m) = (\underline{w},\underline{v}) + (r,q) , \qquad \underline{w},r \in V ,$$

then

$$(3.10) \qquad \|\underline{z},m\|_{2,1} \le c_7 \|\underline{v},q\|_{0,1} .$$

PROOF OF THEOREM 3. Let $\underline{v},q \in L^2(\Omega)^2 \times H_0^1(\Omega)$ and let $\underline{z},m \in V$ satisfy the boundary value problem (3.9). Then

$$(3.11) \qquad \begin{aligned} &(\underline{e},\underline{v}) + (E,q) \\ &= B_0(\underline{e},E;\underline{z},m) \\ &= -\varepsilon(\varphi_h p_h, m) + R((\underline{e}\cdot\nabla)\underline{e},\underline{z}) \end{aligned}$$

Now

$$(3.12) \qquad \begin{aligned} -(\varphi_h p_h, m) &= -(\varphi_h p, m) + (\varphi_h(p-p_h), m) \\ &\le 2c_1 h\|p\|_1\|m\|_1 + \|E\|_0 \|m\|_0 \\ &\le (2c_1 h\|p\|_1 + d)\|m\|_1 , \end{aligned}$$

and

$$R((\underline{e}\cdot\nabla)\underline{e},\underline{z}) \le c_3 R d^2 \|\underline{z}\|_1 .$$

Hence, Theorem 3 follows from (3.10) ∎

4. SMOOTHING ESTIMATES. In this section, we recall some estimates and relations for convolution with

$$K(x) = K(x_1,x_2) = \begin{cases} (2h)^{-2} & \text{if } |x_1|,|x_2| \le h , \\ 0 & \text{otherwise.} \end{cases}$$

The proofs of these results can be found in [1]. We note that

$$\int_{R^2} K(x)dx = 1 .$$

We extend the notation in section 2 by defining for $\Omega_* \subset \Omega$,

$$\|\xi\|^2_{\Omega_*} = \int_{\Omega_*} \xi^2 dx , \qquad \xi \in L^2(\Omega_*) ,$$

$$\|\xi\|^2_{\Omega_*,k} = \sum_{|\alpha|\le k} \|D^\alpha \xi\|^2_{\Omega_*} , \qquad \xi \in H^k(\Omega_*) ,$$

$$\|\xi\|^2_{\Omega_*,-k} = \sup_{\eta \in H_0^k(\Omega_*)} \frac{(\xi,\eta)}{\|\eta\|_{\Omega_*,k}} , \qquad \xi \in L^2(\Omega_*) .$$

For $x \in \Omega$ such that $\mathrm{dist}(x,\partial\Omega) > H$ (recall that $H = \sqrt{2}\,h$), we note that $K*\xi(x)$ is well-defined and

$$K*\xi(x) = \int_{x_1-h}^{x_1+h} \int_{x_2-h}^{x_2+h} \xi(y_1,y_2)dy_1dy_2 \quad .$$

Now it is easy to check that if $\text{dist}(x,\partial\Omega) > H$, then

$$\frac{\partial}{\partial x_1} K*\xi(x) = \int_{x_2-h}^{x_2+h} \left[\frac{\xi(x_1+h\ ,y_2) - \xi(x_1-h,y_2)}{2h}\right] dy_2$$

$$= L_1*\partial_{1,h}\xi(x)$$

where

$$L_1(x) = \begin{cases} \delta(x_1)/2h \ , & -h \leq x_2 \leq h \ , \\ 0 \ , & \text{otherwise} \ . \end{cases}$$

Here, $\delta(x_1)$ is the Dirac delta function. Similar definitions and relations hold for

$$\frac{\partial}{\partial x_2} K*\xi(x) \quad .$$

We have the following lemmas [1].

LEMMA 2. <u>If</u> $\Omega_* \subset\subset \Omega$ <u>and</u> $\text{dist}(\Omega_*,\partial\Omega) > H$, <u>then</u>

$$(4.1) \qquad \|\xi - K*\xi\|_{\Omega_*} \leq c_8 h^2 \|\xi\|_2 \ , \qquad \xi \in H^2(\Omega) \quad .$$

LEMMA 3. <u>If</u> $\Omega_0 \subset\subset \Omega_1 \subset \Omega$, <u>then</u>

$$(4.2) \qquad \|\xi\|_{\Omega_0} \leq c_9(\|\xi\|_{\Omega_1,-1} + \sum_{i=1}^{2} \|\frac{\partial}{\partial x_i} \xi\|_{\Omega_1,-1})$$

LEMMA 4. <u>If</u> $\Omega_0 \subset\subset \Omega_1 \subset \Omega$, <u>then for</u> $H < \text{dist}(\Omega_0,\partial\Omega_1)$ <u>we have</u>

$$(4.3) \qquad \|L_i*\xi\|_{\Omega_0,-1} \leq c_{10}\|\xi\|_{\Omega_1,-1} \ , \qquad i = 1,2 \ ,$$

$$\|K*\xi\|_{\Omega_0,-1} \leq c_{10}\|\xi\|_{\Omega_1,-1} \quad .$$

Now let $\Omega_0 \subset\subset \Omega_1 \subset\subset \Omega_2 \subset\subset \Omega$ and suppose that $\text{dist}(\Omega_1,\partial\Omega_2) > H$. We have that

$$(4.4) \qquad \|p-K*p_h\|_{\Omega_0} \leq \|p-K*p\|_{\Omega_0} + \|K*(p-p_h)\|_{\Omega_0}$$

$$\leq c_8h^2\|p\|_2 + \|K*(p-p_h)\|_{\Omega_0}$$

and from Lemma 3 and Lemma 4 we have

$$(4.5)\qquad \|K*(p-p_h)\|_{\Omega_0} \le c_9(\|K*(p-p_h)\|_{\Omega_1,-1} + \sum_{i=1}^{2} \|\frac{\partial}{\partial x_i} K*(p-p_h)\|_{\Omega_1,-1})$$
$$= c_9(\|K*(p-p_h)\|_{\Omega_1,-1} + \sum_{i=1}^{2} \|L_i*\partial_{i,h}(p-p_h)\|_{\Omega_1,-1})$$
$$\le c_9c_{10}(\|p-p_h\|_{\Omega_2,-1} + \sum_{i=1}^{2} \|\partial_{i,h}(p-p_h)\|_{\Omega_2,-1}) \quad .$$

Thus,

$$(4.6)\qquad \|p-K*p_h\|_{\Omega_0} \le c_8h^2\|p\|_2 + c_9c_{10}\|p-p_h\|_{\Omega_2,-1} + c_9c_{10}\sum_{i=1}^{2}\|\partial_{i,h}(p-p_h)\|_{\Omega_2,-1} \quad .$$

Now by **Theorem** 3

$$(4.7)\qquad \|p-p_h\|_{-1} \le c_6(\epsilon h + \epsilon^2)$$

The next section will be devoted to establishing that

$$(4.8)\qquad \|\partial_{i,h}(p-p_h)\|_{\Omega_2,-1} = O(\epsilon h + \epsilon^2) \quad .$$

5. ESTIMATES FOR DIFFERENCE QUOTIENTS. The results in this section were motivated by the interior estimates for difference quotients for Ritz-Galerkin methods developed in [7]. If $\Omega_1 \subset\subset \Omega$ and $\text{dist}(\Omega_1, \partial\Omega) > H$, then by (2.7)

$$(5.1)\qquad \hat{B}_0(\partial_{1,h}\underline{e}, \partial_{1,h}E;\ \underline{z}, m) = L_1(\underline{z}) + L_2(m) \quad , \qquad \underline{z},\ m \in V(\Omega_1) \quad ,$$

where

$$V(\Omega_1) = \{\underline{z}, m \in V(\Omega) \mid \underline{z} \equiv \underline{0}\ ,\ m \equiv 0 \ \text{ for } x \in \Omega/\Omega_1\} \quad ,$$

$$\hat{\underline{y}}(x_1,x_2) = [\underline{y}(x_1+h,x_2) + \underline{y}(x_1-h,x_2)]/2 \quad ,$$

$$\hat{B}_0(\underline{w},r;\underline{z},m) = \tilde{B}(\underline{w},r;\underline{z},m) + R((\hat{\underline{u}}\cdot\nabla)\underline{w} + (\underline{w}\cdot\nabla)\hat{\underline{u}}, \underline{z}) \quad ,$$

$$L_1(\underline{z}) = R((\partial_{1,h}\underline{u}\cdot\nabla)\hat{\underline{e}} + (\hat{\underline{e}}\cdot\nabla)\partial_{1,h}\underline{u} - (\hat{\underline{e}}\cdot\nabla)\partial_{1,h}\underline{e} - (\partial_{1,h}\underline{e}\cdot\nabla)\hat{\underline{e}}, \underline{z}) \quad ,$$

$$L_2(m) = -\epsilon(\varphi_h\ \partial_{1,h}p, m) \quad .$$

In order to estimate $\partial_{1,h}e$ and $\partial_{1,h}E$, we prove an interior estimate for for a solution $\underline{w} \in H^1(\Omega_1)^2$, $r \in L^2(\Omega_1)$ of

$$(5.2) \qquad \hat{B}_0(\underline{w},r;\underline{z},m) = L_1(\underline{z}) + L_2(m), \quad \underline{z},m \in V(\Omega_1).$$

First, we extend the definition of negative norms to vector-valued functions and we define for $\Omega_* \subset\subset \Omega$,

$$\|\underline{w},r\|_{\Omega_*,-\ell,-m} = \|\underline{w}\|_{\Omega_*,-\ell} + \|r\|_{\Omega_*,-m} .$$

Further, we define for the linear operators, L_i, the norms

$$\|L_1\|_{\Omega_*,-2} = \sup_{\underline{z} \in H_0^2(\Omega_*)^2} \frac{L_1(\underline{z})}{\|\underline{z}\|_{\Omega_*,2}},$$

$$\|L_2\|_{\Omega_*,-1} = \sup_{m \in H_0^1(\Omega_*)} \frac{L_2(m)}{\|m\|_{\Omega_*,1}} .$$

LEMMA 5. <u>Suppose that</u> $\Omega_0 \subset\subset \Omega_1 \subset\subset \Omega$, $H < \mathrm{dist}(\Omega_1,\partial\Omega)$, <u>and that</u> $\underline{w} \in H^1(\Omega_1)^2$, $r \in L^2(\Omega_1)$ <u>satisfy</u> (5.2). <u>Then</u>

$$(5.3) \qquad \|\underline{w},r\|_{\Omega,0,-1} \le c_{11}(\|\underline{w},r\|_{\Omega_1,-1,-2} + \|L_1\|_{\Omega_1,-2} + \|L_2\|_{\Omega_1,-1}) .$$

PROOF. Let $\omega \in C_0^\infty(\Omega_1)$ with $\omega \equiv 1$ on Ω_0. Then

$$\|\underline{w},r\|_{\Omega_0,0,-1} \le \|\omega\underline{w},\omega r\|_{\Omega_1,0,-1}$$

$$= \sup_{\substack{\underline{v},q \in L^2(\Omega)^2 \times H_0^1(\Omega) \\ \underline{v},q \equiv 0 \text{ on } \Omega\setminus\Omega_1}} \frac{(\omega\underline{w},\underline{v}) + (\omega r,q)}{\|\underline{v},q\|_{0,1}} .$$

Now we can extend $\underline{u}$ to $\mathbb{R}^2$ so that $\underline{u} \in C^\infty(\mathbb{R}^2)^2$. Then $\hat{B}_0$ is defined on $V \times V$ and we have the <u>a priori</u> estimate

$$(5.4) \qquad \|\underline{z},m\|_{2,1} \le c_{12}\|\underline{v},q\|_{0,1}$$

for h sufficiently small for solutions to

$$(5.5) \qquad \hat{B}_0(\underline{w},r;\underline{z},m) = (\underline{w},\underline{v}) + (r,q), \qquad \underline{w},r \in V .$$

So,

$$(\omega\underline{w},\underline{v}) + (\omega r,q) = \hat{B}_0(\omega\underline{w},\omega r;\underline{z},m)$$

$$= \hat{B}_0(\underline{w},r;\omega\underline{z},\omega m) + I(\underline{w},r;\underline{z},m)$$

where

$$I(\underline{w},r;\underline{z},m) = (\underline{w}(\nabla\omega),\nabla\underline{z}) - (\underline{w},\nabla\cdot(\underline{z}\nabla\omega))$$
$$- R((\hat{\underline{u}}\cdot\nabla)\omega\underline{w},\underline{z})$$
$$+ (r\nabla\omega,\underline{z}) + (\underline{w},(\nabla\omega)m) \quad .$$

Thus,

$$(\omega\underline{w},\underline{v}) + (\omega r,q)$$
$$= L_1(\omega\underline{z}) + L_2(\omega m) + I(\underline{w},r;\underline{z},m)$$
$$\leq \|L_1\|_{\Omega_1,-2}\|\omega\underline{z}\|_2 + \|L_2\|_{\Omega_1,-1}\|\omega m\|_1$$
$$+ c_{12}\|\underline{w},r\|_{\Omega_1,-1,-2}\|\underline{z},m\|_{2,1}$$

The result (5.3) now follows from (5.4) and (5.5) ∎

We next apply Lemma 5.1 to (5.1). It then follows that

$$\|\partial_{1,h}\underline{e},\partial_{1,h}E\|_{\Omega_0,0,-1}$$
$$\leq c_{12}\|\partial_{1,h}\underline{e},\partial_{1,h}E\|_{\Omega_1,-1,-2}$$
$$(5.6) \qquad + c_{11}(\|L_1\|_{\Omega_1,-2} + \|L_2\|_{\Omega_1,-1})$$
$$\leq c_{12}\|\underline{e},E\|_{0,-1}$$
$$+ c_{11}(\|L_1\|_{\Omega_1,-2} + \|L_2\|_{\Omega_1,-1}) \quad .$$

It thus remains to estimate $\|L_1\|_{\Omega_1,-2}$ and $\|L_2\|_{\Omega_1,-1}$. Now

$$\|L_1\|_{\Omega_1,-2} \leq c_{13}R(\|\underline{e}\| + \|\underline{e}\|_1^2)$$

and

$$L_2(m) = -\epsilon(\varphi_h\,\partial_{1,h}p,m) + \epsilon(\varphi_h\,\partial_{1,h}E,m)$$
$$= -\epsilon(\varphi_h\partial_{1,h}p,m) - \epsilon(\varphi_h E,\partial_{1,h}m)$$
$$\leq [c_{13}\epsilon h\|p\|_2 + \epsilon\|E\|]\|m\|_{\Omega_1,1} \quad ,$$

so,

$$\|L_2\|_{\Omega_1,-1} \leq c_{13}\epsilon h\|p\|_2 + \epsilon\|E\| \quad .$$

Hence,

$$(5.7)\qquad \|\partial_{1,h}\underline{e},\partial_{1,h}E\|_{\Omega_0,0,-1} \le c_{14}\|\underline{e},E\|_{0,-1} + c_{13}R(\|\underline{e}\|+\|\underline{e}\|_1^2) + c_{13}\,\varepsilon h\|p\|_2 + \varepsilon\|E\| \le c_{15}(\varepsilon h+\varepsilon^2)\ .$$

Finally, we note that (5.7) includes the result (4.8), and (4.7) and (4.8) can be substituted in (4.6) to yield Theorem 4.

6. THE FINITE ELEMENT VARIABLE SIGN PENALTY METHOD. We will now describe briefly how to implement the variable sign penalty approximation with a finite element method. Let

$$\mathcal{V} = \{\underline{v}\in H^1(\Omega)^2 \mid \underline{v}\equiv 0 \quad\text{on}\quad \Gamma_1\}$$

and let $\mathcal{V}_h$ be a finite dimensional subspace of $\mathcal{V}$. Further, let $\{\Theta_i\}_{i=1}^N$ be a set of open subsets of Ω such that

$$\Theta_i\cap\Theta_j = \emptyset\ ,\qquad \text{if}\quad i\neq j\ ,$$

$$\overline{\Omega} = \bigcup_{i=1}^N \overline{\Theta}_i\ .$$

We then define the finite dimensional subspace of $L^2(\Omega)$,

$$\mathcal{M}_h = \{\tilde{q}\in L^2(\Omega) \mid \tilde{q}|_{\Theta_i} = c_i \quad\text{for}\quad i=1,\dots,N \quad \text{where } c_i \text{ is a constant}\}\ .$$

The solution $\underline{u}_h\in\mathcal{V}$, $p_h\in L^2(\Omega)$ to (2.5) satisfies

$$(6.1)\qquad 2(\underline{\underline{D}}(\underline{u}_h),\nabla\underline{v}) + R((\underline{u}_h\cdot\nabla)\underline{u}_h,\underline{v}) - (p_h,\nabla\cdot\underline{v}) = (\underline{f},\underline{v})\ ,\qquad \underline{v}\in\mathcal{V}\ ,$$

$$(6.2)\qquad (\nabla\cdot\underline{u}_h,q) + \varepsilon(\varphi_h p_h,q) = 0\ ,\qquad q\in L^2(\Omega)\ .$$

The finite element solution $\tilde{\underline{u}}_h\in\mathcal{V}_h$, $\tilde{p}_h\in\mathcal{M}_h$ to (6.1) - (6.2) satisfies

$$(6.3)\qquad 2(\underline{\underline{D}}(\tilde{\underline{u}}_h),\nabla\tilde{\underline{v}}) + R((\tilde{\underline{u}}_h\cdot\nabla)\tilde{\underline{u}}_h,\tilde{\underline{v}}) - (\tilde{p}_h,\nabla\cdot\tilde{\underline{v}}) = (\underline{f},\tilde{\underline{v}})\ ,\qquad \tilde{\underline{v}}\in\mathcal{V}_h$$

$$(6.4)\qquad (\nabla\cdot\tilde{\underline{u}}_h,\tilde{q}) + \varepsilon(\varphi_h\tilde{p}_h,\tilde{q}) = 0\ ,\qquad \tilde{q}\in\mathcal{M}_h\ .$$

Now suppose that $\varphi_h\in\mathcal{M}_h$ and $\varphi_h(x)\neq 0$ for every $x\in\Omega$. Then we can let $\tilde{q} = \chi_{\Theta_i}$, where χ_{Θ_i} is the characteristic function for the set Θ_i, in (6.4) to obtain

(6.5) $\int_{\Theta_i} \nabla\cdot\tilde{\underline{u}}_h\,dx + \varepsilon\,\varphi_h(\Theta_i)\tilde{p}_h(\Theta_i)m(\Theta_i)$

where $\tilde{q}(\Theta_i)$ denotes the constant value of $\tilde{q}\in\mathcal{M}_h$ on Θ_i and $m(\Theta_i)$ is the measure of Θ_i. Thus,

$$\tilde{p}_h(\Theta_i) = -\frac{1}{\varepsilon\,\varphi_h(\Theta_i)m(\Theta_i)}\int_{\Theta_i}(\nabla\cdot\tilde{\underline{u}}_h)\,dx\ , \qquad i = 1,\ldots,N\ . \tag{6.6}$$

We can then use (6.6) to eliminate $\tilde{p}_h$ from (6.3) to obtain a nonlinear system of equations for $\tilde{\underline{u}}_h$. If $P_h : L^2(\Omega)\to\mathcal{M}_h$ denotes the projection onto $\mathcal{M}_h$, then we obtain

$$2(\underline{\underline{D}}(\tilde{\underline{u}}_h), \nabla\tilde{\underline{v}}) - R((\tilde{\underline{u}}_h\cdot\nabla)\,\tilde{\underline{u}}_h, \tilde{\underline{v}}) + (\tfrac{1}{\varepsilon}\,P_h(\varphi_h^{-1}\nabla\cdot\tilde{\underline{u}}_h), \nabla\cdot\tilde{\underline{v}})$$
$$= (\underline{f},\tilde{\underline{v}})\ , \qquad \tilde{\underline{v}}\in\mathcal{V}_h$$

Spaces $\mathcal{V}_h$ which have been useful in computation can be found in [3,8]. Also, see [5,6] for more details on the finite element implementation of the variable sign penalty approximation. We present the results of a computation for the driven cavity problem

$$\begin{aligned}
&-2\nabla\cdot\underline{\underline{D}} + \nabla p = \underline{f}\ , && 0<x_1,\ x_2<1\ ,\\
&\nabla\cdot\underline{u} = 0\ , && 0<x_1,\ x_2<1\ ,\\
&\underline{u}(0,x_2) = \underline{u}(1,x_2) = \underline{0}\ , && 0<x_2<1\ ,\\
&\underline{u}(x_1,0) = \underline{0}\ , && 0<x_1<1\ ,\\
&\underline{u}(x_1,1) = (1,0), && 0<x_1<1\ ,
\end{aligned}$$

where continuous, bilinear elements have been used for the velocity field and piecewise constant elements have been used for the pressure field. Here φ_h is taken to be the "checkerboard" function (see figure 1).

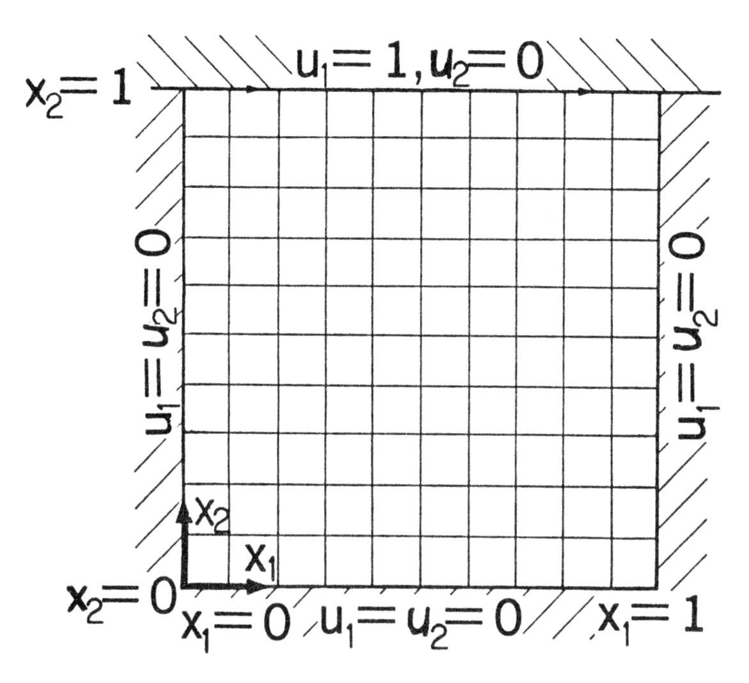

FIGURE 2. Mesh for driven cavity problem.

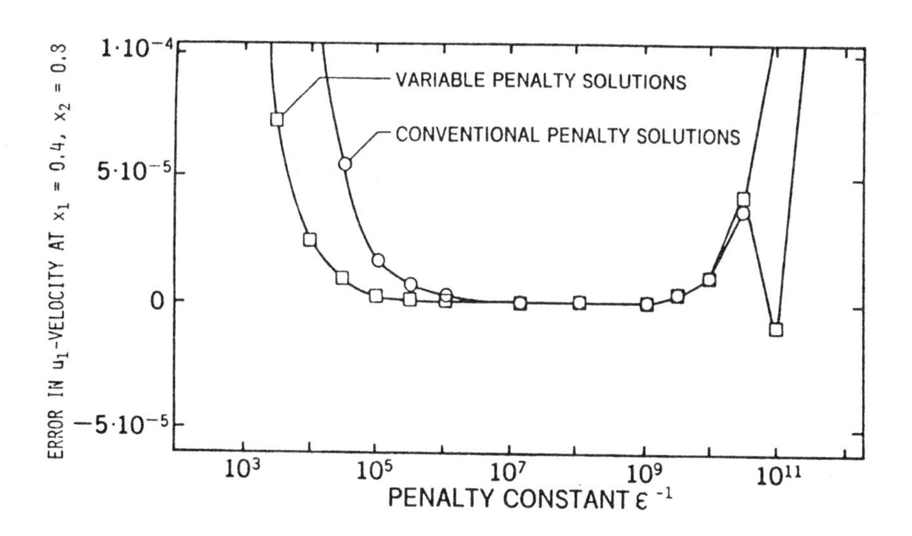

FIGURE 3. Penalty error at a representative point of the driven-cavity problem with elements of equal area. Computation done on CDC Cyber 74 (48 bit floating point numbers).

BIBLIOGRAPHY

1. J. Bramble and A. Schatz, "Higher order local accuracy by averaging in the finite element method". Math. Comput., 31 (1977), 94-111.

2. I. Fried, "Influence of Poisson's Ratio on the Condition of the Finite Element Stiffness Matrix". Int. J. Solids and Structures, 9 (1973), 323-329.

3. V. Girault and P.-A. Raviart, Finite Element Approximation of the Navier-Stokes Equations, Lecture Notes in Mathematics, No. 749, Springer-Verlag, Berlin, 1979.

4. T. Hughes, W. Liu, and A. Brooks, "Finite element analysis of incompressible viscous flows by the penalty function formulation". J. Comp. Physics, 30 (1979), 1-60.

5. H. Kheshgi and L. Scriven, "Finite element analysis of incompressible viscous flow by a variable penalty function method". Proceedings of the winter annual meeting of ASME, Phoenix, Arizona, 1982.

6. H. Kheshgi and L. Scriven, "Variable penalty method for finite element analysis of incompressible Stokes flow". Int. J. Num. Meth. Fluids, to appear.

7. J. Nitsche and A. Schatz, "Interior estimates for Ritz-Galerkin methods". Math. Comput., 28 (1974), 937-958.

8. R. Temam, Navier-Stokes Equations, North Holland, Amsterdam, Second Edition, 1979.

DEPARTMENT OF CHEMICAL ENGINEERING AND MATERIAL SCIENCE
UNIVERSITY OF MINNESOTA
MINNEAPOLIS, MINNESOTA 55455

SCHOOL OF MATHEMATICS
UNIVERSITY OF MINNESOTA
MINNEAPOLIS, MINNESOTA 55455

Contemporary Mathematics
Volume 17, 1983

INITIAL BOUNDARY VALUE PROBLEMS FOR THE EQUATIONS OF MOTION OF COMPRESSIBLE VISCOUS FLUIDS

Akitaka MATSUMURA[1] and Takaaki NISHIDA[2]

ABSTRACT We present a survey of recent results on the solutions to the equations of motion of compressible viscous and heat-conductive fluids.

1. INTRODUCTION. The motion of compressible viscous and heat-conductive fluids is described by the system of five equations for the density ρ, the velocity $u = (u^1, u^2, u^3)$ and the temperature θ:

$$
(1)\quad \begin{cases} \rho_t + (\rho u^j)_{x_j} = 0 \\ u^i_t + u^j u^i_{x_j} + \frac{1}{\rho} p_{x_i} = \frac{1}{\rho}\{\mu(u^i_{x_j} + u^j_{x_i}) + \mu' u^\kappa_{x_\kappa}\delta^{ij}\}_{x_j} + f^i, \quad i = 1,2,3, \\ \theta_t + u^j\theta_{x_j} + \frac{\theta p_\theta}{\rho c} u^j_{x_j} = \frac{1}{\rho c}\{(\kappa\theta_{x_j})_{x_j} + \psi\}, \end{cases}
$$

where p is the pressure, μ is the viscosity coefficient, μ' is the second coefficient of viscosity, κ is the coefficient of heat conduction, c is the specific heat at constant volume, all of which are known functions of ρ and θ, and ψ is the dissipation function.

We survey existence theorems of solution local or global in time to the initial (- boundary) value problem for the system (1). We use the following notations for the function spaces.

1980 Mathematics Subject Classification. 35Q99, 76N10

[1]Supported in part by the University of Wisconsin, Mathematics Research Center.

[2]Supported in part by National Science Foundation, the University of Wisconsin, Mathematics Research Center and the University of California, Mathematical Sciences Research Institute.

0271-4132/83/0000-1520/$03.00

$\mathscr{B}^{m+\sigma}(\Omega)$ = { $u(x)$: u and $\partial_x^\alpha u$, $|\alpha| \leqslant m$, are Hölder continuous functions on $\Omega \subset \mathbb{R}^n$, $n = 1$, 2 or 3, with the Hölder exponent $\sigma \in (0,1)$ }.

$\mathscr{B}^{\ell+\sigma/2,m+\sigma}(Q_T)$ = { $u(t,x)$: u and $\partial_t^j \partial_x^\alpha u$, $0 \leqslant j + |\alpha| \leqslant \ell = m$ or $0 \leqslant 2j + |\alpha| \leqslant 2\ell = m$, are Hölder continuous functions on $Q_T = [0,T] \times \Omega$ with the Hölder exponent $\sigma/2$ and σ with respect to t and x respectively. }

$H^m(\Omega)$ = { $u(x)$: u and $\partial_x^\alpha u \in L_2(\Omega)$, $|\alpha| \leqslant m$. }.

$\mathscr{C}^\ell(0,T;\ H^m(\Omega))$ = { $u(t,x)$: $u(t)$ and $\partial_t^j u(t)$, $0 \leqslant j \leqslant \ell$, are continuous functions of $t \in [0,T]$ with values in $H^m(\Omega)$. }

$L_2(0,T;\ H^m(\Omega))$ = { $u(t,x)$: $u(t)$ is a L_2 - function of $t \in [0,T]$ with values in $H^m(\Omega)$. }

2. LOCAL EXISTENCE. Let us consider the initial (- boundary) value problems for the system (1) under the following conditions.

(A.1) p, c, μ, μ' and κ are smooth functions of $\rho > 0$ and $\theta > 0$, and $\mu > 0$, $\mu' + \frac{2}{3}\mu \geqslant 0$ and $\kappa > 0$ for $\rho, \theta > 0$.

The local existence theorems have very good shape. The initial value problems for the system (1) are solved locally in time by Nash [17] and Itaya [2] in a general situation. Assume that the initial data

$$(2) \quad \begin{cases} \rho_0(x) \in \mathscr{B}^{1+\sigma}(\Omega),\ u_0(x),\ \theta_0(x) \in \mathscr{B}^{2+\sigma}(\Omega), \\ \rho_0(x),\ \theta_0(x) \geqslant \text{constant} > 0,\ x \in \Omega \subset \mathbb{R}^n,\ n = 1,\ 2 \text{ or } 3. \end{cases}$$

<u>THEOREM 1</u>. Under the assumption (A.1) there exists $T > 0$ such that there is the unique solution ρ, u, θ of (1) (2), $\Omega = \mathbb{R}^n$, and

$$\begin{cases} \rho \in \mathscr{B}^{1+\sigma/2,1+\sigma}(Q_T),\ u,\ \theta \in \mathscr{B}^{1+\sigma/2,2+\sigma}(Q_T), \\ \rho,\ \theta > 0 \quad \text{on } Q_T. \end{cases}$$

The proof is delicate and messy, but the idea is simple, i.e., the linearized system of equations for u and θ can be treated as a parabolic one and the linearized equation for ρ as a single hyperbolic one, and then the iteration or the fixed point theorem can be used to solve the nonlinear system. There is an outline of proof of the local existence theorem in a survey article by Solonnikov - Kazhikhov [21]. On the other hand Vol'pert-Khudjaev [26] obtained local existence theorems in the class of H^ℓ functions for a general system containing (1).

The first initial boundary value problem (1) (2) with the boundary condition

$$(3) \qquad u\big|_{\partial\Omega} = 0 \ , \quad \theta\big|_{\partial\Omega} = \theta_1(t,x),$$

is solved locally in time by Tani [22]. The statement of the result is the same as Theorem 1 if we take Ω the domain being considered and the compatibility conditions are satisfied. More general boundary conditions than (3) may be treated similarly. Especially the free surface problems are also solved locally in time by Tani in [23] for the one phase problem and in [24] for the two phase problem in the class of Hölder continuous functions. They require a lot of analysis e.g. to verify Lopachinskii conditions.

The one phase problem bounded by a free surface is also solved in H^ℓ space by Secchi and Valli [20].

3. GLOBAL EXISTENCE. ONE SPACE DIMENSION. The global in time existence theorems are at present very restricted compared to the local one. The polytropic viscous gas is treated on the finite interval I by Kazhikhov-Shelukhin [12].

<u>Theorem 2</u> Assume that

$$p = R\rho\theta \text{ and } R,\ \mu,\ \kappa \text{ are positive constants.}$$

Consider the initial data (2) with $\Omega = I$ and the boundary condition

$$u(t,\cdot)\big|_{\partial I} = \frac{\partial\theta(t,\cdot)}{\partial x}\Big|_{\partial I} = 0$$

that satisfy the compatibility conditions. Then there exists the unique global in time solution $\rho \in \mathcal{B}^{1+\sigma/2,1+\sigma}(Q_T)$, u, $\theta \in \mathcal{B}^{1+\sigma/2,2+\sigma}(Q_T)$ for any $T > 0$ and ρ, $\theta > 0$.

This and another boundary value problem on the finite interval are discussed in [21]. The Dirichlet boundary condition for the temperature is not contained. The piston problems are solved globally by Itaya [4] for the isothermal viscous model gas and for the barotropic viscous model gas.

The Cauchy problem is solved globally under more restrictions. Itaya [3] obtained a global solution of isothermal viscous gas for general initial data. Kanel' [5] got a global solution of barotropic viscous gas for H^1 initial data by energy method. There are several results on the global existence for the initial (- boundary) value problems in [6], [9], [10] and [18] under certain assumptions on the equations of states of gases or on the smallness of the initial data.

4. GLOBAL EXISTENCE. THREE SPACE DIMENSION. If the initial data and the external force are small, there are some global existence theorems in the two and three space dimensions. In this case the Cauchy problem is easier to treat. Let us assume the condition (A.1) and the following

(A.2) $\frac{\partial p}{\partial \rho} > 0$ in a neighborhood of the constant state $\{\rho = \bar{\rho},\ u = 0,\ \theta = \bar{\theta}\}$.

Assume that the initial data $(\rho_0, u_0, \theta_0)(x)$ satisfy

$$\rho_0 - \bar{\rho},\ u_0,\ \theta_0 - \bar{\theta} \in H^{\ell}(\Omega),\qquad \rho_0,\ \theta_0 \geq \text{constant} > 0,\ \ell = 3 \text{ or } 4. \tag{4}$$

Let us denote the space of solution by

$$\begin{aligned} X^{\ell}(t_1,t_2;\ E) = \{(\rho,u,\theta)(t,x):\ & \rho \in \mathcal{C}^{0}(t_1,t_2;\ H^{\ell}),\ D\rho \in L_2(t_1,t_2;\ H^{\ell-1}),\\ & \rho_t \in \mathcal{C}^{0}(t_1,t_2;\ H^{\ell-1}) \cap L_2(t_1,t_2;\ H^{\ell-1}),\\ & u,\ \theta \in \mathcal{C}^{0}(t_1,t_2;H^{\ell}),\ D(u,\theta) \in L_2(t_1,t_2;\ H^{\ell}),\\ & u_t,\ \theta_t \in \mathcal{C}^{0}(t_1,t_2;\ H^{\ell-2}) \cap L_2(t_1,t_2;\ H^{\ell-1}),\\ & N(T_1,t_2) \leq E\ \} \end{aligned}$$

for $0 \leq t_1 \leq t_2 \leq \infty$ and for some constant E, where $H^{\ell} = H^{\ell}(\Omega)$, Ω is a domain in $\mathbb{R}^n$, $n = 2$ or 3. Here the norm $N(t_1,t_2)$ is defined by

$$\begin{aligned} N^2(t_1,t_2) = & \sup_{t_1 \leq t \leq t_2} \{\|(\rho,u,\theta)(t)\|_{\ell}^2,\ \|\rho_t(t)\|_{\ell-1}^2,\ \|(u_t,\theta_t)(t)\|_{\ell-2}^2 \\ & + \int_{t_1}^{t_2} (\|D\rho(s)\|_{\ell-1}^2 + \|\rho_t(s)\|_{\ell-1}^2 + \|D(u,\theta)(s)\|_{\ell}^2 + \|(u_t,\theta_t)(s)\|_{\ell-1}^2) ds, \end{aligned}$$

where $\|\ \|_{\ell}$ is the norm of $H^{\ell}(\Omega)$.

Theorem 3 Consider the initial value problem (1) (4), $\Omega = \mathbb{R}^n$, n = 2 or 3, under the conditions (A,1)(A,2). Let the initial data have the finite norm E_ℓ, $\ell=4$, i.e.,

$$E_\ell = \|\rho_o-\bar{\rho},\ u_o,\ \theta_o-\bar{\theta}\|_\ell + \|\rho_o-\bar{\rho},\ u_o,\ \theta_o-\bar{\theta}\|_{L_1} < \infty.$$

Then there exist positive constants ϵ_o and $C_o < \infty$ such that if $E_\ell < \epsilon_o$, then the problem (1)(4) has the unique solution ρ, u, θ in the large,

$$(\rho(t) - \bar{\rho},\ u(t),\ \theta(t) - \bar{\theta}) \in X^\ell(0,\infty;\ C_oE_\ell),$$

and it has the decay rate

$$\|\rho(t) - \bar{\rho},\ u(t),\ \theta(t) - \bar{\theta}\|_2 \leqslant C_oE_\ell/(1+t)^{n/4}.$$

In particular if μ, μ' and κ do not depend on ρ, then the above assertion holds for $\ell = 3$ also.

The proof [14] consists of a combination of the linear decay rate by spectral analysis and the energy estimate. Using Theorem 3 an asymptotic relation between the equation (1) of compressible viscous fluids and the Boltzmann equation of rarefied gas dynamics as $t \longrightarrow \infty$ is investigated in [8].

The initial boundary value problems in the halfspace, an interior domain and an exterior domain are treated similarly.

Theorem 4 Let Ω be the halfspace, an interior domain or an exterior domain of a bounded region. Let us consider the boundary condition.

$$(5) \qquad u\Big|_{\partial\Omega} = 0 \text{ and } \theta\Big|_{\partial\Omega} = \bar{\theta} \text{ or } \frac{\partial\theta}{\partial n}\Big|_{\partial\Omega} = 0.$$

There exist positive constants ϵ and C such that if $\delta \equiv \|\rho_o - \bar{\rho},\ u_o,\ \theta_o - \bar{\theta}\|_3 < \epsilon$, then the initial boundary value problem (1) (4) (5) has the unique global solution $(\rho,\ u,\ \theta)$ in time,

$$(\rho(t) - \bar{\rho},\ u(t),\ \theta(t) - \bar{\theta}) \in X^3(0,\ \infty;\ C\delta),$$

and it satisfies

$$(6) \qquad \sup\Big|\rho(t,\cdot) - \bar{\rho},\ u(t,\cdot),\ \theta(t,\cdot) - \bar{\theta}\Big| \longrightarrow 0$$

as $t \longrightarrow \infty$. In the case of interior domain the decay rate (6) is exponential in t.

The proof [15] [16] uses energy methods and applies also the Cauchy problem and gives the global solution without the decay rate and without the assumption of L_1 - ness of the initial data. When the domain Ω is the halfspace or the exterior domain, we do not know the exact decay rate of the solution as $t \longrightarrow \infty$ as Theorem 3. To obtain that the spectral analysis for the linearized equation on the domain or the nice energy estimates are necessary. cf. [13]

The exterior problem with the velocity at infinity $u(\infty) \neq 0$, i.e., the flow past an object problem is open even for small $u(\infty)$. In the case of Boltzmann equation it is solved for small $u(\infty)$ by means of detailed linear spectral analysis by Ukai-Asano [25].

The stability of traveling shock wave solutions in one space-dimension is open. cf. [19]

There is an interesting result about a limit of the vanishing viscosity i.e., $\mu \longrightarrow 0$, for the one-dimensional barotropic model gas equation by DiPerna [1].

The system of equations for the electro-magneto fluid dynamics are considered on the global solutions by Kawashima-Okada [11] and by Kawashima [7].

References

[1] DiPerna, R. , Convergence of approximate solutions to conservation laws, preprint (1982).

[2] Itaya, N., On the Cauchy problem for the system of fundamental equations describing the movement of compressible viscous fluid, Kōdai Math. Sem. Rep., 23 (1971) 60-120.

[3] ...., A survey on two model equations for compressible viscous fluid, J. Math. Kyoto Univ. 19 (179) 293-300.

[4] ...., Piston problems of model equations for compressible viscous fluid, preprint (1982).

[5] Kanel', Ya. I., On a model system of equations for one-dimensional gas motion, Diff. Eq. 4 (1968) 374-380.

[6] ...., On the Cauchy problem for the equations of gas dynamics with viscosity. Sib. Math. J. 20 (1979) 208-218.

[7] Kawashima, S., Smooth global solutions for two-dimensional equations of electro-magneto-fluid dynamics, preprint (1982).

[8] ...., Matsumura, A. and Nishida, T., On the fluid-dynamical approximation to the Boltzmann equation at the level of the Navier-Stokes equation, Comm. Math. Phys. 70 (1979) 97-124.

[9] .... and Nishida, T., Global solutions to the initial value problem for the equations of one-dimensional motion of viscous polytropic gases, J. Math. Kyoto Univ. 21 (1981) 825-837.

[10] ... and Okada, M., The initial and initial boundary value problems for the equations of one-dimensional motion of compressible viscous fluids, to appear in J. Math. Kyoto Univ.

[11] .... and ...., Smooth global solutions for the one-dimensional equations in Magneto-hydrodynamics, preprint (1982).

[12] Kazhikhov, A. V. and Shelukhin, V. V., The unique solvability in the large with respect to time of initial-boundary value problems for one-dimensional equations of a viscous gas, J. Appl. Math. Mech. 41(1977) 273-282.

[13] Matsumura, A., An energy method for the equations of motion of compressible viscous and heat-conductive fluids, Univ. of Wisconsin - Madison, MRC Tech. Summary Rep. # 2194 (1981).

[14] .... and Nishida, T., The initial value problem for the equations of motion of compressible viscous and heat-conductive fluids, Proc. Japan Acad. Ser. A, 55 (1979) 337-342.

[15] .... and ...., Initial boundary value problems for the equations of motion of general fluids, Proc. 5th Intern. Symp. Comp. Methods Appl. Sci. and Eng'g, Dec. 1981 INRIA, France.

[16] .... and ...., Initial boundary value problems for the equations of motion of compressible viscous and heat-conductive fluids, to appear in Comm. Math. Phys.

[17] Nash, J., Le problème de Cauchy pour les équations différentielles d'un fluide général. Bull. Soc. Math. France 90 (1962) 487-497.

[18] Padula, M., Existence and continuous dependence for solutions to the equations of a one-dimensional model in gas dynamics, Meccanica, J. of A.I.ME.T.A., 17 (1981) 128.

[19] Pego, R., Linearized stability for extreme shock profiles for systems of consevation laws, Rep. #3, Center Pure Appl. Math. Univ. Calif. Berkeley (1982).

[20] Secchi, P. and Valli, A., A free boundary problem for compressible viscous fluids, Tech. Rep. Libera Univ. Trento (1982).

[21] Solonnikov, V.A. and Kazhikhov, A.V., Existence theorems for the equations of motions of a compressible viscous fluid, Ann. Rev. Fluid Mech. 13 (1981) 79-95.

[22] Tani, A., On the first initial-boundary value problem of compressible viscous fluid motion, Publ. RIMS Kyoto Univ. 13 (1977) 193-253.

[23] ...., On the free boundary value problem for compressible viscous fluid motion, J. Math. Kyoto Univ. 21 (1981) 839-859.

[24] ...., Two phase free boundary value problem for compressible viscous fluid motion, preprint (1982).

[25] Ukai, S. and Asano, K., Stationary solutions of the Boltzmann equation for a gas flow past an obstacle, I Exitence, II Stability, preprints (1982).

[26] Vol'pert, A.I. and Khudjaev, S.I., On the Cauchy problem for composite systems of nonlinear differential equations, Math. USSR Sbornik 16 (1972) 514-544.

DEPARTMENT OF APPLIED MATHEMATICS AND PHYSICS
KYOTO UNIVERSITY
KYOTO 606 JAPAN

DEPARTMENT OF MATHEMATICS
KYOTO UNIVERSITY
KYOTO 606 JAPAN

Contemporary Mathematics
Volume 17, 1983

CONVERGENCE OF FINITE DIFFERENCE APPROXIMATIONS TO NONLINEAR PARABOLIC SYSTEMS

T. Nishida and J. Smoller*

1. We consider the system of equations

$$(1)\qquad \begin{aligned} v_t - u_x &= \varepsilon v_{xx} \\ u_t + p(v)_x &= \mu u_{xx}, \qquad \varepsilon,\mu > 0, \end{aligned}$$

where $(x,t) \in \mathbb{R} \times \mathbb{R}_+$, with initial data

$$(2)\qquad (v,u)(x,0) = (v_0(x), u_0(x)), \quad x \in \mathbb{R}.$$

The function $p(v)$ satisfies $p' < 0$ and $p'' > 0$, thus in the case $\varepsilon = \mu = 0$, the system (1) reduces to the familiar "p-system", which includes, as a special case, the isentropic gas dynamics equations, (see [2,9,11,12,15]).

The problem (1), (2) with $\varepsilon = \mu > 0$ has been studied by Kanel', [5], and the full compressible Navier-Stokes equations, in three space dimensions, was considered by Matsumura and Nishida, [10]. These authors use the "energy" method to prove the existence of a (small) global solution. They require that the data functions, (2), lie in some Sobolev space H^s, where $s > 1$. This means that the data is small at infinity, and excludes some natural choices. For example one cannot consider "Riemann problem" data, nor can one take data near a travelling wave which approximates a shock wave.

We study (1), (2) by the method of finite differences. The class of difference schemes which we consider includes the Lax-Friedrichs scheme, [6,7]. We assume that the data has finite total variation, (not necessarily small), and that the difference approximants for $n\Delta t \le T$ all lie in a region of the form $v \ge C(T)$, where $C(T)$ doesn't depend on the mesh parameters. Under these conditions, we are able to obtain variation-norm bounds on the difference approximants, and to thereby prove that the entire sequence of difference approximants converges to a unique classical solution in $t > 0$. As a consequence of a theorem of D. Hoff, [4], we show that if $\varepsilon = \mu > 0$,

*Research supported by NSF Contract MCS 80-02337

the difference scheme converges to a unique global classical solution provided only that the data is of class BV.

In the final section, we consider some more general "viscosity matrices", and make some remarks about the "vanishing viscosity method."

2. We choose Δt, $\Delta x > 0$, and divide the upper-half plane by the two families of lines $x = n\Delta x$, $t = k\Delta t$, $n \in \mathbb{Z}$, $k \in \mathbb{Z}_+$, and use the notation $f_n^k = f(n\Delta x, k\Delta z)$. The difference schemes we consider are defined by

$$(3)\qquad \begin{aligned} &\frac{v_n^k - v_n^{k-1}}{\Delta t} - \frac{u_{n+1}^k - u_{n-1}^k}{2\Delta x} = \varepsilon \frac{v_{n+1}^k - 2v_n^k + v_{n-1}^k}{\Delta x^2} \\ &\frac{u_n^k - u_n^{k-1}}{\Delta t} + \frac{p(v_{n+1}^k) - p(v_{n-1}^k)}{2\Delta x} = \mu \frac{u_{n+1}^k - 2u_n^k + u_{n-1}^k}{\Delta x^2}. \end{aligned}$$

Observe that if $\varepsilon = \mu = \Delta x^2/2\Delta t$, we recover the Lax-Friedrichs scheme.

Writing $U_n^k = u_n^k - u_{n-1}^k$, $P_n^k = p(v_n^k) - p(v_{n-1}^k)$, $V_n^k = v_n^k - v_{n-1}^k$, $\tau = \Delta t/\Delta x$, $\gamma = \mu\Delta t/\Delta x^2$, and assumming $\beta = (\Delta x^2/\mu\Delta t) - 2 \geq 0$, and $\beta' = \Delta x^2/\varepsilon\Delta t - 2 \geq 0$, we have

$$(4)\qquad U_n^k = \gamma(U_{n+1}^{k-1} + \beta U_n^{k-1} + U_{n-1}^k) - \frac{\tau}{2}(P_{n+1}^{k-1} - P_{n-1}^{k-1}).$$

By iteration, we get, for $0 < \ell \leq k$,

$$(5)\qquad U_n^k = \sum_{|m| \leq \ell} \alpha_m^\ell U_{m+n}^{k-\ell} - \tau \sum_{j=1}^{\ell} \sum_{m=0}^{j} A_m^j (P_{n+m}^{k-j} - P_{n-m}^{k-j}),$$

where

$$(6)\qquad \begin{aligned} &\alpha_m^\ell = \gamma^\ell \sum_{j=0}^{\ell} \binom{\ell}{j} \beta^{\ell-j} \begin{pmatrix} j \\ \frac{1}{2}(j-m) \end{pmatrix}, \quad |m| \leq \ell \\ &A_m^\ell = \frac{1}{2}\gamma^{\ell-1} \sum_{j=0}^{\ell-1} \binom{\ell-1}{j} \beta^{\ell-j-1} \frac{m}{j+1} \begin{pmatrix} j+1 \\ \frac{1}{2}(j+1-m) \end{pmatrix}, \quad m \leq \ell. \end{aligned}$$

Using these formulas, and assuming that the difference approximants lie in a region $v \geq C(T)$, $T = k\Delta t$, we can obtain the estimate

$$\sum_{n \in \mathbb{Z}} |U_n^k| \leq C_0 \left(1 + C'\sqrt{\frac{k\Delta t}{(2+\beta)\mu}}\right) \exp \frac{C_1 k\Delta t}{\sqrt{(2+\beta)(2+\beta')\varepsilon\,\mu}},$$

where C_0 and C_1 are independent of Δx, Δt, and k. Since a similar estimate holds for $\sum_{n\in\mathbb{Z}} |v_n^k|$, these give the desired bounds on the total variation of the approximants. These are used to prove an estimate of the form

$$\sum_{n\in\mathbb{Z}} \{|v_n^k - v_n^p| + |u_n^k - u_n^p|\}\Delta x \leq C_2[\{(k-p)\Delta t\}^{1/3} + (k-p)\Delta t] ,$$

for $k > p$, where C_2 is independent of Δx and Δt.

These estimates enable us to prove the following theorem.

THEOREM 1. Suppose that $v_0(x)$, and $u_0(x)$ have bounded total variation, and $\inf v_0(x) \geq \delta > 0$. Let $T > 0$ be given, and assume that for $k\Delta t \leq T$, $n \in \mathbb{Z}$, the solutions of (3) satisfy $v_n^k \geq C_T$ where C_T depends only on T and the data. If

(7) $$\max_{v\geq C_T} \sqrt{-p'(v)} \leq \Delta x/\Delta t, \quad \text{and} \quad \max(\varepsilon,\mu) \leq \Delta x^2/2\Delta t ,$$

then the entire set of difference approximations converges to a unique classical solution of (1), (2).

In the case $\varepsilon = \mu > 0$, Hoff, [4], has shown that the difference approximants lie in a region $v \geq \delta'$, where δ' is independent of T, provided that

(8) $$\Delta x \leq 2\varepsilon/\max_{v\geq\delta'} \sqrt{-p'(v)} .$$

We thus obtain the following corollary.

COROLLARY 2. Consider the equations (1), (2) where $\varepsilon = \mu > 0$, and assume that u_0, and v_0 are of class BV. If the mesh parameters satisfy (7) and (8), then the conclusions of the theorem are valid.

3. We describe two important ingredients in our method. The first is how to obtain formulas (6). To this end, we first define the "translation" operator S by $SU_n^k = U_{n+1}^k$, $S^{-1}U_n^k = U_{n-1}^k$, and $SP_n^k = P_{n+1}^k$, $S^{-1}P_n^k = P_{n-1}^k$. In terms of S, (4) becomes

$$U_n^k = (S+\beta+S^{-1})U_n^{k-1} - \frac{\tau}{2}(S-S^{-1})P_n^{k-1} .$$

This formula is easy to iterate, and gives

$$U_n^k = \gamma^\ell(S+\beta+S^{-1})^\ell U_n^{k-\ell} - \frac{\tau}{2}\sum_{j=1}^{\ell}\gamma^{j-1}(S+\beta+S^{-1})^{j-1}(S-S^{-1})P_n^{k-j} .$$

The operators $(S+\beta+S^{-1})^k$ can be simplified by using a binomial expansion This lets us easily calculate the desired expressions. We note that α_m^ℓ is a linear combination of random-walk coefficients with unrestricted boundaries, while A_m^ℓ is a linear combination of random-walk coefficients with absorbing boundaries; see [3].

The second important point is the following inequality; namely

(9) $$\sum_{m=0}^{\ell} A_m^{\ell} \leq C/\sqrt{\ell}$$

where C is independent of ℓ. In case $\beta = 0$, (9) is proved using Stirlings approximation. If $\beta \neq 0$, we show that (9) holds provided that

$$\frac{C}{(\beta+2)} \sum_{j=0}^{\ell-1} \binom{\ell-1}{j} \left(\frac{\beta}{\beta+2}\right)^{\ell-j-1} \left(\frac{2}{\beta+2}\right)^{j+1} \frac{1}{\sqrt{j+1}} \leq C_1/\sqrt{\ell} ,$$

and this latter inequality, in turn, is proved using Bernstein polynomials.

4. The results in the previous sections are contained, with complete proofs, in our paper [12]. We wish to make here a few new remarks which are of relevance to the above discussion.

In [1], the following system was considered:

(10) $$\begin{aligned} v_t - u_x &= \varepsilon v_{xx} + \varepsilon u_{xx} \\ u_t + p(v)_x &= \varepsilon v_{xx} + 2\varepsilon_{xx} \end{aligned}$$

where $p'(v) = \operatorname{Arctan} v - M$, and $M > \pi/2$ is chosen so large that $M + 1 - \operatorname{Arctan} v_1 > [10(M - \operatorname{Arctan} v_2]^{1/2}$ for every v_1 and v_2. For this system, it was shown that no shock-wave solution of

$$v_t - u_x = 0 , \quad u_t + p(v)_x = 0$$

can be a limit of travelling wave solutions of (10) as $\varepsilon \to 0$. This indicates that the vanishing viscosity method for this system may not work; indeed computer calculations show this to be the case; see [13]. We wish here to make a few remarks which may shed some light on the problem.

First, we rewrite (10) in the form

(11) $$U_t + F(U)_x = \varepsilon B U_{xx} ,$$

where $U = (v,u)$, $F = (-u,p(v))$ and

$$B = \begin{pmatrix} 1 & 1 \\ 1 & 2 \end{pmatrix} .$$

If $P^{-1}BP = \Lambda$, a diagonal matrix, then writing $U = PV = (w,z)$, the equations (11) become

$$V_t + P^{-1}F'(PV)PV_x = \varepsilon \Lambda V_{xx} ,$$

or if we define

$$A(V) = P^{-1}F(PV) = \frac{1}{\sqrt{5}} [(w-\lambda_+ z-p(z+w), \lambda_- w-z+p(z+w)]$$

we may write the system in conservation form:

(12) $$V_t + A(V)_x = \varepsilon \Lambda V_{xx} .$$

In component form, this system can be written as

$$w_t + r(w,z)_x = \varepsilon\lambda_1 w_{xx}$$
$$z_t + s(w,z)_x = \varepsilon\lambda_2 z_{xx}$$

where ∇r and ∇s are all bounded functions. Thus if we consider the difference approximation

(13) $$\frac{w_n^k - w_n^{k-1}}{\Delta t} + \frac{r(w_{n+1}^{k-1}, z_{n+1}^{k-1}) - r(w_{n-1}^{k-1}, z_{n-1}^{k-1})}{2\Delta x} = \varepsilon\lambda_1 \frac{w_{n+1}^{k-1} - 2w_n^{k-1} + w_{n-1}^{k-1}}{\Delta x^2}$$
$$\frac{z_n^k = z_n^{k-1}}{\Delta t} + \frac{s(w_{n+1}^{k-1}, z_{n+1}^{k-1}) - s(w_{n-1}^{k-1}, z_{n-1}^{k-1})}{2\Delta x} = \varepsilon\lambda_2 \frac{z_{n+1}^{k-1} - 2z_n^{k-1} + z_{n-1}^{k-1}}{\Delta x^2}$$

we can use our above methods to show that the system (10) has a unique globally defined classical solution, which is the limit of the above difference approximation. Note that this result is easily capable of generalization; in fact, we have the following theorem.

Theorem 3. Consider the system

$$U_t + f(U)_x = BU_{xx} , \quad (x,t) \in \mathbb{R} \times \mathbb{R}_+ ,$$

with initial data $U(x,0) = U_0(x)$, $x \in \mathbb{R}$, where B is a positive symmetric matrix, and the entries of the matrix $f'(U)$ are all bounded. If U_0 is of class BV, then this problem has a unique globally defined classical solution which is of class BV on each line $t > 0$. The solution is obtainable as a limit of difference approximations.

The proof follows just as in the above special case; namely, we diagonalize B to obtain a system which we can write in the form (12), and we then use the difference approximations (13).

Thus, from Theorem 3, we see that (10) has a unique solution U for each $\varepsilon > 0$, provided that the data is of class BV. We are interested in the behavior of the set $\{U_\varepsilon\}$, as $\varepsilon \to 0$. Now in a recent paper, [2], DiPerna, has shown that, subject to some other conditions, the method of "compensated compactness" can be used to study this limit. One of the hypotheses in his argument is that there exist an entropy-entropy flux pair, (η,q), (see [8]), which is compatible with B in the sense that

$$H(\eta)B \geq 0 , \tag{14}$$

where $H(\eta)$ denotes the hessian matrix of η. We shall show that for the system (10), condition (14) <u>cannot</u> be achieved.

In fact, we shall consider a somewhat more general case; namely, let $q(u)$ be any bounded function with $q'(u) \neq 0$ for all u. Define p by

$$p'(u) = \begin{cases} q(u) - M , & \text{if } q'(u) > 0 \\ -q(u) - M , & \text{if } q'(u) < 0 , \end{cases}$$

where $M > 0$ is chosen so large that $p'(u) < 0$ for all u. Let

$$D = \begin{pmatrix} a & b \\ b & c \end{pmatrix}$$

be a positive matrix, so that $ac > b^2$, and $a,c > 0$, and consider the system

$$U_t + F(U)_x = \varepsilon D U_{xx} , \tag{15$_\varepsilon$}$$

where $U = (v,u)$, and $F(U) = (-u,p(v))$. (Note that (10) satisfies these conditions.)

Now in order that an entropy-entropy flux pair (η,q) exist for $(15)_0$, it is necessary that η be a solution of the (linear) equation

$$\eta_{vv} + p'(v)\eta_{uu} = 0 .$$

We use this equation and compute:

$$H(\eta)D = \begin{pmatrix} \eta_{vv} & \eta_{uv} \\ \eta_{uv} & \eta_{uu} \end{pmatrix} \begin{pmatrix} a & b \\ b & c \end{pmatrix} = \begin{pmatrix} -p'\eta_{uu} & \eta_{uv} \\ \eta_{uv} & \eta_{uu} \end{pmatrix} \begin{pmatrix} a & b \\ b & c \end{pmatrix}$$

$$= \begin{pmatrix} -ap'\eta_{uu} + b\eta_{uv} & -p'\eta_{uu}b + c\eta_{uv} \\ a\eta_{uv} + b\eta_{uu} & b\eta_{uv} + c\eta_{uu} \end{pmatrix} .$$

This last matrix is positive semi-definite iff

$$[(a+c)\eta_{uv} + b(1-p')\eta_{uu}]^2 \leq 4[(b\eta_{uv}+c\eta_{uu})(b\eta_{uv}-ap'\eta_{uu})] ,$$

or

$$(a+c)^2\eta_{uv}^2 + 2(a+c)b(1-p')\eta_{uv}\eta_{uu} + b^2(1-p')^2\eta_{uu}^2$$

$$\leq 4b^2\eta_{uv}^2 - 4p'ac\eta_{uu}^2 + 4(bc-abp')\eta_{uv}\eta_{uu} .$$

We shall show that given D, we can violate this for large enough M. For this, it suffices to show that the quadratic form

$$[(a+c)^2 - 4b^2]\eta_{uv}^2 + [2b(a+c)(1-p')-4(bc-abp')]\eta_{uv}\eta_{uu} + [b^2(1-p')^2+4p'ac]\eta_{uu}^2$$

is positive definite. This form can be simplified to

$$[(a+c)^2-4b^2]\eta_{uv}^2 + 2b(a-c)(1+p')\eta_{uu}\eta_{uv} + [b^2(1-p')^2+4p'ac]\eta_{uu}^2 ,$$

and the associated matrix is

$$\begin{bmatrix} (a+c)^2 - 4b^2 & b(a-c)(1+p') \\ b(a-c)(1+p') & b^2(1-p')^2 + 4p'ac \end{bmatrix} .$$

(Note that $(a+c)^2 - 4b^2 \geq (a+c)^2 - 4ac = (a-c)^2 \geq 0$). Now this matrix is positive definite iff

$$(16) \qquad 4b^2(a-c)^2(1+p')^2 < 4[(a+c)^2 - 4b^2][b^2(1-p')^2 + 4p'ac] .$$

Since $p' = O(M)$ as $M \to \infty$, we see that (16) holds for large M if

$$b^2(ac-b^2) > 0 .$$

This latter inequality is true however, iff $b \neq 0$, since we know that $ac - b^2 > 0$.

REFERENCES

1. Conley, C., and J. Smoller, Viscosity matrices for two-dimensional nonlinear hyperbolic systems, Comm. Pure Appl. Math., 23, (1970), 867-884.

2. DiPerna, R., Convergence of approximate solutions to conservation laws, (preprint).

3. Feller, W., An Introduction to Probability Theory and Its Application, Vol. 1, (2nd ed.), J. Wiley: New York, (1957).

4. Hoff, D., A finite difference scheme for a system of two conservation laws with artificial viscosity, Math. Comp., 33, (1979), 1171-1193.

5. Kanel', Ya., On some systems of quasilinear parabolic equations, Zh. Vych., Mat. i Mat. Fiz., 6, (1966), 446-477; Eng. transl. in: USSR Comp. Math and Math. Phys., 6, (1966), 74-88.

6. Lax, P., Weak solutions of nonlinear hyperbolic equations and their numerical computation, Comm. Pure Appl. Math., 7, (1954), 159-193.

7. _______, Hyperbolic systems of conservation laws, II, Comm. Pure Appl. Math., 10, (1957), 537-566.

8. _______, Shock waves and entropy, in: Contributions to Nonlinear Functional Analysis, ed. by E. Zarantonello, Academic Press: New York, (1971), 603-634.

9. Liu, T., and J. Smoller, On the vacuum state for the isentropic gas dynamics equations, Adv. Appl. Math., 1, (1980), 345-359.

10. Matsummura, A., and T. Nishida, The initial-value problem for the equations of motion for compressible viscous, and heat conducting fluids, Proc. Japan Acad., 55, Ser. A, (1979), 337-342.

11. Nishida, T., Global solution for an initial-voundary value problem of a quasilinear hyperbolic system, Proc. Japan. Acad., 44, (1968), 142-646.

12. Nishida, T., and J. Smoller, Solutions in the large for some nonlinear hyperbolic conservation laws, Comm. Pure Appl. Math., 26, (1973), 183-200.

13. _______, A class of convergent finite difference schemes for certain nonlinear parabolic systems, Comm. Pure Appl. Math., to appear.

14. Smoller, J., and M. Taylor, Wave front sets and the viscosity method, Bull. A.M.S. 79, (1973), 431-436.

15. Zhang Tong, and Guo Yu-Fa, A class of initial-value problems for systems of aerodynamic equations, Acta Math. Sinica, 15, (1965), 386-396; Engl. transl. in: Chinese Math., 7, (1965), 90-101.

DEPARTMENT OF MATHEMATICS
KYOTO UNIVERSITY
KYOTO, JAPAN

DEPARTMENT OF MATHEMATICS
THE UNIVERSITY OF MICHIGAN
ANN ARBOR, MI 48109

Contemporary Mathematics
Volume 17, 1983

LINEARIZED STABILITY OF EXTREME SHOCK PROFILES FOR SYSTEMS OF CONSERVATION LAWS WITH VISCOSITY

Robert L. Pego[1]

A shock profile is a smooth traveling wave solution $\phi((x - st)/\mu)$ of a system of conservation laws with added dissipation,

(1) $$u_t + f(u)_x = \mu u_{xx}, \quad t \geq 0,\ x \in \mathbf{R},\ u \in \mathbf{R}^m$$

such that $\phi(\xi) \to u_L$ as $\xi \to -\infty$, $\phi(\xi) \to u_R$ as $\xi \to +\infty$. Here u_L and u_R are given constant states which must satisfy the Rankine Hugoniot conditions

(2) $$f(u_L) - f(u_R) - s(u_L - u_R) = 0$$

and hence determine a weak solution

(3) $$u(x,t) = \begin{cases} u_L & x < st \\ u_R & x > st \end{cases}$$

(called a shock wave when the entropy condition, below, is satisfied) of the inviscid system

(4) $$u_t + f(u)_x = 0 .$$

When dissipation terms are neglected, many equations of mathematical physics take the form (4), for which smooth initial data typically exhibits breakdown in finite time, and discontinuous weak solutions, including shocks, must be admitted. The dissipation term in (1) is to be considered a simple model for more complicated mechanisms in physical systems.

1980 Mathematics Subject Classifications: 35K55, 35L65, 35B35

[1]Sponsored by the United States Army under Contract No. DAAG29-80-C-0041. This material is based upon work supported by the National Science Foundation under Grant No. MCS-7927062, Mod. 2.

0271-4132/83/0000-1397/$02.75

We are concerned with the stability, to perturbations in initial data, of weak (small amplitude) shock profiles of a particular family. We assume the system (4) is strictly hyperbolic, i.e., the matrix $df(u)$ has m distinct real eigenvalues $\lambda_1(u) < \cdots < \lambda_m(u)$ with corresponding eigenvectors $r_k(u)$, $k = 1$ to m. We assume these eigenvalues are genuinely nonlinear in the sense of Lax, so that $\nabla\lambda_k \cdot r_k(u) \neq 0$, and normalize r_k so $\nabla\lambda_k \cdot r_k(u) \equiv 1$. The weak solution (3) is called a k-shock if Lax's entropy condition is satisfied:

$$\lambda_k(u_L) > s > \lambda_k(u_R)$$
$$\lambda_{k+1}(u_R) > s > \lambda_{k-1}(u_L) \;. \tag{5}$$

Foy [2] first showed that shock profiles, as above, exist for weak k-shocks. For u_R fixed, a one parameter family of weak k-shocks exists, with $u_L = \hat{u}^k(\varepsilon)$ where $\frac{1}{2}\varepsilon = s - \lambda_k(u_R)$, for $\varepsilon > 0$ small, satisfying

$$\hat{u}^k(0) = u_R \qquad \frac{d\hat{u}^k}{d\varepsilon}(0) = r_k(u_R) \;.$$

An elegant proof is given by Conlon [1]. Also see Lax [5].

Our approach to stability is based on Sattinger's criterion of linearized stability for traveling waves of parabolic systems [7]. Scale equation (1) so $\mu = 1$, and change to the moving frame $\xi = x - st$. Equation (1) becomes

$$u_t = u_{\xi\xi} - (df(u) - s)u_\xi \;. \tag{7}$$

Writing $u(\xi,t) = \phi(\xi) + v(\xi,t)$, the perturbation v satisfies a nonlinear evolution equation

$$v_t = L(v) \;. \tag{8}$$

Linearizing at $v = 0$, we get

$$v_t = L'v = v_{\xi\xi} - ((df(\phi) - s)v)_\xi \;. \tag{9}$$

Any translate $\phi(\xi + \gamma)$ is a stationary solution of (7), so it follows that $L'\phi_\xi = 0$. So zero is an eigenvalue of L', and it seems that linear analysis can yield only marginal stability at best. However, Sattinger showed that if, on a suitable function space, zero is a <u>simple</u> eigenvalue of L' and the rest of the spectrum lies in $\{\mathrm{Re}\,\lambda < -\delta\}$ for some $\delta > 0$, then "orbital asymptotic stability" obtains: a slightly perturbed traveling wave asymptotically approaches a nearby translate.

An important feature of Sattinger's analysis is the use of spatially weighted norms for function spaces to "push" the essential spectrum (the spectrum aside from isolated points of finite multiplicity) of L' to the left. In our situation, in fact, it is easy to verify that the essential spectrum on unweighted L^p spaces includes the origin (cf. [4]). Our main result is to exhibit, for weak <u>extreme</u> shock profiles (ε sufficiently small and $k = 1$ or m), a weighted space on which L' satisfies Sattinger's criterion of linearized stability.

Let us now state our result for 1-shock profiles (treat m-shock profiles by reflection, $x \to -x$). By a linear change of coordinates, we may assume

$$df(u_R) = \operatorname{diag}(\lambda_1(u_R), \ldots, \lambda_m(u_R)) .$$

We introduce the weighted space

$$(C_u)^m_c = \{u : \mathbf{R} \to \mathbf{R}^m \mid u^1 \cosh cx \in C_u,\ u^j e^{-cx} \in C_u,\ j = 2, \ldots m\}$$

with norm

$$\|u\|_{\infty,c} = \max\{\|u^1 \cosh cx\|_\infty, \|u^j e^{-cx}\|_\infty, j = 2, \ldots, m\} .$$

Here C_u is the space of bounded uniformly continuous functions on $\mathbf{R}$ under the sup norm.

THEOREM. Fix $u_R \in \mathbf{R}^m$ and suppose $\lambda_1(u)$ is genuinely nonlinear. Consider the linearized equation (9), $v_t = L'v$, for the evolution of perturbations of the 1-shock profile $\phi(\xi; \varepsilon)$. Fix c, $0 < c < \frac{1}{2}$. Then there exists β, $0 < \beta < c(1 - c)$, such that if $\varepsilon > 0$ is sufficiently small, the spectrum of L' on the space $(C_u)^m_{\frac{1}{2}\varepsilon c}$ consists of a simple isolated eigenvalue at the origin and a part which lies in a sector strictly contained in the left half plane,

$$S_\alpha(-\tfrac{1}{4}\varepsilon^2\beta) = \{\lambda \in \mathbf{C} \mid \operatorname{Re}(\lambda + \tfrac{1}{4}\varepsilon^2\beta) \le -(\cos\alpha)|\lambda + \tfrac{1}{4}\varepsilon^2\beta|\}$$

where α depends on ε, $0 < \alpha < \pi/2$.

Unfortunately, this result does not immediately yield nonlinear stability by Sattinger's theorem, for our weight fails to satisfy two of his requirements. In particular, (a) the weight is not a <u>scalar</u> function, and (b) it is not bounded below in some components, so the nonlinear terms in (8) may fail to be continuous on the weighted space. The latter may be a serious problem, as illustrated by the following example: It is well-known that the solution of the equation

$$u_t = u_{xx} + |u|^{p-1}u$$

with any positive initial data blows up in finite time, for $p \le 2$ [Fu]. Change to a moving frame $\xi = x - ct$ and linearize the equation about the zero state, obtaining

$$v_t = Av = v_{\xi\xi} + cv_\xi \,.$$

The spectrum of A on a space with weight $e^{1/2\, c\xi}$,

$$B_c = \{u : \mathbf{R} \to \mathbf{R} \mid e^{1/2\, c\xi} u(\xi) \in C_u\}$$

is the same as the spectrum of $e^{-1/2\, c\xi} A e^{1/2\, c\xi}$ on $B_0 = C_u$. But

$$e^{-1/2\, c\xi} A e^{1/2\, c\xi} w = w_{\xi\xi} - \frac{1}{4} c^2 w$$

has spectrum $(-\infty, -\frac{1}{4} c^2]$ on C_u. So we have strict linearized stability, but nonlinear instability. Despite this situation, the theorem above, one of few of its type for systems, does provide information about the eigenvalues and spectrum of L' which may prove valuable.

A detailed proof of the theorem above will appear elsewhere, but a brief sketch is appropriate: Foy's proof of the existence of the shock profile gives an asymptotic description of it as $\varepsilon \to 0$; in fact,

$$\phi(\xi;\varepsilon) = u_R + \frac{1}{2}\varepsilon \;\psi(\frac{1}{2}\varepsilon\xi;\varepsilon)$$

where

$$\psi(x;\varepsilon) = \phi^B(x) r_1(u_R) + O(\varepsilon)$$

and

$$\phi^B(\eta) = 1 - \tanh \frac{1}{2}\eta$$

is the shock profile solution of Burgers' equation

$$u_t + (\frac{1}{2} u^2)_x = u_{xx} \tag{10}$$

(for $u_L = 2$, $u_R = 0$; and $s = 1$). Introduce appropriate scaled variables

$$x = \frac{1}{2}\varepsilon\xi, \quad \frac{1}{2}\varepsilon u = v, \quad \hat{t} = \frac{1}{4}\varepsilon^2 t$$

and write

$$u = \begin{pmatrix} u^1 \\ \hat{u} \end{pmatrix}, \quad \text{where } \hat{u} = (u^2,\dots,u^m)^t \,.$$

Now rewrite (9) in these variables in block form relative to this decomposition, neglecting terms of order ε, which turn out to be insignificant:

$$u_t = \hat{L}' u = \begin{bmatrix} L^B & M^1 \\ \hat{M} & L^T + M^T \end{bmatrix} \begin{pmatrix} u^1 \\ \hat{u} \end{pmatrix} . \tag{11}$$

Here

$$\hat{M}u^1 = \hat{A}(\phi^B u^1)_x \qquad \begin{array}{l} M^1\hat{u} = A^1(\phi^B\hat{u})_x \\ M^T\hat{u} = A^T(\phi^B\hat{u})_x \end{array}$$

where

$$\begin{pmatrix} 1 & A^1 \\ \hat{A} & A^T \end{pmatrix} = d^2 f \cdot r_1(u_R) \quad \text{is a constant matrix.}$$

The main terms in (11) are

$$L^B u^1 = u^1_{xx} - ((\phi^B - 1)u^1)_x$$

$$(L^T\hat{u})^j = u^j_{xx} - \left(\frac{\lambda_j - \lambda_1}{1/2\,\varepsilon}\right)u^j_x .$$

It is no surprise that L^B corresponds to the linearized L' for the shock profile $\phi^B(x - t)$ of Burgers' equation (10). It is well-known that on spaces with weight $\cosh \frac{1}{2}x$, L^B satisfies the linearized stability criterion we are concerned with ([6], [7]). So one would like to "uncouple" the first component of (11), and show that the other components only "contribute" spectrum lying strictly in the left half plane. However, the time evolution for a component of L^T yields the solution of the heat equation in a frame moving at constant velocity. No weight which is bounded below will yield an exponential decay rate for such solutions (necessary if the spectrum lies strictly in the left half plane). However, for an extreme shock, the solutions in each component are convected in the same direction at increasingly high velocity as $\varepsilon \to 0$. So one may expect that a weight e^{cx}, decaying in the direction of convection, could yield exponential decay for solutions of $\hat{u}_t = L^T\hat{u}$ at a rate which improves as $\varepsilon \to 0$. Thus as $\varepsilon \to 0$, the spectrum of L^T should move to the left, and for λ fixed, the resolvent $(\lambda - L^T)^{-1}$ should decay in norm. But then the resolvent equation $(\lambda - \hat{L}')u = f$ can be solved as if it were diagonally dominant, by Gaussian elimination.

The discussion above explains our choice of weight. Some care is required, however, since the weight is not scalar. We let

$$W_c(x) = \begin{bmatrix} \cosh cx & 0 \\ 0 & e^{-cx} \end{bmatrix} \quad \text{and} \quad z = W_c u .$$

Equation (11) is transformed into

$$(12) \qquad z_t = \hat{L}'_c z = W_c \hat{L}' W_c^{-1} z = \begin{bmatrix} w_c L^B w_c^{-1} & w_c M^1 e^{cx} \\ e^{-cx} \hat{M} w_c^{-1} & e^{-cx}(L^T + M^T) e^{cx} \end{bmatrix} \begin{pmatrix} z^1 \\ \hat{z} \end{pmatrix}$$

where $w_c(x) = \cosh cx$. After this scaling and transformation, the theorem we stated is equivalent to:

THEOREM. Fix c, $0 < c < \frac{1}{2}$. Then there exists β, $0 < \beta < c(1-c)$, such that if ε is sufficiently small, then the spectrum of $\hat{L}'_c$ on $(C_u)^m$ consists of a simple eigenvalue at the origin and a part in the sector

$$S_\alpha(-\beta) = \{\lambda \in \mathbf{C} \mid \mathrm{Re}(\lambda + \beta) \leq -(\cos \alpha)|\lambda + \beta|\}$$

lying strictly in the left half plane, where α may depend on ε, $0 < \alpha < \pi/2$.

Because the weight is not scalar, one must verify that the off-diagonal terms in (12) have bounded coefficients. It suffices in our situation to show that

$$\sup_x |\psi(x;\varepsilon)e^{2cx}| < C \quad \text{independent of } \varepsilon .$$

This yields the strict requirement $c < \frac{1}{2}$. As a consequence, we must estimate the resolvent of $(\cosh cx)L^B(\cosh cx)^{-1}$ (also $e^{-cx}L_j^T e^{cx}$), which is not in self adjoint form if $c \neq \frac{1}{2}$. (Sattinger used only weights which transformed the scalar equation to self adjoint form.) But an advantage is that the weights $\cosh cx$, $0 < c < \frac{1}{2}$, are not as restrictive as the weight $\cosh \frac{1}{2}x$. So for instance, applying Sattinger's theorem we find that the profile ϕ^B of Burgers' equation is stable to sufficiently small perturbations that decay exponentially as $|x| \to \infty$, regardless of the rate. Similar results might hold for other types of scalar equations considered by Sattinger.

BIBLIOGRAPHY

1. J. Conlon, "A theorem in ordinary differential equations with an application to hyperbolic conservation laws", Advances in Math. 35 (1980), 1-18.

2. L. Foy, "Steady state solutions of conservation laws with viscosity terms", Comm. Pure Appl. Math. 17 (1964), 177-188.

3. H. Fujita, "On the blowing up of the solution of the Cauchy problem $u_t = \Delta u + u^{\alpha+1}$", J. Fac. Sci. Univ. Tokyo, Sec. 1, 13 (1966), 109-124.

4. D. Henry, Geometric Theory of Semilinear Parabolic Equations, Lec. Notes in Math. v. 840, Springer-Verlag, New York, 1981.

5. P. D. Lax, "Hyperbolic systems of conservation laws, II", Comm. Pure Appl. Math. 10 (1957), 537-566.

6. L. A. Peletier, Asymptotic stability of traveling waves, in Instability of Continuous Systems, IUTAM Symposium (1969), 418-422.

7. D. H. Sattinger, "On the stability of waves of nonlinear parabolic systems", Advances in Math. 22 (1976), 312-355.

MATHEMATICS RESEARCH CENTER
UNIVERSITY OF WISCONSIN-MADISON
MADISON, WI 53706

Contemporary Mathematics
Volume 17, 1983

ON SOME "VISCOUS" PERTURBATIONS OF QUASI-LINEAR FIRST ORDER HYPERBOLIC SYSTEMS ARISING IN BIOLOGY

Michel Rascle

ABSTRACT. The following problem arises in Biology :
Find $u \geqslant 0$ and v such that

$$(\mathscr{P}_{\mu,\nu}) \quad \begin{cases} u_t - \mu\, u_{xx} - (uv)_x = 0 \\ v_t - \nu\, v_{xx} - u_x = 0 \end{cases} \quad \text{in} \quad \Omega \times]0,T[\ , \ \Omega \subseteq \mathbb{R}$$

with suitable initial (and boundary) conditions, and $\mu > 0$, $\nu \geqslant 0$.

This problem is a "viscous" perturbation of $(\mathscr{P}_{0,0})$, which is hyperbolic for $u \geqslant 0$. In a previous work, we studied the case $\mu > 0$, $\nu = 0$, Ω bounded, and we proved for this problem the existence and the uniqueness of a smooth solution, by using a convex entropy of $(\mathscr{P}_{0,0})$.

Here, we prove the same results for $\mu = \nu > 0$, $\Omega = \mathbb{R}$. In this case, the previous method fails, and we must use an argument of invariant regions for the problem $(\mathscr{P}_{\mu,\mu})$. These regions, defined by the Riemann Invariants of $(\mathscr{P}_{0,0})$, are unbounded, but nevertheless, they enable us to prove the global existence of the solution.

1. INTRODUCTION AND NOTATIONS. The following problem is a particular case (see Nossal [7]) of a general mathematical model proposed by Keller and Segel [4] to describe the dynamics of populations of so-called chemotactic Bacteria, i.e. of Bacteria which are attracted by the gradient of concentration of some chemical substance :

Find two functions u and U such that

$$(\mathscr{P}) \quad \begin{cases} \dfrac{\partial u}{\partial t} - \mu\, \Delta u - \operatorname{div}(u\, \nabla U) = 0 & \text{in} \quad Q_T = \Omega \times]0,T[& (1) \\ \dfrac{\partial U}{\partial t} - \nu\, \Delta U = u \quad , \quad u \geqslant 0 & \text{in} \quad Q_T & (2) \\ \dfrac{\partial u}{\partial n} = \dfrac{\partial U}{\partial n} = 0 & \text{on} \quad \Sigma_T = \Gamma \times]0,T[& (3) \\ u(\cdot,0) = u_0(\cdot) \geqslant 0 & \text{in} \quad \Omega & (4) \\ U(\cdot,0) = U_0(\cdot) \equiv 0 & \text{in} \quad \Omega & (5) \\ (\mu > 0 \ \text{and} \ \nu \geqslant 0 \ \text{are (fixed) constants}) \end{cases}$$

1980 Mathematics Subject Classification. 35L65, 35K55, 92A15.

0271-4132/83/0000-1398/$03.50

where Ω is an open connected set in $\mathbb{R}^N$ $(N \leq 3)$, with smooth boundary Γ, div denotes the divergence of a vector field, and $\frac{\partial}{\partial n}$, ∇ , Δ are respectively the normal derivative, the gradient and the Laplace operator. The function u_0 is known, u is the concentration of Bacteria, and $U = -k \ln s/s_0$ where s is the concentration of substrate, $s_0(\cdot) = s(\cdot,0) \equiv s_0$ (s_0 and k positive constants).

The difficulty is that $(\mathcal{P})$ is a <u>"viscous" perturbation of a first order quasi-linear hyperbolic system</u>. In [8], we studied the case <u>$\nu = 0$</u>, and we proved for this problem the local existence and uniqueness of the solution in a space of Hölder-continuous functions (and its local $\mathcal{C}^\infty$ regularity if the initial data are smooth enough and satisfy compatibility conditions at any order) <u>and its global existence if Ω is bounded in $\mathbb{R}$ $(N = 1)$</u>. In [9] we extended these results to the (more complicated) Keller-Segel model.

Here, we are going to give a similar result in a somewhat different situation : <u>we prove the global existence and uniqueness of the solution of $(\mathcal{P})$ if $\Omega = \mathbb{R}$</u> (i.e. N = 1 and Ω <u>unbounded</u>) and <u>$\mu = \nu$</u>. Let us explain why the method used in [8] cannot work here. Let us set $v = U_x = \frac{\partial U}{\partial x}$, and let us write $(\mathcal{P})$ under the following form :

$$(\mathcal{P}')\begin{cases} u_t - \mu\, u_{xx} - (uv)_x = 0 & \text{in} \quad Q_T = \mathbb{R} \times]0,T[\qquad (6)\\ v_t - \nu\, v_{xx} - u_x = 0 & \text{in} \quad Q_T \qquad (7)\\ u(\cdot,0) = u_0(\cdot) & \text{in} \quad \mathbb{R} \qquad (8)\\ v(\cdot,0) = v_0(\cdot)\ (\equiv 0) & \text{in} \quad \mathbb{R} \qquad (9)\end{cases}$$

which is clearly a "viscous" perturbation of

$$(*)\begin{cases} u_t - (uv)_x = 0 & \\ v_t - u_x = 0 & \text{in} \quad Q_T\\ u(\cdot,0) = u_0(\cdot) & \\ v(\cdot,0) = v_0(\cdot) & \text{in} \quad \mathbb{R}\end{cases}$$

This problem is hyperbolic for $u \geq 0$ (and even for $v^2 + 4u \geq 0$), and strictly hyperbolic for $u \geq 0$ and $(u,v) \neq (0,0)$. It admits a <u>convex entropy</u>

$$\eta(u,v) = \frac{v^2}{2} + u(\ln u - 1) \geq -1 \tag{10}$$

associated with the entropy flux

$$q(u,v) = -uv \ln u$$

In [8], we used this fact to obtain a crucial *a-priori* estimate by multiplying (6) and (7) by $\eta_u = \frac{\partial \eta}{\partial u}$ and $\eta_v = \frac{\partial \eta}{\partial v}$ respectively, by integrating on $\Omega \times]0,t[$, and by using (10), which implies

$$\int_\Omega (\eta(u,v))(x,t)dx \geq -\operatorname{meas}(\Omega) \tag{11}$$

Here, $\Omega = \mathbb{R}$, and therefore (11) is not interesting. Moreover, we cannot replace η by $\eta+1$ because, in the physical problem, u_0 has a compact support, which implies

$$\int_{\mathbb{R}}[(\eta(u_0,v_0))(x) + 1]\, dx = +\infty$$

Hence, we can no longer use this entropy of (*), but, in the case where $\mu = \nu$, we can use its Riemann invariants w,z. Their expressions are really horrific, but it is easy to see that they define in the phase plane (u,v) a family of regions

$$\mathcal{R} = \{(u,v)/w(u,v) \leqslant c_1 \ , \ z(u,v) \leqslant c_2 \ , \ u \geqslant 0\}$$

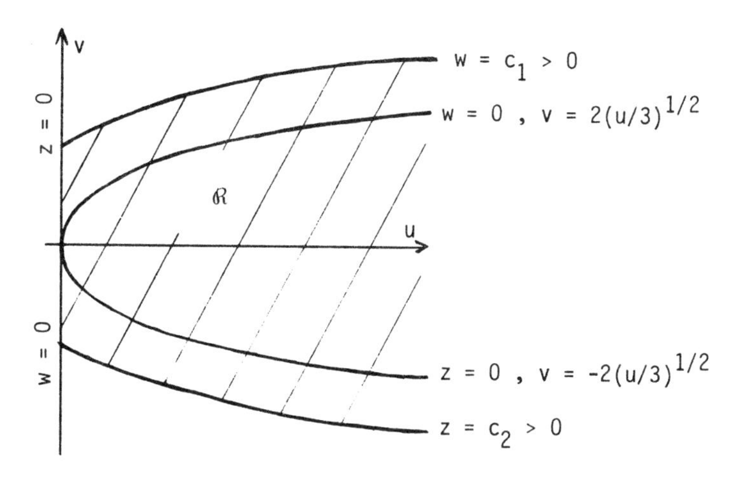

and that, for all c_1, c_2, $\mathcal{R}$ is convex, unbounded, and satisfies

$$\begin{cases} \exists\, c \ , \ c' \text{ (which depend on } c_1 \ , \ c_2) \text{ such that} \\ \forall\, (u,v) \in \mathcal{R} \ , \ |v| \leqslant c\, u^{1/2} + c' \end{cases} \tag{12}$$

Now, if $\nu = \mu$, it is known - see Chueh, Conley and Smoller [2] - that such a region $\mathcal{R}$ is invariant for $(\mathscr{P}')$, and we shall see below that, although $\mathcal{R}$ is unbounded, (12) is sufficient to prove the global existence of the solution.

We are going to use some classical spaces. Let k be a positive integer, $1 \leqslant q \leqslant +\infty$, and let $[\ell]$ denote the integer part of a non-integer positive real number ℓ, let $\bar{\Omega}$ and $\overline{Q_t}$ be respectively the closures of Ω and of $Q_t = \Omega \times]0,t[$. We consider the following spaces (see the classical book of Ladyzenskaja, Solonnikov and Ural'ceva, hereafter referred to as L.S.U. [5]) :

$$W_q^k(\Omega) = \{u \in L^q(\Omega)/D^s u = D_x^s u = \partial^{|s|}u/\partial x_1^{s_1} \ldots \partial x_N^{s_N} \in L^q(\Omega) \ , \quad \forall\, s/|s| = \sum_{i=1}^{N} s_i \leqslant k\}$$

where $s = (s_1,\dots,s_N)$ is a multi-integer and the derivatives are taken in the sense of the Distributions

$$W_q^{2,1}(Q_t) = \{u \in L^q(Q_t) / \frac{\partial u}{\partial t} , D_x^s u \in L^q(Q_t) , \forall s/|s| \leqslant 2\}$$
$$H^k(\Omega) = W_2^k(\Omega) , H^{2,1}(Q_t) = W_2^{2,1}(Q_t) .$$

All these spaces are equipped with their usual norm.

$C^\ell(\overline{\Omega})$ is the space of functions u such that their derivatives $D_x^s u$ are continuous and bounded in $\overline{\Omega}$ for $0 \leqslant |s| \leqslant [\ell]$, and are Hölder-continuous of exponent $(\ell-[\ell])$ for $|s| = [\ell]$. Let $|\cdot|_\Omega^{(\ell)}$ the classical norm in this space (see again L.S.U. [5] , where this space is denoted by $H^\ell(\overline{\Omega})$).

$C^{\ell,\ell/2}(\overline{Q_t})$ is the space of functions u such that their derivatives $D_t^r D_x^s u$ are continuous and bounded in $\overline{Q_t}$ for $0 \leqslant 2r + |s| \leqslant [\ell]$, are Hölder-continuous with respect to x (resp. t) of exponent $(\ell-[\ell])$ for $2r+|s| = [\ell]$ (resp. of exponent $(\ell - 2r - |s|)/2$ for $2r + |s| = [\ell]$ or $[\ell] - 1$). Let $|\cdot|_{Q_t}^{(\ell)}$ be the classical norm on this space.

Finally, if $1 \leqslant p < +\infty$ and if X is a Banach space, $L^p(0,t;X)$ is the space of measurable functions u of [0,t] in X such that $(\int_0^t \|u(\tau)\|_X^p \, d\tau)^{1/p} < +\infty$, with the usual modification for $p = +\infty$.

2. GLOBAL EXISTENCE OF THE SOLUTION FOR $\mu = \nu$ AND $\Omega = \mathbb{R}$. To simplify the notations, we may assume that $\mu = \nu = 1$. First we consider the more general case where the second member of (1) and the initial data (5) are any given functions f and U_0. Let $\mathscr{P} = \mathscr{P}(T, f, u_0, U_0)$ be the associated problem. We consider the problem $(\mathscr{P})$ rather than the (equivalent) problem $(\mathscr{P}')$.

THEOREM 1.

i) Suppose that for some $\ell > 0$ *(ℓ non integer), we have :*

$$(u_0 , U_0) \in (H^1(\mathbb{R}) \cap C^{\ell+2}(\mathbb{R}))^2 \qquad (14)$$

and

$$f \in L^2(Q_T) \cap C^{\ell,\ell/2}(\overline{Q_T}) \qquad (15)$$

then the problem $\mathscr{P}(t, f, u_0, U_0)$ *admits, at least locally in time (i.e. for all* t *in a time - interval* $\mathscr{I}$ *, with* $\mathscr{I} = [0,t^*[$ *,* $0 < t^* \leqslant T$ *or* $\mathscr{I} = [0,T]$*) a unique solution* (u,U) *such that*

$$\forall t \in \mathscr{I} , (u,U) \in (H^{2,1}(Q_t) \cap C^{\ell+2,\ell/2+1}(\overline{Q_t}))^2 \qquad (16)$$

ii) if the assumptions are true for all non-integer positive ℓ*, then*

$$\forall t \in \mathscr{I} , (u,U) \in (C^\infty(\overline{Q_t}))^2$$

iii) suppose moreover that f *is positive, that* $\ell > 1$*, and that*

$$\forall x \in \mathbb{R} , (u_0(x) , v_0(x)) = (u_0(x) , \frac{\partial U_0}{\partial x}(x)) \in \mathscr{R}$$

where $\mathcal{R}$ is one of the regions in $\mathbb{R}^2$ which was previously defined by the Riemann invariants w,z *then*

$$\forall\ (x,t) \in \mathbb{R} \times \mathcal{I}\ ,\ (u(x,t)\ ,\ v(x,t)) = (u(x,t)\ ,\ \frac{\partial U}{\partial x}(x,t)) \in \mathcal{R} \tag{17}$$

i.e. $\mathcal{R}$ is invariant for $(\mathcal{P}')$, and therefore (12) is satisfied for all (x,t) in $\mathbb{R} \times \mathcal{I}$, where the constants c, c' *are independent of* t.

SKETCH OF THE PROOF.

1) As in Rascle [8] , but in a different way, we use a <u>fixed point method</u>. Let us define

$$K = K(\alpha,\tau) = \{\varphi \in L^2(Q_\tau)/\|\varphi\|_{L^2(Q_\tau)} \leqslant \alpha\} \qquad \tau > 0\ ,\ \alpha > 0$$

$\mathcal{T} = \mathcal{T}(\tau,\ f,\ u_0\ ,\ U_0)\ :\ \varphi \mapsto U = \mathcal{T}(\varphi)$ solution of

$$\begin{cases} \dfrac{\partial U}{\partial t} - \Delta U = \varphi & \text{in} \quad Q_\tau \\ U(\cdot,0) = U_0(\cdot) \end{cases} \tag{18}$$

$\mathcal{S} = \mathcal{S}(\tau,\ f,\ u_0\ ,\ U_0)\ :\ U \mapsto u = \mathcal{S}(U)$ solution of

$$\begin{cases} \dfrac{\partial u}{\partial t} - \Delta u - \nabla U\cdot\nabla u - (\Delta U)u = f & \text{in} \quad Q_\tau \\ u(\cdot,0) = u_0(\cdot) \end{cases} \tag{19}$$

Now, it is easy to prove that

<u>$\forall\ \alpha > 0\ ,\ \forall\ \beta \in [0,1[\ ,\ \exists\ \tau_0 > 0\ ,\ \forall\ \tau \leqslant \tau_0\ ,\ \mathcal{S} \circ \mathcal{T}$ is a contraction of $K(\alpha,\tau)$ in itself, of rate less than or equal to β</u> . (20)

<u>Let us prove (the first part of) (20)</u>. For all φ in $K(\alpha,\tau)$, it is easy to see that $U = \mathcal{T}(\varphi)$ satisfies

$$\begin{aligned} \|\nabla U\|_{W(0,\tau)} &= \|\nabla U\|_{L^\infty(0,\tau\ ;\ L^2(\mathbb{R}))\ \cap\ L^2(0,\tau\ ;\ H^1(\mathbb{R}))} \leqslant \\ &\leqslant c(\|U_0\|_{H^1(\mathbb{R})} + \|\varphi\|_{L^2(Q_\tau)}) \leqslant c(\|U_0\|_{H^1(\mathbb{R})} + \alpha) = c_1(\alpha) \end{aligned} \tag{21}$$

(multiply both members of (2) by $(-\Delta U)$, and integrate by parts).
Here and in the sequel, c denotes any constant independent of τ, for all τ in $[0,T]$.

Now, by use of the classical Embedding Theorems, we obtain

$$\begin{cases} \|\nabla U\|_{L^r(0,\tau\ ;\ L^q(\mathbb{R}))} \leqslant c\|\nabla U\|_{W(0,\tau)} \leqslant c\ c_1(\alpha) \\ \forall\ (q,r)/\ \dfrac{1}{r} + \dfrac{1}{2q} = \dfrac{1}{4}\ ,\quad r \in [4\ ,\ +\infty]\ ,\quad q \in [2\ ,\ +\infty] \end{cases} \tag{22}$$

(see L.S.U. [5] , where the proof of this result can be easily extended to the unbounded case $\Omega = \mathbb{R}$). Let us choose $q = 3$, $r = 12$; we obtain

$$\|\nabla U\|_{L^3(Q_\tau)} \leqslant \tau^{1/4} \|\nabla U\|_{L^{12}(0,\tau\ ;\ L^3(\mathbb{R}))} \leqslant c\ c_1(\alpha)\ \tau^{1/4} \tag{23}$$

We consider now (19) as a linear problem to the unknown function u, under the general form

$$\begin{cases} \dfrac{\partial u}{\partial t} - \mu\ \Delta u + a_1(x,t)\ \dfrac{\partial u}{\partial x} + a(x,t)u = f(x,t) \\ u(\cdot,0) = u_o(\cdot) \end{cases} \tag{24}$$

with

$$a_1(x,t) = -\ \nabla U(x,t)\ ;\ a(x,t) = -\ \Delta U(x,t) \tag{25}$$

We use again the general Theorems (L.S.U. p. 341) where we choose

$$\begin{cases} a_1 \in L^r(Q_\tau) \quad , \quad r = 3 = \max(q,N+2) \quad , \quad N = 1 \quad , \quad q = 2 \\ a \in L^s(Q_\tau) \quad , \quad s = 2 = \max(q,(N+2)/2) \\ u_o \in W_q^{2-2/q}(\mathbb{R}) = H^1(\mathbb{R}) \ , \\ f \in L^q(\mathbb{R}) \quad , \quad q = 2\ . \end{cases} \tag{26}$$

We obtain :

$$\|u\|_{H^{2,1}(Q_\tau)} \leqslant c(\tau, \|a_1\|_{L^r_{loc}(Q_\tau)} + \|a\|_{L^s_{loc}(Q_\tau)}) \left[\|u_o\|_{H^1(\mathbb{R})} + \|f\|_{L^2(Q_\tau)}\right]$$

where :

$$\|a_1\|_{L^r_{loc}(Q_\tau)} = \sup_{\substack{\omega \subset \mathbb{R} \\ \text{meas}(\omega) = 1}} \|a_1\|_{L^r(\omega \times]0,\tau[)} \leqslant \|a_1\|_{L^r(Q_\tau)}$$

and where c is an increasing function of its arguments (see Rascle [8]).

Hence we obtain :

$$\|u\|_{H^{2,1}(Q_\tau)} \leqslant c(T, c\ c_1(\alpha)T^{1/4} + c\ c_1(\alpha))\ \|u_o\|_{H^1(\mathbb{R})} + \|f\|_{L^2(Q_T)} = c_2(\alpha) \tag{27}$$

Then we write :

$$u(x,t) = u_o(x) + \int_0^t \frac{\partial u}{\partial t}\ (x,s)ds$$

from which it is easy to deduce :

$$\begin{aligned} \|u\|^2_{L^2(Q_\tau)} &\leqslant \tau^2 \left\| \frac{\partial u}{\partial t} \right\|^2_{L^2(Q_\tau)} + 2\tau\ \|u_o\|^2_{L^2(\mathbb{R})} \\ &\leqslant \tau^2\ \|u\|^2_{H^{2,1}(Q_\tau)} + 2\tau\ \|u_o\|^2_{L^2(\mathbb{R})} \\ &\leqslant \tau^2\ c_2(\alpha) + 2\tau\ \|u_o\|^2_{L^2(\mathbb{R})} \leqslant \alpha^2 \quad \text{for } \tau \text{ small enough.} \end{aligned} \tag{28}$$

The first assertion of (20) follows, and the second would be proved in a similar way. Hence (20) is proved.

Obviously, (20) implies the existence and the uniqueness of the solution of $(\mathscr{P})$ in the closed convex set $K(\alpha,\tau)$, for all $\tau \leqslant \tau_0$, i.e. on a "small" interval of time.

Next, the regularity of this solution (u, U), which is (independently) proved below, will imply

$$(u, U) \in C^0([0,\tau_0] ; H^1(\mathbb{R})) \tag{29}$$

then we can use the same fixed point method to extend this solution on another interval $[\tau_0, \tau_1]$, and so on, and, finally, to define this solution on an interval $\mathscr{I} = [0,t^*[$ $(0 < t^* \leqslant T)$ or $\mathscr{I} = [0,T]$.

Now, suppose that there exist two solutions of $(\mathscr{P})$, then it is easy to check that they are identical on a set of times which is both open and closed : this proves the uniqueness of the solution of $(\mathscr{P})$. (Observe that, *a priori*, (20) implied the uniqueness of this solution in the set $K(\alpha,\tau)$, for $\tau \leqslant \tau_0$, but not necessarily in the whole space).

2) Now, let us study the regularity of this solution (u,U) on $[0,t]$, $t \in \mathscr{I}$. First, by (18) (where, now, $\varphi = u$) and (19), we have

$$(u, U) \in (H^{2,1}(Q_t))^2$$

which implies, by the Embedding Theorems (see L.S.U. [5], p. 342) :

$$u \in C^{m,m/2}(\overline{Q_t}), \quad \text{with} \quad m = 2-(N+2)/q = 1/2, \quad N = 1, \quad q = 2 > (N+2)/2, \quad q \neq N+2 \tag{30}$$

And now, we can apply the classical Theorems (see again L.S.U. [5], p. 320) to obtain better and better estimates on the regularity of the solution :

We first come back to the Problem (18), for which (14) and (29) imply (now $\varphi = u$) :

$$U \in C^{p+2,p/2+1}(\overline{Q_t}), \quad p = \min(\ell,1/2) \tag{31}$$

In turn, (31) implies :

$$(\nabla U, \Delta U) \in (C^{p,p/2}(\overline{Q_t}))^2 \tag{32}$$

which, in view of (14), (15) and (19), implies :

$$u \in C^{p+2,p/2+1}(\overline{Q_t}) \tag{33}$$

Then, if $0 < \ell < 1/2$, $p = \ell$ and (i) is proved. If not, using alternately (18) and (19), we can improve more and more the regularity of the solution, and we can thus prove (i) and (ii).

3) Let us prove (iii). If $\ell > 1$, u and $v = U_x$ are smooth enough to apply the Theorem of Chueh, Conley and Smoller [2], i.e. to apply the Maximum Principle to the functions $w(u(x,t),v(x,t))$ and $z(u(x,t),v(x,t))$, where w and z are the previously defined Riemann invariants of (*). We just observe

that, if f is positive, then the vector-field

$$(x,t) \mapsto (f(x,t),0)$$

is pointing-inside at the boundary of the region $\mathcal{R}$. Then (12) is satisfied for all (x,t) in $\mathbb{R} \times \mathcal{I}$. Thus Theorem 1 is proved.

THEOREM 2.

We suppose now :

$$u_0 \in L^1(\mathbb{R}) \cap L^2(\mathbb{R}) \tag{34}$$

$$\exists\, U_0 \in H^1(\mathbb{R}) / v_0 = \frac{\partial U_0}{\partial x} \tag{35}$$

$$f = 0 \; (^1) \tag{36}$$

$$\forall\, x \in \mathbb{R}\ ,\ (u_0(x), v_0(x)) \in \mathcal{R} \tag{37}$$

then the Problem ($\mathcal{P}'$) *admits a unique solution* (u,v), *<u>defined on any bounded interval</u>* [0,T] , *in the space*

$$(L^\infty(0,T\ ;\ L^1(\mathbb{R})) \cap W(0,T)) \times Z(0,T)$$

where :

$$W(0,T) = L^\infty(0,T\ ;\ L^2(\mathbb{R})) \cap L^2(0,T\ ;\ H^1(\mathbb{R})) \tag{38}$$

and :

$$Z(0,T) = \{v / \exists\, U \in L^\infty(0,T\ ;\ H^1(\mathbb{R})) \cap H^{2,1}(Q_T) / v = U_x\} \tag{39}$$

Moreover, we have :

$$(u(x,t),v(x,t)) \in \mathcal{R} \quad \text{a.e. in } Q_T \ (\text{whence } u \geq 0 \quad \text{a.e. in } Q_T) \tag{17}$$

and :

$$\left(\frac{\partial u}{\partial t}, \frac{\partial v}{\partial t}\right) \in (L^2(0,T\ ;\ H^{-1}(\mathbb{R})))^2 \tag{40}$$

where $H^{-1}(\mathbb{R})$ *is the dual space of* $H^1(\mathbb{R})$. *Finally, if* u_0 *and* U_0 *satisfy* (14), *then* u *and* U *(given by* (39)*) satisfy* (16) *for all time* t *in* [0,T] .

SKETCH OF THE PROOF.

1) First, we suppose that u_0 and U_0 are smooth, i.e. satisfy also (14), with $\ell > 1$, and we write the main *a priori* estimates which will prove the <u>global existence</u> of the solution (i.e. $\mathcal{I} = [0,T]$).

<u>First estimate</u> :

We integrate (1) on $\mathbb{R} \times]0,t[$. As u is positive, we obtain :

---

(1) The same result would be true for any positive f in $L^2(Q_T) \cap L^1(Q_T)$, but we limit ourselves to this case which was the one of the initial Problem ($\mathcal{P}$). We just introduced f in Theorem 1 because Chueh, Conley and Smoller's Theorem uses an argument of perturbation of the second members of the equations (1) and (2).

$$\forall\, t \in \mathscr{I} \;,\; \|u(t)\|_{L^1(\mathbb{R})} = \|u_0\|_{L^1(\mathbb{R})} = \|u\|_{L^\infty(0,t\,;\,L^1(\mathbb{R}))} \tag{41}$$

(i.e. the total bacterial population is constant)

Second estimate :

We multiply both members of (1) by u, and we integrate (by parts) on $\mathbb{R} \times]0,t[$. We obtain :

$$\frac{1}{2}\,\|u(t)\|^2_{L^2(\mathbb{R})} + \int_0^t \int_{\mathbb{R}} (\nabla u)^2(x,\tau)dx\, d\tau = \frac{1}{2}\,\|u_0\|^2_{L^2(\mathbb{R})} + I(t) \tag{42}$$

where :

$$I(t) = -\int_0^t \int_{\mathbb{R}} u\, \nabla U\cdot\nabla u\, dx\, d\tau = -\int_0^t \int_{\mathbb{R}} uv\, \nabla u\, dx\, d\tau \tag{43}$$

Now, by (17), we have :

$$|I(t)| \leqslant \int_0^t \int_{\mathbb{R}} u\, (c\, u^{1/2} + c')|\nabla u| dx\, d\tau \leqslant$$

$$\leqslant c\|u\|_{L^2(0,t\,;\,L^\infty(\mathbb{R}))}\cdot\|u^{1/2}\|_{L^\infty(0,t\,;\,L^2(\mathbb{R}))}\cdot\|\nabla u\|_{L^2(Q_t)} + c'\|u\|_{L^2(Q_t)}\,\|\nabla u\|_{L^2(Q_t)} \tag{44}$$

On the other hand, we can use the same Embedding Theorem as in (22), with $q = +\infty$, $r = 4$, and Hölder inequality, to obtain, in view of (41) :

$$|I(t)| \leqslant (c\, t^{1/4}\, \|u_0\|^{1/2}_{L^1(\mathbb{R})} + c' t^{1/2})\|u\|^2_{W(0,t)} \tag{45}$$

Next, we can choose t such that :

$$c\, t^{1/4}\, \|u_0\|^{1/2}_{L^1(\mathbb{R})} + c' t^{1/2} \leqslant 1/4 \tag{46}$$

Then (42) and (45) imply :

$$\|u\|^2_{W(0,t)} \leqslant 2\, \|u_0\|^2_{L^2(\mathbb{R})} \tag{47}$$

Now, by (41) and (46), we can obtain the same inequality on the interval <u>of same length</u> [t,2t] , and so on, until we have exhausted the interval [0,T]. Thus we have proved :

$$u \in W(0,T) \tag{48}$$

(and it would be very difficult to obtain this crucial estimate without using the invariant regions $\mathcal{R}$).

Third estimate :

Obviously (35) and (48) imply :

$$U \in L^\infty(0,T\,;\,H^1(\mathbb{R})) \cap H^{2,1}(Q_T) \tag{49}$$

Then (48) and (49) imply (40). Next, if u_0 and U_0 satisfy (14), by the same arguments as in Theorem 1, (49) implies (16) for all time t in [0,T] , i.e. globally in time.

2) Thus Theorem 2 is proved if u_0 and v_0 are smooth. Now if it is not the case, we can approximate them by two sequences of smooth initial data (u_0^n) and (v_0^n), to which we associate the corresponding solutions u^n and $v^n = \frac{\partial U^n}{\partial x}$ of Problem ($\mathscr{P}'$).

Then, for each n, the previous *a priori* estimates are valid, with constants which do not depend on n, and we can apply the classical compactness method (see e.g. Lions [6]) : we extract subsequences of (u^n), (v^n) which converge weakly in suitable Sobolev spaces, next, using Rellich Theorem on any bounded sub-domain of $\mathbb{R}$, and the compactness Theorem of Aubin (see [1]), we can pass to the limit, even in the non-linear term, and prove that (u,v) is a solution - actually, the solution - in the sense of the Distributions, of Problem ($\mathscr{P}'$) in the space $W(0,T) \times Z(0,T)$. Hence Theorem 2 is proved.

To conclude, it seems interesting, and quite natural, to observe the prominent part of the hyperbolic properties of (*) - the entropy η in [8] , the Riemann invariants (w,z) here - in the resolution of Problem ($\mathscr{P}$) (or of the equivalent Problem ($\mathscr{P}'$)). Finally, we have just considered here this Problem with a fixed viscosity coefficient μ. The problem of the convergence when $\mu \to 0$ is still open, because in this case the recent Theorem of Di Perna [3] does not work.

BIBLIOGRAPHY

1. J.P. Aubin, Un Théorème de compacité, C.R. Acad. Sc., t. 256 (1963), 5042-5044.

2. K.N. Chueh, C.C. Conley, and J.A. Smoller, Positively Invariant Regions for Systems of Non Linear Equations, Indiana Univ. Math. J., 26, 2 (1977), 373-392.

3. R.J. Di Perna, Convergence of Approximate Solutions to Conservation Laws, to appear.

4. E.F. Keller, and L.A. Segel, Traveling Bands of Chemotactic Bacteria : A Theoretical Analysis, J. Theor. Biol., 30 (1971), 235-248.

5. O.A. Ladyzenskaja, V.A. Solonnikov, and N.N. Ural'Ceva, Linear and quasi-linear Equations of Parabolic Type, Amer. Math. Soc. Translations, Vol. 23, A.M.S., Providence, R.I. (1968).

6. J.L. Lions, Quelques méthodes de résolution des problèmes aux limites non linéaires, Dunod-Gauthier-Villars, Paris, 1969.

7. R. Nossal, Boundary Movement of Chemotactic Bacterial Populations, Math. Biosc., 13 (1972), 397-406.

8. M. Rascle, Sur une équation intégro-différentielle non linéaire issue de la Biologie, J. Diff. Eq., 32, 3 (1979), 420-453.

9. M. Rascle, On a system of non-linear strongly coupled Partial Diff. Equations arising in Biology, Proc. Internat. Conference on Ordinary and Partial Diff. Eq., Dundee 1980, Lecture Notes in Math., Springer, 846, 290-298.

DEPARTEMENT DE MATHEMATIQUES
UNIVERSITE DE ST ETIENNE, 42023 ST ETIENNE CEDEX (FRANCE)

Contemporary Mathematics
Volume 17, 1983

SYSTEMS OF CONSERVATION LAWS WITH COINCIDING SHOCK AND RAREFACTION CURVES*

BLAKE TEMPLE

1. INTRODUCTION

Systems of conservation laws which have coinciding shock and rarefaction curves arise in the study of oil reservoir simulation, nonlinear wave motion in elastic strings, as well as in multicomponent chromatography [1, 4, 5, 6, 9, 11, 12]. These systems have many interesting features. The Riemann problem for these equations can be explicitly solved in the large, and wave interactions have a simplified structure, even in the presence of a nonconvex flux function. For this reason, these systems represent some of the few examples for which the Cauchy problem has been solved for arbitrary data of bounded variation. Also, hyperbolic degeneracies appear in each of these systems. In the present paper we are concerned with locating the class of equations that exhibit the phenomenon of coinciding shock and rarefaction curves. For $n \times n$ systems, we give necessary and sufficient conditions for a shock curve to coincide with a rarefaction curve. We use these general results to write down explicitly the class of 2×2 conservation laws which have shock and rarefaction curves that coincide.

A system of conservation laws in one space dimension is a set of partial differential equations of the form

$$U_t + F(U)_x = 0. \tag{1A}$$

Here $-\infty < x < \infty$, $t \geq 0$, and U and F are vector valued functions, $U = (u_1,\ldots,u_n) \equiv U(x,t)$, $F(U) = (f_1(U),\ldots,f_n(U))$. The Cauchy problem is the natural problem to pose for system (1), and it is commonly known that discontinuities can form in the solutions of (1). For this reason we look for weak solutions $U(x,t)$; i.e., solutions that satisfy the following integral equation [7] for any smooth function $\psi(x,t)$ with compact support:

$$\iint_{\substack{-\infty<x<+\infty \\ t\geq 0}} U\psi_t + F(U)\psi_x + \int_{-\infty}^{\infty} U(x,0)\psi(x,0)dx = 0. \tag{1B}$$

*This research supported by an N.S.F. Mathematical Sciences Postdoctoral Research Fellowship, Grant #MC580-17157.

0271-4132/82/0000-1013/$03.50

Two important systems of conservation laws arise in applications, and have been studied in [1, 4, 5, 6, 9, 11, 12].

$$\begin{aligned} u_t + \{u\phi(u,v)\}_x &= 0 \\ v_t + \{v\phi(u,v)\}_x &= 0. \end{aligned} \tag{2}$$

$$\begin{aligned} u_t + \left\{\frac{u}{1+u+v}\right\}_x &= 0 \\ v_t + \left\{\frac{\kappa v}{1+u+v}\right\}_x &= 0. \end{aligned} \tag{3}$$

System (2) arises in problems of oil reservoir simulation, as well as in elasticity theory [4, 6, 9, 12]. For example, in the reservoir simulation problem [4, 6], u is the saturation of water in the reservoir and v is the concentration of a polymer in the water, so that $0 \leq u \leq 1$, $0 \leq v \leq 1$. The system is determined by specifying the function $\phi(u,v)$, but the structure of the solutions is determined by qualitative properties of ϕ which can be verified experimentally.

System (3) arises in the study of two component chromatography [1, 5, 11]. Here u and v are transformations of the concentrations of the two solutes, and x and t are transformations of the actual space and time variables (see Aris & Amundson [1], pp. 268). The domains of the variables u and v can be taken to be $u \geq 0$, $v \geq 0$, and $\kappa \in (0,1)$ is determined by adsorption properties of the stationary phase.

Systems (2) and (3) are remarkable because for both systems, the shock curves and rarefaction curves coincide. This leads us to study the phenomenon of coinciding shock and rarefaction curves in general. To say this precisely, assume first that system (1) is hyperbolic; i.e., that the eigenvalues (wave speeds) of dF (the matrix defined by $F(U)_x = dF \cdot U_x$) are real, but not necessarily distinct. Let λ denote an eigenvalue, and R a corresponding eigenvector of dF. We call (λ,R) a "characteristic family" or "characteristic field" for system (1) if $\lambda(U)$, $R(U)$ are defined and C^3 in some neighborhood N of U-space. Let $S \subset N$ be the integral curve of R through some point $U_0 \in N$. S is called the λ-rarefaction curve of U_0 in N. Rarefaction curves are the one-dimensional sets that smooth solutions to system (1) can take values on. For example, if the range of a smooth solution $U(x,t)$ of system (1) lies on a one-dimensional curve in U-space, then that curve must be a rarefaction curve. An analagous one-dimensional curve in U-space applies to the study of discontinuous solutions of (1). The Hugoniot locus of a point U_0 is defined to be the set of points U such that

$$\sigma[U] = [F(U)] \tag{4}$$

for some scalar $\sigma = \sigma(U,U_0)$, $[U] = U - U_0$, $[F(U)] = F(U) - F(U_0)$. A state U_1 is in the Hugoniot locus of U_0 if and only if the discontinuous function

$$U(x,t) = \begin{cases} U_0 & \text{for } x < \sigma t \\ U_1 & \text{for } x > \sigma t \end{cases} \tag{5}$$

satisfies the weak form (1B) of system (1) [2, 7]. Under very general conditions, there corresponds to each family (λ, R) a one parameter subset of the Hugoniot locus of U_0 that has C^2 contact with the integral curve of R at U_0 [7]. This is called the λ-shock curve. (Often this term is reserved for that portion of the curve that determines the physically acceptable solutions in (5).)

DEFINITION. We say that the λ-shock curve coincides with the λ-rarefaction curve on S if the Hugoniot locus of each point on S, contains S.

In [13] we prove Theorem 1 (details are omitted here) which states that the λ-shock curve and λ-rarefaction curve coincide on S if and only if either S is linear in U-space or λ is constant on S; and this occurs if and only if the equations reduce to a scalar conservation law on S. We call (λ, R) a "contact" family if λ is constant on each integral curve of R, and we call (λ, R) a "line" family if each integral curve of R is a straight line in U-space. In the next section we derive the class of 2×2 equations that have either a contact or a line family. (In this case we let $U \equiv (u,v)$ and for convenience we assume that a contact family satisfies $\nabla\lambda = (\frac{\partial\lambda}{\partial u}, \frac{\partial\lambda}{\partial v}) \neq 0$ in N, and that a line family satisfies $\nabla q \neq 0$ in N, where $q(u,v)$ is the slope $\frac{dv}{du}$ of the integral curve of R through (u,v). Weaker assumptions can be made). All of the characteristic families in systems (2) and (3) are then seen to be either line or contact families. In this way the phenomenon of coinciding shock and rarefaction curves is observed from the explicit form of the equations.

2. COINCIDING SHOCK AND RAREFACTION CURVES FOR 2×2 SYSTEMS

Consider an arbitrary system of 2×2 conservation laws.

$$\begin{aligned} u_t + f(u,v)_x &= 0 \\ v_t + g(u,v)_x &= 0, \end{aligned} \tag{2.1}$$

where we take $U \equiv (u,v)$, $F = (f,g)$. We now locate the class of such 2×2 equations that have either a contact or a line field in a region N of U-space. These are generically the only fields that have coinciding shock and rarefaction curves, as indicated by Theorem 1.

We let $q = q(u,v)$ denote a Riemann invariant for a contact or line field in N. A Riemann invariant for a family (λ,R) is a function which is constant on the integral cuves of R. We assume that $\nabla q \neq 0$, and because it suffices to prove our results locally, we always assume that $q = \text{const.}$ determines a unique integral curve of R in N, which has a finite slope $\frac{dv}{du}$.

First assume that $q(u,v)$ is the wave speed of a contact field for a system of 2×2 equations, $\nabla q \neq 0$. The function q is thus a Riemann invariant for the contact field given by $\lambda = q$, $R = (-q_v,q_u) \equiv \nabla q^{\perp}$. Since (λ,R) is a contact field in N, we have by Theorem 1 that the λ-shock curves coincide with the λ-rarefaction curves in N, so the curve $q = \text{const.}$ must be contained within the Hugoniot locus of every point on that curve. Thus by (4), if $[q] \equiv q(U) - q(U_0) = 0$, then also

$$\begin{aligned} \sigma[u] &= [f], \\ \sigma[v] &= [g], \end{aligned} \tag{2.2}$$

and in the contact case, $\sigma \equiv \sigma(U,U_0) = q(U_0)$ (c.f. [7]). Conversely, if (2.2) holds in N when $[q] = 0$ for some smooth function $q(u,v)$, $\nabla q \neq 0$, then $(\sigma(U_0,U_0), \nabla q^{\perp}(U_0))$ must be a characteristic field for dF. To see this, note that (2.2) implies that

$$\begin{aligned} \sigma(U_0,U_0) &= \lim_{U\to U_0} \frac{[f]}{[u]} = f_u + f_v \frac{dv}{du} \\ &= \lim_{U\to U_0} \frac{[g]}{[v]} = g_v + g_u \frac{du}{dv}, \end{aligned}$$

where the vector $(1, \frac{dv}{du})$ is parallel to $\nabla q(U_0)$. Therefore, we can verify that

$$dF\cdot(1, \tfrac{dv}{du})^{tr} = \sigma(U_0,U_0)(1, \tfrac{dv}{du}). \tag{2.3}$$

Thus the statement that (2.2) holds with $\sigma = q$ when $[q] = 0$ in N, is equivalent to the statement that $(\lambda,R) = (q,\nabla q^{\perp})$ is a contact field in N. But (2.2) holds when $[q] = 0$ if and only if

$$\begin{aligned} [f - uq] &= 0, \\ [g - vq] &= 0, \end{aligned} \tag{2.4}$$

when $[q] = 0$; and (2.4) holds if and only if $f = uq + F(q)$ and $g = vq + G(q)$ for some smooth functions F and G. We have the following theorem:

THEOREM 2. A system of 2×2 conservation laws (2.1) has a contact field in a domain N of uv-space if and only if f and g satisfy

$$\begin{aligned} f(u,v) &= uq + F(q), \\ g(u,v) &= vg + G(q), \end{aligned} \tag{2.5}$$

in N, for some smooth functions q, F and G, $\nabla q \neq 0$. In this case $q(u,v) = \lambda$ where λ is the wave speed of the contact family.

Next assume that (λ,R) is a line family for a system of 2×2 equations defined in a region N of U-space. Let $q(u,v)$ be the slope $\frac{dv}{du}$ of the integral curve of R through the point $(u,v) \in N$, $\nabla q \neq 0$. The function q is a Riemann invariant of R. Moreover, q is a smooth nonconstant solution to Burger's equation

$$q_u + q\, q_v = 0 \tag{2.6}$$

since ∇q is orthogonal to the vector $(1,q)$ at every point in N. Since (λ,R) is a line field, Theorem 1 again implies that a curve defined by $q = \text{const.}$ contains the Hugoniot locus of each point on that curve. Thus, when $[q] = 0$,

$$\sigma[u] = [f]$$
$$\sigma[v] = [g].$$

Dividing we obtain that, when $[q] = 0$,

$$q = \frac{[v]}{[u]} = \frac{[g]}{[f]}, \tag{2.7}$$

so that

$$[fq - g] = 0. \tag{2.8}$$

By (2.3) this is equivalent to the statement that (λ,R) is a line field in N. But (2.8) holds if and only if

$$g = fq + H(q)$$

for some smooth function H. We have proven the following theorem:

THEOREM 3. A system of 2×2 conservation laws has a line family in N if and only if f and g satisfy

$$g = fq + H(q) \tag{2.9}$$

in N, for some smooth function H, where $q(u,v)$ is a smooth solution to Burger's equation (2.6) with $\nabla q \neq 0$. In this case q is a Riemann invariant for the line family.

Theorem 2 applies to system (2) with $F = G \equiv 0$, and thus $(\phi,\nabla\phi^{\perp})$ must be a contact family for system (2). Since system (2) also satisfies

$$g(u,v) = v\phi(u,v) = \frac{v}{u} f(u,v),$$

and $\frac{v}{u}$ is a smooth solution to Burger's equation in $u > 0$, $v > 0$, we have by Theorem 3 that $q(u,v) = \frac{v}{u}$ is the Riemann invariant of a linear family for system (2). Moreover, we can use Theorems 2 and 3 to locate the class of 2×2 equations that have both a line and a contact field; i.e., we say that system (1) has both a contact and a line field in N if system (1) has two Riemann invariants q and p that satisfy Theorem 3 and 4 respectively, such that u and v are smooth functions of (p,q) off a closed set of measure 0 in N. By Theorems 2 and 3,

$$vq + G(q) = uqp + F(q)P + H(p). \tag{2.10}$$

Formally differentiating (2.10) with respect to u holding p fixed yields

$$v + G'(q) = up + F'(q)p \tag{2.11}$$

since $\frac{\partial}{\partial u} v(u,p) = p$ because p is a smooth solution of (2.6). Differentiating (2.10) with respect to q holding p fixed gives

$$\frac{\partial}{\partial q} V(p,q) + v + G'(q) = up + \frac{\partial}{\partial q} u(p,q)qp + F'(q)p. \tag{2.12}$$

Therefore, substituting (2.11) into (2.12) we obtain

$$\frac{\partial}{\partial q} v(p,q) = p \frac{\partial}{\partial q} u(p,q). \tag{2.13}$$

Now differentiate (2.11) with respect to q holding p fixed and obtain

$$\frac{\partial}{\partial q} v(p,q) + G''(q) = P \frac{\partial}{\partial q} u(p,q) + F''(q)p, \tag{2.14}$$

which by (2.13) is

$$G''(q) = F''(q)p. \tag{2.15}$$

Finally, differentiating (2.15) with respect to p holding q fixed, we conclude

$$F''(q) = G''(q) = 0, \tag{2.16}$$

or

$$F(q) = aq + c, \quad G(q) = bq + d \tag{2.17}$$

for some constants a, b, c, d.

The assumptions made in (2.11) to (2.17) are that v is a differentiable function of $u(p,q)$, $\frac{\partial}{\partial u} q(u,p) \neq 0$, and $q \neq 0$. But by (2.6) these must hold off a closed set of measure 0 in N, so (2.17) must hold everywhere in N. Moreover since the addition of a constant to the flux functions f and g in (2.1) does not affect the solutions, we can take $c = d = 0$. Substituting (2.17) into (2.11) then gives

$$p = p(u,v) = \frac{v + b}{u + q},$$

and the constraint in (2.10) yields

$$H(p) = -bp. \tag{2.18}$$

We have thus proven the following corollary:

COROLLARY 1. System (2.1) has both a line and a contact field if and only if

$$f = (u + a)q, \qquad g = (v + b)q, \tag{2.19}$$

for some smooth function $q = q(u,v)$, and some constants a and b. In this case q is the wave speed of the contact family, and

$$p = \frac{v + b}{u + a}. \tag{2.20}$$

Now consider an arbitrary 2×2 system that has two distinct line families. By Theorem 3, there exist two distinct solutions p and q of Burger's equation such that (2.9) holds; i.e., such that

$$g = fp + H_1(p), \tag{2.21}$$

$$g = fq + H_2(q), \tag{2.22}$$

for some smooth functions H_1 and H_2. Equating (2.10) and (2.11) gives

$$f = \frac{H_1(p) - H_2(q)}{q - p}.$$

This proves the following corollary:

COROLLARY 2. System (2.1) has two line families if and only if

$$f = \frac{H_1(p) - H_2(q)}{q - p}, \qquad g = \frac{gH_1(p) - pH_2(q)}{q - p}, \tag{2.23}$$

where p and q are smooth solutions of (2.6).

For system 3 one can verify that

$$\frac{u}{1+u+v} = \frac{H(p) - H(q)}{q - p}, \quad \frac{Kv}{1+u+v} = \frac{gH(p) - pH(q)}{q - p}, \tag{2.24}$$

where

$$H(z) = \frac{z - \kappa z}{z + 1}, \tag{2.25}$$

and p,q are the two solutions of Burger's equation which satisfy

$$uz^2 + \{\kappa(u+1) - (v+1)\}z - \kappa v = 0 \tag{2.26}$$

in z, and are smooth in $u > 0$, $v > 0$. This verifies that system (3) has a pair of line families with integral curves given by $p = \text{const.}$ and $q = \text{const.}$

Finally, note that a system of 2×2 equations generically has two characteristic families. The functions f and g given in (2.5) and (2.9) involve explicitly a Riemann invariant of one family. The following result, which is easily verified, determines both characteristic families given a Riemann invariant of one family. This general result is simple and explains many of the calculations in [4, 12].

THEOREM 4. Let $q = q(u,v)$ be a Riemann invariant for a system of 2×2 conservation laws (2.1), and let $\overline{f}(u,q) = f(u,v)$ and $\overline{g}(v,q) = g(u,v)$ be smooth. Then (λ_i, R_i), $i = 1,2$ are characteristic fields for the system, where

$$\begin{aligned} \lambda_1 &= \overline{f}_u, R_1 = \nabla q^{\perp} \\ \lambda_2 &= \overline{f}_u + \overline{f}_q q_u + \overline{g}_q q_v, R_2 = (\overline{f}_q, \overline{g}_q). \end{aligned} \tag{2.27}$$

ACKNOWLEDGEMENTS

I would like to thank Professor James Glimm for calling my attention to the coincidence of shock and rarefaction curves in the chromatography equations. I also thank Dr. Eli Isaacson for many helpful discussions.

BIBLIOGRAPHY

[1] Aris, R. and Amundson, N., Mathematical Methods in Chemical Engineering, vol. 2, Prentice-Hall, Inc., Englewood Cliffs, New Jersey.

[2] Courant, R. and Friedricks, K. O., Supersonic Flow and Shock Waves, Wiley New York, 1948.

[3] Glimm, J., "Solutions in the large for nonlinear hyperbolic systems of equations", Comm. Pure Appl. Math., 18 (1965), 697-715.

[4] Isaacson, E., "Global solution of a Riemann problem for a non-strictly hyperbolic system of conservation laws arising in enhanced oil recovery", J. Comp. Phys., to appear.

[5] Helfferich, F. and Klein, G., Multicomponent Chromatography, Marcel Dekker, Inc., New York, 1970.

[6] Keyfitz, B. and Kranzer, H., "A system of non-strictly hyperbolic conservation laws arising in elasticity theory", Arch. Rat. Mech. Anal., 72 (1980).

[7] Lax, P. D., "Hyperbolic systems of conservation laws", II, Comm. Pure Appl. Math., 19 (1957), 537-566.

[8] Lax, P. D., "Shock waves and entropy", in Contributions to Nonlinear Functional Analysis", ed. E. H. Zarantonello, Academic Press, New York, 1971, 603-634.

[9] Liu, T. P. and Wang, C. H., "On a hyperbolic system of conservation laws which is not strictly hyperbolic", MRC Technical Summary Report #2184, December 29, 1980.

[10] Peaceman, D. W., Fundamentals of Numerical Reservoir Simulation, Elsevier North-Holland, Inc., 52 Vanderbilt Avenue, New York, N.Y. 10017.

[11] Rhee, H. and Aris, R. and Amundson, N. R., "On the theory of multicomponent chromatography", Phil. Trans. Roy. Soc., A267, 419 (1970).

[12] Temple, Blake, "Global solution of the cauchy problem for a class of 2×2 non-strictly hyperbolic conservation laws", Advances of Appl. Math., to appear.

[13] Temple, Blake, "Nonlinear conservation laws with invariant submanifolds". To appear in Trans. Amer. Math. Soc.

DEPARTMENT OF MATHEMATICAL PHYSICS
THE ROCKEFELLER UNIVERSITY
NEW YORK, NEW YORK 10021

Contemporary Mathematics
Volume 17, 1983

COMPUTATIONAL SYNERGETICS AND MATHEMATICAL INNOVATION: WAVES AND VORTICES*

Norman J. Zabusky

1. INTRODUCTION

My goal is to show with concrete examples how, by the judicious use of computers, we can penetrate into new areas and discover linkages to diverse areas of mathematics unforeseen by our forbears. With insight obtained from numerous solutions, often displayed naturally by graphs and cinemas, we may be liberated from the prejudices of our conservative and sometimes misguided mathematical intuitions.

I use the word "synergetic" here to mean the enhancement in the rate and depth of mathematical understanding through the combined use of analysis and computer simulation. Synergetic was first used in this context by S. M. Ulam. My primary example is the Fermi-Pasta-Ulam (FPU) problem and the discovery of solitons. The second example is from two dimensional inviscid vortex dynamics, where concrete analytic-mathematical developments are still in their infancy. For further details and references to the literature on waves and solitons see [1].

Almost everyone using computers has experienced instances where computational results have sparked new insights. The range covered is large: from uncovering mistakes in formal derivations or calculations; to suggesting combinations of parameters with which to make asymptotic expansions and thereby obtain equations which are analytically tractable; and finally to shining the light of inspiration into areas which have been thought devoid of new concepts or new fundamental truths. The last, namely, the heuristic use of computers to obtain and display the results of numerical experiments where parameters are varied and solutions returned in a rapid interactive fashion is the mode of working that I will elaborate.

In 1946 John von Neumann asked: "To what extent can human reasoning in the sciences be more efficiently replaced by mechanisms?" and: "What phases

*Work supported by the Mathematics Division of the Army Research Office, Contract DAAG-29-82-K-0069 and the Fluid Dynamics Branch of the Office of Naval Research, Contract N00014-81-C-0531.

0271-4132/82/0000-1048/$02.50

of pure and applied mathematics can be furthered by the use of large-scale, automatic computing instruments?" He noted that:

> Our present analytical methods seem unsuitable for the solution of the important problems arising in connection with non-linear partial differential equations and, in fact, with virtually all types of non-linear problems in pure mathematics. The truth of this statement is particularly striking in the field of fluid dynamics. Only the most elementary problems have been solved analytically in this field... .
>
> The advance of analysis is, at this moment, stagnant along the entire front of non-linear problems. That this phenomenon is not of a transient nature but that we are up against an important conceptual difficulty... no decisive progress has been made against time... which could be rated as important by the criteria that are applied in other, more successful (linear!) parts of mathematical physics... .
>
> ...we conclude by remarking that really efficient high-speed computing devices may, in the field of non-linear partial differential equations as well as in many other fields which are now difficult or entirely denied of access, provide us with those heuristic hints which are needed in all parts of mathematics for genuine progress... . This should ultimately lead to important analytical advances.

As described below, several pioneering steps have been taken. But this is just the beginning of a mind-augmenting revolution that inexpensive and robust computing will allow the prepared investigator.

2. WAVES

In the early fifties, Fermi, Pasta and Ulam used the Los Alamos MANIAC I computer to obtain solutions for the one-dimensional nonlinear lattice. Their intuition led them to the hypothesis:

<u>H1</u>. Given a one-dimensional lattice with fixed boundary conditions and with identical nearest neighboring masses coupled by identical nonlinear springs. If the energy of the system is initially in a long-wavelength state, the energy will eventually be shared "equally" among all the degrees of freedom of the system.

In fact, they observed the contrary, namely <u>near</u>-recurrence to the initial state and only the highest modes participating significantly in the dynamics. This led Kruskal and me to a second hypothesis:

H2. The near-recurrence phenomenon and the detailed modal history can be explained by a continuum model with periodic boundary conditions and progressive wave initial conditions.

This was verified computationally and, for the "cubic" lattice, Kruskal derived a Korteweg-de Vries (KdV) equation. We continued with the hypothesis:

H3. The Korteweg-de Vries equation

$$u_t + uu_x + \delta^2 u_{xxx} = 0 \tag{1}$$

can describe propagation of small-but-finite amplitude long waves on a lossless "cubic" lattice excited by a progressive wave initial condition.

This was verified by solving (1) numerically, and Figure 1 shows the trajectory of $\max_x u(x,t)$ over an interval of time where "soliton" interactions are indicated.

The modified KdV equation (u^2u_x replaces uu_x) had similar properties and both equations seemed to have a large number of corresponding conservation law forms. That is, both equations can be written in the form $T_t^{(n)} + X_x^{(n)} = 0$ where $T^{(n)}$ and $X^{(n)}$ were polynomial functions of u and its spatial derivatives, and $n = 1,2,\ldots$.

This milieu of computational and analytical results led R. Miura to prove the theorem.

THEOREM. If v is a solution of the modified KdV equation

$$Q(v) \equiv v_t - 6v^2v_x + v_{xxx} = 0 \tag{2}$$

then

$$u = v^2 + v_x \tag{3}$$

is a solution of the KdV equation

$$P(u) \equiv u_t - 6uu_x + u_{xxx} = 0. \tag{4}$$

Eq. (3) is a Ricatti equation which can be transformed to a Sturm-Liouville problem $[\psi_{xx} + (\lambda - u(x,t))\psi = 0]$ and the problem is exactly linearized.

Much beautiful mathematics followed this pioneering step, including: identifying and solving the Toda lattice, the first of many exactly solvable discrete systems; an operator approach based on the invariance of λ and the conservation laws (P. D. Lax); the KdV can be written as a Hamiltonian system (C. Gardner, L. Fadeev, V. E. Zakharov); an ingenious direct method (R. Hirota); application to the nonlinear Schrödinger equation, a first step in generalization (V. Zakharov and A. B. Shabat); and an inverse (or isospectral) transform method for dealing with a wide class of nonlinear partial differential equations (M. Ablowitz, D. Kaup, A. Newell, H. Segur). In the last decade the exponential surge of beautiful mathematics and applications to natural phenomena has continued. For example, their have been several studies to investigate the validity of the KdV equation as a model for shallow-water waves. The solitary wave and soliton phenomena were first carefully observed by J. Scott Russel in 1834 for waves generated in canals.

3. GRAPHICS

As our analytical insight matured, the character of the graphical representation became focussed on particular phenomena. FPU plotted waveforms $y_n(t)$ and modal energies. When it became clear that progressive waves with periodic boundary conditions contained the same effect, we began tracking trajectories of waveform extrema; this proved informative, for it allowed us to see phase shifts arising from localized entity interactions.

For example, this "Riemann-invariant" filtering was applied to a cubic lattice with periodic boundary conditions that was strongly excited with a low-mode progressive wave.

The oblique projections of $\psi(x,t)$ developed by the Manchester group (Bullough, Eilbeck, Caudrey) have provided useful and artistically pleasing global summaries of nonlinear wave phenomena. For example, Fig. 2 gives $\psi(x,t)$ for the interaction of two solitons of the nonlinear Schrödinger equations. The phase shift of the right-going soliton is evident. Figure 3 shows $\psi(r,t)$, solutions of the cylindrically symmetric sine-Gordon equation with a localized impurity current $j(r)$,

$$\psi_{rr} + r^{-1}\psi_r - \psi_{tt} - j(r)\sin\psi = 0, \tag{5}$$

where

$$j(r) = 1 + a\ \mathrm{sech}^2(r-R_1).$$

Exhibited are nearly two periods of a permanent oscillation of a cylindrically symmetric "pulson." The pulson "falls" toward the origin and is reflected by the localized impurity current at R_1. If one adds a damping term, the oscillation decays. However, if one switches on a constant bias (that is, a constant is added to the right side of (5)) for an appropriate time interval a tunable periodic oscillation is obtained.

The use of color and cinema can greatly enhance the perception of small unexpected phenomena. They also provide other "parameters" to aid the retention and communication process. They enhance the mind's ability to recall and correlate old and new results. For problems of waves and fluctuating or turbulent fluids it would be desirable to develop algorithms to track and plot extremal lines, areas and volumes in one, two, and three dimensions. These diagrams will probably be "noisy". Spatial and temporal correlations should be used to obtain histories of the location of small but important regions when they are hidden in a busy background. Color, shading, and alternate view angles will facilitate the recognition of these events.

4. VORTICES

Many computational studies of 2D Euler equations

$$\zeta_t + \psi_x \zeta_y - \psi_y \zeta_x = 0, \quad \Delta\psi = -\zeta$$

have been made in the last two decades [2,3,4,5]. Point-vortex interaction studies go back to Rosenhead and others in the 1930's. In the last two decades finite-difference, spectral and vortex-in-cell studies have also elucidated dynamical phenomena. The contour dynamics (CD) approach [6], because of its clarity and precision, has recently opened the door to a deeper mathematical understanding of inviscid phenomena.

In CD we assume that the velocity field is created by piecewise-constant localized regions of vorticity or FAVR's. Using the Green's function, we convert 2D integrals into 1D contour integrals. A multitude of "free-boundary" steady states, that we call "V-states", have been found [7,8]. Some singly-connected rotating states and one-quarter of the translating states is shown in Fig. 4. The linear stability of "wake-like" periodic configurations of vorticity have been investigated and the role of finite-area is being clarified [9,10]. Dynamical studies have been made, including merger (or "pairing") of isolated vortex regions (see Fig. 5, and reference [11]) and the coaxial scattering of translating V-states [12]. These processes occur in the laboratory and ocean when "coherent" vortex states interact.

For problems that are ill-posed when dissipation is neglected, we have introduced a tangential regularization procedure [13]. For dissipative regularization, the equations have the form

$$\dot{\underset{\sim}{x}} = \hat{\dot{\underset{\sim}{x}}} + \nu \underset{\sim}{x}_{ss},$$

where s is the arc length and $\hat{\dot{x}}$ is the advective part that arises from contour integration (for the Euler equations [6]) or by solving integral equations (for ionospheric plasma clouds or deformable dielectrics [14]).

5. CONCLUSIONS

To me it is clear that the soliton paradigm validates von Neumann's foresight. Not only has it engendered much new activity in pure and applied mathematics, but it provides a new concentual basis for applications in diverse areas of physics. For realistic systems that are near-integrable, the soliton concept provides an economy-of-thought in posing problems and obtaining solutions.

E. Chargaff, in his review of nucleic acid research, notes "It is in general true of every scientific discovery that the road means more than the goal. But only the latter appears in ordinary scientific papers." This paper is an attempt to show concretely that the analytical-computational synergetic approach is a mode of working that is applicable generally in the natural sciences. It requires good analysis and good computation. But it also requires good graphics and other modes of computer expression.

REFERENCES

1. N. J. Zabusky, "Computational synergetics and mathematical innovation", J. Comput. Phys., 43 (1981), 195-249.

2. N. J. Zabusky, "Coherent structures in fluid dynamics", in The Significance of Nonlinearity in the Natural Sciences, A. Perlmutter and L. F. Scott, eds., Plenum Press, New York, 1977, 145-206.

3. P. G. Saffman and G. R. Baker, "Vortex interactions", Ann. Rev. Fluid Mech., 11 (1979), 95-122.

4. A. Leonard, "Vortex methods for flow simulation", J. Comput. Phys., 37 (1980), 289-335.

5. H. Aref, "Integrable, chaotic and turbulent vortex motion in two-dimensional flows", Ann. Rev. Fluid Mech., 15 (1983). (To be published.)

6. N. J. Zabusky, M. H. Hughes and K. V. Roberts, "Contour dynamics for the Euler equations in two dimensions", J. Comput. Phys., 30 (1979), 96-106.

7. G. S. Deem and N. J. Zabusky, "Vortex waves: Stationary "V-states", interactions, recurrence and breaking", Phys. Rev. Lett., 40 (1979), 859-862.

8. H. M. Wu, E. A. Overman, II and N. J. Zabusky, "Steady-state solutions of the Euler equations in two dimensions: Rotating and translating V-states with limiting cases. I. Numerical algorithms and results", J. Comput. Phys., (1983). (Submitted for publication.)

9. P. G. Saffman and J. C. Schatzman, "Properties of a vortex street of finite vortices", SIAM J. Sci. Stat. Comput., 2 (1981), 285-295.

10. P. G. Saffman and J. C. Schatzman, "Stability of a vortex street of finite vortices", J. Fluid Mech., 117 (1982), 171-185.

11. E. A. Overman II and N. J. Zabusky, "Evolution and merger of isolated vortex structures", Phys. Flds., 25 (1982), 1297-1305.

12. E. A. Overman, II and N. J. Zabusky, "Coaxial scattering of Euler equation translating V-states via contour dynamics," J. Fluid. Mech., 125, (1982), 187-202.

13. N. J. Zabusky and E. A. Overman, II, "Regularization of contour dynamical algorithms. I. Tangential regularization", J. Comput. Phys., (1983). (Accepted for publication.)

14. E. A. Overman, II, N. J. Zabusky and S. L. Ossakow, "Ionospheric plasma clouds dynamics via regularized contour dynamics. I. Stability and nonlinear evolution of one contour models," Phys. Flds. Also See E. A. Overman, II and N. J. Zabusky, Phys. Rev. Lett., 45 (1980), 1693.

INSTITUTE OF COMPUTATIONAL MATHEMATICS AND APPLICATIONS
DEPARTMENT OF MATHEMATICS AND STATISTICS
UNIVERSITY OF PITTSBURGH
PITTSBURGH, PENNSYLVANIA 15261

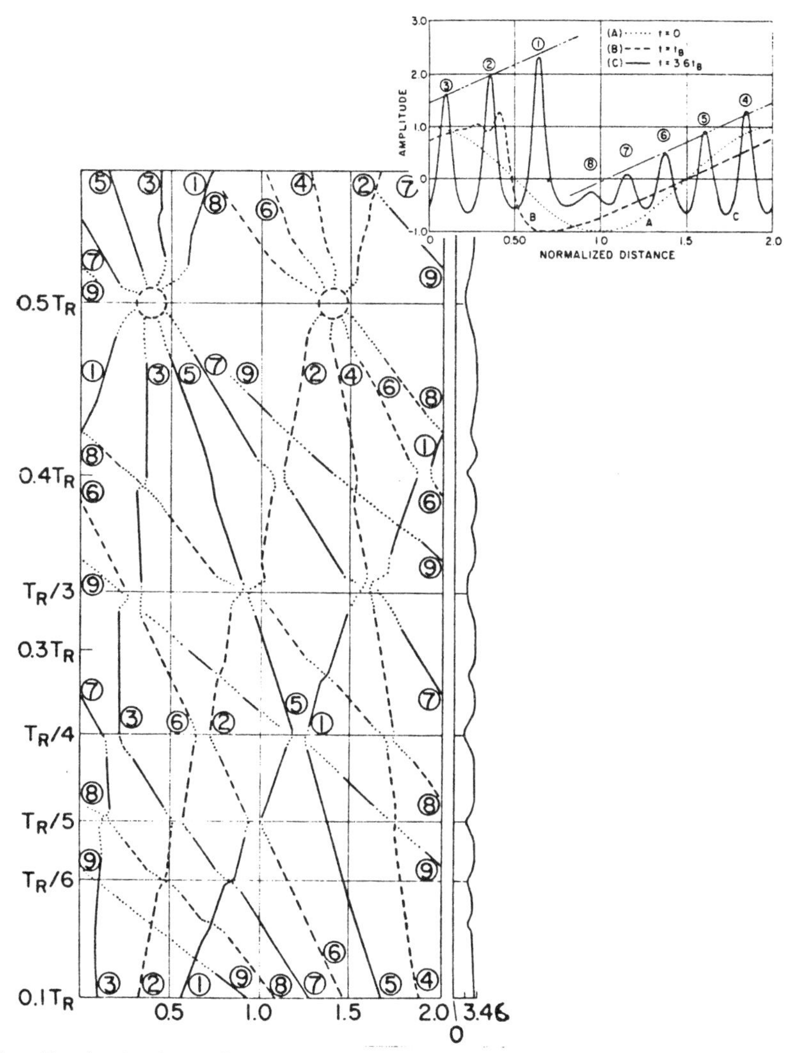

Figure 1: Trajectories of maxima of the Korteweg-de Vries equation (δ = 0.022) on a space time diagram beginning at $t = 0.1t_R = 3.04t_B$. The insert shows the waveforms $u(x,t)$ at three times.

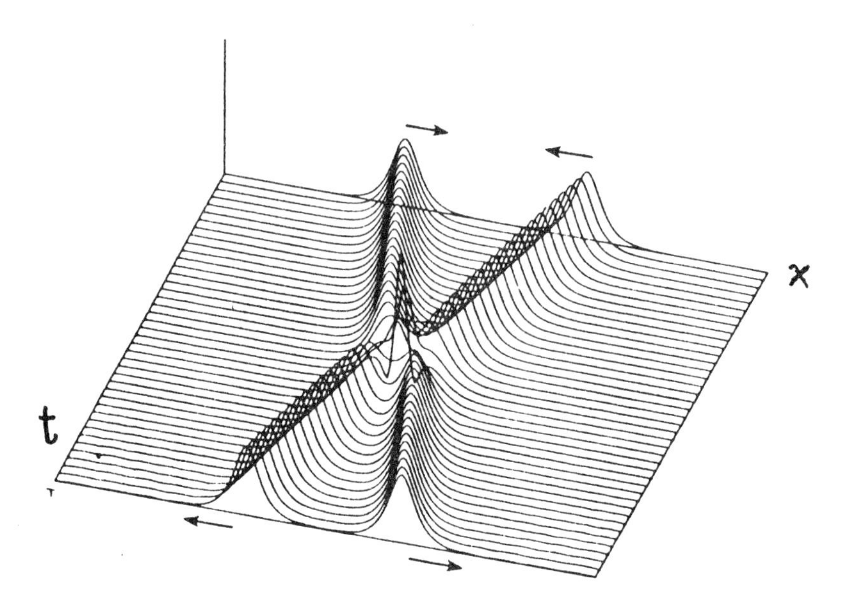

Figure 2: Oblique projections $\psi(x,t)$ of two-soliton interactions. Solutions of the nonlinear Schrödinger equation $i\psi_t + \psi_{xx} + \kappa\psi|\psi|^2 = 0$.

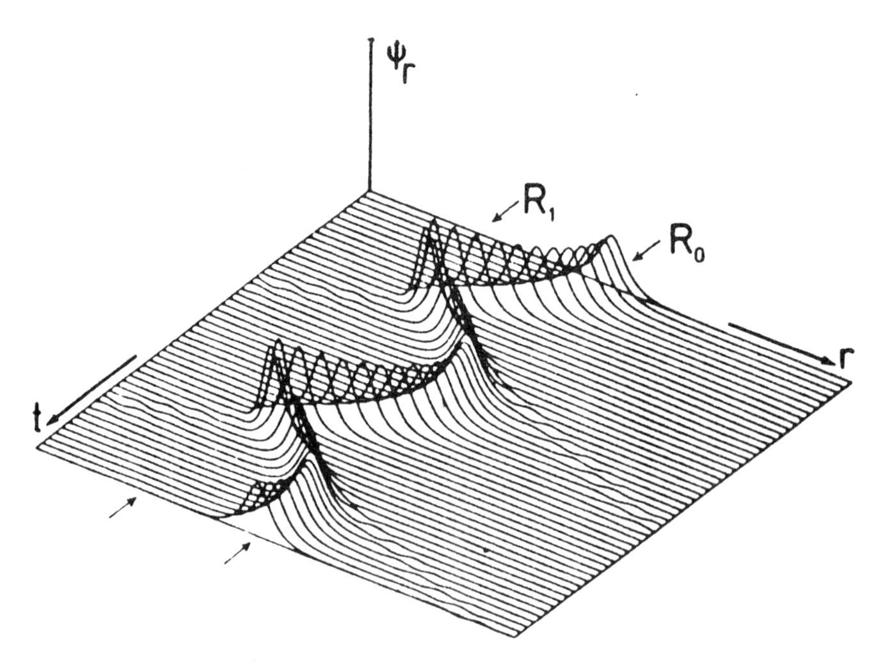

Figure 3: Solutions $\partial_r\psi(r,t)$ of azimuthally symmetric sine-Gordon equation with a current impurity $j(r)$, $\psi_{rr} + r^{-1}\psi_r - \psi_{tt} - j(r)\sin\psi = 0$. Exhibited are oscillations of a pulson which has a "ballistic" behavior.

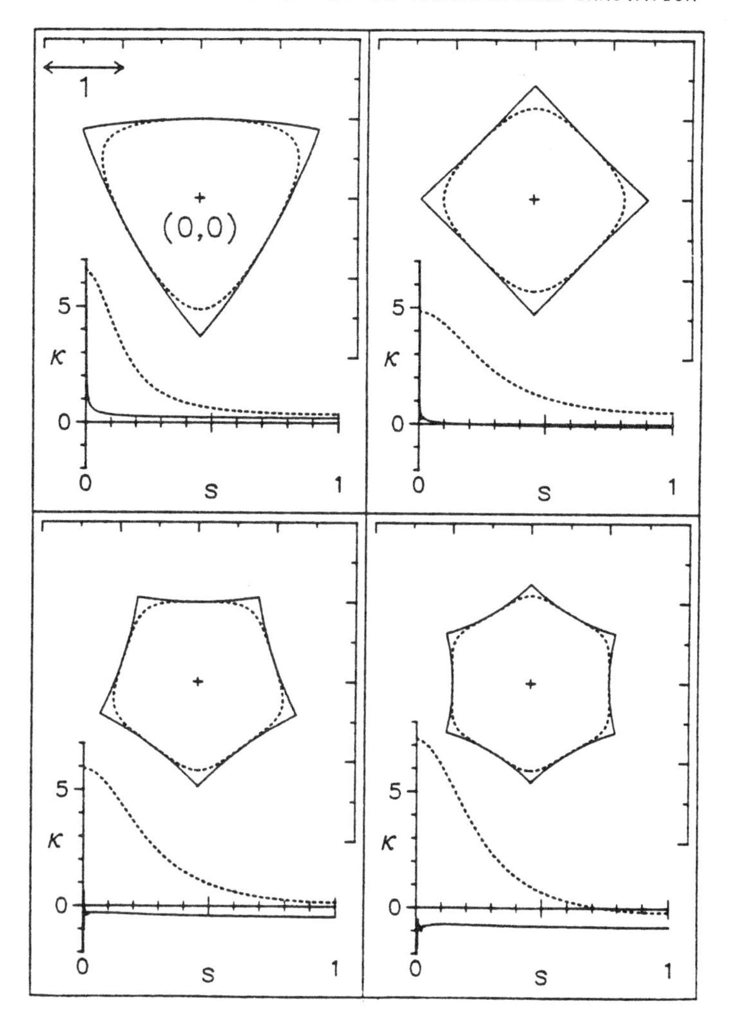

Figure 4a: Rotating V-states and their curvatures $\kappa(s)$, for $3 < m < 6$. The solid curves are the limiting V-states. (See Reference 8.)

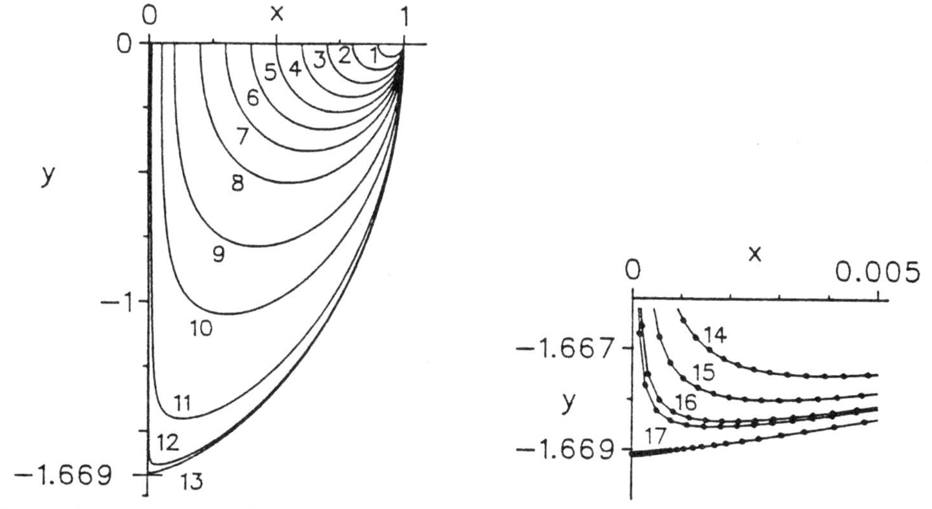

Figure 4b: A sector (1/4) of the translating V-states. (See Reference 8.)

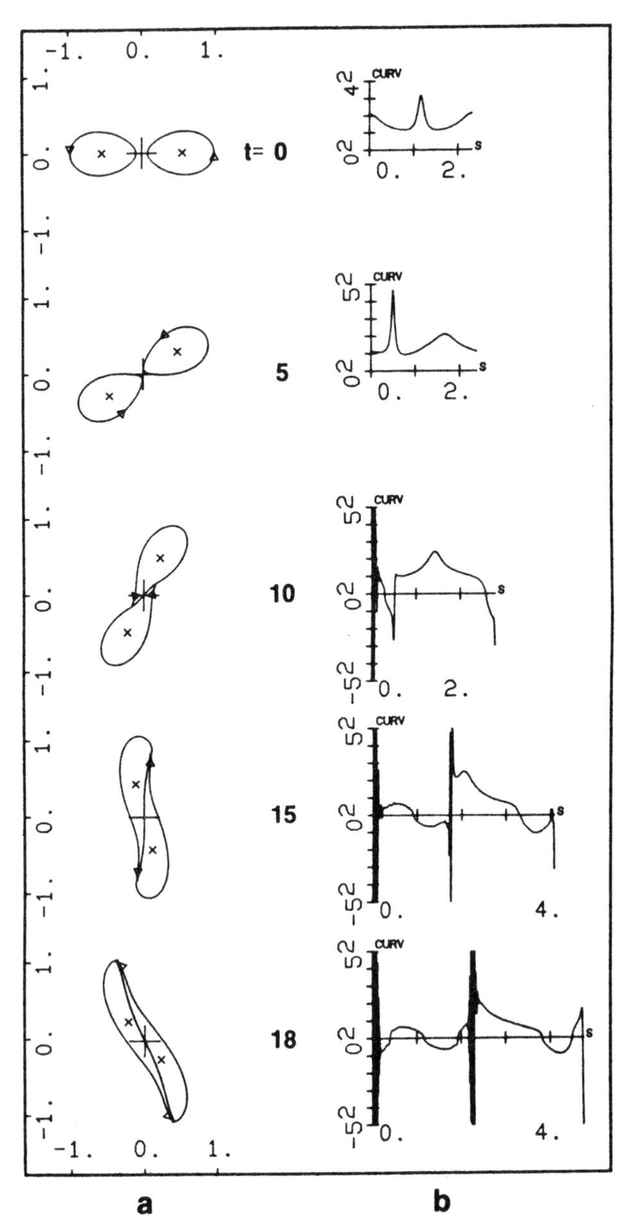

Figure 5a: Merger of inwardly perturbed corotating V-state for $0 \leq t \leq 18$. (a) Physical space. (b) Curvature vs. arc length. (Figure 7 of reference 11.)

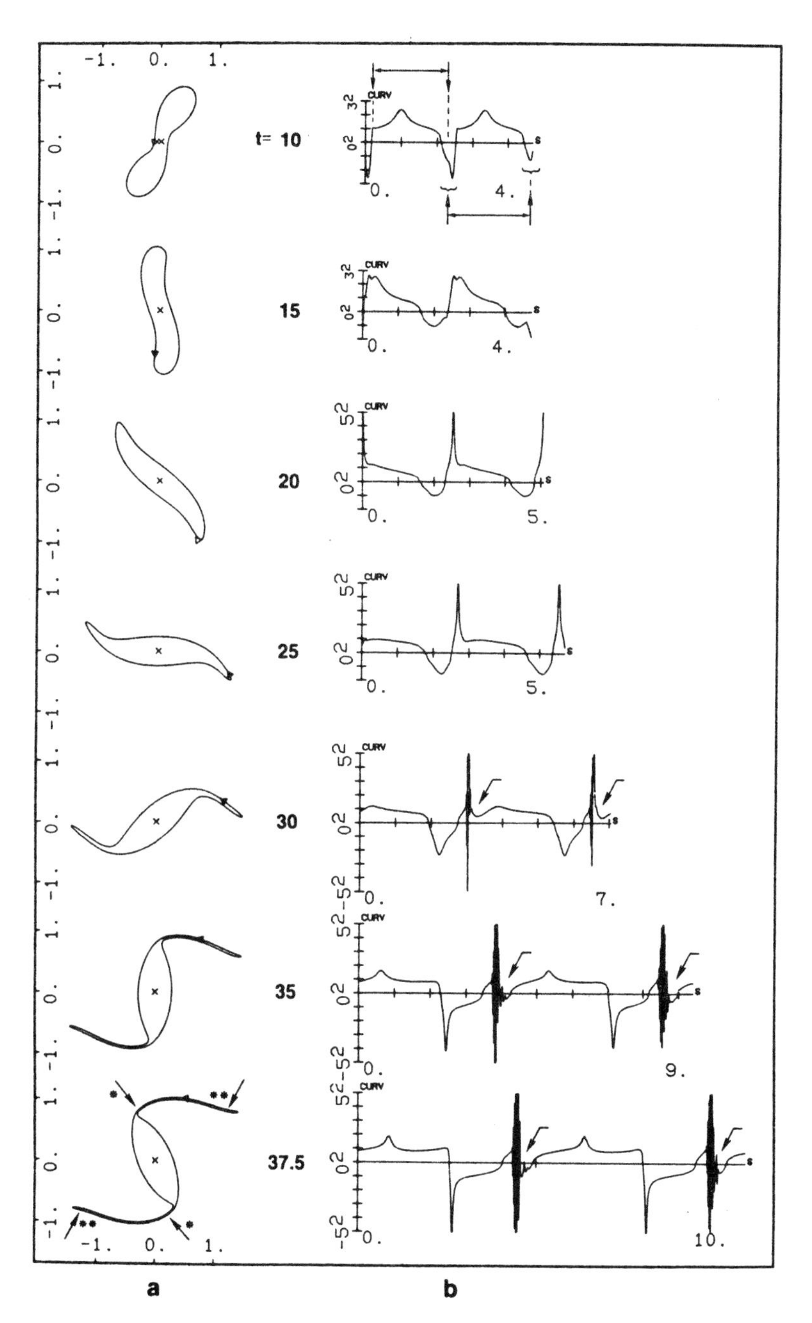

Figure 5b: Continued evolution of merged vortex regions as a singly-connected region for $10 \le t \le 37.5$. (a) Physical space. (b) Curvature vs. arc length. (Figure 9 of reference 11.)

Contemporary Mathematics
Volume 17, 1983

REACTION AND DIFFUSION IN CARBON COMBUSTION*

Neal R. Amundson

ABSTRACT. The combustion of a single particle of non-porous carbon surrounded by a stagnant boundary layer is a complex problem in reaction and diffusion. The diffusion is through the boundary layer in which there is reaction between carbon monoxide and oxygen. There is reaction at the surface and heat generated or absorbed which must be conducted through the boundary layer. This paper treats a simple model which exhibits solution multiplicity, quenching, and lack of ignition. Bounding solutions are also exhibited which may be generalized to more complicated models.

1. INTRODUCTION. In this paper we will present the simplest non-trivial model for combustion of an impervious carbon particle surrounded by a stagnant boundary layer, a problem in diffusion and reaction. Suppose we consider a particle of radius a surrounded by a boundary layer of external radius b. Outside the boundary layer the ambient atmosphere consists of oxygen in an inert, say nitrogen. The concentration of oxygen in the ambient is specified as well as the ambient temperature. Oxygen (O_2) must diffuse from the ambient through the boundary layer where when it reaches the carbon surface (C) reacts to form carbon monoxide (CO). The CO then diffuses outwards from the surface and meets O_2 with which it reacts to form carbon dioxide (CO_2). The CO_2 diffuses both towards the carbon surface and to the ambient. At the surface it reacts to form more CO. At the surface heat is generated by the O_2 reaction and absorbed by the CO_2 reaction. In the boundary layer heat is generated by the CO oxidation. Thus we have three diffusing species and three reactions

$$\text{I:}\quad C + \tfrac{1}{2}O_2 \to CO, H_1$$

$$\text{II:}\quad C + CO_2 \to 2CO, H_2$$

$$\text{III:}\quad CO + \tfrac{1}{2}O_2 \to CO_2, H_3$$

H_i stands for the heat generated or absorbed. Let g_1, g_2, g_3, g_4 stand, respectively, for the mass fractions of CO_2, CO, O_2, and the temperature. The reaction rates of the above reactions are given by

$$R_1 = k_1 e^{-\alpha_1/g_4} g_3$$

*35K55

$$R_2 = k_2 e^{-\alpha_2/g_4} g_3$$

$$R_3 = k_3 e^{-\alpha_3/g_4} g_3^{\frac{1}{2}}$$

The k_i and α_i are positive constants. The reaction-diffusion conservation equations for the boundary layer are

$$\frac{1}{r^n}\frac{\partial}{\partial r}(r^n \rho_f D_i \frac{\partial g_i}{\partial r}) + \nu_i M_i k_3 e^{-\alpha_3/g_4} g_2 g_3^{\frac{1}{2}} = \frac{\partial g_i}{\partial \theta} \ , \ a<r<b \ , \ \theta>0 \qquad (1)-(4)$$

where n = 0, 1, or 2 for a slab, cylinder, or sphere respectively. The D_i, i = 1,2,3 are molecular diffusion coefficients, $D_4 = K/c_f$, a thermal diffusivity, ρ_f the fluid density (assumed constant), M_i the molecular weights of the three species and $\nu_1 = 1$, $\nu_2 = -1$, $\nu_3 = -\frac{1}{2}$, $\nu_4 = H_3/c_f\rho_f$.

Although there is no reaction within the particle there will be heat effects at the surface necessitating consideration of the conduction equation for the solid carbon

$$\frac{1}{r^n}\frac{\partial}{\partial r}(r^n K_s \frac{\partial g_4}{\partial r}) = c_s\rho_s \frac{\partial g_4}{\partial \theta} \ , \ 0<r<a \qquad (5)$$

At the particle surface we write the boundary and compatibility conditions as,

$$\left.\begin{aligned} \rho_f D_3 \frac{\partial g_3}{\partial r} &= \tfrac{1}{2}k_1 e^{-\alpha_1/g_4} g_3 && (6)\\ -\rho_f D_2 \frac{\partial g_2}{\partial r} &= k_1 e^{-\alpha_1/g_4} g_3 + 2k_2 e^{-\alpha_2/g_4} g_1 && (7)\\ \rho_f D_1 \frac{\partial g_1}{\partial r} &= k_2 e^{-\alpha_2/g_4} g_1 && (8) \end{aligned}\right\} \text{ at } r = a+$$

$$g_4(a-,\theta) = g_4(a+,\theta) \qquad (9)$$

$$-K_s \frac{\partial g_4}{\partial r})_{a-} + K_f \frac{\partial g_4}{\partial r})_{a+} + k_1 e^{-\alpha_1/g_4} g_3 H_1 - k_2 e^{-\alpha_2/g_4} g_1 H_2 - \beta(g_4^{\ 4} - g_{4b}^{\ 4}) = 0 \ , \ r=a+ \qquad (10)$$

$$-\frac{\rho_s}{M_c}\frac{da}{d\theta} = k_1 e^{-\alpha_1/g_4} g_3 + k_2 e^{-\alpha_2/g_4} g_1 \ , \ r=a+ \qquad (11)$$

At the boundary r = a we have $g_i = g_{is}$. Equations (6), (7), (8) reflect only that the species which diffuse to the surface must react. Equation (10) reflects the generation of heat by reaction, the losses by conduction away from the boundary, and radiation from the surface to an ambient temperature g_{4b}. Equation (11) describes the loss of carbon by reaction and relates it to the changing particle size. We have uncoupled the changing value of a from the conservation equations (1)-(4), (5). Not doing this would considerably complicate the problem. We must specify the conditions at r = b and these will be taken as $g_i = g_{ib}$, constants. Initial conditions at $\theta = 0$ are also essential if complete solutions are to be generated. We assume that the problem is well posed and that solutions exist. Our aim is to establish some of the solution structure and to illustrate the solution pathology for systems substantially simpler than that written above. During the burning of a carbon particle there are many transient processes occurring and it would be desirable to sort these out since the solution of the whole system is not only difficult but probably unnecessary as well for realistic process parameters.

2. THE ISOTHERMAL PARTICLE AND LIMITING SOLUTIONS. In order to simplify the model we will assume that the particle conductivity is very high which is a good assumption at high temperatures for carbon. The particle then has a uniform temperature which varies with time. The new model consists of equations (1)-(9), (11), and a modified equation (10) becomes

$$c_s\rho_s \frac{\partial g_4}{\partial\theta} = k_1 e^{-\alpha_1/g_4} g_3 H_1 - k_2 e^{-\alpha_2/h_4} g_1 H_2 + K_f \left.\frac{\partial g_4}{\partial r}\right)_{a+} - \beta(g_4{}^4 - g_{4b}{}^4),\quad r=a+,\ g_4=g_{4s} \tag{12}$$

Here g_{4s} is the uniform particle temperature, and we have assumed that the particle temperature and size are the meaningful time dependent processes, all others having smaller time constants. In order to determine the solution structure it is instructive and interesting to consider some limiting quasi-steady state solutions. There are indeed no steady state solutions but we can freeze the particle size and assume that at that instant the system has a steady state solution. It is later a matter of checking to determine if this procedure has any validity. Suppose for simplicity we take the slab geometry for illustration and write a set of quasi-steady state equations. These will be

$$\frac{d}{dx}\left(\rho_f D_i \frac{dg_i}{dx}\right) + \nu_i M_i k_3 e^{-\alpha_3/g_4} g_2 g_3{}^{\frac{1}{2}} = 0 \qquad (13)\text{-}(16)$$

Equations (6), (7), (8), (9) are the same. Equation (10) becomes with radiation neglected

$$k_1 e^{-\alpha_1/g_4} g_3 H_1 - k_2 e^{-\alpha_2/g_4} g_1 H_2 + K_f \left.\frac{dg_4}{dx}\right)_{a+} = 0\ ,\quad r=a\ ,\quad g_4=g_{4s} \tag{17}$$

and we take at $x = b$

$$g_1 = 0\ ,\quad g_2 = 0\ ,\quad g_3 = g_{3b}\ ,\quad g_4 = g_{4b} \tag{18}$$

Now one can generate numerical solutions for equations (6)-(8) and (13)-(18) but it is instructive, and, as we shall show, very useful to consider three limiting cases. What we will show is that these three limiting solutions bound the quasi-steady state solutions.

LIMITING SOLUTION A. In this case we assume that there is <u>no</u> reaction in the boundary layer; i.e., the rate of reaction III is zero, $k_3 = 0$, so that no CO_2 is present and the only reaction at the surface is I. Under these circumstances the model is, for a slab,

$$\frac{d^2 g_i}{dx^2} = 0\ ,\quad i=1,2,3,4$$

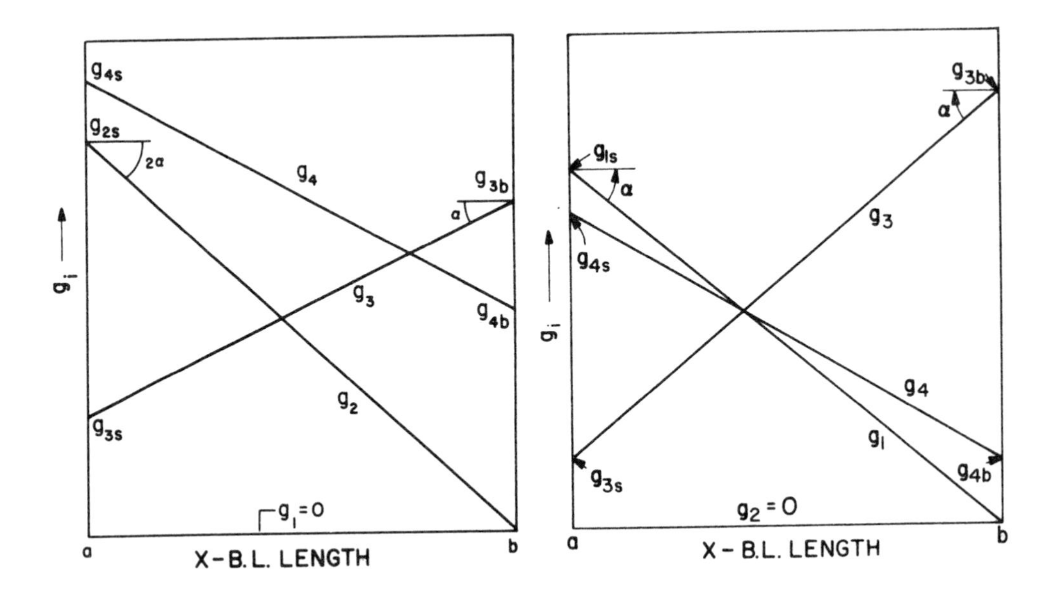

Figure 1. Limiting solution A.

Figure 2. Limiting solution B.

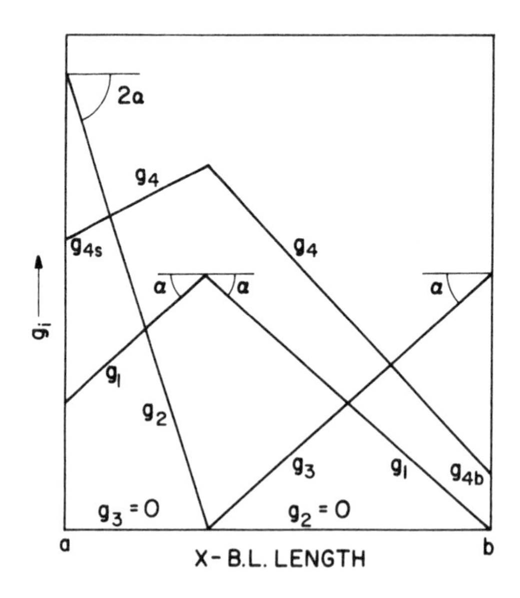

Figure 3. Limiting solution C.

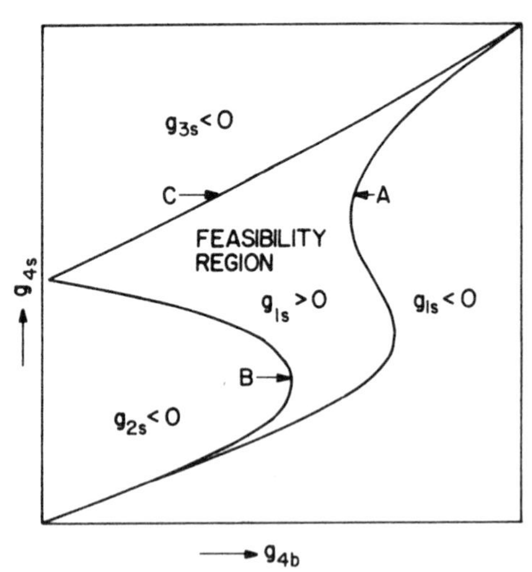

Figure 4. Loci of limiting solutions and feasibility region.

$$\left.\begin{aligned} \rho_f D_3 \frac{dg_3}{dx} &= \tfrac{1}{2}k_1 e^{-\alpha_1/g_4} g_3 \\ -\rho_f D_3 \frac{dg_2}{dx} &= k_1 e^{-\alpha_1/g_4} g_3 \\ K_f \frac{dg_4}{dx} &= -k_1 e^{-\alpha_1/g_4} g_3 H_1 \end{aligned}\right\} \;, \quad x=a \;, \quad (g_i = g_{is})$$

$$g_1 = 0 \;, \; g_2 = 0 \;, \; g_3 = g_{3b} \;, \; g_4 = g_{4b} \;, \quad x=b$$

For geometric simplicity assume that $D_2 = D_3$. The profiles of g_i in the boundary layer will be straight lines with $g_1 \equiv 0$. Shown in Figure 1 are the profiles. The slope of the g_3 curve must be half that of the g_2 curve since the flux of O_2 must be half the flux of CO. The exact computation of the profiles is a purely algebraic problem once the necessary parameters have been specified. The parameter g_{4b} is an interesting one to vary and having specified it (and g_{3b}) one can compute algebraically a locus of g_{4s} versus g_{4b} which is shown as curve A in Figure 4. For very high g_{4b} the curve becomes asymptotic to a line $g_{4s} = g_{4b} + A$ where A is the adiabatic temperature rise for reaction I, and for very low g_{4b} the curve is asymptotic to a line of slope unity. For some values of the parameters the curve may be monotonic increasing.

LIMITING SOLUTION B. In this case we assume as before that there is no reaction in the boundary layer but at the surface only CO_2 is produced by a combination of all three reactions. The model for this case is

$$\frac{d^2 g_i}{dx^2} = 0 \;, \quad i=1,2,3,4$$

$$\rho D_3 \frac{dg_3}{dx} = k_1 e^{-\alpha_1/g_4} g_3 + k_2 g_1 e^{-\alpha_2/g_4} \;, \quad x=a$$

$$-\rho D_1 \frac{dg_1}{dx} = +k_2 g_1 e^{-\alpha_2/g_4} + k_1 g_3 e^{-\alpha_1/g_4} \;, \quad x=a$$

$$K \frac{dg_4}{dx} = \tfrac{1}{2}k_1 g_3 e^{-\alpha_1/g_4} H_1 - k_2 g_1 e^{-\alpha_2/g_4} H_2 + (k_1 g_3 e^{-\alpha_1/g_4} + k_2 g_1 e^{-\alpha_2/g_4}) H_3 \;, \quad x=a$$

$$g_1 = 0 \;, \; g_2 = 0 \;, \; g_3 = g_{3b} \;, \; g_4 = g_{4b} \quad \text{at} \quad x=b$$

Again all of the profiles are straight lines so the problem is algebraic once the parameters have been specified. Figure 2 shows the schematic profiles while the locus of g_{4s} versus g_{4b} with all paramters held fixed except g_{4b} is shown as curve B in Figure 4. For low values of g_{4b} the curve becomes asymptotic to $g_{4s} = g_{4b}$ as in the previous case while for high temperatures the curve will intersect the next limiting solution.

LIMITING SOLUTION C. In this case one assumes that the CO and O_2 react instantaneously at a position ξ where the stoichiometry is exactly met, that is, where the flux of O_2 inwards is one half of the flux of CO outwards. This position then becomes a variable to be determined. The model can be written then as

$$\frac{d^2 g_i}{dx^2} = 0 \ , \ x \neq \xi \ , \ i=1,2,3,4 \ , \ a<x<b$$

$$\left.\begin{aligned} -\rho_f D_2 \frac{dg_2}{dr} &= 2k_2 e^{-\alpha_2/g_4} g_1 \\ \rho_f D_1 \frac{dg_1}{dx} &= k_2 e^{-\alpha_2/g_4} g_1 \\ K_f \frac{dg_4}{dx} - k_2 e^{-\alpha_2/g_4} g_1 H_2 &= 0 \end{aligned}\right\} , \ x=a \ , \ g_i = g_{is}$$

$$-\rho_f D_2 \left.\frac{dg_2}{dx}\right)_{\xi-} = 2\rho_f D_3 \left.\frac{dg_3}{dx}\right)_{\xi+}$$

$$g_2 = 0 \ , \ \xi<x<b$$

$$g_3 = 0 \ , \ a<x<\xi$$

$$K_f \left.\frac{dg_4}{dx}\right)_{\xi-} - K_f \left.\frac{dg_4}{dx}\right)_{\xi+} + \rho_f D_2 \left.\frac{dg_2}{dx}\right)_{\xi-} H_3 = 0$$

$$g_4(\xi-) = g_4(\xi+)$$

$$g_1 = 0 \ , \ g_2 = 0 \ , \ g_3 = g_{3b} \ , \ g_4 = g_{4b} \quad \text{at} \quad x=b$$

This time the profiles will be piecewise straight lines and the problem a purely algebraic one. We hold all parameters fixed but allow g_{4b} to be varied developing a locus of steady state solutions. A particular steady state solution is shown in Figure 3 while the locus (g_{4s}, g_{4b}) is shown as curve C in Figure 4. We can compute easily from the above the asymptote at high temperatures and this is the same as for Limiting Solution A. At low temperatures the curve will intersect the locus for Limiting Solution B.

In Figure 4 we show the loci of the three Limiting Solutions and these bound a region of the g_{4s}, g_{4b} plane. It will be shown in the next section that the solutions of the quasi-steady state problem must lie inside this region.

3. THE QUASI-STEADY STATE SOLUTIONS. Examination of equations (13)-(16) reveals that the reaction rate term may be eliminated to give three simple ordinary differential equations which may be integrated to give three algebraic equations. Let

$$\frac{\nu_i M_i}{\rho_f D_i} = \alpha_i$$

then

$$\frac{d^2 g_i}{dx^2} + \alpha_i k_3 e^{-\alpha_3/g_4} g_2 g_3^{\frac{1}{2}} = 0$$

and

$$\alpha_i \frac{d^2 g_j}{dx^2} = \alpha_j \frac{d^2 g_i}{dx^2}$$

$$\alpha_i \frac{dg_j}{dx} - \alpha_j \frac{dg_i}{dx} = A_{ij}$$

$$\alpha_i(g_j - g_{js}) - \alpha_j(g_i - g_{is}) = A_{ij}x$$

The A_{ij} may be determined from the boundary conditions at $x = a+$ and they will depend upon the values of the g_i's at $x = a+$, say g_{is}. We have three independent algebraic equations involving g_i and g_{is} for $i=1,2,3,4$. Substitution of $x = b$ in these equations relates the four g_i's evaluated in the vicinity of the surface to the four values specified in the ambient at $x = b$. We can then write the following system of equations.

$$\frac{d^2 g_1}{dx^2} + \alpha_1 k_3 e^{-\alpha_3/g_4} g_2 g_3^{\frac{1}{2}} = 0 \tag{19}$$

$$\rho_f D_1 \frac{dg_1}{dx} = k_2 e^{-\alpha_2/g_4} g_1 \;, \quad x=a \tag{20}$$

$$\alpha_i(g_j - g_{js}) - \alpha_j(g_i - g_{is}) = A_{ij}x \tag{21}$$

$$\alpha_i(g_{jb} - g_{js}) - \alpha_j(g_i - g_{ib}) = A_{ij}b \tag{22}$$

$$g_i = g_{ib} \;, \quad x=b \tag{23}$$

Such a system can be solved by a marching method by assuming a value for g_{1s} at $x = a$ and marching to $x = b$ to g_{1b}. This method is fraught with problems for a variety of reasons. The exponent one half on g_3 means that g_3 may become zero and stay there over some interval say $a < x < \eta < b$. The system is extremely stiff. In order to circumvent these difficulties to some extent we adopt the following ruse. The three algebraic equations involve g_{is} and g_{ib}. Suppose in these three equations we set $g_{1s} = 0$. Then the three equations contain only g_{2s}, g_{3s}, g_{4s} and it turns out algebraically we can solve to obtain

$$g_{1s}(g_{4s}, g_{1b}, g_{2b}, g_{3b}, g_{4b}) = 0$$

This is much simpler than anticipated and results in an explicit solution for g_{4b} in terms of g_{4s} and the other g_{ib};

$$g_{4b} = F_1(g_{4s}) \;, \quad \text{for } g_{1s} = 0$$

We can repeat this procedure setting in term $g_{2s} = 0$ and $g_{3s} = 0$ obtaining

$$g_{4b} = F_2(g_{4s}) \;, \quad g_{2s} = 0$$

$$g_{4b} = F_3(g_{4s}) \;, \quad g_{3s} = 0$$

These three functions turn out to be exactly the same as those giving the loci of the three limiting solutions A, B, C. It is trivial to show that the loci of the three functions separate regions where g_{is}, $i=1,2,3$ are positive on one side and negative on the other. Inside the region all $g_{is} > 0$ while outside

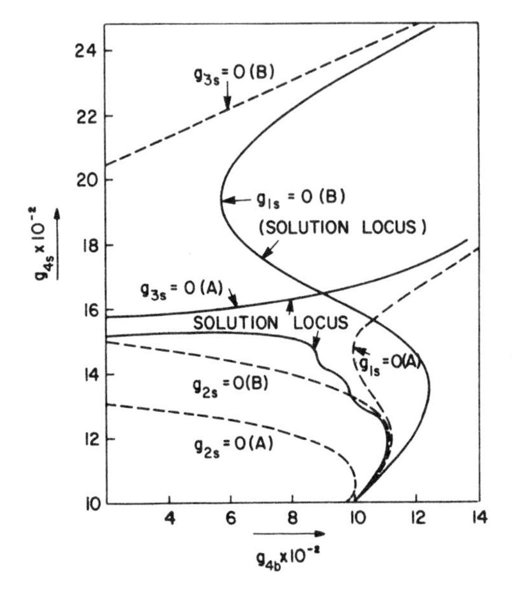

← Figure 5. Limiting solutions, feasibility region, and quasi-steady state solution loci for a large particle (500 μm) (A) and a small particle (50 μm) (B).

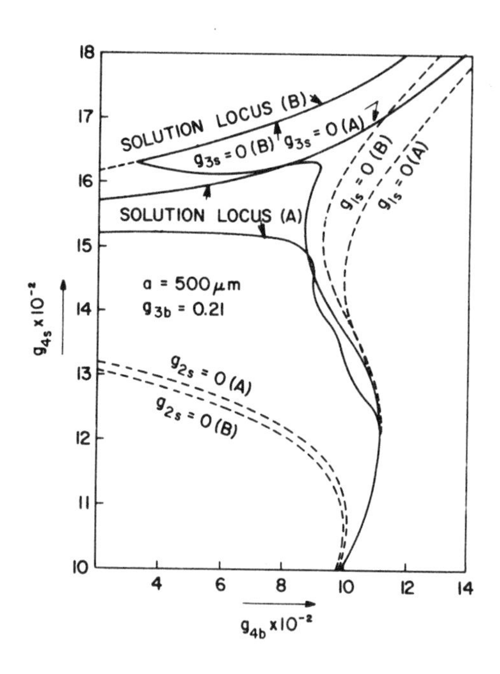

Figure 6. Limiting solutions, feasibility region, and quasi-steady state solution loci for two different boundary layer thicknesses: A, twice the particle radius and B, equal to the particle radius.

one is negative. Thus it follows that all solutions of the quasi-steady state problem must be inside this feasibility region, which is defined not only by the limiting solutions A, B, and C but by the three functions above. (Figure 4)

We are now in a position to compute the loci of solutions for it is apparent that beginning with a problem in either of the two asymptotic regions gives an easy starting solution and it is then possible to move from problem to problem in a systematic way. This sounds somewhat easier than it is but this is not the place to discuss the numerical difficulties of obtaining the whole solution locus. We stress here that we exhibit only the quasi-steady particle temperature versus the ambient temperature loci.

Let us now consider the numerical solution of the system given by equations (19)-(23). We treat the cases here where $g_{1b} = 0$, $g_4 = g_{4b}$, $g_{2b} = 0$, and $g_3 = g_{3b}$ at $r = b$ for a variety of parameters. Our aid is to show only the solution structure and some of the interesting properties. These results will all be for a spherical geometry.

In Figure 5 we show the feasibility region for two cases (A) and (B), the only difference being the particle size. For region B the particle is of radius 50 μm and we note that the locus of solutions is coincident with the limiting solution for which no $CO_2(g_1)$ is formed. Region A is for a particle of 500 μm radius and the locus meanders through its feasibility region but ultimately coincides with the limiting solution for which $g_3 = 0$ at the surface. Note that for these two cases there will be three solutions in general for each specified value of g_{4b}, a lower unignited state, an intermediate probably unstable state, and the ignited state.

Figure 6 shows the variation in feasibility regions as well as the solution loci when the boundary layer thickness is varied. Region A is for a boundary layer thickness twice that of the particle radius while region B is for a boundary layer thickness equal to that of the particle radius. We observe that for the thicker boundary layer there may be five solutions for a given ambient temperature while for the thinner one there will, in general, be three except at very low and very high temperatures.

There are other ways to present the information. For example one might compute quasi-steady state solutions for a number of particle sizes but with all other conditions held fixed. Figure 7 shows these loci for three different ambient temperatures (g_{4b}). The numbers 1 through 7 on the ordinate represent the particle temperature for a 50 μm radius particle at the ambient temperatures shown. SS 1,2,3 represent the three particle temperatures corresponding to g_{4b} = 1000 K, etc. The decreasing particle size runs from 50 μm to zero. Again we note the solution multiplicity. The structure shown in Figure 8 is much richer where we show plots of particle temperature versus particle size for a

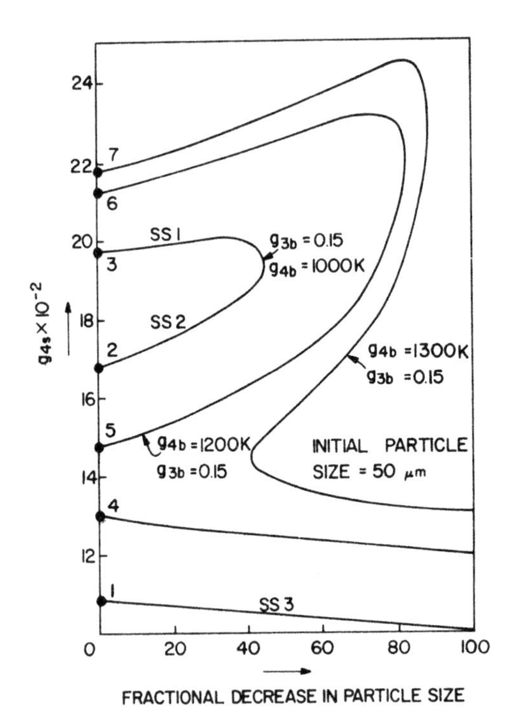

← Figure 7. Quasi-steady state solution loci for three different ambient temperatures as a function of particle size.

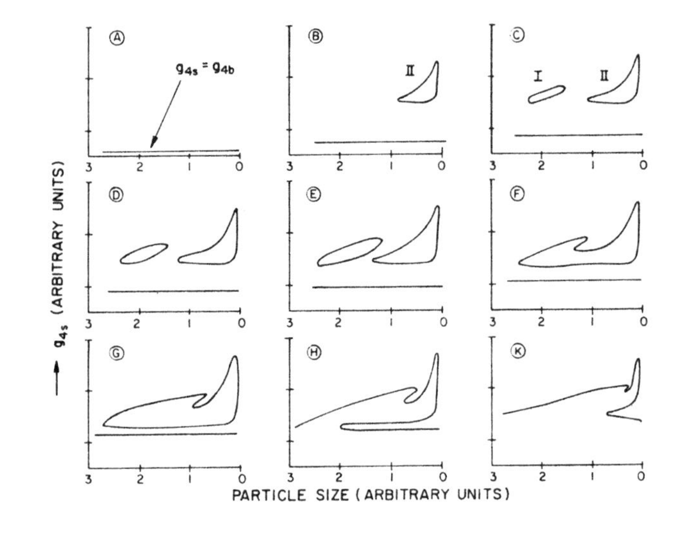

Figure 8. Quasi-steady state solution loci as a function of particle size for a sequence of ambient temperatures.

sequence of ambient temperatures. As we proceed from A to K the ambient temperature g_{4b} is gradually increased. For low ambient temperatures the particle is not ignited for any particle size. As the temperature is increased the unignited state persists and an isola appears; at higher temperatures a second isola appears and these eventually merge, disappearing into a single locus with solution multiplicity.

4. TRANSIENT SOLUTIONS. The curves presented above are all for the quasi-steady state problem described by equations (19)-(23). The simplest kind of transient to consider as a first model is that in which the changing temperature of the particle as well as its size reduction is included but with the time derivatives in the boundary layer itself set equal to zero.

Such a system is described by equations (19)-(23), (11) and (12) along with suitable initial conditions. The radiation term may be included or not. Inclusion of the radiation term of course also changes the computation of the feasibility region. The solution of this system if the quasi-steady state has any validity should follow the same trajectories whenever there is a quasi-steady state solution. In Figure 9 which corresponds to the curves marked SS 1-2-3 in Figure 8 the transient profiles are shown in solid lines and we note that if the initial condition of the particle is the same as the initial quasi-steady state solution the transient does indeed follow the quasi-steady state solution except for the intermediate state which therefore must be unstable. We note that the particle may quench if started on the upper branch or between the upper branch and the intermediate branch and will not ignite if started below the intermediate branch. We note the same sort of phenomenon occurring in Figures 10 and 11, that is, quenching and lack of ignition. The intermediate branch in these cases is not approached by any trajectory.

If a conventional stability analysis is made or if the transients described above are performed then one can compute regions of instability in the plane of g_{4s} and g_{4b} to give Figure 12. This is a plot for different ambient values of g_{3b} and shows that most of the states are unstable, there being only a high temperature ignited state and a low temperature unignited state.

While this transient model is the simplest one for such a problem, more sophistocated models for realistic physical parameters give results not substanially different than those shown above. For less realistic physical parameters the computations of course give a much richer structure, and varying the choice of initial conditions produces profiles with great pathology.

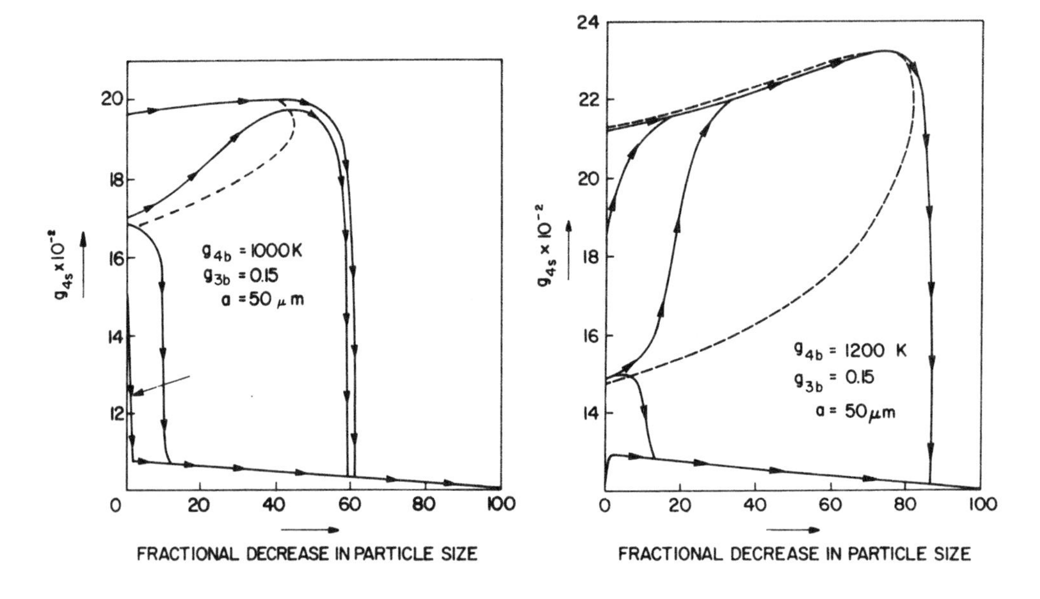

Figure 9. Transient solution for 1000 K case shown in Figure 7.

Figure 10. Transient solution for 1200 K case shown in Figure 7.

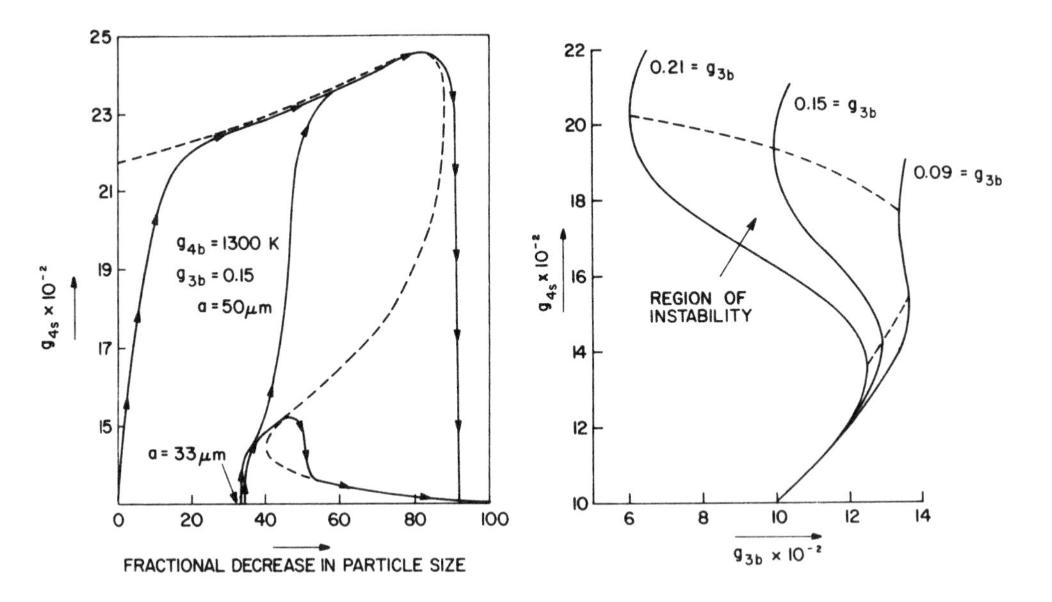

Figure 11. Transient solution for 1300 K case shown in Figure 7.

Figure 12. Instability region for quasi-steady state solutions for three different oxygen concentrations in the ambient.

BIBLIOGRAPHY

1. Hugo S. Caram and Neal R. Amundson, "Diffusion and Reaction in a Stagnant Boundary Layer", Ind. Eng. Chem. Fund., 16 (1977), 171-181.

2. Christos Georgakis, John Congalidis, and Yam Yee Lee, "Physical Interpretation of the Feasibility Region in the Combustion of Char by Use of the Single and Double Film Theories", Ind. Eng. Chem. Fund., 19 (1980), 98-103.

3. Eduardo Mon and Neal R. Amundson, "Diffusion and Reaction in a Stagnant Boundary Layer", 2. Ind. Eng. Chem. Fund., 17 (1978), 313-321; 3. Ind. Eng. Chem. Fund., 18 (1979), 162-168; 4. Ind. Eng. Chem. Fund., 19 (1980), 243-250.

4. S. Sundaresan and Neal R. Amundson, "Diffusion and Reaction in a Stagnant Boundary Layer", 5. Ind. Eng. Chem. Fund., 19 (1980), 344-351; 6. Ind. Eng. Chem. Fund., 19 (1980), 351-357; 7. Am. Inst. Chem. E. J., 27 (1981), 679-686.

DEPARTMENTS OF CHEMICAL ENGINEERING AND MATHEMATICS
UNIVERSITY OF HOUSTON
HOUSTON, TEXAS 77004

Current Address: Same as above.

Contemporary Mathematics
Volume 17, 1983

BIFURCATION FROM COLLINEAR SOLUTIONS TO A REACTION-DIFFUSION SYSTEM

David L. Barrow and Peter W. Bates

ABSTRACT. This paper extends results reported in [1] on spatially periodic solutions to a system of two reaction-diffusion equations. The system $u_t = u_{xx} + k\phi_0(\rho)u - \varepsilon\phi_1(\beta)v$, $v_t = v_{xx} + \varepsilon\phi_1(\beta)u + k\phi_0(\rho)v$, where $u = \rho \cos \beta$, $v = \rho \sin \beta$, $\phi_0(\rho) = k(1-p)$, $\phi_1(\beta)$ is smooth and 2π-periodic is shown in [1] to have, among others, families of linearly dependent ("collinear") 2π-periodic equilibrium solutions when $\varepsilon = 0$. Here we consider a larger class of equations, and then show that if ϕ_1 satisfies certain conditions, there are traveling wave solutions bifurcating from these families for small $|\varepsilon|$. We believe these solutions are interesting because computer simulations suggest they are central to understanding the qualitative properties of the flow induced by the equations.

1. INTRODUCTION.

We have recently studied a model system of reaction-diffusion equations with the objective of describing the qualitative properties of its solutions, in the belief that this model may be a prototype for reaction-diffusion systems modelling chemical and biological phenomena having "excitable" dynamics. By excitable we mean, loosely, that a small perturbation from a certain rest state will decay directly to that state, but a larger perturbation will result in a large excursion before returning to the rest state. The system $(1.1)_\varepsilon$ is a slight modification of one proposed by J. M. Lasry as a model that may have properties sufficiently similar to the Hodgkin-Huxley equations of nerve conduction that its study will lead to a better understanding of that system.

We reported in the paper [1] on the existence and stability of (spatially) 2π-periodic solutions to the system

$$(1.1)_\varepsilon \qquad \begin{aligned} u_t &= u_{xx} + k(1-\rho)u - \varepsilon\phi(\beta)v \\ v_t &= v_{xx} + k(1-\rho)v + \varepsilon\phi(\beta)u, \end{aligned}$$

where $u = \rho \cos \beta$, $v = \rho \sin \beta$, k and ε are parameters with $k > 0$, and ϕ is a smooth 2π-periodic function changing sign at 0 and $\beta_0 \in (0,2\pi)$ with $\varphi'(0) < 0$. Thus, the vector field determined by the nonlinearity has a

1980 Mathematics Subject Classification. 35K60.

radial component directed toward the circle $u^2 + v^2 = 1$, and a rotational component which changes direction and vanishes on the rays $\beta = 0$ and β_0. For this vector field, (1,0) is stable node, $(\cos \beta_0, \sin \beta_0)$ is a saddle point, (0,0) is an unstable node, and these are the only critical points.

This paper will describe an extension of the results reported in [1]. In order to set the stage for these new results, and to show the rich structure of the solution set of the system $(1.1)_\varepsilon$, we first summarize the contents of [1].

The approach taken in [1] was to first look for all steady state solutions to (1.1) when $\varepsilon = 0$, for all values of $k > 0$, and then to "continue" these solutions to small $|\varepsilon|$. Hence, we first consider solutions in $(C^2(S^1))^2$ to

$$(1.2) \qquad \begin{aligned} u_{xx} + k(1 - \rho)u &= 0 \\ v_{xx} + k(1 - \rho)v &= 0 \end{aligned} .$$

We combine several results of [1] into the following:

THEOREM A. All solutions of (1.2) satisfy $u^2 + v^2 \leq 1$. For all $k > 0$, (1.2) has the constant solutions (u_0,v_0), where $u_0^2 + v_0^2 = 0$ or 1, and these are the only solutions if $0 < k \leq 1$. If (u,v) solves (1.2), so does any rotate and translate of (u,v), i.e., so does $(\tilde{u}(x),\tilde{v}(x)) = (u(x-h)\cos \theta - v(x-h)\sin \theta,\ u(x-h)\sin \theta + v(x-h)\cos \theta)$ for any h and θ; moreover, every C^2 2π-periodic solution to (1.2) is a rotate and translate of a solution in $X = \{(u,v) \in (C^2(S^1))^2 : u \text{ is even, } v \text{ is odd}\}$. For each non-zero integer j, if $k > j^2$ there are solutions in X of the form $(u(x),v(x)) = (\rho_0 \cos jx,\ \rho_0 \sin jx)$, where $\rho_0 = 1 - j^2/k$. In addition, there is secondary bifurcation (with bifurcation parameter k) of solutions in X from each "j^{th} branch" of these solutions, at the points $k = 5j^2 - n^2$, $n = 1, 2, \ldots, 2|j| - 1$. Finally, for each positive integer j, if $k > j^2$ there are solutions in X of the form $(u(x),v(x)) = (\bar{u}(jx),0)$, where $\bar{u}$ has precisely 2 zeroes.

We will call the solutions $(\rho_0 \cos jx,\ \rho_0 \sin jx)$ and $(\bar{u}(jx),0)$ <u>circular</u> and <u>collinear</u> solutions, respectively, because of their images in the u-v plane. We conjecture that the solutions described in Theorem A represent the totality of solutions to (1.2).

We next consider the case when $\varepsilon \neq 0$. Motivated by computer simulations of $(1.1)_\varepsilon$, which indicate that the circular solutions continue to traveling wave solutions when $\varepsilon \neq 0$, we seek solutions to $(1.1)_\varepsilon$ of the form $u(x - ct), v(x - ct)$, c constant. Hence, we seek solutions $(u(z),v(z))$ of

$$u'' + cu' + k(1 - \rho)u - \varepsilon\phi(\beta)v = 0 \tag{1.3}$$
$$v'' + cv' + k(1 - \rho)v + \varepsilon\phi(\beta)u = 0.$$

We prove the following in [1]:

THEOREM B. Let j be a nonzero integer and let $k > j^2$ be fixed such that $k \neq 5j^2 - n^2$, $n = 1, 2, \ldots, 2|j| - 1$. Then there exists $\varepsilon_k > 0$ such that for $|\varepsilon| < \varepsilon_k$, $(1.1)_\varepsilon$ has a unique (up to translation by the independent variable z) 2π-periodic traveling wave solution with winding number j in a C^2 neighborhood of each circular solution to (1.2) having winding number j. The speed of the traveling wave is given by

$$c = -(\varepsilon/2\pi j)\int_{-\pi}^{\pi}\phi(x)\,dx + o(\varepsilon).$$

REMARK: Note that the sign of the wave speed c is determined by that of the winding number j, for given φ and small ε.

Because of the special form of the system $(1.1)_o$, the linearization of the right side of that equation at a circular solution is a constant coefficient second order differential operator when the polar coordinates (ρ, β) are used for the dependent variables (u,v). Hence, we were able to compute the spectrum of that operator and study the stability of the circular solutions. Moreover, we used a perturbation argument and extended the stability result to the case $\varepsilon \neq 0$, i.e., to the traveling waves of Theorem B (which includes the circular solutions as a special case). The result is (using the notation X^α for the fractional power of a Banach space X [cf. 4]):

THEOREM C. Let j be a nonzero integer, let $k > j^2$ and let $(u(x,t), v(x,t)) = \psi(x - ct)$ be a traveling wave solution of $(1.1)_\varepsilon$ as in Theorem B. Then, if $k > 5j^2 - 1$ and ε is sufficiently small, ψ is stable in the following sense: for $0 \leq \alpha < 1$ let $\|\cdot\|_\alpha$ be the norm on $((C(S^1))^2)^\alpha$ and let (u,v) be the 2π-periodic solution of $(1.1)_\varepsilon$ with initial data $\psi + w$. Then for $\|w\|_\alpha$ sufficiently small there exists a real number h such that

$$\|(u(x,t), v(x,t)) - \psi(x - ct + h)\|_\alpha \leq Ke^{-\omega t}$$

where K, ω are independent of w. If $k < 5j^2 - 1$, then for all small ε, ψ is linearly unstable.

In order to motivate the next section, the main result of this paper, we will conclude this section with a brief discussion of some numerical simulations we made. We simulated the parabolic system $(1.1)_o$ with initial data $(\bar{u}, v_0)$, a small perturbation of a collinear solution $(\bar{u},0)$. If v_0 was of one sign, the solution tended to a constant on the circle $u^2 + v^2 = 1$. On the other hand, if v_0 was such that the initial data encircled the origin once in the $u - v$ plane, then the solution tended either to a constant, as above, or to a circular solution $\rho(x) = 1 - 1/k$, the winding number being $+1$ or -1 dependending on the winding number of the initial data. This led us to raise the following question: when $\varepsilon \neq 0$, is there a (traveling wave or equilibrium) solution near a collinear solution? If so, then that solution is probably on the boundary of the domains of attraction of certain other solutions; that is, we suspect, on the basis of the simulations mentioned above, that arbitrarily small perturbations of such a solution could tend either to extinction $(u = 1,\ v = 0)$, or to propogation in either direction (i.e., the stable traveling waves of Theorem C). The next section answers the above existence question in the affirmative if ϕ satisfies certain conditions.

2. BIFURCATION FROM A CURVE OF COLLINEAR SOLUTIONS.

A version of the theorem we will prove in this section can be proved for the original system $(1.1)_\varepsilon$ but, for technical reasons related to a lack of smoothness of the right side $(1.1)_\varepsilon$, the proof is more involved than the one we will present, and the result isn't as strong - uniqueness isn't proved. Consequently, we make a slight modification and consider the system

$$(2.1)_\varepsilon \qquad \begin{aligned} u_t &= u_{xx} + k(1 - \rho)u - \varepsilon\eta(\rho)\phi(\beta)v \\ v_t &= v_{xx} + k(1 - \rho)v + \varepsilon\eta(\rho)\phi(\beta)u, \\ &u \text{ and } v \ \ 2\pi\text{-periodic in } x, \end{aligned}$$

where $\eta \geq 0$ is a C^1 function which is positive except at $\rho = 0$, where $\eta(0) = 0$.

For fixed $k > 1$ (as we will assume throughout this section) let $(\bar{u},0)$ be the collinear solution mentioned in the last section. That is, $\bar{u}$ is the unique 2π-periodic even solution to

$$(2.2) \qquad u'' + k(1 - |u|)u = 0$$

having two zeroes and satisfying $\bar{u}(0) > 0$. (For the construction of $\bar{u}$, see [1]; it may be helpful to think of $\bar{u}(x)$ as $\cos x$, since the two have the same symmetries). The rotational invariance of equilibrium solutions to $(2.1)_0$ implies that $(u,v) = (\bar{u}\cos\tau,\ \bar{u}\sin\tau)$ is a "circle" of steady state solutions to $(2.1)_0$. We will show, using [3, Theorem 1.7] on bifurcation from simple eigenvalues, that, if ϕ satisfies certain hypotheses, there are traveling wave solutions to $(2.1)_\varepsilon$ for small ε bifurcating from this circle of collinear solutions at certain values of τ.

LEMMA 2.1. For u in $C^2(S^1)$, define

$$L_1u = u'' + k(1 - 2|\bar{u}|)u$$
$$L_2u = u'' + k(1 - |\bar{u}|)u.$$

Let the eigenvalues of L_1 and L_2 be denoted by $\lambda_0 > \lambda_1 \geq \lambda_2 \geq \cdots$ and $\mu_0 > \mu_1 \geq \mu_2 \geq \ldots$, respectively. Then

(i) $\lambda_1 = 0$ and $N(L_1) = \text{span}\{\bar{u}'\}$

(ii) $\mu_2 = 0$ and $N(L_2) = \text{span}\{\bar{u}\}$,

where N denotes null space.

PROOF. Let

$$G(u,v) = \begin{pmatrix} u'' + k(1-\rho)u \\ v'' + k(1-\rho)v \end{pmatrix}$$

for (u,v) in $(C^2(S^1))^2$. An easy calculation gives the Fréchet derivative

$$G'(\bar{u},0)\begin{pmatrix} u \\ v \end{pmatrix} \equiv A\begin{pmatrix} u \\ v \end{pmatrix} = \begin{pmatrix} L_1u \\ L_2v \end{pmatrix} .$$

Now since $G(\bar{u}(\cdot+h),0) = 0$ for all h, we have

$$0 = \frac{d}{dh} G(\bar{u}(\cdot+h),0) \Big|_{h=0} = A\begin{pmatrix} \bar{u}' \\ 0 \end{pmatrix}, \text{ so}$$

$\bar{u}' \in N(L_1)$. Similarly, $G(\cos\tau\bar{u}, \sin\tau\bar{u}) = 0$ for all τ implies

$$0 = \frac{d}{d\tau} G(\bar{u}\cos\tau, \bar{u}\sin\tau) = A\begin{pmatrix} 0 \\ \bar{u} \end{pmatrix}, \quad \text{so}$$

$\bar{u} \in N(L_2)$. The following are based on standard Sturm-Liouville theory (see for example, [2, Theorem 3.1]):

a. Since $1 - |\bar{u}| > 1 - 2|\bar{u}|$, $\mu_i > \lambda_i$, $i = 0, 1, 2, \ldots$.

b. Since $\bar{u}'$ has exactly two zeroes, either $\lambda_1 = 0$ or $\lambda_2 = 0$.

c. Since $\bar{u}$ has exactly two zeroes, either $\mu_1 = 0$ or $\mu_2 = 0$.

d. For $n = 0, 1, \ldots$, $\lambda_{2n+1} < \lambda_{2n}$ and $\mu_{2n+1} < \mu_{2n}$.

Now a - c immediately imply that $\lambda_1 = \mu_2 = 0$. Hence, $\lambda_2 < \mu_2 = 0$ by a., so $N(L_1)$ is spanned by $\bar{u}'$. Finally, $\mu_3 < \mu_2 = 0 = \lambda_1 < \mu_1$ by d. and a., so $N(L_2) = \text{span}\{\bar{u}\}$, completing the proof of Lemma 2.1.

We are interested in traveling wave solutions to $(2.1)_\varepsilon$, so we let $z = x - ct$ and seek C^2, 2π-periodic solutions $(u(z), v(z))$ to

$$\begin{aligned} &u'' + cu' + k(1-\rho)u - \varepsilon\eta(\rho)\phi(\beta)v = 0 \\ (2.3)\qquad &\\ &v'' + cv' + k(1-\rho)v + \varepsilon\eta(\rho)\phi(\beta)u = 0 . \end{aligned}$$

As mentioned above, we will show bifurcation of solutions to (2.3) from the curve of collinear solutions $(u,v) = (\bar{u}\cos\tau, \bar{v}\sin\tau)$ at certain values of τ depending on ϕ. It will be convenient to first make a translation in order to study bifurcation at $\tau = 0$, i.e., from the point $(u,v) = (\bar{u},0)$, and for this purpose the following lemma will be useful:

LEMMA 2.2. If (u,v) satisfies (2.3), then $(u\cos\theta - v\sin\theta,\ u\sin\theta + v\cos\theta)$ satisfies (2.3) with $\phi(\beta)$ replaced by $\phi^{\theta}(\beta) \equiv \phi(\beta + \theta)$.

The proof is straightforward and will be omitted.

We will prove bifurcation by applying the following result [3, Theorem 1.7]:

THEOREM D. Let X, Y be Banach spaces, V a neighborhood of 0 in X and

$$F: (-1,1) \times V \to Y$$

have the properties

(a) $F(\tau,0) = 0$ for $|\tau| < 1$,

(b) The partial derivatives F_{τ}, F_{x} and $F_{\tau x}$ exist and are continuous,

(c) $N(F_x(0,0))$ and $Y/R(F_x(0,0))$ are one-dimensional.

(d) $F_{\tau x}(0,0)x_0 \notin R(F_x(0,0))$, where

$$N(F_x(0,0)) = \text{span}\{x_0\}.$$

If Z is any complement of $N(F_x(0,0))$ in X, then there is a neighborhood U of $(0,0)$ in $\mathbb{R} \times X$, an interval $(-a,a)$, and continuous functions $g: (-a,a) \to \mathbb{R}$, $h: (-a,a) \to Z$ such that $g(0) = 0$, $h(0) = 0$ and

$$F^{-1}(0) \cap U = \{(g(\alpha), \alpha x_0 + \alpha h(\alpha)) : |\alpha| < a\} \cup \{(\tau,0) : (\tau,0) \in U\}.$$

REMARK: The proof of Theorem D only uses the continuity of $F_{\tau x}(\tau,0)$ in τ, and this is what we will establish in applying it.

In the interest of clarity, we will begin the proof of the main result of this section before stating it precisely. First, suppose that for some $\theta \in [0,2\pi)$, ϕ satisfies the conditions

$$\text{(2.4)} \qquad \begin{aligned} &\text{(a)} \quad \phi(\theta) = -\phi(\theta + \pi) \\ &\text{(b)} \quad \phi'(\theta) \neq -\phi'(\theta + \pi) \end{aligned}$$

Let $W = \{(u,v,c,\varepsilon) \in (C^2(S^1))^2 \times \mathbb{R} \times \mathbb{R} : u'(0) = 0\}$. The purpose of the condition $u'(0) = 0$ is to eliminate the translation invariance of solutions. Next, for real τ let $w(\tau)$ in W be $w(\tau) = (\bar{u}\cos\tau, \bar{u}\sin\tau, 0, 0) \equiv (u(\tau), v(\tau), 0, 0)$. For (u,v,c,ε) in W let

$$G(u,v,c,\varepsilon) = \begin{pmatrix} u'' + k(1-\rho)u + cu' - \varepsilon\eta(\rho)\phi^{\theta}(\beta)v \\ v'' + k(1-\rho)v + cv' + \varepsilon\eta(\rho)\phi^{\theta}(\beta)u \end{pmatrix},$$

where θ is as in (2.4) and ϕ^{θ} is the translate of ϕ defined earlier. Hence

$$(2.5)\qquad \begin{aligned} &(a)\quad \phi^\theta(0) = -\phi^\theta(\pi)\ , \text{ and} \\ &(b)\quad (\phi^\theta)'(0) \neq -(\phi^\theta)'(\pi)\ . \end{aligned}$$

Let $X = \{(u,v,c,\varepsilon) \in W\colon v \perp \bar{u}'\}$, where $f \perp g$ means $\int_{-\pi}^{\pi} f(s)g(s)ds = 0$. For $x \in X$, define $F(\tau,x) = G(w(\tau) + x)$. Let $Y = (C(S^1))^2$. We now proceed to show that the hypotheses of Theorem D are satisfied. $F(\tau,0) = 0$ for all τ since $(u,v) = (\bar{u}\cos\tau, \bar{u}\sin\tau)$ satisfies $(2.1)_o$. A calculation gives (with ' denoting differentiation with respect to z)

$$(2.6)\qquad F_x(\tau,0) = \frac{\partial F(\tau,0)}{\partial(u,v,c,\varepsilon)} =$$

$$= \begin{bmatrix} \frac{d^2}{dz^2} + k(1-\rho(\tau)) - \frac{ku^2(\tau)}{\rho(\tau)}, & -\frac{ku(\tau)v(\tau)}{\rho(\tau)}, & u'(\tau), & -\eta(\rho(\tau))\phi^\theta(\beta(\tau))v(\tau) \\ -\frac{ku(\tau)v(\tau)}{\rho(\tau)}, & \frac{d^2}{dt^2} + k(1-\rho(\tau)) - \frac{kv^2(\tau)}{\rho(\tau)}, & v'(\tau), & \eta(\rho(\tau))\phi^\theta(\beta(\tau))u(\tau) \end{bmatrix}.$$

where $\rho(\tau)$, $\beta(\tau)$ are given by

$$(2.7)\qquad u(\tau) + iv(\tau) = \bar{u}\cos\tau + i\bar{u}\sin\tau = \bar{u}e^{i\tau} = \rho(\tau)e^{i\beta(\tau)}\ ,\ \text{i.e.,}$$

$$(2.8)\qquad \rho(\tau) = |\bar{u}| \quad\text{and}\quad \beta(\tau) = \tau + \operatorname{Arg}(\bar{u})\ .$$

In the above we have suppressed explicit reference to the fact that all functions also depend on z. Now since $u(0) = \bar{u}$ and $v(0) = 0$, we have

$$(2.9)\qquad F_x(0,0)(u,v,c,\varepsilon) = \begin{pmatrix} \tilde{L}_1 u + c\bar{u}' \\ \tilde{L}_2 v + \varepsilon\eta(|\bar{u}|)\phi^\theta(\beta(0))\bar{u} \end{pmatrix},$$

where $\tilde{L}_1 = L_1|_{u'(0) = 0}$ and $\tilde{L}_2 = L_2|_{v \perp \bar{u}}$.

PROPOSITION 2.3: $N(F_x(0,0)) = \operatorname{span}\{(0,y,0,1)\} \equiv \operatorname{span}\{x_0\}$, where $y = \tilde{L}_2^{-1}(\eta(|\bar{u}|)\phi^\theta(\beta(0))\bar{u})$.

PROOF: $\tilde{L}_1 u + c\bar{u}' = 0$ if and only if (iff) $L_1 u = -c\bar{u}'$, $u \in C^2(S^1)$ and $u'(0) = 0$. Now by the Fredholm alternative $L_1 u = -c\bar{u}'$ has a solution iff $-c\bar{u}' \perp N(L_1) = \operatorname{span}(\bar{u}')$ (by Lemma 2.1). The last condition is true if and only if $c = 0$, in which case $u = \alpha\bar{u}'$ for some α. But now the requirement $u'(0) = 0$ implies $\alpha = 0$, so $u \equiv 0$.

$\tilde{L}_2 v + \varepsilon\eta(|\bar{u}|)\phi^\theta(\beta(0))\bar{u} = 0$ iff $L_2 v = -\varepsilon\eta(|\bar{u}|)\phi^\theta(\beta(0))\bar{u}$, $v \in C^2(S^1)$ and $v \perp \bar{u}$. Again by the Fredholm alternative, the equation has a solution iff the right side is orthogonal to $N(L_2)$, i.e., to $\bar{u}$. This is true if $\varepsilon = 0$, but then the condition $v \perp \bar{u}$ would imply $v = 0$. If $\varepsilon \neq 0$, the condition for solvability is $\eta(|\bar{u}|)\phi^{\theta}(\beta(0))\bar{u} \perp \bar{u}$, which is true by (2.5,a), since $\beta(0) = \mathrm{Arg}(\bar{u}) = 0$ if $|z| < \pi/2$ and $\beta(0) = \pi$ if $\pi/2 < |z| < \pi$. The assertion of Proposition 2.3 now follows.

PROPOSITION 2.4: $R(F_x(0,0)) = \{(f,g) \in (C(S^1))^2 : g \perp \bar{u}\}$.

PROOF: Let $(f,g) \in (C(S^1))^2$. Then $\tilde{L}_1 u + c\bar{u}' = f$ iff $L_1 u = f - c\bar{u}'$, $u \in C^2(S^1)$ and $u'(0) = 0$. The equation $L_1 u = f - c\bar{u}'$ has a solution iff $(f - c\bar{u}') \perp \bar{u}'$, and this condition is true for $c = (f,\bar{u}')/(\bar{u}',\bar{u}')$ (where (f,g) denotes $\int_{-\pi}^{\pi} f(s)g(s)ds$).

Similarly, $\tilde{L}_2 v + \varepsilon\eta(|\bar{u}|)\phi^{\theta}(\beta(0))\bar{u} = g$ has a solution iff $g \perp \bar{u}$, since $\varepsilon\eta(|\bar{u}|)\phi^{\theta}(\beta(0))\bar{u} \perp \bar{u}$ by (2.5,a). This proves the proposition.

PROPOSITION 2.5: $F_{\tau x}(0,0)x_0 \notin R(F_x(0,0))$.

PROOF: We must show that the second component of $F_{\tau x}(0,0)x_0$, denoted $F^{(2)}_{\tau x}(0,0)x_0$, is not orthogonal to $\bar{u}$. Now

$$F^{(2)}_{\tau x}(0,0)x_0 = \frac{\partial}{\partial\tau} F_x^{(2)}(\tau,0)x_0\Big|_{\tau=0}, \text{ and}$$

$$F_x^{(2)}(\tau,0)x_0 = \left[\frac{d^2}{dz^2} + k(1 - \rho(\tau)) - \frac{kv^2(\tau)}{\rho(\tau)}\right] y + \eta(\rho(\tau))\phi^{\theta}(\beta(\tau))u(\tau).$$

We next differentiate with respect to τ, using (from (2.7) and (2.8)) $\frac{\partial}{\partial\tau}u(\tau) = -v(\tau)$, $\frac{\partial}{\partial\tau}v(\tau) = u(\tau)$, $\frac{\partial}{\partial\tau}\rho(\tau) = 0$, $\frac{\partial\beta(\tau)}{\partial\tau} = 1$. Hence,

$$F_{\tau x}^{(2)}(\tau,0)x_0 = -\frac{2kv(\tau)u(\tau)}{\rho(\tau)} y + \eta(\rho(\tau))(\phi^{\theta})'(\beta(\tau))u(\tau) + \eta(\rho(\tau))\phi^{\theta}(\beta(\tau))v(\tau),$$

so that

$$F^{(2)}_{\tau x}(0,0)x_0 = \eta(\rho(0))\phi^{\theta\prime}(\beta(0))\bar{u} \quad .$$

Condition (2.5,b) guarantees that this is not orthogonal to $\bar{u}$, proving Proposition 2.5.

We have now verified (a), (c) and (d) of Theorem D. The continuity of F_τ and F_x follow from the fact that the nonlinear terms in $(2.1)_\varepsilon$, considered as functions of the real variables u and v, are C^1. The continuity of $F_{\tau x}(\tau,0)$ in τ is seen by differentiating $F_x(\tau,0)$ with respect to τ. The discontinuity of $\beta(\tau)$ (in z) has been smoothed out by the function η. Hence, by Theorem D and Lemma 2.2, we have established the first part of the following:

THEOREM 2.6. Suppose that ϕ in $(2.1)_\varepsilon$ satisfies (2.4). Then there is a neighborhood U of $(\bar{u}\cos\theta, \bar{u}\sin\theta, 0, 0)$ in $X = \{(u,v,c,\varepsilon) \in (C^2(S^1))^2 \times \mathbb{R} \times \mathbb{R} : u'(0) = 0\}$, an interval $I = (-a,a)$, and continuous functions $g: I \to \mathbb{R}$, $h_1, h_2: I \to C^2(S^1)$, $h_3, h_4: I \to \mathbb{R}$ such that $g(0) = 0$, $h_i(0) = 0$, $i = 1, \ldots, 4$ and the only solutions to (2.3) (i.e., the only traveling wave solutions to $(2.1)_\varepsilon$) in U are $(u,v,c,\varepsilon) = (\bar{u}\cos\tau, \bar{u}\sin\tau, 0, 0)$ and

$$(u,v,c,\varepsilon) = (\bar{u}\cos(\theta + g(\alpha)) + \alpha(y\sin\theta + h_1(\alpha)),$$

$$\bar{u}\sin(\theta + g(\alpha)) + \alpha(y\cos\theta + h_2(\alpha)),\ \alpha h_3(\alpha),\ \alpha + \alpha h_4(\alpha)),\quad |\alpha| < a,$$

where y is defined in Proposition 2.3. In addition, if (2.4,a) doesn't hold, the only solutions in X to (2.3) near $(u,v,c,\varepsilon) = (\bar{u}\cos\theta, \bar{u}\sin\theta, 0, 0)$ must be of the form $(\bar{u}\cos\tau, \bar{u}\sin\tau, 0, 0)$.

PROOF. We have only to prove the last statement. Suppose (2.4,a) does not hold. Then, by examining the proof of Propositions 2.3 and 2.4, we see that $F_x(0,0)$ is an isomorphism, and the result follows from the implicit function theorem.

REMARK. If (u,v,c,ε) solves (2.3), then so does $(\tilde{u},\tilde{v},-c,\varepsilon)$, where $\tilde{u}(z) = u(-z)$, $\tilde{v}(z) = v(-z)$. Hence, by uniqueness, it follows that $c = 0$ in the above theorem.

BIBLIOGRAPHY

1. D. L. Barrow and P. W. Bates, "Bifurcation and Stability of Periodic Traveling Waves for a Reaction-Diffusion System", J. Diff. Eq. (to appear).

2. E. A. Coddington and N. Levinson, Theory of Ordinary Differential Equations, McGraw-Hill, New York, 1955.

3. M. G. Crandall and P. H. Rabinowitz, "Bifurcation from Simple Eigenvalues", J. Functional Analysis, 8(1971), 321-340.

4. D. Henry, Geometric Theory of Semilinear Parabolic Equations, Lecture Notes in Math., vol. 840, Springer-Verlag, New York, 1981.

DEPARTMENT OF MATHEMATICS
TEXAS A&M UNIVERSITY
COLLEGE STATION, TEXAS 77843

Contemporary Mathematics
Volume 17, 1983

TRAVELING WAVE SOLUTIONS TO REACTION DIFFUSION SYSTEMS MODELING COMBUSTION

Henri Berestycki, Basil Nicolaenko, and Bruno Scheurer

ABSTRACT. We consider the deflagration wave problem for a compressible reacting gas, with species involved in a single step chemical reaction. In the limit of small Mach numbers, the one-dimensional traveling wave problem reduces to a system of reaction-diffusion equations. Thermomechanical coefficients are temperature dependant. Existence is proved by first considering the problem in a bounded domain, and taking an infinite domain limit. In the singular limit of high activation energy in the Arrhenius exponential term, we prove strong convergence to a limiting free boundary problem (discontinuity of the derivatives on the free boundary).

0. INTRODUCTION

We consider the deflagration wave problem for a compressible reacting gas, with one reactant involved in a single step chemical reaction. In the limit of small Mach numbers, the one-dimensional traveling wave problem reduces to a system of two reaction-diffusion equations (cf. Section 1 for review of the basic flame equations). We assume both heat conductivity and diffusion coefficients temperature-dependent. For sake of clarity, we will first develop our methods on a simpler scalar case (corresponding to a Lewis number equal to one, cf. Section 1.4). The renormalized model is:

$$\begin{cases} -(k(u)u')' + cu' = g(u) \quad \text{on } \mathbb{R} \\ u(-\infty) = 0 \qquad u(+\infty) = 1 \end{cases} \tag{0.1}$$

where u is the renormalized temperature, $0 \leq u \leq 1$, $k(u)$ is a C^1 function of u, which is strictly positive; $g(u)$ is a renormalized reaction term such that $g(u) \equiv 0$ on $[0,\theta)$ and $g(u) > 0$ on $(\theta,1)$ where θ is some ignition temperature $(0 < \theta < 1)$. Moreover, $g(1) = 0$. c is the unknown mass flux of the wave.

This research supported by the Center for Nonlinear Studies, Los Alamos National Laboratory. Work also performed under the auspices of the U.S. Department of Energy under contract W-745-ENG-36 and contract KC-04-02-01, Division of Basic and Engineering Sciences.

Next, we will investigate the system

$$\begin{cases} -u'' + cu' = f(u)v & \text{on } \mathbb{R} \\ -\Lambda v'' + cv' = -f(u)v & \text{on } \mathbb{R} \\ u(-\infty) = 0 \qquad u(+\infty) = 1 \\ v(-\infty) = 1 \qquad v(+\infty) = 0 \ , \end{cases} \tag{0.2}$$

where u is again the renormalized temperature $(0 \leq u \leq 1)$, v is a renormalized reactant concentration $(0 \leq v \leq 1)$ and $f(u)$ as the same properties as $g(u)$, except that now $f(1) > 0$. In (0.2), Λ is taken constant again for simplicity.

However, our methods trivially extends to the full system:

$$\begin{cases} -(\lambda(u)u')' + c\ \kappa(u)u' = Q\ f(u)v \\ -(\mu(u)v')' + c\ v' = -f(u)v \end{cases} \tag{0.3}$$

where $\lambda(u)$ is the heat conductivity, $\mu(u)$ is the reactant diffusion coefficient and $\kappa(u)$ the variable specific heat of the mixture. Finally, Q is the heat of reaction and is a given positive constant.

The above equations (0.1), (0.2), (0.3) are nonlinear eigenvalue problems for c. Existence for the systems is proved by first considering the problem on a bounded domain. This allows the reduction of the corresponding problem to a fixed point formulation w.r.t. the triplet (u,v,c). Then the usual Leray Schauder degree gives the existence in a bounded domain. Taking an infinite domain limit completes the proof. Obtaining strictly positive lower and upper bounds for c, independent of the size of the domain, is crucial for the above.

Considerable amount of work has been performed on the above equations in the asymptotic limit of infinite activation energy in the Arrhenius reaction term (see [4] and the bibliography there). Equivalently, f in (0.2) is now allowed to depend on ε (proportional to the reciprocal of the activation energy); $f_\varepsilon(u)v$, in (0.2), formally behaves as $\delta(u-1)$ when $\varepsilon \to 0$, where δ is the Dirac function centered at zero. Such formal asymptotic limits have not been rigorously established from a mathematical point of view. In this paper, we prove strong convergence of the traveling waves to singular limit free boundary problems, with discontinuous derivatives. The plan of the paper is the following:

1. The basic equations of the premixed 1-dimensional laminar flame.

2. The scalar equation case.

3. The case of the system of equations.

This communication is a shorter version of a full paper to appear [2] with several proofs shortened or deleted. Our methods are also generic to the case of two reactants (that is, a system of 3 equations) and multistep branching reactions. In the latter case, detailed asymptotic analysis has been done by P. C. Fife and B. Nicolaenko [7,8].

§1. THE BASIC EQUATIONS OF THE PREMIXED ONE-DIMENSIONAL LAMINAR FLAME

1.1. REACTIVE FLOW EQUATIONS IN ONE-DIMENSION.

We summarize the equations of a chemically reacting mixture; this is essentially a compressible, heat-conducting viscous fluid with the added complexity of species diffusion and source terms representing the chemical reaction. Let ρ be the total mass density, T the temperature, p the hydrostatic pressure and $\underline{v}$ the mass-average velocity. The reacting mixture is considered to be made of N fluids whose separate densities are ρY_i $(i = 1,2,\ldots,N)$; here the Y_i are mass fractions of species i, with molecular mass m_i:

$$\sum_{i=1}^{N} Y_i = 1 \quad . \tag{1.1}$$

In what follows, we will investigate in detail the case of a single reactant A which yields a global product P, in a one-dimensional geometry:

$$A \rightarrow P \quad . \tag{1.2}$$

The total mass density satisfies the equation

$$\frac{D\rho}{Dt} + \rho \frac{\partial v}{\partial x} = 0 \quad , \tag{1.3}$$

where $D/Dt \equiv \partial/\partial t + v\frac{\partial}{\partial x}$ is the convective derivative.

The balance law for momentum is the same as for nonreactive flows, that is:

$$\rho \frac{Dv}{Dt} + \frac{\partial p}{\partial x} = \frac{4}{3} \frac{\partial}{\partial x} (\kappa\frac{\partial v}{\partial x}) \quad , \tag{1.4}$$

where κ is the dynamic viscosity coefficient and external forces on the mixture have been neglected. Energy conservation is expressed by:

$$\rho\, c_p \frac{DT}{Dt} - \frac{\partial}{\partial x}\left(\lambda \frac{\partial T}{\partial x}\right) = Q\,\omega + \frac{Dp}{Dt} + \frac{4}{3}\kappa\left(\frac{\partial v}{\partial x}\right)^2 , \tag{1.5}$$

where $c_p(T)$ is the specific heat at constant pressure, $\lambda(T)$ the coefficient of thermal conductivity, Q the chemical heat release of the reaction (1.2); ω measures the rate at which reaction (1.2) is proceeding:

$$\omega = B(T)\,\frac{\rho Y}{m}\exp\left(-\frac{E}{RT}\right) , \tag{1.6}$$

where E is the activation energy of the reaction (a constant); in some sense, E/R is the temperature below which the reaction is negligible; R is the perfect gas constant, and B(T) has a weak dependence on T. (1.6) encompasses both the law of mass action and Arrhenius kinetics [4].

The continuity equation for the mass fraction Y of reactant A with molecular mass m is:

$$\rho \frac{DY}{Dt} - \frac{\partial}{\partial x}\left(\mu \frac{\partial Y}{\partial x}\right) = - m\,\omega , \tag{1.7}$$

where $\mu(T,Y)$ is the diffusion coefficient. From (1.1), the mass concentration of the combustion product P is 1-Y, which enables its elimination. Finally, the equation of state for a perfect gas yields a supplementary constitutive law:

$$p = R\,\rho\,T . \tag{1.8}$$

1.2. <u>APPROXIMATION OF COMBUSTION</u>.

A flame is a low-speed wave whose Mach number M_o is $\ll 1$. As an immediate consequence the variations in pressure are also small, i.e.:

$$p = p_c + \delta p , \quad \delta p = O(M_o^2) ,$$

so that we may set

$$p = p_c = \text{Const}$$

everywhere except in the momentum equation (1.4). The momentum equation now only controls small "flow-induced" variations δp in pressure [4] and uncouples from the remaining equations. Similarly, in the energy equation (1.5), one can neglect $\frac{Dp}{Dt} \equiv \frac{D}{Dt}\,\delta p$ and $(\frac{\partial v}{\partial x})^2 = O(M_o^2)$. Finally, in the combustion approximation, the Eqs. (1.3)-(1.7) reduce to:

$$\begin{cases} \frac{D\rho}{Dt} + \rho\,\frac{\partial v}{\partial x} = 0 \quad , \\ \rho\, c_p\,\frac{DT}{Dt} - \frac{\partial}{\partial x}(\lambda\,\frac{\partial T}{\partial x}) = +Q\omega \quad , \\ \rho\,\frac{DY}{DT} - \frac{\partial}{\partial x}\,(\mu\,\frac{\partial}{\partial x}Y) = -m\omega \quad , \end{cases} \tag{1.9}$$

together with the equation of state (1.8). Note that the combustion approximation does not imply a constant density approximation in one dimension.

1.3. THE FLAME FRONT EQUATIONS FOR A SINGLE REACTANT.

To study a one-dimensional flame moving with constant velocity $V_o > 0$ to the left, it is appropriate to write (1.9) in the frame of reference of an observer moving at this same speed. Let $\xi = x + V_o t$ be the observer's space variable, and prime denote the differentiation with respect to ξ, then (1.9) becomes:

$$V_o \rho' + (\rho\, v)' = 0 \quad , \tag{1.10a}$$

$$c_p(\rho\, V_o + \rho\, v)T' - (\lambda T')' = Q\omega \quad , \tag{1.10b}$$

$$(\rho\, V_o + \rho\, v)Y' - (\mu\, Y')' = -m\omega \quad ; \tag{1.10c}$$

(1.10a) integrates to yield:

$$\rho(V_o + v) = \text{Const} \equiv c \quad , \tag{1.11a}$$

where c is the mass flux; then the energy and species equations uncouple from the rest of the system:

$$c\, c_p T' - (\lambda T')' = +Q\omega \tag{1.11b}$$

$$c\, Y' - (\mu\, Y')' = -m\omega \quad ; \tag{1.11c}$$

the following boundary conditions apply to (1.11): at $\xi = -\infty$, the mixture is cold and unburned:

$$\begin{aligned} T &= T(-\infty) \\ Y &= Y(-\infty) \quad ; \end{aligned} \tag{1.12a}$$

at $\xi = +\infty$ the reactant is burned out and

$$Y(+\infty) = 0 \quad ; \tag{1.12b}$$

$T(+\infty)$ is determined through Rankine-Hugoniot-like conditions obtained by integrating (1.11) from $-\infty$ to $+\infty$ and injecting (1.12a-b):

$$T(+\infty) = T(-\infty) + \frac{Q}{c_p m} Y(-\infty) \quad , \tag{1.12c}$$

where we have assumed c_p independent from T.

In Section 3 of the paper, we shall consider a renormalized version of (1.11-1.12):

$$\begin{cases} c\, u' - u'' = v\, f(u) \\ c\, v' - \Lambda\, v'' = -\, v\, f(u) \end{cases} \tag{1.13}$$

where u is the renormalized temperature, v the renormalized concentration of reactant, and f(u) the renormalized formulation of ω/Y. For simplification, λ, μ and c_p are assumed constant. The renormalized boundary conditions are:

$$u(-\infty) = 0 \quad , \quad v(-\infty) = 1 \quad ,$$

$$u(+\infty) = 1 \quad , \quad v(+\infty) = 0 \quad .$$

1.4. THE CASE OF LEWIS NUMBER EQUAL TO 1.

Consider again the (unnormalized) single reactant system (1.11). If the Lewis number

$$Le = \frac{\lambda}{\mu c_p} \tag{1.14}$$

is such that

$$Le \equiv 1 \quad , \forall\, T \quad ,$$

we can eliminate the concentration Y, through the use of the "Shvab-Zeldovich variable" [4]:

$$Z = Q\frac{Y}{m} + c_p T \quad ; \tag{1.15}$$

Z satisfies a purely convective diffusive equation

$$c Z' - (\mu Z')' = 0 \quad ,$$

whose only bounded solution is

$$Z = \text{Constant} = Z(+\infty) \quad ; \tag{1.17a}$$

since the reactant Y is depleted at $+\infty$, $Y(+\infty) = 0$ and

$$Z \equiv c_p T(+\infty) \quad , \tag{1.17b}$$

where $T(+\infty)$ is the adiabatic flame temperature:

$$T(+\infty) \equiv \frac{Q}{c_p}\frac{Y(-\infty)}{m} + T(-\infty) \quad . \tag{1.18}$$

Injecting the identity

$$Y \equiv \frac{mc_p}{Q}(T(+\infty) - T) \tag{1.19}$$

into equation (1.11b) we obtain the reduced scalar equation:

$$c T' - (\frac{\lambda}{c_p} T')' = \rho B(T)(T(+\infty) - T) \exp(-\frac{E}{RT}) \quad . \tag{1.20}$$

In Section 2, we will investigate the normalized version of (1.20)

$$c u' - (k(u)u')' = g(u) \quad , \tag{1.21}$$

where

$$g(u) \equiv (1-u) f(u) \quad ,$$

f(u) defined in (1.13).

1.5. CONSIDERATIONS ON THE IGNITION TEMPERATURE.

Consider the flame equations (1.11). If we fix boundary conditions at $\xi = -\infty$, i.e., $T(-\infty)$ and $Y(-\infty)$, then the problem is improperly posed, since

$$\exp\left(\frac{-E}{RT(-\infty)}\right) \neq 0$$

and there exist no bounded solutions to (1.11). The origin of the difficulty is clear: this formulation requires the mixture to react all the way in from $\xi = -\infty$, so that by the time finite ξ is reached, the combustion would be complete [4]; this is the "cold-boundary" difficulty, on which a considerable amount of ingenuity has been spent [9,11]. To resolve it, we classically modify the reaction term ω through the introduction of an <u>ignition temperature</u> θ^c, such that $\omega \equiv 0 \ \forall \ T < \theta^c$; this is equivalent to replacing ω by $H(T - \theta^c)\,\omega$, where H is the Heaviside function. It has been proven [9] that if one takes a sequence θ_i^c such that $\theta_i^c \to T(-\infty)$, c_i converges to some limit c_∞ from below. However, this lack of universal significance for c disappears as the activation energy E becomes large; the high activation energy asymptotics developed in Section 2 completely circumvent this difficulty by yielding limiting formulas for c independent of θ^c [4].

§2. THE SCALAR EQUATION CASE

2.1. EXISTENCE AND UNIQUENESS RESULT.

The goal of this section is to study the problem:

Find a function $u:\mathbb{R} \to (0,1)$ and $c \in \mathbb{R}$ satisfying

$$\begin{cases} -(k(u)u')' + cu' = g(u) \text{ on } \mathbb{R} \\ u(-\infty) = 0 \quad u(+\infty) = 1 \ . \end{cases} \tag{2.1}$$

It has been shown, in Section 1.4, that (2.1) corresponds to a reduction to a scalar equation if the Lewis number is one. The more general situation where $k(u) \not\equiv 1$ is investigated here as a model for the more complex systems of the following section. In (2.1), c is an unknown of the problem (representing the mass flux), so we are looking for a nonlinear eigenvalue problem. Many earlier works have assessed the question of existence for problems of the type (2.1). A good survey of the question with a discussion of the literature can be found in the notes of P. C. Fife [5,6]. We should mention, specifically devoted to combustion, the works of Ya. I. Kanel' [10], W. Johnson and W. Nachbar [9]. In the context of population genetics, the paper of D. G. Aronson and H. F. Weinberger [1] is also relevant. All these papers use a "phase plane" approach to solve (2.1). Here, using a more analytical approach, namely a "shooting method", we solve (2.1) with more general hypotheses than the former works. Precisely, the function g will satisfy the following conditions:

$$g : [0,1] \to \mathbb{R}^+ \quad (\text{i.e., } g \geq 0) \quad . \tag{2.2}$$

There is some θ, (ignition temeprature) $0 < \theta < 1$, for which

$$g \equiv 0 \text{ on } [0,\theta) \text{ and } g > 0 \text{ on } (\theta,1). \text{ Moreover } g(1) = 0. \tag{2.3a}$$

$$g \text{ is Lipschitz continuous on } [\theta,1]. \tag{2.3b}$$

The nonlinear coefficient k will satisfy:

$$k : [0,1] \to \mathbb{R} \text{ is a } C^1 \text{ function and } k(s) \geq \alpha > 0,\ 0 \leq s \leq 1. \tag{2.4}$$

The smoothness assumption on k is the minimal one in order to work with classical solutions. Indeed, u will be of class C^2 on $\mathbb{R}$, excepting the points where $u = \theta$. We are now ready to state the

<u>*THEOREM 2.1*</u>. *Assuming the condition (2.2) to (2.4), there exists* $u:\mathbb{R} \to [0,1]$ *and* $c > 0$ *solution of (2.1). The function u is of class* C^1 *and strictly monotone increasing on* $\mathbb{R}$, *of class* C^2 *on* $\mathbb{R}-\{x_o\}$ *with* $u(x_o) = \theta$. *Furthermore, u and c are unique (up to a translation of the origin).*

<u>REMARK 2.2</u>. It is also possible to precise the (exponential) rate of decay of $u(x)$ as x goes to infinity. □

The existence part of this theorem will not be proved here (see [2]). We just notice that a shooting argument is possible due to the following:

<u>LEMMA 2.3</u>. Suppose (u,c) is a solution of (2.1). Then, $c > 0$, $0 < u < 1$, $u' > 0$ and $\lim_{x\to\pm\infty} u'(x) = 0$.

Indeed, to solve (2.1) is equivalent to solving a problem on $\mathbb{R}^+$:

$$\begin{cases} -(k(u)u')' + cu' = g(u) \text{ on } \mathbb{R}^+ \\ u(o) = \theta \quad u'(0) = c\theta\, k(\theta)^{-1} \\ u(+\infty) = 1 \quad . \end{cases} \tag{2.5}$$

2.2. <u>PROOF OF UNIQUENESS</u>.

It suffices to prove the uniqueness for the problem (2.5), equivalent to (2.1) (Lemma 2.3). We proceed by contradiction, introducing a "hodograph variable." Precisely, u (solution of (2.5)) being strictly increasing, we define a function $x:[\theta,1) \to \mathbb{R}^+$ by

$$x(s) = u^{-1}(s) \quad . \tag{2.6}$$

Clearly, this function is continuous, strictly increasing and $x(\theta) = 0$, $x(1) = +\infty$. Then, set

$$z(s) = \frac{d}{ds}\, x(s) \quad .$$

A simple calculation shows that the problem (2.5) becomes (now $\frac{d}{ds} = {}'$):

$$\begin{cases} -\left(\dfrac{k(s)}{z(s)}\right)' + c = g(s)\, z(s) \quad , \ \theta \leq s \leq 1 \\ \\ z(\theta) = \dfrac{k(\theta)}{c\theta} \qquad z(1) = +\infty \end{cases} \tag{2.7}$$

noting that $z(s) > 0 \quad \theta \leq s < 1$.

So, suppose (u_1,c_1) and (u_2,c_2) are two solutions of (2.5) with $c_1 < c_2$. The corresponding functions necessarily satisfy $z_1 > z_2$ on $[\theta,1]$. If not, for some σ, $\theta < \sigma < 1$, we have:

$$z_1(s) > z_2(s) \quad , \forall s \in [\theta,\sigma) \text{ and } z_1(\sigma) = z_2(\sigma) \quad . \tag{2.8}$$

Hence, using the equation (2.7), we obtain:

$$-\left(\frac{k(\sigma)}{z_1(\sigma)} - \frac{k(\sigma)}{z_2(\sigma)}\right) = c_2 - c_1 > 0 \quad .$$

Thus letting $w(s) = \dfrac{k(s)}{z_1(s)} - \dfrac{k(s)}{z_2(s)}$, we get $w'(\sigma) < 0$ which contradicts $w(s) < 0$, $\forall s \in [\theta,\sigma)$ and $w(\sigma) = 0$ consequence of (2.8). Now the integration of (2.7) over $[\theta,1)$ for z_1 and z_2 yields:

$$c_1 - c_2 = \int_\theta^1 g(s)(z_1(s) - z_2(s))\, ds \tag{2.9}$$

But $c_1 < c_2$, $z_1 > z_2$ on $[0,1]$, and $g(s) > 0$ on $(\theta,1)$ by (2.3). Thus (2.9) is impossible and $c_1 = c_2$; $u_1 = u_2$ follows from the uniqueness of the initial value problem associated to (2.5) ($\theta \leq u(x) < 1$ on $\mathbb{R}^+$). The proof is complete.

2.3. ASYMPTOTIC ANALYSIS.

Now we assume that g depends on a parameter $\varepsilon > 0$ (essentially the inverse of the activation energy) and we let $g = g_\varepsilon$. The corresponding solutions of (2.1) will be denoted by $(u_\varepsilon, c_\varepsilon)$. We also set:

$$K_\varepsilon(u) = \int_o^u k(s)\, g_\varepsilon(s)\, ds \tag{2.10}$$

The function g_ε depends on ε in the following way:

For each $\varepsilon > 0$, g_ε verifies conditions (2.2) - (2.4) with a fixed θ, $0 < \theta < 1$. (2.11)

$\exists\, \theta_\varepsilon$, $\theta \leq \theta_\varepsilon \leq 1$ such that $\theta_\varepsilon \nearrow 1$ when $\varepsilon \searrow 0$ and

$$\lim_{\varepsilon \searrow 0} \max_{\theta \leq s \leq \theta_\varepsilon} g_\varepsilon(s) = 0 \tag{2.12}$$

$$\lim_{\varepsilon \searrow 0} K_\varepsilon(1) \equiv m > 0 \text{ and } m < +\infty \tag{2.13}$$

These hypotheses include, in particular, the Arrhenius Law described in the first section. We state:

THEOREM 2.4. Assuming the conditions (2.11)-(2.13), the unique solution $(u_\varepsilon, c_\varepsilon)$ to (2.1) such that $u_\varepsilon(0) = \theta$, when $\varepsilon \searrow 0$, converges strongly in $H^1(\mathbb{R}) \times \mathbb{R}$ and $C^0(\mathbb{R}) \times \mathbb{R}$ to (u_o, c_o). Moreover, $c_o = \sqrt{2m}$ and u_o is the unique (continuous) solution of the problem:

$$\begin{gathered} -(k(u)u')' + c_o u' = c_o\, \delta_{x=\bar{x}} \\ u(-\infty) = 0 \qquad u(+\infty) = 1 \qquad u(0) = \theta\ , \end{gathered} \tag{2.14}$$

where $\delta_{x=\bar{x}}$ is the Dirac measure at the point $\bar{x}$ and $\bar{x}$ is uniquely determined by the condition $u(o) = \theta$.

REMARK 2.5. The theorem and its proof give a rigorous justification for the heuristic "internal layer analysis" by which this type of asymptotic limit was usually obtained. We now sketch the proof of the theorem which consists of five parts.

PROOF:

1) A priori estimates.

Multiplying the equation in (2.1) successively by 1, u_ε, $k(u_\varepsilon)u'_\varepsilon$ and integrating from 0 to x we obtain:

$$-k(u_\varepsilon(x))\, u'_\varepsilon(x) + c_\varepsilon u_\varepsilon(x) = \int_0^x g_\varepsilon(u_\varepsilon)\, dx \tag{2.15}$$

$$-k(u_\varepsilon(x))\, u_\varepsilon(x)\, u'_\varepsilon(x) + \tfrac{1}{2}c_\varepsilon(u_\varepsilon(x)^2 + \theta^2) + \int_0^x k(u_\varepsilon)\, u'^2_\varepsilon dx = \int_0^x g_\varepsilon(u_\varepsilon)\, u_\varepsilon\, dx \tag{2.16}$$

$$-\tfrac{1}{2}k(u_\varepsilon(x))^2\, u'_\varepsilon(x)^2 + \tfrac{1}{2}c_\varepsilon^2\, \theta^2 + c_\varepsilon \int_0^x k(u_\varepsilon)\, u'^2_\varepsilon\, dx = K_\varepsilon(u_\varepsilon(x)) \quad . \tag{2.17}$$

Moreover, since $0 < u_\varepsilon < 1$, $u_\varepsilon(+\infty) = 1$ and $g_\varepsilon(u_\varepsilon) \geq 0$, letting $x \to +\infty$ in these identities, we deduce easily upper and lower bounds for c_ε, independently of ε (hypothesis 2.13):

$$\sqrt{2K_\varepsilon(1)} \leq c_\varepsilon \leq \frac{1}{\theta}\sqrt{2K_\varepsilon(1)} \quad . \tag{2.18}$$

2) Let us define $x_\varepsilon > 0$ by $u_\varepsilon(x_\varepsilon) = \theta_\varepsilon$, for each $\varepsilon > 0$. We claim that:

$$x_\varepsilon < \frac{1-\theta}{\alpha} \equiv x_o \quad , \alpha > 0 \quad . \tag{2.19}$$

Since

$$\int_0^{x_\varepsilon} u'_\varepsilon\, dx = u_\varepsilon(x_\varepsilon) - u_\varepsilon(0) \leq 1-\theta \quad ,$$

it is sufficient to prove $u'_\varepsilon(x) \geq \alpha > 0$ on $[0,x_\varepsilon]$. From (2.17), since K_ε is nondecreasing, we get, if $\beta \equiv \underset{0\leq s\leq 1}{\text{Max}}\ k(s)$:

$$-\frac{\beta^2}{2}\ u'_\varepsilon(x)^2 + \tfrac{1}{2}\ c_\varepsilon^2\theta^2 \leq K_\varepsilon(\theta_\varepsilon) \quad , \ 0 < x \leq x_\varepsilon \quad .$$

So, with (2.18), we obtain

$$\frac{\beta^2}{2}\ u'_\varepsilon(x)^2 \geq K_\varepsilon(1)\ \theta^2 - K_\varepsilon(\theta_\varepsilon)$$

and the lower bound follows, for ε small enough, from (2.12) and (2.13).

3) $\lim\limits_{\varepsilon\searrow 0} c_\varepsilon = \sqrt{2m} \equiv c_o$.

From (2.12) and (2.19) we get, as $\varepsilon \to 0$:

$$0 < \int_o^{x_\varepsilon} g_\varepsilon(u_\varepsilon)\ dx \leq \frac{1-\theta}{\alpha}\ \max_{\theta\leq s\leq\theta_\varepsilon} g_\varepsilon(s) \to 0 \quad .$$

Then (2.15), written at $x = x_\varepsilon$, implies:

$$\lim_{\varepsilon\searrow 0} \{k(\theta_\varepsilon)\ u'_\varepsilon(x_\varepsilon) - c_\varepsilon\ \theta_\varepsilon\} = 0 \qquad (2.20)$$

But, multiplying (2.1) by $k(u_\varepsilon)u'_\varepsilon$ then integrating from x_ε to $+\infty$, shows that:

$$\tfrac{1}{2}k(\theta_\varepsilon)^2\ u'_\varepsilon(x_\varepsilon)^2 \leq K_\varepsilon(1) - K_\varepsilon(\theta_\varepsilon) \quad . \qquad (2.21)$$

Combining (2.20) and (2.21) yields

$$\overline{\lim_{\varepsilon\downarrow 0}}\ c_\varepsilon \leq \sqrt{2m}$$

and (2.18) yields

$$\underline{\lim_{\varepsilon\downarrow 0}}\ c_\varepsilon \geq \sqrt{2m} \quad .$$

Therefore,

$$\lim_{\varepsilon \downarrow 0} c_\varepsilon = \sqrt{2m} \equiv c_o \quad .$$

4) On $\mathbb{R}^-$, due to (2.3), the convergence of u_ε to u_o solution of the problem:

$$-(k(u_o)\, u_o')' + c_o u_o' = 0 \qquad \text{on } \mathbb{R}^-$$

$$u_o(0) = \theta \qquad u_o'(0) = \frac{c_o\theta}{k(\theta)}$$

is straightforward. Moreover, on $[x_\varepsilon, +\infty)$, $\theta_\varepsilon \leq u_\varepsilon(x) < 1$; so (2.19) shows that $u_\varepsilon \to u_o$ in the C^o norm on $[x_o,+\infty)$, where $u_o(x) \equiv 1 \ \forall\, x \geq x_o$.

5) It remains to study the limit of u_ε on the bounded interval $[0,x_o]$. Letting $x \to +\infty$ in (2.17) shows that $\int_0^{+\infty} u_\varepsilon'(x)^2\, dx$ and thus the $H^1((0,x_o))$ - norm of u_ε remains bounded independently of ε. Using (2.19) and the compact embedding $H^1((0,x_o)) \hookrightarrow C^o([0,x_o])$ there exists a subsequence $\varepsilon_n \downarrow 0$ such that

$$\begin{cases} x_{\varepsilon_n} \to \bar{x} \leq x_o \\ u_{\varepsilon_n} \to u_o \text{ in } C^o([0,x_o]) \quad . \end{cases} \tag{2.22}$$

Since $\theta_\varepsilon \uparrow 1$, we know that $u_o(x) = 1$, $\forall\, x \geq \bar{x}$. But, by (2.12), $g_{\varepsilon_n}(u_{\varepsilon_n}(x))$ converges uniformly to 0 on any interval $[0,\bar{x} - \delta]$ with $\delta > 0$. Therefore, the equation shows that u_o is of class C^2 on $\mathbb{R} -(\{\bar{x}\})$ and verifies:

$$\begin{cases} -(k(u_o)u_o')' + c_o\, u_o' = 0 \text{ on } (-\infty,\bar{x}] \\ u_o(0) = \theta \qquad u'(0) = \dfrac{c_o\theta}{k(\theta)} \end{cases} \tag{2.23}$$

One checks easily, that $\bar{x}$ is uniquely determined by $u_o(0) = \theta$. Thus u_o is also unique and (2.22) holds for any subsequence. The $H^1(\mathbb{R})$ convergence is straightforward and (2.23), together with $u_o(x) \equiv 1$ for $x \geq \bar{x}$, implies (2.14). The proof is complete. □

§3. THE CASE OF THE SYSTEM OF EQUATIONS

3.1. EXISTENCE RESULT ON A BOUNDED DOMAIN.

The goal of this section is to study the problem:

Find two functions $u,v:(-a,+a) \to \mathbb{R}$ and $c \in \mathbb{R}$ satisfying:

$$\begin{cases} -u'' + cu' = f(u)v & \text{on } (-a,+a) \\ -\Lambda v'' + cv' = -f(u)v & \text{on } (-a,+a) \\ -u'(-a) + cu(-a) = 0 & u(a) = +1 \\ -\Lambda v'(-a) + cv(-a) = c & v(a) = 0 \\ u(0) = \theta \end{cases} \tag{3.1}$$

where $0 < \theta < 1$, $\Lambda > 0$ are given and $a > 0$. Letting $a \to +\infty$ will yield the needed existence result on $\mathbb{R}$ (see § 3.2); the problem is also useful for some numerical calculations. The assumptions on f are similar to those on g (see § 2). Precisely:

$$f:[0,1] \to \mathbb{R}^+ \tag{3.2}$$

There is some θ, $0 < \theta < 1$, for which $f \equiv 0$ on $[0,\theta)$ and $f > 0$ on $(\theta,1]$ (3.3)

f is Lipschitz continuous on $[\theta,1]$. (3.4)

$$G(1) \equiv \int_\theta^1 f(s)(1-s)\, ds < +\infty \quad .(3.5)$$

THEOREM 3.1. Assuming the conditions (3.2) to (3.5), there exists $u,v : (-a,+a) \to (0,1)$ and $c > 0$ solution of (3.1). The function u (resp. v) is of class C^1 and strictly monotone increasing (resp. decreasing).

We will not give the complete proof of this theorem (see [2]), but just show how to reduce the problem (3.1) to a fixed point equation then solvable using Leray-Schauder degree. Let us emphasize again that c is part of the unknowns in (3.1). The setting for the fixed point equation is as follows. Let us set $I_a = (-a,+a)$ and define $X = C^1(\bar{I}_a)^2 \times \mathbb{R}$, equipped with the natural norm. For (u,v,c) given in X, and $\tau \in [0,1]$, let (U,V) be the solution of the problem:

$$\begin{cases} -U'' + cU' = \tau\, f(u)v & \text{on } I_a \\ -\Lambda V'' + cV' = -\tau\, f(u)v & \text{on } I_a \\ -U'(-a) + cU(-a) = 0 & U(a) = +1 \\ -\Lambda V'(-a) + cV(-a) = c & V(a) = 0 \end{cases} \tag{3.6}$$

One easily checks that U,V are in $C^1(\bar{I}_a)$ and that the application:

$$K_\tau : X \to X : (u,v,c) \to (U,V,c - u(0) + \theta) \tag{3.7}$$

is compact for any $\tau \in [0,1]$. Then, setting $F_\tau = I - K_\tau$, to solve the problem (3.1) is equivalent to finding a fixed point of K_τ or a zero of F_τ. The main step in the proof of the theorem 3.1 is the:

PROPOSITION 3.2: Assuming (3.2) to (3.5), there exists positive and finite constants R, $\underline{c}$, $\bar{c}$ such that:

$$F_\tau(\partial\Omega) = (I - K_\tau)(\partial\Omega) \neq 0 \qquad 0 \leq \tau \leq 1$$

where K_τ is defined by (3.7) and

$$\Omega = \{(u,v,c) \in X \,\big|\, \|u\|_{C^1(\bar{I}_a)} < R \quad , \|v\|_{C^1(\bar{I}_a)} < R \quad , \underline{c} < c < \bar{c}\} \quad .$$

Moreover deg $(F_\tau,\Omega,0) = -1$, for $0 \leq \tau \leq 1$.

REMARK 3.3. The solutions u,v of (3.1) are in fact in $W^{2,\infty}(I_a)$. □

REMARK 3.4. Moreover, we show the existence of R, $\underline{c}$, $\bar{c}$ independent of a, which will allow to pass to the limit $a = +\infty$ (see § 3.2). □

3.2. EXISTENCE RESULT ON $\mathbb{R}$.

We now consider the problem:

Find two functions $u,v:\mathbb{R} \to \mathbb{R}$ and $c \in \mathbb{R}$ satisfying ($0 < \theta < 1$ and $\Lambda > 0$ are given):

$$\begin{cases} -u'' + cu' = f(u)v & \text{on } \mathbb{R} \\ -\Lambda v'' + cv' = -f(u)v & \text{on } \mathbb{R} \\ u(-\infty) = 0 \quad u(+\infty) = 1 \\ v(-\infty) = 1 \quad v(+\infty) = 0 \\ u(0) = \theta \quad . \end{cases} \tag{3.8}$$

THEOREM 3.5. Assuming the conditions (3.2) to (3.5), there exist $u,v:\mathbb{R} \to (0,1)$ and $c > 0$ solutions of (3.8). Moreover the functions u,v are in $W^{2,\infty}(\mathbb{R})$ and u (resp. v) is strictly monotone increasing (resp. decreasing).

REMARK 3.6. We conjecture the uniqueness of the solution of (3.8). □

A solution (u,v,c) of (3.8) is obtained as a limit, as $a \to +\infty$, of a solution of (3.1). This supposes adequate a priori estimates independent of a, for any solution, noted (u_a,v_a,c_a) of (3.1).

We refer to [2] for details. Here we just sketch the proof of the crucial step, that is the bounds on c_a.

PROPOSITION 3.7. Let $a_o > 0$ fixed. Then, if (u_a,v_a,c_a) is any solution, c_a satisfies :

$$\left\{\frac{2G(1)}{|\lambda-1|+1}\right\}^{\frac{1}{2}} \leq c_a \leq \mathrm{Max}\left\{\mathrm{Max}\left\{\sqrt{\frac{2M}{\theta}},\ 2M\right\},\ -\frac{\mathrm{Log}\ \theta}{a_o}\right\}, \tag{3.9}$$

for any $a \geq a_o$. Here $M \equiv \sup_{\theta \leq s \leq 1} f(s)$.

PROOF: Adding the two equations in (3.1) one easily shows the pointwise inequalities $|u + v-1| \leq |\Lambda-1|v$ and $|u + v-1| \leq \frac{|\Lambda-1|}{\Lambda}(1-u)$ from which one deduces $\frac{1}{|\Lambda-1|+1}(1-u) \leq v$. Now, from the equation $-u'' + cu' = f(u)v$, multiplied successively by 1, u, u' and integrated from 0 to +a, one gets (using $f(u)\,u\,v \geq 0$ and $u \leq 1$) after some calculations:

$$\frac{c_a^2}{2} \geq \int_0^a f(u)vu'\,dx \quad .$$

The lower bound on c_a follows then from the lower bound on v since $f(u)u' \geq 0$. For the upper bound, one introduces the unique solution $\bar{u}$ of the problem:

$$\begin{cases} -\bar{u}'' + c\bar{u}' = MH & \text{on } (-a,+a) \\ -\bar{u}'(-a) + c\bar{u}(-a) = 0 & \bar{u}(a) = 1 \end{cases}$$

where H is the Heaviside function at $x = 0$. It is easy to show $u_a(x) \leq \bar{u}(x)$ on $(-a,+a)$. In particular $\theta = u_a(0) \leq \bar{u}(0)$. The upper bound follows then from an explicit calculation of $\bar{u}(0)$. □

3.3 ASYMPTOTIC ANALYSIS.

We are now looking for an analysis similar to the one of Section 2.3. So we assume f depending on a parameter $\varepsilon > 0$ and set $f = f_\varepsilon$. The dependency on ε is as follows:

For each $\varepsilon > 0$, f_ε verifies conditions (3.2) - (3.4) with a fixed θ, $0 < \theta < 1$. (3.10)

$\exists\, \theta_\varepsilon$, $\theta \le \theta_\varepsilon \le 1$ such that $\theta_\varepsilon \uparrow 1$ when $\varepsilon \downarrow 0$ and

$$\lim_{\varepsilon \downarrow 0} \ \max_{\theta \le s \le \theta_\varepsilon} f_\varepsilon(s)(1-s) = 0 \tag{3.11}$$

$$\lim_{\varepsilon \downarrow 0} \int_\theta^1 f_\varepsilon(s)(1-s)\, ds \equiv \lim_{\varepsilon \downarrow 0} G_\varepsilon(1) \equiv m < +\infty \ . \tag{3.12}$$

REMARK 3.8. The function $s \to f_\varepsilon(s)(1-s)$ will play the role of g_ε in Section 2.3. □

In what follows, for each $\varepsilon > 0$, $(u_\varepsilon, v_\varepsilon, c_\varepsilon)$ will be a solution of (3.8), where $f = f_\varepsilon$.

THEOREM 3.9. Assuming the conditions (3.10) to (3.12), there exists a sequence $\{\varepsilon_n\}$ *with* $\varepsilon_n \downarrow 0$ *such that* $(u_{\varepsilon_n}, v_{\varepsilon_n}, c_{\varepsilon_n})$ *converge strongly in* $H^1(\mathbb{R})^2 \times \mathbb{R}$ *and* $C^0(\mathbb{R})^2 \times \mathbb{R}$ *to* (u_o, v_o, c_o). *Moreover,* u_o, v_o *are solutions of the problem*

$$\begin{cases} -u'' + c_o u' = c_o\, \delta_{x=\bar{x}} \\ -\Lambda v'' + c_o v' = -c_o\, \delta_{x=\bar{x}} \\ u(-\infty) = 0 \qquad u(+\infty) = 1 \\ v(-\infty) = 1 \qquad v(+\infty) = 0 \end{cases}$$

where $\bar{x}$ *is uniquely determined by the condition* $u(0) = \theta$. *The proof generalizes the proof of the Theorem 2.4, and we refer to [2].*

CONCLUSION: We have just given here a brief account of the methods used in the work [2]. These methods are fairly flexible and apply to more complex systems than (3.8) also of interest in the study of deflagration waves. For example, one can introduce nonlinear diffusion coefficients in (3.8), as for problem (2.1). One can also consider reaction terms in (3.8) of the form $f(u)v^n$ (this corresponds to a simple n^{th} order reaction). Finally, one can also study reactive flow associated to a reaction $A + B \to P$. The system of interest is now:

$$-(D(U)U')' + c\,U' = F(U)\,K \quad ,$$

on $\mathbb{R}$ where $U:\mathbb{R} \to \mathbb{R}^3$, $F:\mathbb{R}^3 \to \mathbb{R}$ and K is a given vector of $\mathbb{R}^3$. These problems are worked out in [2]. More general systems are studied in [3].

BIBLIOGRAPHY

1. D. G. Aronson and H. F. Weinberger, "Multidimensional Nonlinear diffusion arising in Population Genetics," Advances in Math 30 (1978), 33-76.

2. H. Berestycki, B. Nicolaenko, and B. Scheurer, "Mathematical analysis and asymptotic limits of the premixed laminar flame."

3. H. Berestycki, B. Nicolaenko, and B. Scheurer, in preparation.

4. J. Buckmaster and G. S. S. Ludford, "Theory of Laminar Flames," Cambridge University Press, 1982.

5. P. C. Fife, "Mathematical aspects of reacting and diffusing systems," Lecture Notes in Biomathematics 28, Springer-Verlag (1979).

6. P. C. Fife, "Propagating Fronts in Reactive Media," Nonlinear Problems: Present and Future, Mathematics Studies 61 (1982), 267-285 (ed. A. Bishop, D. Campbell, B. Nicolaenko) North-Holland.

7. P. C. Fife and B. Nicolaenko, "The singular perturbation approach to flame theory with chain and competing reactions," to appear in Proceedings Conference on Differential Equations, Dundee, (1982), Springer-Verlag Lecture Notes in Mathematics.

8. P.C. Fife and B. Nicolaenko, "Asymptotic flame theory with complex chemistry," Proceedings of this conference.

9. W. E. Johnson and W. Nachbar, "Laminar flame theory and the steady linear burning of a monopropellant," Arch. Rat. Mech. Anal. 12 (1963), 58-91. See also: W. E. Johnson, "On a first order boundary value problem for laminar flame theory," Arch. Rat. Mech. Anal. 13 (1963), 46-54.

10. Ya. I. Kanel', "On the stabilization of solutions of the Cauchy problem for the equations arising in the theory of combustion," Mat. Sbornik 59 (1962), 245-288.

11. F. Williams, "Combustion theory," Addison-Wesley, 1963.

HENRI BERESTYCKI
CENTER FOR NONLINEAR STUDIES, MS-B258
LOS ALAMOS NATIONAL LABORATORY
LOS ALAMOS, NM 87545

Current Address:
Analyse Numérique
Universite Pierre et Marie Curie
Paris, France

BASIL NICOLAENKO
CENTER FOR NONLINEAR STUDIES, MS-B258
LOS ALAMOS NATIONAL LABORATORY
LOS ALAMOS, NM 87545

BRUNO SCHEURER
CENTER FOR NONLINEAR STUDIES, MS-B258
LOS ALAMOS NATIONAL LABORATORY
LOS ALAMOS, NM 87545

Current Address:
CENTRE d'ETUDES DE LIMEIL AND UNIVERSITE PARIS-SUD (ORSAY)
BOITE POSTALE 27
94190 VILLENEUVE ST. GEORGES
FRANCE

Contemporary Mathematics
Volume 17, 1983

STABLE EQUILIBRIA WITH VARIABLE DIFFUSION

Michel Chipot[+] and Jack K. Hale[*]

ABSTRACT. For a scalar nonlinear parabolic equation in one space dimension with homogeneous Neumann boundary conditions, criteria are given on the diffusion coefficient to ensure that the stable equilibrium solutions are constant functions regardless of the nonlinearities. The Dirichlet problem is also discussed.

Consider the equation

$$u_t = (au_x)_x + f(u), \qquad 0 < x < 1, \tag{1}$$

with homogeneous Neumann conditions

$$u_x = 0 \quad \text{at} \quad x = 0,\ x = 1, \tag{2}$$

where $a \in C^2[0,1]$ is a positive function and $f \in C^2(\mathbb{R})$. System (1), (2) defines a local dynamical system in $H^1(0,1) = W^{1,2}(0,1)$ (see Henry [8]). Furthermore, the ω-limit set of any bounded orbit is exactly one equilibrium point; that is, a solution of the equations

$$(au_x)_x + f(u) = 0, \qquad 0 < x < 1 \tag{3}$$

$$u_x = 0 \quad \text{at} \quad x = 0,\ x = 1 \tag{4}$$

(see Matano [11], Hale and Massatt [6], Zelenyak [15]).

1980 Mathematics Subject Classification. 35K55

[+]This research has been supported in part by the Air Force Office of Scientific Research under contract #AF-AFOSR 81-0198.

[*]This research has been supported in part by the Air Force Office of Scientific Research under contract #AF-AFOSR 81-0198, in part by the National Science Foundation under contract #MCS 79-05774-05 and in part by the U.S. Army Research Office under contract #ARO-DAAG-29-79-C-0161.

0271-4132/82/0000-0761/$02.25

The first result is the following.

THEOREM 1. If $a''(x) \leq 0$ on $[0,1]$, then every nonconstant equilibrium solution of (1),(2) is unstable.

For $a(x) = c > 0$, where c is a constant, this result was proved by Chafee [2]. Note that the conclusion in the theorem is valid for every function f. Matano [13] has given an example (1),(2) for which there are stable nonconstant equilibrium solutions. The function f can be chosen to be a cubic with three simple zeros and the function a is ≥ 1 on intervals $[0,\alpha],[\beta,1]$ and $\leq \varepsilon$ on $[\gamma,\delta]$, $\alpha < \gamma < \delta < \beta$, and ε is sufficiently small. Numerical examples indicated the same property can occur with $a \geq 1$ on $[0,\alpha]$ and $\leq \varepsilon$ on $[\beta,1]$ with $\alpha < \beta$. These examples show clearly that more work is needed to determine the optimal conditions on a for the conclusions of the theorem to hold.

For a more general parabolic equation in several space dimensions in a bounded domain Ω and the diffusion coefficients constant, Casten and Holland [1], Matano [12] have shown that Theorem 1 is true if Ω is convex. The authors have had no success in extending this result in a significant way with variable diffusion coefficients. If Ω is not convex and the diffusion coefficients are constants, nonconstant equilibrium solutions can occur (see Matano [12], Hale and Vegas [7], Vegas [14]) for certain nonlinear functions f. If the diffusion coefficient in (1) is constant and f is a function of $u(\cdot,t)$, for example, $f(u(x,t)) = g(u(x,t), \int_0^1 \alpha(x)u(x,t)dx)$, then stable nonequilibrium solutions can occur (see Chafee [3]). A similar situation was considered in several space dimensions and nonconvex domains by Keyfitz and Kuiper [9]. For constant diffusion in (1) and f replaced by $s(x)f(u)$, Fleming [5], (see also Henry [8]) has considered how the existence of stable nonconstant solutions depend on $s(x)$. All of these papers should be reconsidered with variable diffusion.

There is an analogue of Theorem 1 for the Dirichlet problem.

THEOREM 2. Consider the equation (1) with Dirichlet boundary conditions

$$u = 0 \quad \text{at} \quad x = 0,\ x = 1.$$

If $a''(x) \leq 0$ on $[0,1]$ and $v(x)$ is a nonconstant equilibrium solution such that $v_x = 0$ at two points in $(0,1)$, then v is unstable.

For $a(x) = c > 0$, this result was proved by Chafee and Infante [4], Maginu [10].

<u>Proof of Theorem 1.</u> If v is a solution of (3),(4) and

$$\mathcal{A}(\phi) = \int_0^1 [a\phi_x^2 - f'(v)\phi^2]dx,$$

then the first eigenvalue λ_1 of the operator

$$Lu = -(au_x)_x - f'(v)u$$

is given by

$$\lambda_1 = \min\{\mathcal{A}(\phi) : \phi \in H^1(0,1), |\phi|_{L^2(0,1)} = 1\}.$$

Furthermore, the equilibrium solution v of (1),(2) is unstable if $\lambda_1 < 0$ (see, for example, Henry [8]). Consequently, it is sufficient to show that the hypotheses of the theorem imply $\lambda_1 < 0$ if v is not a constant and this will be the case if we show that $\mathcal{A}(av_x) < 0$ if v is not a constant. We have

$$\begin{aligned}\mathcal{A}(av_x) &= \int_0^1 [a[(av_x)_x]^2 - f'(v)(av_x)^2]dx \\ &= \int_0^1 af^2(v)dx - \int_0^1 f(v)_x a\, av_x dx \\ &= \int_0^1 af^2(v)dx + \int_0^1 f(v)a(av_x)_x dx + \int_0^1 f(v)a'av_x dx \\ &= -\int_0^1 (av_x)_x a'(av_x)dx\end{aligned}$$

To compute the last expression, we integrate by parts to obtain

$$\int_0^1 (av_x)_x a'(av_x)dx = -\int_0^1 av_x a'' av_x dx - \int_0^1 av_x a'(av_x)_x dx$$

and, thus,

$$\int_0^1 (av_x)_x a'(av_x)dx = -\frac{1}{2}\int_0^1 a''(av_x)^2 dx.$$

Therefore,

$$\mathcal{A}(av_x) = \frac{1}{2}\int_0^1 a''(av_x)^2 dx \le 0 \tag{5}$$

since we have assumed $a'' \le 0$.

If $\mathscr{A}(av_x) < 0$, then $\lambda_1 < 0$ and the solution v is unstable. Thus, suppose $\lambda_1 \geq 0$. From (5), this implies $\mathscr{A}(av_x) = 0$ which implies $\lambda_1 = 0$. Since the first eigenvalue of L is simple, there is a ϕ satisfying $L\phi = 0$, $\phi_x = 0$ at $x = 0$, $x = 1$, $|\phi|_{L^2(0,1)} = 1$. Also, $\mathscr{A}(av_x) = 0$ implies there is a constant c such that $av_x = c\phi$ on $[0,1]$. If $c \neq 0$, then $v_x = 0$ at $x = 0$, $x = 1$ and $a > 0$ imply $\phi = 0$ at $x = 0$, $x = 1$. But this would imply $\phi = 0$ on $[0,1]$ which is a contradiction. Thus, $c = 0$ and $av_x = 0$, $v_x = 0$ and $v =$ constant. This proves the theorem.

Proof of Theorem 2. Suppose v is an equilibrium solution, $v_x = 0$ at $x = \alpha$, $x = \beta$, $\alpha,\beta \in (0,1)$. Then v is an equilibrium solution of the Neumann problem on the interval $[\alpha,\beta]$. As in the proof of Theorem 1, one shows that the first eigenvalue λ_1 of the linear variational operator is negative if v is not a constant. By the characterization of λ_1 as a minimum of the functional $\mathscr{A}(u)$, it follows that λ_1 for the Dirichlet problem on $[0,1]$ is negative. This proves the theorem.

After this paper was written, the authors learned that E. Yanagida "Stability of stationary distributions in a space dependent population growth process" (to appear in J. Math. Biol.) has obtained more general results than the above. In fact, if $a = b^2$, then Yanagida proves that it is sufficient to have $b'' \leq 0$. The proofs are somewhat different and the more general result is obtained from the above argument after showing that $\mathscr{A}(bv_x) = \int_0^1 b^3 b_{xx} v_x^2 \leq 0$. Yanagida also proves the converse in the sense that if $b_{xx} > 0$ for some x, then there is an f such that (1),(2) has a nonconstant stable equilibrium.

BIBLIOGRAPHY

1. R.G. Casten and C.J. Holland, Instability results for a reaction diffusion equation with Neumann boundary conditions. J. Differential Equations 27 (1978), 266-273.

2. N. Chafee, Asymptotic behavior of a one-dimensional heat equation with homogeneous Neumann boundary conditions. J. Differential Equations 18 (1975), 111-134.

3. N. Chafee, The electric ballast resistor: homogeneous and nonhomogeneous equilibria. In Nonlinear Differential equations: Invariance Stability and Bifurcation. P. de Mottoni and L. Salvadori eds. Academic Press(1981), 161-173.

4. N. Chafee and E.F. Infante, A bifurcation problem for a nonlinear partial differential equation of parabolic type. J. Math. Anal. & Appl. 4 (1974), 17-37.

5. W. Fleming, A selection-migration model in population genetics. J. Math. Biol. 2 (1975), 219-233.

6. J.K. Hale and P. Massatt, Asymptotic behavior of gradient-like systems. Univ. of Fla. Symp. Dyn. Systems, II, Academic Press, 1982.

7. J.K. Hale and J. Vegas, A nonlinear parabolic equation with varying domain. Arch. Rat. Mech. Ana. To appear.

8. D. Henry, Geometric Theory of Semilinear Parabolic Equations, Lecture Notes in Math., Vol. 840, Springer-Verlag, 1981.

9. B.L. Keyfitz and H.J. Kuiper, Bifurcations resulting from changes in domain in a reaction diffusion equation. J. Differential Eqns. To appear.

10. K. Maginu, Stability of stationary solutions of one-dimensional semilinear parabolic partial differential equation, J. Math. Anal. Appl. 3(1978), 224-243.

11. H. Matano, Convergence of solutions of one-dimensional semilinear parabolic equations. J. Math. Kyoto University 18(1978), 221-227.

12. H. Matano, Asymptotic behavior and stability of solutions of semilinear diffusion equations. Res. Inst. Math. Sci., Kyoto 15(1979), 401-454.

13. H. Matano, Nonincrease of lap number of a solution for a one dimensional semilinear parabolic equation. Pub. Fac. Sci. Univ. Tokyo. To appear.

14. J. Vegas, Bifurcations caused by perturbing the domain in an elliptic equation. J. Differential Eqns. To appear.

15. T.I. Zelenyak, Stabilization of solutions of boundary value problems for a second order parabolic equation with one space variable. Differential Equations 4(1968), 17-22.

Michel Chipot

On leave during the academic year 1981-82 from:

Université de Nancy I
UER. Sc. Mathematiques
Case off. 239-54506 Vandoeuvre les Nancy-Cedex
FRANCE

Visitor to:

Lefschetz Center for Dynamical Systems
Division of Applied Mathematics
Brown University
Providence, Rhode Island 02912

Jack K. Hale
Lefschetz Center for Dynamical Systems
Division of Applied Mathematics
Brown University
Providence, Rhode Island 02912

Contemporary Mathematics
Volume 17, 1983

DIFFUSION AND THE PREDATOR-PREY INTERACTION: STEADY STATES WITH FLUX AT THE BOUNDARIES

E. D. Conway[1]

ABSTRACT. We study the system describing two constituents in a predator-prey relationship that are undergoing simple diffusion in a bounded one-dimensional medium. Under homogeneous Dirichlet boundary conditions we study the dependence of the number and character of steady states upon the length of the interval and the relative rates of diffusion of each constituent. We show that certain features of this relationship are to be found in a broad class of models.

§1. INTRODUCTION. In [6], Conway, Gardner and Smoller studied a particular predator-prey type interaction in the presence of diffusion. It is the purpose of this paper to extend that study to a fairly general class of predator-prey interactions and, in particular, to show that certain features discovered in that earlier study are to be found in many of the models discussed in the ecological literature. This is important because these elementary two-species ecological models are at best caricatures of realistic situations and hence, only those features which are shared by large classes of models can be significant.

We shall study equations of the form

$$\text{(1.1a)}\qquad \begin{cases} u_t = u_{xx} + uM(u,v) \\ v_t = dv_{xx} + vN(u,v) \end{cases} \qquad t > 0\ , \quad |x| < L$$

subject to homogenous boundary conditions of Dirichlet type,

$$\text{(1.2)}\qquad \begin{cases} u(\pm L,t) = 0 = v(\pm L,t)\ , \quad t > 0\ , \text{ for } d > 0 \\ u(\pm L,t) = 0\ , \quad t > 0\ , \text{ for } d = 0. \end{cases}$$

1980 Mathematics Subject Classification. 35K50, 35K55.
[1]Partially supported by a COR Grant from Tulane University.

The variables u and v may be thought of as linear population densities of two interacting species continuously distributed along a one-dimensional medium. Both boundary conditions model situations in which the medium is in contact at each end with a region which is hostile to both organisms. (We refer to Okubo [14] for a general discussion, with bibliography, of continuum models of populations.) We have chosen the length scale so as to set the diffusion coefficient for u equal to 1 and thus the parameter d is a measure of the rate of diffusion of v relative to that of u.

We have written the interaction term, i.e. the vector field $(uM(u,v),uN(u,v))$, in Kolmogorov form (cf. [10]). If M and N are smooth then the positive quadrant $[u \geqslant 0, v \geqslant 0]$ is invariant for (1.1a)-(1.2) i.e. if u and v are non-negative for $t = 0$ then they remain so for $t > 0$ (cf. Chueh, Conley and Smoller [3] for a discussion of this notion). If

$$\frac{\partial M}{\partial v} \equiv M_v < 0 \quad \text{and} \quad \frac{\partial N}{\partial u} \equiv N_u > 0$$

then the interaction is said to be of predator-prey type (cf May [10] and Magnard-Smith [12] for a discussion of this). Most model interactions discussed in the ecological literature are special cases of the Kolmogorov form. To make contact with this literature we change notation:

$$(1.3) \qquad \begin{aligned} f(u) &\equiv uM(u,0) \ , \quad \phi(u,v) \equiv f(u) - uM(u,v) \\ g(v) &\equiv vN(0,v) \ , \quad \psi(u,v) \equiv vN(u,v) - g(v) \end{aligned}$$

Equations (1.1a) now become

$$(1.1) \qquad \begin{cases} u_t = u_{xx} + f(u) - \phi(u,v) \\ v_t = dv_{xx} + g(r) + \psi(u,v) \end{cases} \qquad t > 0 \ , \quad |x| < L$$

The interaction will be of predator-prey type if $\phi_v > 0$ and $\psi_u > 0$. We also see that

$$(1.4) \qquad \phi(0,v) = \phi(u,0) = \psi(0,v) = \psi(u,0) = 0.$$

We see then that f describes the growth of the prey (whose density is u) when there is no predator ($v \equiv 0$) and g describes the growth of the predator when the prey is absent.

Assumptions concerning $g(v)$. Unless the predator has alternative prey, the function g is usually assumed to be negative and, in fact, the most common form is $g(v) = -\mu v$ for some constant $\mu > 0$. We shall always assume

$$(G) \quad g(0) = 0 \ , \quad g'(0) = -\mu < 0 \ , \quad g(v) \leq -\mu v$$

Assumptions concerning $f(u)$. When there is no predator ($v \equiv 0$) the growth of the prey species is governed by

$$u_t = u_{xx} + f(u).$$

We assume there is a limitation to growth expressed by $f(K) = 0$, $f(u) < 0$ for $u > K$, where K is a constant referred to as the carrying capacity of the environment. We choose units for u so that $K = 1$. Since $f(0) = 0$ we have the following minimal assumptions satisfied by most realistic models

$$(1.5) \qquad f(0) = 0 \ , \quad f(1) = 0 \ , \quad f(u) < 0 \quad \text{for} \quad u > 1.$$

Beyond these restrictions the models appearing in the ecological literature differ in the way f behaves near $u = 0$. We concentrate on three classes characterized by the shape of the graph of $f(u)/u$, the per capita growth rate.

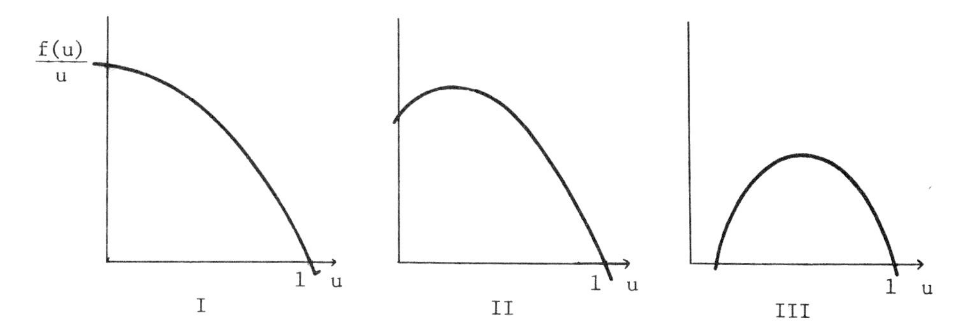

FIGURE 1: $f(u)/u$ vs u for three types of f

In case of type I, the function $f(u)/u$ is monotone decreasing for $0 < u < 1$ while in cases of type II or III it has a single local maximum. An example of a function of type I is the classic logistic $f(u) = au(1 - u)$. Even when f is not quadratic, we shall refer to type I as "logistic type" because, as we show in §2, non-negative solutions for all type I function f have the same qualitative behavior as in the case of $f(u) = u(1 - u)$. Both types II and III exhibit the Allee effect of a fall-off in per capita growth rate at low population densities. The very strong form of fall-off in type III has been termed "asocial" by Philip [15].

Each of the above types can be realized by a cubic polynomial and we shall assume f to be of this form:

(F) $$f(u) = u(Bu + A)(1 - u), \qquad B + A > 0.$$

The restriction that $B + A$ be positive insures that $f'(1) < 0$. Any such functon f will be of one of the three types indicated above depending upon which of the following conditions is satisfied.

(F I) $A > B \geq 0$

(F II) $B > A > 0$

(F III) $A < 0$ but $A + B > 0$.

Figure 2 indicates examples of each situation.

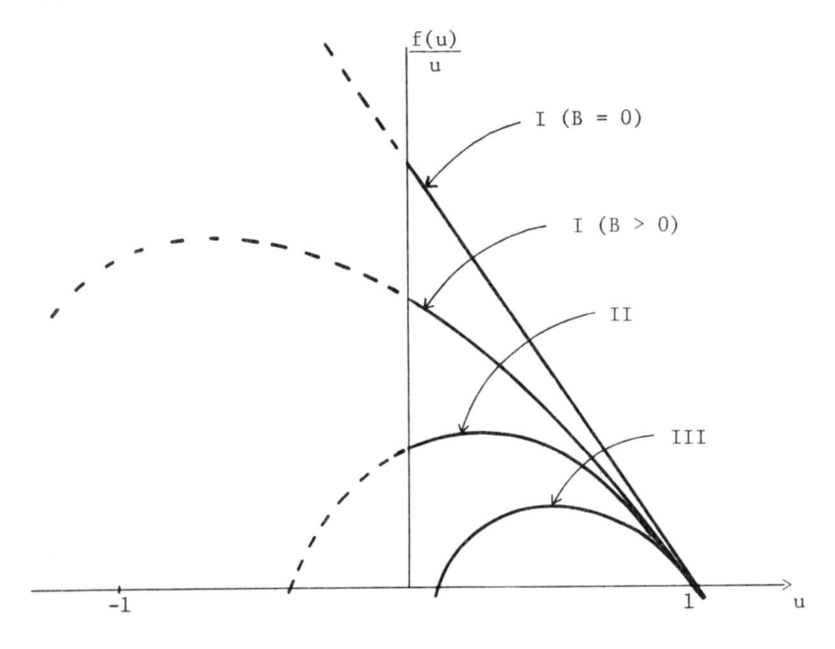

FIGURE 2: Examples of each type of per capita growth

We shall show that the bifurcation of steady states differs considerably from type to type.

REMARK. We restrict ourselves to polynomials of degree no more than three because we use results of Smoller and Wasserman [19]. Similar results are clearly true in more generality but I know of no published work to this effect.

<u>Assumptions Concerning ϕ and ψ.</u> Some of our results require only minimal restrictions on ϕ and ψ while others require more stringent requirements.

Our minimal assumption is

(1.6) $$0 < \frac{\phi(u,v)}{uv} = O(1) \quad \text{and} \quad 0 < \frac{\psi(u,v)}{uv} = O(1) \quad \text{as} \quad u,v \to 0.$$

The boundedness of these ratios follows from (1.4) if ϕ and ψ are C^2.

In [11], May lists the following common examples:

$$cuv \; ; \; cvu^p, \; 0 < p < 1 \; ; \; \frac{cuv}{u + d}, \quad \frac{cu^2v}{u^2 + d^2}, \quad cv(1 - \exp(-au)).$$

Notice that each of these is of the form

$$(1.7) \qquad vR(u) \quad \text{where} \quad R(0) = 0 \; , \quad R'(u) > 0.$$

Any such function has the property that the bound in (1.6) can be chosen to be independent of v and we need ψ to satisfy such a condition:

$$(1.8) \qquad \psi(u,v) \leq Kuv \quad \text{for} \quad 0 \leq u \leq M \; , \quad v > 0$$

where M can be any number greater than 1 since, as will be shown, all solutions will satisfy $\limsup u(x,t) \leq 1$ as $t \to \infty$.

Up to this point we have not fully used the properties of ϕ and ψ that reflect the predator-prey nature of the interaction nor have we ruled out some rather vacuously trivial cases. The following restrictions take us a few step further.

$$(1.9) \qquad \begin{aligned} &a) \quad \psi_{uv}(u,v) > 0 \\ &b) \quad \frac{\partial}{\partial v}\left(\frac{\psi(u,v)}{v}\right) \leq 0 \\ &c) \quad \psi_v(1,0) > \mu \end{aligned}$$

where the parameter $\mu = g'(0)$.

If ψ is C^2 then from (1.4) and the fact that $\psi > 0$ it would follow that $\psi_{uv} \geq 0$. Condition (1.9) a) is a slight strengthening of this. We can think of g/v as the normalized (i.e. per capita) death rate for v and ψ/v as the normalized growth rate. Condition (1.9) b) then is seen to require that the normalized growth rate does not increase with v. This is certainly a reasonable assumption in many instances. In any case, notice that a) and b) are automatically satisfied by all the above mentioned examples and by any function of the type (1.7). The final condition, (1.9) c), rules out the uninteresting case where the predator would die out even if the prey were at saturation (Notice that $\psi_v(1,0)$ is the limiting value of the normalized growth rate as $v \to 0$ and $u = 1$ while μ is the limiting death rate.)

When ϕ and ψ satisfy

$$(1.10) \qquad \psi(u,v) = k\phi(u,v) \; , \qquad k > 0 \; ,$$

we shall use the term Rosenzweig-MacArthur equations.

REMARK. In our earlier study [6] we studied the case of $g(v) = -\mu v - \varepsilon v^2$, $\phi(u,v) = uv$, $\psi = k\phi$ and f either of class I with $B = 0$ or of class III. In other words, we studied the system

$$(1.11)\qquad \begin{aligned} u_t &= u_{xx} + f(u) - uv \\ v_t &= dv_{xx} + v[-\varepsilon v + k(u - \gamma)] \end{aligned}$$

where $\gamma = \mu/k$.

§2 STEADY STATES WITH NO PREDATOR ($v \equiv 0$)

The equations (1.1)-(1.2) admit solutions of the form $(u,0)$ where u is a solution of

$$(2.1)\qquad u'' + f(u) = 0\ , \quad |x| < L$$

$$(2.2)\qquad u(\pm L) = 0.$$

As is usual in these studies, we analyze (2.1)-(2.2) by a study of the phase plane of (2.1). The qualitative nature of the phase plane depends upon the "type" of the function f. We are concerned only with non-negative solutions hence we shall be interested only in the half-plane (u,u') , $u \geq 0$. Utilizing the fact that $\frac{1}{2}(u')^2 + F(u)$ is invariant on the flow defined by (2.1) (where $F'(u) = f(u)$), a standard argument shows that the positive part of the phase planes for (2.1) are as in Figure 3.

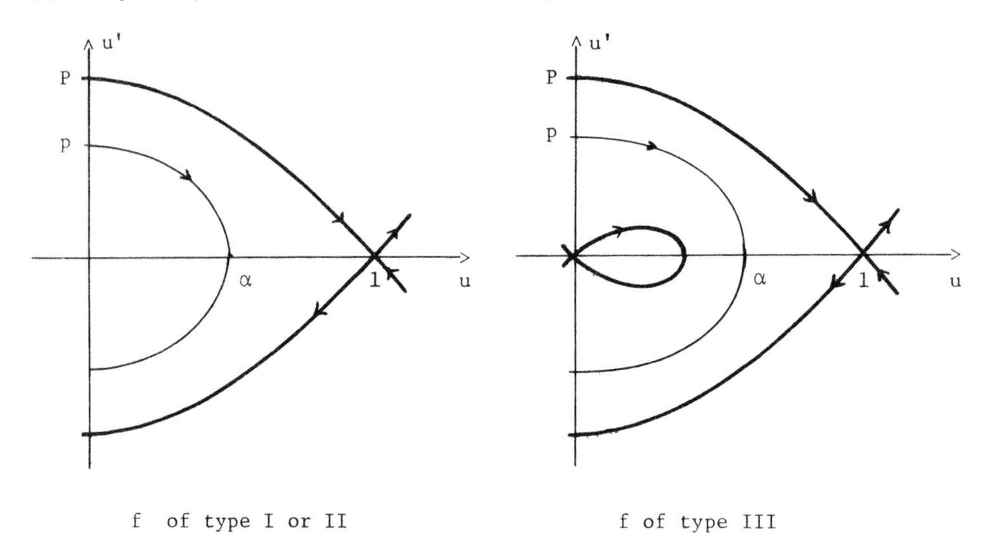

FIGURE 3

In all cases, we see that solutions of (2.1) with $u(-1) = 0$ are unbounded if $p = u'(-L) > P$ where P is some constant depending upon f. If $0 < p < P$ then the solution remains positive for a finite interval of length $2T(p)$ and $u(-L + 2T(p)) = 0$. Such a solution of (2.1) satisfies the boundary condition

(2.2) when and only when $T(p) = L$. The nature of T as a function of p (the "time map") has been analyzed by Smoller and Wasserman in [19]. Applying their results and methods we obtain the graphs of T in each of the three cases. They are portrayed in Figure 4.

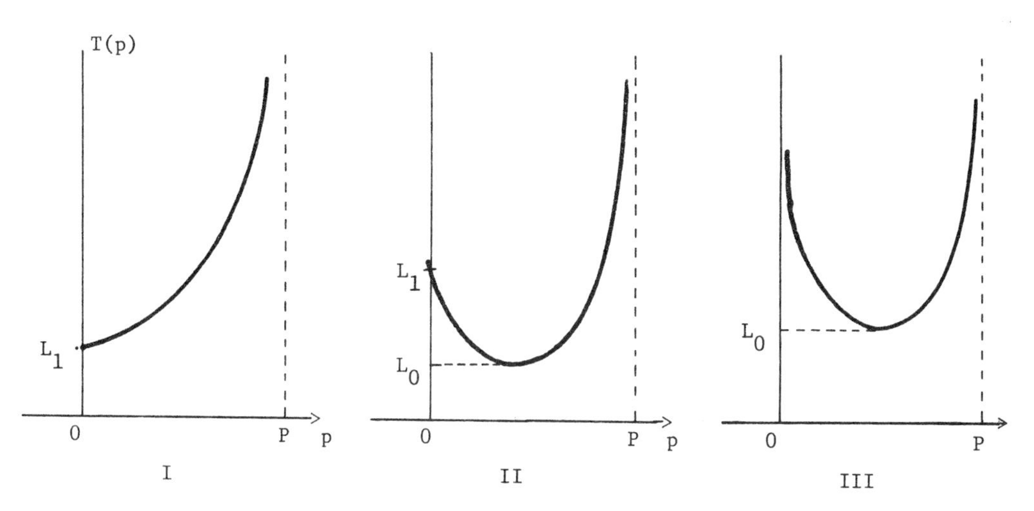

FIGURE 4: Graphs of $T(p)$, $0 < p < P$, for each type f.

The numbers L_0 and L_1 depend only upon f and, in fact, $L_1 = \pi/2\sqrt{f'(0)} = \pi/2\sqrt{A}$.

Figure 4 enables us to determine the number of non-negative solutions of (2.1)-(2.4) for any value of L. We shall also be interested in the stability character of these solutions considered as steady state solutions (independent of t) of

(2.3) $$u_t = u_{xx} + f(u) , \quad t > 0 , \quad |x| < L.$$

The following is a summary of the results we need from the work of Smoller, Conley, Tromba and Wasserman ([4],[5],[18],[19]; see also [17]).

Type I. For $L < L_1$, $u_0 \equiv 0$ is a global attractor for all non-negative solutions of (2.3)-(2.2). I.e. if u is a solution of (2.3)-(2.2) and $u \geq 0$ then $u(\cdot,t) \to 0$ as $t \to \infty$ uniformly in $|x| \leq L$. For $L > L_1$, $u_0 \equiv 0$ is unstable and there is a strictly positive steady state, u_1 , which is a global attractor for non-negative solutions of (2.3)-(2.2). I.e. $u_1(x) > 0$ for $x > 0$, u_1 is a solution of (2.1)-(2.2) and u_1 is the uniform limit, as $t \to \infty$, of all non-negative solutions of (2.3)-(2.2) except the solution $\equiv 0$.

<u>Type II</u>. For $L < L_0$, $u_0 \equiv 0$ is a global attractor for non-negative solutions of (2.3)-(2.2). For $L_0 < L < L_1$, there are three steady states: $u_0 \equiv 0$, $u_2(x) > u_1(x) > 0$ for $|x| < L$. u_0 and u_2 are stable. u_1 has a one-dimensional unstable manifold and, in fact, there are heteroclinic solutions $v_0(x,t)$ and $v_2(x,t)$ which <u>connect</u> u_1 to u_0 and u_2 respectively i.e.

$$\lim_{t\to-\infty} v_i = u_1 , \quad \lim_{t\to+\infty} v_i = u_i , \quad i = 1,2.$$

For $L_1 < L$, $u_0 \equiv 0$ is unstable and u_2 is a global attractor for all non-negative, nontrivial solutions.

<u>Type III</u>. For $L < L_0$, the situation is as in Type II, $L < L_0$.
For $L_0 < L$, the situation is as in Type II, $L_0 < L < L_1$.

We now turn to the question of the stability character of $(u_0,0) \equiv (0,0)$ and $(u_i,0)$ considered as steady states of the system (1.1)-(1.2).

§3. THE CASE OF TYPE I FUNCTIONS f.

In our earlier paper [6] we studied the special system (1.11)-(1.2). For $f(u) = Au(1 - u)$ (i.e. Type I, $B = 0$.) The asymptotic behavior of non-negative solutions depends upon the values of d and L. We divide the $d - L$ plane into the three regions as in Figure 5.

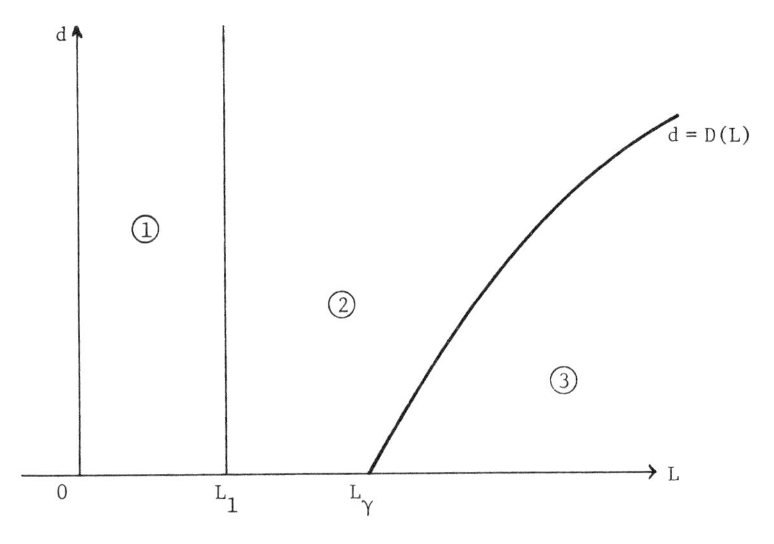

FIGURE 5. The $d - L$ Plane

The number L_1 is as in the preceding section while L_γ and the monotone increasing function $D(L)$ depend upon f and γ. In region ①, $(u_0,0) \equiv (0,0)$ is a global attractor for all non-negative solutions of (1.11)-(1.2). In region ②, $(u_1,0)$ is a global attractor for all non-negative solutions except the solution $(0,0)$ which is now an unstable steady state. In region ③, both $(0,0)$ and $(u_1,0)$ are unstable and, near $d = D(L)$, there is a unique stable steady state (u_2,v_2) with both u_2 and v_2 positive. We were unable to prove that (u_2,v_2) exists throughout region ③ and that it is a global attractor. (Computer simulation seems to indicate that this is so.)

In this section, we show that basically the same results hold for all f in Type I and for quite broad classes of functions g, ϕ and ψ. It is, perhaps, particularly surprising that for an intermediate range of interval lengths L, the solution $(u_1,0)$ is a global attractor for virtually any reasonable functions ϕ and ψ as along as $g(v) \leq -\mu v$. But let us turn now to the details.

THEOREM 1. Let f be of Type I and g satisfy condition (G). Let ϕ and ψ be smooth nonnegative functions satisfying condition (1.6) but assume also that ψ satisfies (1.8).

Then

a) for $0 < L < L_1 = \pi/2\sqrt{f'(0)}$ all non-negative solutions of (1.1)-(1.2) converge uniformly to $(0,0)$

and

b) there is an $L_2 > L_1$ which depends upon f, μ and K such that for $L_1 < L < L_2$ all non-negative solutions of (1.1)-(1.2) except $(0,0)$ converge uniformly to $(u_1,0)$ where $u_1(x;L)$ is the unique positive solution of (2.1)-(2.2).

REMARK. If $v(x,t)$ is known to be uniformly bounded, by M say, then we need not assume condition (1.8). In that case, however, L_2 would also depend upon M. See Alikakos [1] for a discussion of the question of uniform bounds for v.

Proof. We see that

$$u_t - u_{xx} - f(u) = -\phi(u,v) \leq 0 = U_t - U_{xx} - f(U)$$

where U is the solution of the scalar equation (2.3) having the same initial and boundary values as does u. Since $f'(z)$ is bounded from above on $z \geq 0$ it follows from the maximum principle (cf. [2], Proposition 2.1) that

(3.1) $$u(x,t) \leq U(x,t) \quad \text{for} \quad t > 0 , \quad |x| < L .$$

If $L < L_1$ it follows from the results of §2 that U and thus u converge unformly to zero. Therefore, from (1.8) we see that there is a $T > 0$ such that for $t > T$ we have $\phi(u,v) \leq \frac{\mu}{2} v$ for $|x| < L$. From this and condition (G) it follows that for $t > T$ we have

$$v_t - dv_{xx} + \frac{\mu}{2} v = g(v) + \mu v + \phi(u,v) - \frac{\mu}{2} v \leq 0 = V_t + \frac{\mu}{2} V$$

where

$$V(t) = [\max_{|x| \leq L} v(T,x)] e^{-\frac{\mu}{2}t} .$$

Since $0 = v(\pm L,t) < V(t)$ it again follows from the maximum principle that $v(x,t) \leq V(t)$ for all $t > T$ and $|x| \leq L$. This completes the proof of part a).

If $L > L_1$ it follows from §2 that $U(x,t) \to u_1(x;L)$ uniformly in $|x| < L$ as $t \to \infty$. From (3.1) it then follows that

$$\limsup_{t\to\infty} u(x,t) \leq u_1(x;L) \tag{3.2}$$

Now, from Figures 3 and 4 we see that $u_1(x;L) \leq u_1(0;L)$ and that $u_1(0;L)$ is monotone increasing in L. We can then define L_2 by

$$u_1(0;L_2) = \frac{\mu}{K} \tag{3.3}$$

where K is as in (1.8). For $L_1 < L < L_2$ define β by

$$u_1(0;L) = \frac{\mu}{K} - \frac{\beta}{K} .$$

From (3.2) we see that there is a T_β such that for $t > T_\beta$ we have

$$u(x,t) \leq u_1(x;L) + \frac{1}{2}\frac{\beta}{K} \leq u_1(0;L) + \frac{1}{2}\frac{\beta}{K} = \frac{\mu}{K} - \frac{1}{2}\frac{\beta}{K} .$$

From (1.8) and condition (G) we thus see that for $t > T_\beta$,

$$\begin{aligned} v_t - dv_{xx} + \frac{1}{2}\beta v &= g(v) + \phi(u,v) + \frac{1}{2}\beta v \\ &\leq -\mu v + Kuv + \frac{1}{2}\beta v \\ &\leq -\mu v + Kv(\frac{\mu}{K} - \frac{1}{2}\frac{\beta}{K}) + \frac{1}{2}\beta v \\ &= 0 = V_t + \frac{1}{2}\beta V \end{aligned}$$

where, as before,

$$V(t) = [\max_{|x| \leq L} v(x,T_\beta)] e^{-\frac{1}{2}\beta t} ,$$

so that once again $v(x,t) \leq V(t) \to 0$ as $t \to \infty$. From (1.6) it then follows that for every $\varepsilon > 0$ there is a $T_\varepsilon > 0$ such that $\phi(u(x,t),v(x,t)) < \varepsilon u(x,t)$ for $t > T_\varepsilon$ and $|x| \leq L$. Thus, for $t > T_\varepsilon$ we have

$$(3.4) \qquad \begin{aligned} u_t - u_{xx} - f(u) + \varepsilon u &= \varepsilon u - \phi(u,v) \\ &> 0 = W^\varepsilon_t - W^\varepsilon_{xx} - f(W^\varepsilon) + \varepsilon W^\varepsilon \end{aligned}$$

where W^ε is the solution of

$$(3.5) \qquad \begin{aligned} &W_t = W_{xx} + f(W) - \varepsilon W \qquad t > T_\varepsilon, \quad |x| < L \\ &W(\pm L,t) = 0, \qquad t > T_\varepsilon \\ &W(x,T_\varepsilon) = u(x,T_\varepsilon), \qquad |x| < L. \end{aligned}$$

Now $f_\varepsilon(z) = f(z) - \varepsilon z$ is a cubic polynomial which, for all sufficiently small ε, gives rise to the same phase plane for

$$(3.6) \qquad z'' + f_\varepsilon(z) = 0$$

as does f for (2.1). In particular, for $L > L_1^\varepsilon = \pi/2\sqrt{A - \varepsilon}$ there is a unique positive solution, $w^\varepsilon(x;L)$, of (3.6) which vanishes at $x = \pm L$. Moreover, we have

$$\lim_{t\to\infty} W^\varepsilon(x,t) = w^\varepsilon(x,;L).$$

Since from (3.4), it follows that $u \geq W^\varepsilon$ for $t > T_\varepsilon$, we can conclude that

$$\liminf_{t\to\infty} u(x,t) \geq w^\varepsilon(x;L).$$

Since $w^\varepsilon \to u_1$ as $\varepsilon \to 0$ we can conclude that

$$\liminf_{t\to\infty} u(x,t) \geq u_1(x;L).$$

Together with (3.2) this completes the proof of the theorem.

If we now assume more concerning the function ψ we can improve the preceding result considerably.

THEOREM 2. Let f, g, ϕ and ψ be as in Theorem 1 but, in addition, assume that ψ satisfies the conditions (1.9). Then there is a monotone increasing function $D(L)$ defined for $L \geq L_\gamma$ where L_γ depends upon f, μ and ϕ and $D(L_\gamma) = 0$ (cf. Figure 5).

a) For $0 < L < L_1$ and $d \geq 0$ all non-negative solutions of the system (1.1)-(1.2) converge uniformly to $(0,0)$ as $t \to \infty$.

b) For $L_1 < L < L_\gamma$, $d \geq 0$, all non-negative solutions converge uniformly to $(u_1,0)$ as in Theorem 1.

c) For $L > L_\gamma$, $d > D(L)$, all non-negative solutions of the system converge uniformly to $(u_1,0)$ as in part b).

d) For $L > L_\gamma$, $d < D(L)$, both $(0,0)$ and $(u_1,0)$ are unstable. For d near $D(L)$ there is a unique stable steady state (u_2,v_2) with $0 < u_2 < u_1$ and $v_2 > 0$ for $|x| < L$.

We begin by noting that part a) is exactly the same as part a) of Theorem 1 and thus needs no further proof. Let us turn then to part b). We define a real number γ by

$$\phi_v(\gamma,0) = \mu. \tag{3.7}$$

This number is well defined because of conditions (1.9) a) and c) and we see that $0 < \gamma < 1$. But, for $L > L_1$, we know that $u_1(0;L)$ is monotone increasing to 1 as $L \to \infty$ so that we can define L_γ by

$$u_1(0;L_\gamma) = \gamma. \tag{3.8}$$

Now, for $L_1 < L < L_\gamma$ we have

$$u_1(0;L) < u_1(0;L_\gamma) = \gamma$$

so that if

$$\beta \equiv \mu - \phi_v(u_1(0;L),0)$$

then we see that $\beta > 0$. As in the proof of Theorem 1 we see that (3.2) is satisfied so that we can find a $T_\beta > 0$ such that $t > T_\beta$ implies

$$\begin{aligned}\phi_v(u(x,t),0) &\leq \phi_v(u_1(x;L),0) + \frac{1}{2}\beta \\ &\leq \phi_v(u_1(0;L),0) + \frac{1}{2}\beta = \mu - \frac{1}{2}\beta\end{aligned} \tag{3.9}$$

thus, we can proceed as in the proof of Theorem 1 to write

$$\begin{aligned}v_t - dv_{xx} + \frac{1}{2}\beta v &= g(v) + \phi(u,v) + \frac{1}{2}\beta v \\ &\leq v[-\mu + \frac{\phi(u,v)}{v} + \frac{1}{2}\beta] \\ &\leq v[-\mu + \phi_v(u,0) + \frac{1}{2}\beta] \\ &\leq 0 = V_t + \frac{1}{2}\beta V\end{aligned}$$

where we have used (1.9) b) and (3.9) and where V is again $M\exp(-\beta t/2)$, M = maximum of $v(\cdot,T_\beta)$. The comparison theorem then yields $v(x,t) \leq V(t) \to 0$ as $t \to \infty$. The proof of part b) is now completed by showing that $\liminf u(x,t) \leq u_1(x;L)$ as $t \to \infty$. This is done just as in the proof of Theorem 1.

The proof of parts a) and d) of the theorem depend upon a spectral analysis of the linearization about $(u_1,0)$ of the elliptic part of (1.1)-(1.2), i.e. of the operator $A(d,L)$ from $C_0^2(-L,+L) \times C_0^2(-L,+L)$ into $C(-L,+L) \times C(-L,+L)$ defined by

$$A(d,L)\begin{bmatrix} w \\ z \end{bmatrix} = \begin{bmatrix} w'' + f'(u_1)w - \phi_v(u_1,0)z \\ dz'' - \mu z + \psi_v(u_1,0)z \end{bmatrix}$$

where we have set $g'(0) = -\mu$ and have taken advantage of the fact that $\phi_u(u,0) = \psi_u(u,0) = 0$ which follows from the smoothness of ϕ and ψ and from (1.6). Because $A(d,L)$ is triangular we are able to determine its spectrum in terms of that of the scalar operators G and H defined on C_0^2 by

$$\begin{cases} G(L): w \longrightarrow w'' + f'(u_1)w \\ H(d,L): z \longrightarrow dz'' + q(x;L)z \end{cases} \tag{3.11}$$

where $q(x;L) = -\mu + \psi_v(u,(x;L),0)$. Concerning these operators we have:

<u>Lemma</u> 1. The spectrum of G consists entirely of eigenvalues $\mu_1 > \mu_2 > \mu_3 > \ldots$. The eigenvalues are real, simple, negative and have no finite point of accumulation.

<u>Lemma 2.</u> For all $L > L_1$ and $d > 0$ the operator $H(d,L)$ has pure point spectrum: $\nu_1 > \nu_2 > \ldots$. The eigenvalues are real, simple and have no finite point of accumulation. For all $L > L_1$ we have

$$\nu_1 \leq q(0;L)$$

and if $L > L_\gamma$ then $\nu_1 > 0$ for all sufficiently small values of $d > 0$.

Except for the negativity of the eigenvalues everything else in Lemma 1 is classical (Cf. [13] and [17]). That $\mu_1 \leq 0$ follows from the stability of u_1 as a steady state of (2.3)-(1.2). That $\mu_1 \neq 0$ is shown in [18]. With the exception of the last statement of Lemma 2, everything there is classical also since $q(0)$ is the maximum value of $q(x)$, $|x| \leq L$. This last is a consequence of the monotonicity in u of ψ_v (condition (1.9)a) and the fact that $u_1(0)$ = maximum value of $u_1(x)$, $|x| \leq L$. To see that $\nu_1 > 0$ for small d we use the classical variational property of the principal eigenvalues,

(3.12)
$$\nu_1 = \sup_{\omega} \frac{-d\int_{-L}^{+L} (\omega')^2 dx + \int_{-L}^{+L} q\omega^2 dx}{\int_{-L}^{+L} \omega^2 dx}$$

For $L > L_\gamma$ we know that $q(0) > 0$ for x in some interval centered at $x = 0$. For any ω with support in this interval we see that the quotient will be positive for all sufficiently small $d > 0$ which yields the final assertion of Lemma 2. We can now describe the spectrum of $A(d,L)$.

Proposition a) For $L > L_1$ and $d > 0$ the operator $A(d,L)$ has pure point spectrum:

$$\Sigma = \{\mu_1, \mu_2, \ldots\} \cup \{\nu_1, \nu_2, \ldots\} .$$

b) For $L > L_1$ and $d = 0$, the spectrum of A has both point and residual components, $\Sigma = \Sigma_p \cup \Sigma_r$, where

$$\Sigma_p = \{\mu_1, \mu_2, \ldots\}$$
$$\Sigma_r = [\alpha, \beta] \backslash \Sigma_p$$
$$\alpha = q(L) = \min q(x) , \quad \beta = q(0) = \max g(x).$$

This is the exact analogue of Theorem 2.7 of [6] and is proved in exactly the same way.

Now returniing to the proof of Theorem 2 we can easily check that ν_1 is monotone decreasing with respect to d and that

$$\nu_1(d,L) = 0$$

defines a function $D(L)$ with the properties indicated in the statement of the theorem. This is done in detail in [6] (cf. proof of Lemma 3.3) and the arguments work just as well in the present context. It is thus clear that for $L > L_\gamma$ and $d > D(L)$ the full spectrum of $A(d,L)$, which is real, is negative so that $(u_1,0)$ is clearly stable. (Cf. [17] for a proof of the principle of linearized stability in this context) To see, as is claimed in part c), that $(u_1,0)$ is a global attractor first note that (3.2) is still valid. Thus, from the monotonicity of ϕ_v (i.e. 1.9 a)), we see that for every $\varepsilon > 0$ there is a $T_\varepsilon > 0$ such that

$$\phi_v(u(x,t),0) \leqslant \phi_v(u_1(x),0) + \varepsilon$$

for $t > T_\varepsilon$. Note also that, as a consequence of (1.9) b) we have

$$\frac{\phi(u,v)}{v} \leqslant \phi_v(u,0).$$

Therefore, for $t > T_\varepsilon$ we have

$$\begin{aligned}\frac{1}{2}\frac{d}{dt}\int_{-L}^{+L}[v(x,t)]^2dx &= d\int_{-L}^{+L} vv_{xx}\,dx + \int_{-L}^{+L}[g(v)v + \phi(u,v)v]\,dx \\ &\leq -d\int_{-L}^{+L}(v_x)^2dx + \int v^2[-\mu + \frac{\phi(u,v)}{v}]\,dx \\ &\leq -d\int (v_x)^2dx + \int v^2[-\mu + \phi_v(u,0)]\,dx \\ &\leq -d\int (v_x)^2dx + \int qv^2dx + \varepsilon\int v^2dx \\ &\leq (\nu_1 + \varepsilon)\int v^2dx\end{aligned}$$

where we again used the variational property (3.12). Since $d > D(L)$, we see that $\nu_1 < 0$ so that if ε is sufficiently small $\nu_1 + \varepsilon$ will also be negative. From Gronwall's inequality we then see that $v \to 0$ in $L^2(-L,+L)$ as $t \to \infty$. From this it follows (cf. [7] or [16]) that v converges uniformly to zero. To complete the proof of part c) we need only show that $\liminf u(x,t) \geq u_1(x)$ and this is done just as in the proof of Theorem 1.

Finally, part d) is proved by employing the standard results (cf. Crandall-Rabinowitz [9]) on bifurcation from a simple eigenvalue. In [6] we worked out the details for the special system (1.11) and these arguments go through, with very little modification, in our more general situation.

§4. THE CASE OF TYPE III FUNCTIONS: ASOCIAL f.

As we saw in §2 the positive phase plane and time map for the scalar stationary equation (2.1)-(2.2) are as indicated in Figure 6.

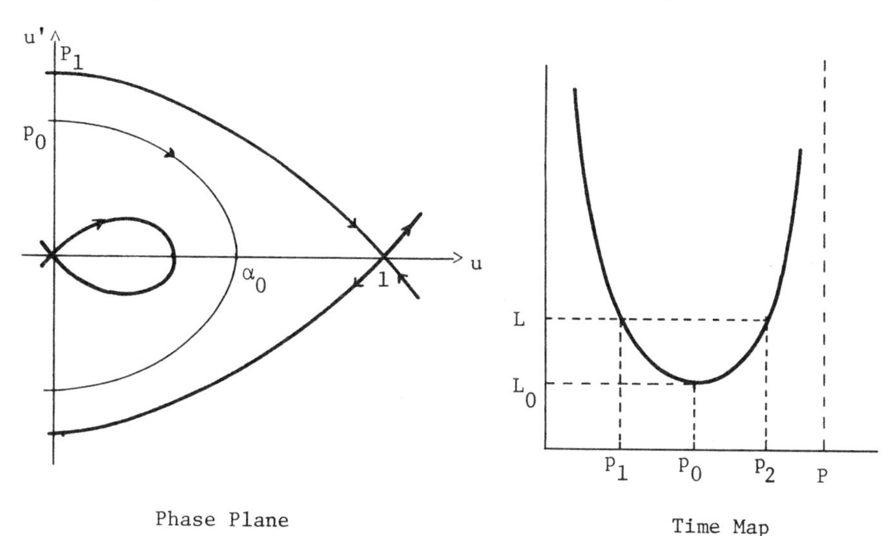

FIGURE 6: f of type III

As L increases we see that at $L = L_0$ there is the sudden appearance of a (large) solution $u_0(x)$ which bifurcates, for $L > L_0$, into two solutions $u_1 < u_2$ where $u_i(-L) = p_i$. We shall let $\alpha_0 = u_0(0)$ = maximum vallue of u_0. Corresponding to u_1 and u_2 there are solutions $(u_1,0)$ and $(u_2,0)$ of the system (1.1)-(1.2). It is easy to see that because u_1 is unstable as a solution of the scalar equation, $(u_1,0)$ is unstable as a solution of the system. Although u_2 is a stable solution of the scalar equation, the stability of $(u_2,0)$ will depend upon the size of u_2 and the quantities d, μ and ϕ appearing in the second equation of (1.1). We shall assume that ϕ satisfies the same conditions as in Theorem 2 notably that (1.9) is satisfied. Thus we can again define γ, as in the proof of Theorem 2, by

$$\phi_v(\gamma,0) = \mu.$$

It will be convenient for us to distinguish two cases: case 1 when $\alpha < \gamma < 1$ and case 2 when $\gamma < \alpha < 1$. The analogues of Figure 5 are as indicated in Figure 7.

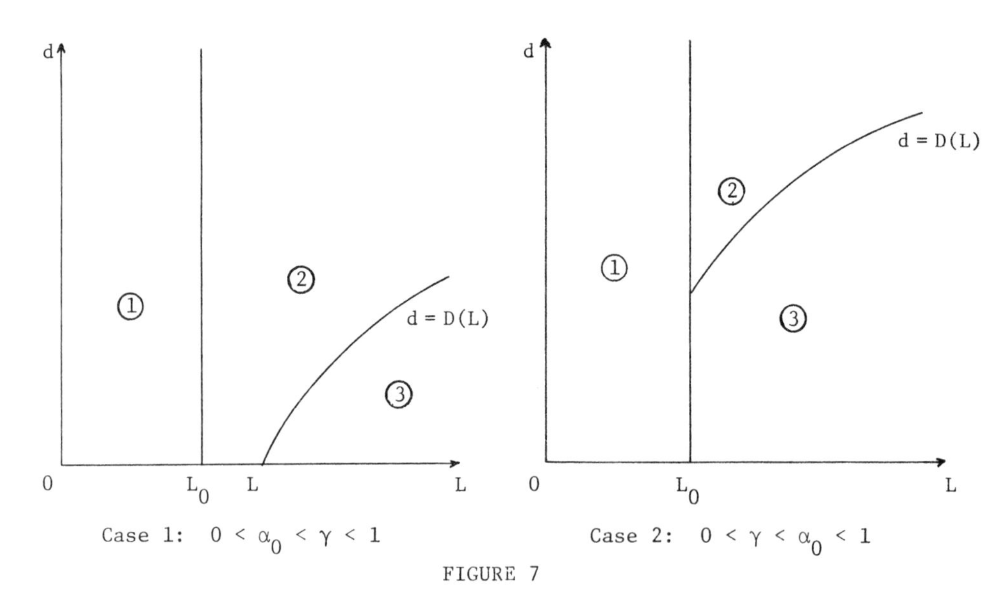

Case 1: $0 < \alpha_0 < \gamma < 1$ Case 2: $0 < \gamma < \alpha_0 < 1$

FIGURE 7

As in §3, $u_2(0;L)$ (= maximum value of $u_2(x;L)$) is monotone increasing as a function of L. Note that $u_2(0;L_0) = u_0(0) = \alpha_0$. In case $\alpha_0 < \gamma$ we define L_γ by

$$u_2(0;L_\gamma) = \gamma.$$

In both cases we define $D(L)$ by solving

$$\nu_1(d,L) = 0$$

for d as a function of L. Here, ν_1 is the principal eigenvalue for the linearization of the "elliptic part" of (1.1) about u_2 just as, in §3, it was the principal eigenvalue for the linearization about u_1. The following theorem summarizes our results in this case.

THEOREM 3. Let (i) f be a cubic polynomial of type III
(ii) g satisfy condition (G)
(iii) ϕ and ψ satisfy (1.6)
(iv) ψ satisfy (1.9)

then a) there is a monotone increasing function $D(L)$ defined for $L \geq L_\gamma$ (case 1) or for $L \geq L_0$ (case 2). In case 1 $D(L_\gamma) = 0$ while in case 2, $D(L_0) = d_0 > 0$ as indicated in Figure 7.

b) For (d,L) in region ①, the solution $(0,0)$ is the uniform limit of all non-negative solutions of (1.1)-(1.2) and thus is the only non-negative steady state.

c) For (d,L) in region ② there are at least three steady states; $(0,0)$, $(u_1,0)$ and $(u_2,0)$. Both $(0,0)$ and $(u_2,0)$ are attractors while $(u_1,0)$ is unstable but has a stable manifold of co-dimension 1.

d) For (d,L) in region ③ both $(u_1,0)$ and $(u_2,0)$ are unstable while $(0,0)$ remains an attractor. For (d,L) sufficiently near the boundary curve $d = D(L)$, there is an additional attractor (u_3,v_3) with $v_3 > 0$ and $u_2 > u_3 > 0$ which has bifurcated from $(u_2,0)$.

The proof of this theorem is sufficiently similar to that of Theorem 2 and to the discussion in §4 of [6] that we shall not give the details.

Note that in regions ② and ③ we are unable to determine completely the stationary states of our system.

§5. THE CASE OF TYPE II FUNCTIONS: ALLEE f.

In this case, the positive phase plane and time map for the scalar stationary equation are as in Figure 8.

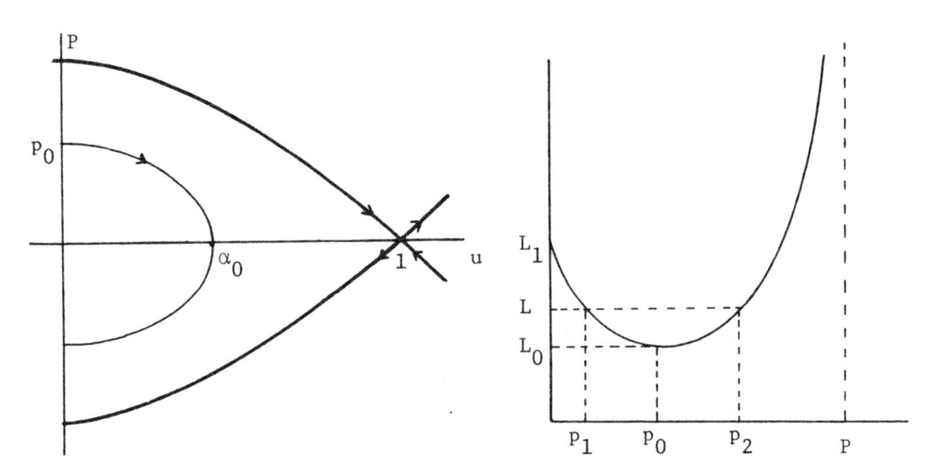

FIGURE 8: Type III f

In this case we have features of both previous types. We have spontaneous bifurcation of a pair of solutions as in type III but $u \equiv 0$ eventually becomes unstable as in type I. The stability of the corresponding steady states for the system, $(0,0)$, $(u_1,0)$ and $(u_2,0)$, can be considered most conveniently if we distinguish three separate cases. If $\alpha_0 < \gamma$ then we consider separately the cases of $L_\gamma > L_1$ and $L_\gamma < L_1$. The third case is defined by $0 < \gamma < \alpha_0$. The corresponding $L - d$ diagrams are pictured in Figure 9.

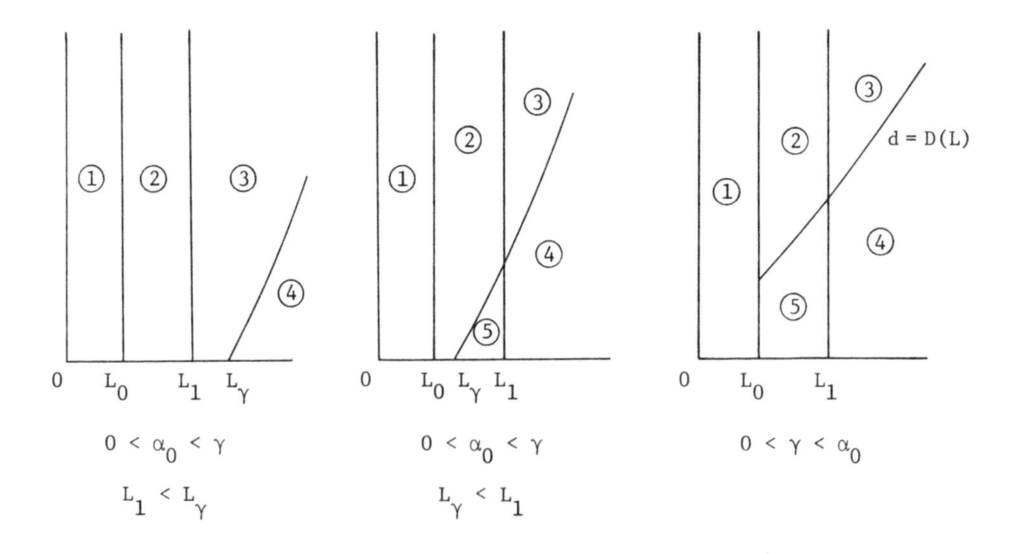

FIGURE 9: $L - d$ planes for type III

If we impose the same restrictions on g, ϕ and ψ as in Theorem 3 then the following is a summary of our results. We will again, for the sake of brevity, not give the details of the arguments.

① In this region $(0,0)$ is the only non-negative steady state and is a global attractor.

② In this region there are three steady states: $(0,0)$, $(u_1,0)$ and $(u_2,0)$. As in the case of type III, both $(0,0)$ and $(u_2,0)$ are attractors while $(u_1,0)$ has a stable manifold fo codimension 1.

③ In this region there are two steady states, $(0,0)$ and $(u_2,0)$. The state $(u_2,0)$ is a global attractor (uniform limit of all non-negative nontrivial solutions) while $(0,0)$ is unstable.

④ Both $(0,0)$ and $(u_2,0)$ are unstable while $(u_1,0)$ no longer exists. For (d,L) near $d = D(L)$ there is an attractor (u_3,v_3) with $v_3 > 0$ and $u_2 > u_3 > 0$ which has simply bifurcated from $(u_2,0)$.

⑤ Both $(u_1,0)$ and $(u_2,0)$ are unstable while $(0,0)$ is an attractor. Again $(u_2,0)$ has bifurcated across $d = D(L)$ so that, near that curve, we have the attractor (u_3,v_3).

In this case of type II functions f it is in regions ②, ④ and ⑤ where we are unable to completely determine the number and character of the steady states.

Except for small values of d we see that as L increases from zero, the point (L,d) passes through ① ② ③ and ④. We thus have the bifurcation diagram pictured in Figure 10.

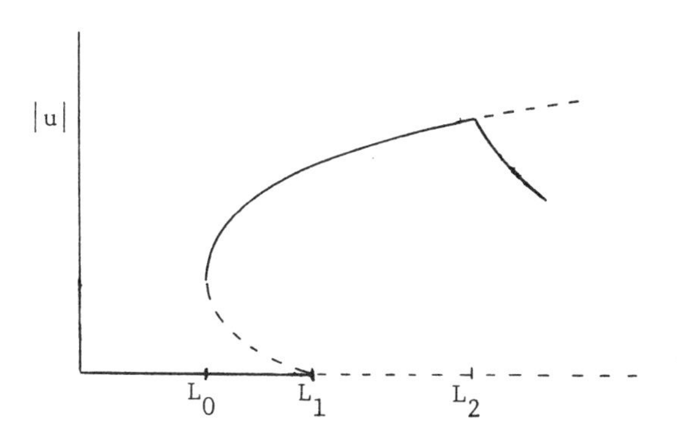

FIGURE 10. Bifurcation Diagram

In that figure the vertical co-ordinate can be interpreted as some measure of the magnitude of the first component of steady states. The value L_2 is determined by $D(L_2) = d$. The dotted lines indicate unstable steady states while the solid lines indicate stable steady states although the curve is defined only locally for $L > L_2$.

REFERENCES

1. N. Alikakos, "L^p bounds of solutions of reaction-diffusion equations," Comm. in Partial Diff. Equations, 4(8), 827-868 (1979).

2. D. G. Aronson and H. F. Weinberger, 1975, Nonlinear diffusion in population genetics, combustion and nerve propagation, in "Proceedings of the Tulane Program in Partial Differential Equations and Related Topics, Lecture Notes in Mathematics 446, Springer, Berlin, 5-49.

3. K. N. Chueh, C. C. Conley, J. A. Smoller, Positively invariant regions for systems of nonlinear diffusion equations, Indiana University Math. J. 26 (1977), pp. 373-391.

4. C. Conley and J. Smoller, "Topological techniques in reaction-diffusion equations," in Biological Growth and Spread, W. Jager, Mathematics, No. 38, Springer-Verlag (1980).

5. C. Conley and J. Smoller, "Remarks on the stability of steady-state solutions of reaction-diffusion equations," in Bifurcation Phenomena in Mathematical Physics and Related Topics, C. Bardos and D. Bessis (Eds.), Reidel Publishing Co. (1980).

6. E. Conway, R. Gardner, and J. Smoller, "Stability and bifurcation of steady-state solutions for predator-prey equations," Advances in Appl. Math. (to appear).

7. E. Conway , D. Hoff and J. Smoller, "Large time behavior of solutions of systems of nonlinear reaction-diffusion equations," SIAM J. Appl. Math. 35, 1-16 (1978).

8. E. Conway and J. Smoller, "Diffusion and predator-prey interaction," SIAM J. Appl. Math. 33 (1977), pp. 673-686.

9. M. Crandall and P. Rabinowitz, "Bifurcation, perturbation of simple eigenvalues, and linearized stability," Arch. Rat. Mech. Anal. 52, 161-180 (1973).

10. R. M. May, Stability and Complexity in Model Ecosystems, Princeton, N. J., (1975) Princeton University Press.

11. R. M. May (editor), Theoretical Ecology, Second Edition, Oxford, Boston (1981); Blackwell Scientific Publication.

12. J. Maynard-Smith, Models in Ecology, Cambridge (1974), Cambridge University Press.

13. V. P. Mikhailov, Partial Differential Equations, MIR Publishers, Moscow, 1978.

14. A. Okubo, Diffusion and Ecological Problems: Mathematical Models, Berlin, Heidelberg, New York (1980) Springer-Verlag.

15. J. R. Philip, "Sociality and sparce populations," Ecology, 38 (1975), 107-111.

16. J. Rauch and J. A. Smoller, "Qualitative theory of the Fitzbugh-Nagumo equations," Advances in Math. 27 (1978) 12-44.

17. J. Smoller, Shock Waves and Reaction Diffusion Equations, Springer-Verlag (to appear).

18. J. Smoller, A. Tromba and A. Wasserman, "Non-degenerate solutions of boundary value problems," J. Nonlinear Anal. 4 (1980), 207-216.

19. J. Smoller and H. Wasserman, "Global bifurcation of steady-state solutions," J. Diff. Eq. 39 (1981), 269-290.

DEPARTMENT OF MATHEMATICS
TULANE UNIVERSITY
NEW ORLEANS, LOUISIANA 70118

Contemporary Mathematics
Volume 17, 1983

ASYMPTOTIC FLAME THEORY WITH COMPLEX CHEMISTRY

Paul C. Fife and Basil Nicolaenko

ABSTRACT

We investigate the structure of laminar flames with general complex chemistry networks in the limit of high activation energy asymptotics. Depending on the specific reaction network and other given thermomechanical data, a wide variety of flame configurations are possible. Here we present a first version of a systematic asymptotic reduction of complex chemistry networks and give criteria to determine the dominant reactions when transport and chemistry are coupled.

1. INTRODUCTION

Mathematically, the subject of this presentation can be considered to be traveling waves for reaction-diffusion systems

$$U_t = DU_{zz} + f(U) \quad , \tag{1.1}$$

where $U \in \mathbb{R}^n$, D is a positive definite transport matrix, and f has the form

$$f(U) = \sum_{j=1}^{m} \omega_j(U)\, K_j \quad .$$

Here ω_j are scalar functions, $K_j \in \mathbb{R}^n$, and each ω_j depends in a certain way on a small parameter ε_j. Typically,

$$\omega_j(U) = \mu_j(\varepsilon_j)\, \psi_j\left(\frac{U-U_j}{\varepsilon_j}\right)$$

for certain reference vectors U_j, constants μ_j, and nonnegative functions ψ_j with $\psi_j(0) = 0$.

Physically, the traveling wave represents a combustion front in a premixed reactive gas. Then the components of U specify the dimensionless

This research partially supported by the Center for Nonlinear Studies, Los Alamos National Laboratory; work also performed under the auspices of the U.S. Department of Energy under contract W-7405-ENG-36 and contract KC-04-02-01, Division of Basic and Engineering Sciences; and partial support was provided by NSF Grant number MCS-8202056.

0271-4132/83/0000-1383/$06.50

temperature and the concentrations of the reactants. There are m chemical reactions involved in the burning process; ω_j are their rates. The smallness of ε_j expresses the fact that these rates depend strongly on temperature (the reactions have high activation energy). Nearly all actual flames have a large reaction network, most of it unknown. Very often it is modeled by a network consisting of a few reactions with high activation energy, and these are the cases in which the present analysis may be useful. Even more commonly, only some of the model reactions have high activation energy; better asymptotic procedures need to be developed for that case.

Traveling waves with velocity c are solutions of the form $U(z,t) = U(z+ct)$; for convenience we keep the same notation U; in a system of coordinates moving with speed c, the traveling wave equation becomes

$$DU'' - \bar{M}\, U' + f(U) = 0 \tag{1.2}$$

where the prime denotes differentiation with respect to $\zeta = z+ct$, and the symbol $\bar{M}$ replaces c. Here $\bar{M}$ is mass flux, which equals velocity if the density is unity. When the density is not taken to be constant, but the pressure _is_ (often a good approximation), the evolution equation is more complicated than (1.1); nevertheless, the problem of determining traveling wave solutions can be reduced to (1.2).

We are concerned with exploiting the smallness of ε_j to obtain approximate wave front solutions. The procedures are linked to the properties of the underlying reaction network, which supplies the vectors K_j. These procedures are well known in the case when the network is modeled by a single reaction $A \to P$. (By way of introduction, this case is reviewed in Sec. 2.) An approach similar to the one in this paper was used in treating two-reaction networks of parallel, sequential, and competing-fuel types in [3] and [4], and it was clear that the extension of this approach to more complex schemes could uncover a baffling array of possibilities. A systematic formalism was called for, and hopefully is provided in part by the present paper.

The methods are those of asymptotic analysis. Roughly speaking, to lowest order this means developing rational procedures for replacing the original problem with a limit problem in which the ε_i do not appear explicitly. The limit problem is expected to yield an approximate solution, and is easy to solve. In the limit, flame sheets appear as free boundaries with jumps in the gradients. A rigorous justification of this expectation has been given recently for the case of a single reaction [1].

Many other parameters besides ε_i enter the problem, for example, the elements of the matrix D and of the unburned state vector of the gas, U_-. Throughout the paper, we assume that their magnitudes do not interfere with the asymptotics. Particularly, we assume that any constants not depending explicitly on ε_i, B_i, or θ_i (as defined in 2.2 and 2.5) are O(1) quantities. This especially refers to elements of D and D^{-1}, reference dimensionless T_j, their differences $T_i - T_j$ and the heat releases Q_j. These restrictions are relaxed, and a more meticulous account of the other parameters is taken in [4].

For background and previous work on asymptotic methods in flame theory, the reader is referred to [2]. A number of papers have been written on the multiple reaction case; see the references in [3].

2. THE CASE OF A SINGLE-REACTION MECHANISM

Although this argument has appeared in the literature many times before, it will be convenient to review here the asymptotic analysis of simple flames, that is, those with a single reaction A → P, hence a single K:

$$DU'' - \bar{M}U' = -\omega(U)K \quad , \; U(-\infty) = U_- \; . \tag{2.1}$$

Here U = (T,Y), the temperature and molar concentration of A, K = (Q,-1), and we take

$$\omega(U) = BY\, e^{-\theta/T} \quad , \tag{2.2}$$

for positive constants B,θ. As $z \to \infty$, the reaction must go to equilibrium, that is,

$$Y(\infty) = 0 \quad . \tag{2.3}$$

Integrating (2.1), we find

$$[DU' - \bar{M}U]_{-\infty}^{\infty} = -\int_{-\infty}^{+\infty} \omega(U(z))\, dz\, K \equiv -\beta\, K \quad ,$$

hence $U(\infty) - U(-\infty) = \alpha K$, $\alpha = \beta/\bar{M}$. Applying (2.3), we find $\alpha = Y_-$, so

$$U(\infty) \equiv U_+ = U_- + Y_- \, K \equiv (T_+, 0) \quad . \tag{2.4}$$

We define $\varepsilon = T_+^2/\theta$, and assume

$$0 < \varepsilon \ll 1 \tag{2.5}$$

In a neighborhood of $T = T_+$, $Y = 0$, we approximate

$$\omega(U) = \frac{\eta}{\varepsilon}\left(\frac{Y}{\varepsilon}\exp[(T - T_+)/\varepsilon]\right) \equiv \frac{\eta}{\varepsilon}\, y\, e^t \quad ,$$

where $u = (t,y) \equiv (U - U_+)/\varepsilon$, and

$$\eta = B\,\varepsilon^2 \exp(-\theta/T_+) \quad .$$

This approximation is obtained by taking the first two terms in the Taylor series expansion of the exponent θ/T in (2.2) about $T = T_+$. The remainder is $O(\varepsilon)$ and is discarded. A careful analysis of this approximation is given in [4].

We next introduce new variables

$$x = \zeta\eta^{\frac{1}{2}} \quad , \quad M = \bar{M}\eta^{-\frac{1}{2}} \tag{2.6}$$

in (2.1) to obtain (differentiation is now with respect to x)

$$D\ddot{U} - M\,\dot{U} = -\frac{1}{\varepsilon}\,\psi\left(\frac{U-U_+}{\varepsilon}\right) K \quad , \tag{2.7}$$

where (recall $u = (U-U_+)/\varepsilon$)

$$\psi(u) = y\, e^t \quad . \tag{2.8}$$

We may position the origin in space at will; we choose to place it where $Y = \varepsilon$ ($y = 1$). Rewriting (2.7) with stretched variables $\xi = x/\varepsilon$, $u = (U-U_+/\varepsilon)$, and discarding terms formally of order ε, we obtain the "flame layer" equation

$$D\,\frac{\partial^2 u}{\partial \xi^2} + \psi(u)\,K = 0 \quad . \tag{2.9}$$

This is to be solved under the boundary conditions $u(\infty) = 0$, $y(0) = 1$. To do so, we observe that (2.9) and the condition $u(\infty) = 0$ imply

$$u = \sigma(\xi)D^{-1}K \tag{2.10}$$

for <u>some scalar function</u> σ. Substituting into (2.9), we have

$$\sigma'' + \psi(\sigma D^{-1}K) = 0 \quad , \; \sigma(\infty) = 0 \quad .$$

This has a first integral

$$(\sigma')^2 - \Omega(\sigma) = 0 \quad ,$$

where $\Omega(\sigma) = 2 \int_\sigma^o \psi(sD^{-1} K)\, ds$. Assuming that $D^{-1}K$ has first component positive and the second negative, we see from (2.8) that $\Omega(-\infty)$ exists. Thus the layer problem has a unique solution, and

$$D\frac{\partial u}{\partial \xi}(-\infty) = \beta\, K \quad ,$$

where $\beta = (\Omega(-\infty))^{\frac{1}{2}}$. Matching with the solution outside the flame layer requires, in the limit $\varepsilon \to 0$, that

$$D\dot{U}(0-) = \beta\, K,\; D\dot{U}(0+) = 0 \quad .$$

Finally, for $x < 0$, $T < T+$ and the strong dependence of ψ on T shows that it can be neglected in (2.7). For $x > 0$, Y is already nearly zero if ε is small, so ψ can again be neglected. What is left is the problem

$$D\ddot{U} - M\dot{U} = 0 \quad , \; x \neq 0 \quad , \tag{2.11}$$

$$[D\dot{U}]_{0-}^{0+} = -\beta\, K \quad , \tag{2.12}$$

$$U(\pm\infty) = U\pm \quad . \tag{2.13}$$

This problem has a unique solution. Integration of (2.11) reveals (from (2.4)),

$$[D\dot{U}]_{-\infty}^{0-} = M[U]_{-\infty}^{0-} = M(U_+ - U_-) = MY_-K \quad .$$

Adding this to (2.12), we have

$$0 = [DU]_{-\infty}^{0+} = MY_{-}K - \beta K \quad .$$

Hence,

$$M = \beta / Y_{-} \quad .$$

And with M known, (2.11) can be integrated backward with initial conditions at $x = 0$ to yield the total profile for $x < 0$. For $x > 0$, the only solution is $U \equiv U_{+}$.

In summary, asymptotics have shown the way to a limit problem (2.11) - (2.13) with nonlinearity replaced by a jump condition (2.12). The coefficient β is known independently from a layer analysis, and $(M, U(x))$ are to be determined. In the original variables, U and $\bar{M}$ are determined by reversing the transformation (2.6).

3. FEASIBLE ALLOCATIONS

We begin by assuming given a reaction network, which is simply a specification of which reactions take place in the combustion process. It is usually symbolized by arrows, such as:

$$A_1 + A_2 \to A_3 \quad , \; 2A_2 + A_3 \to P \quad , \; A_1 \to 2A_1 \quad . \tag{3.1}$$

(In this example, there are three reactions involving three reacting species A_i, and one inert product P.) In general, we symbolize the reactions by R_i, the total number of reactions by m, and the number of species by n.

The vector $Y = (Y_1, \ldots, Y_n)$ signifies mass concentration of the various species, and will be a function of space and time. We will be concerned only with nonnegative concentrations, so restrict Y to the nonnegative cone $\mathcal{Y}$ of n-dimensional (species) space. To simplify the presentation, we will take equal molecular weights for all species.

Associated with each reaction R_j is a <u>reaction vector</u> $V_j \in \mathcal{Y}$. It represents the net change in the number of mass units of the various species during a reaction event of reaction R_j. For example, in the above network (3.1), the three reaction vectors are $(-1, -1, 1)$, $(0, -2, -1)$, and $(1, 0, 0)$. Our basic assumption about these vectors is as follows:

<u>*Assumption 1*</u>. *If* $\sum_{j=1}^{m} \alpha_j V_j = 0$ *and* $\alpha_j \geq 0$, *then* $\alpha_j \equiv 0$. (3.2)

This assumption excludes, for example, reversible reactions, in which $V_2 = -V_1$ (say). More generally, it assures, as we shall see, that equilibrium can only be reached when each reaction rate is zero.

Each reaction R_j is assumed to have a reaction rate $\omega_j \geq 0$ which is a function of Y and of the temperature T.

<u>*Def:*</u> *$Y \in \mathscr{Y}$ is an <u>equilibrium state</u> if, for all T,*

$$\sum_{j=1}^{m} \omega_j(Y,T)\, V_j = 0 \quad . \tag{3.3}$$

It follows from (3.2) that this can only happen if each $\omega_j = 0$.

For each $i = 1, \ldots, m$, we denote by $J(i)$ the set of all indices j of species A_j entering reaction R_i (appearing on the left side of the arrow). Eventually, a specific form will be assumed for the ω_i; but at this point we assume $\omega_i = 0$ if and only if $Y_j = 0$ for some $j \in J(i)$. We symbolize the equilibrium cone for each reaction, and the total equilibrium cone, as follows:

$$\mathscr{E}_i = \{Y \in \mathscr{Y}\colon Y_j = 0 \text{ for some } j \in J(i)\} \quad ,$$

$$\mathscr{E} = \bigcap_{i=1}^{m} \mathscr{E}_i \quad .$$

(This is, of course, a cone imbedded in a set of coordinate hyperplanes.)

For simplicity only we require

<u>*Assumption 2.*</u> *$V_i \notin \mathscr{E}_i$, for all i.*

<u>*Def:*</u> *Given an input state $Y_- \in \mathscr{Y}$, an <u>allocation</u> of Y_- to the network is a set of numbers $\alpha_i \geq 0$, $i = 1, \ldots, m$, such that*

$$Y_- + \sum_{i=1}^{m} \alpha_i V_i \in \mathscr{E} \quad . \tag{3.4}$$

In all the examples below, we normalize the molecular weights to unity.

<u>Ex. 1</u>. In the network (3.1), we have $\mathscr{E} = \{Y = (Y_1, Y_2, Y_3) \geq 0 : Y_1 = 0$, and Y_2 or $Y_3 = 0\}$. Let $Y_- = (1, 1, 1)$. Then the only allocation is $\alpha_1 = 1$,

$\alpha_2 = \frac{1}{2}$, $\alpha_3 = 0$.

Ex. 2: $A_1 \xrightarrow{R_1} 2A_2+P_1$, $A_2 \xrightarrow{R_2} 2A_1+P_2$, $Y_- = (1,1)$. Here $\mathcal{E} = \{0\}$, $V_1 = (-1,2)$, $V_2 = (2,-1)$. The reader may check that (3.2) is satisfied, but that no allocation exists. Thus, the existence of an allocation is not always guaranteed.

Ex. 3:

$$A_1 \xrightarrow{R_1} P_1, \quad A_1 \xrightarrow{R_2} P_2 \quad . \tag{3.5}$$

Here $\mathcal{E} = \{0\}$, Y_- is a scalar, and (α_1,α_2) is an allocation iff $\alpha_1 + \alpha_2 = Y_-$, $\alpha_i \geq 0$. So, of course, allocations are not necessarily unique.

Besides concentrations, we are also concerned with temperature, so use $n + 1$-component state vectors $U = (T, Y_1, \ldots, Y_n) = (T,Y)$. Associated with each reaction is a heat release $Q_i > 0$. We define the $(n + 1)$ component vectors

$$K_i = (Q_i, V_i) \quad ,$$

and shall refer to Q_i as the 0th component of K_i.

Our basic goal, now, is to find approximations to bounded solutions $(U(x), M)$ of the problem

$$D\ddot{U} - M\dot{U} = \sum_{j=1}^{m} \omega_j(U,\theta_j)\, K_j \quad , \; U(-\infty) = U_- = (T_-,Y_-) \quad . \tag{3.6}$$

Here we have indicated the dependence of the rates ω_j on large positive parameters θ_j (activation temperatures). Specifically, we take

$$\omega_i = B_i \prod_{j\in J(i)} Y_j^{\nu(j,i)}\, e^{-\theta_i/T} \tag{3.7}$$

where $B_i > 0$ also might be large or small, and $\nu(j,i)$ is the stoichiometric coefficient of species j on the left side of reaction i.

The procedure will be to construct a suitable reduced problem in which each nonlinear term $\omega_j K_j$ is replaced by a jump condition in $D\dot{U}$ at some point x_j, not known in advance. The reduced problem will be fashioned in terms of "feasible allocations," which we proceed to define.

Suppose an allocation $\alpha = (\alpha_1, \ldots, \alpha_m)$ is given. Let

$$U_+ = U_- + \sum_{j=1}^{m} \alpha_j K_j \quad . \tag{3.8}$$

In particular,

$$T_+ = T_+(\alpha) = T_- + \sum_{j=1}^{m} \alpha_j Q_j \quad . \tag{3.9}$$

Using the constants in (3.7), we define m functions

$$H_j(T) = \varepsilon(\ell n\, B_j - \theta_j/T) \quad , \tag{3.10}$$

where

$$\varepsilon \equiv \min_j \, (T_+^2/\theta_j) \quad . \tag{3.11}$$

We define temperatures T_j, for all j such that $\alpha_j \neq 0$, such that

$$\begin{aligned} &H_j(T_j) \text{ is independent of } j, \text{ and} \\ &\text{Max } T_j = T_+ \quad . \end{aligned} \tag{3.12}$$

(For j such that $\alpha_j = 0$, T_j is not defined.)

Prop. 3.1. *There exists a unique temperature sequence* $\{T_j\}$ *satisfying (3.12).*

Proof. Choose the index k so that $H_k(T_+) = \text{Min}_j\, H_j(T_+)$. We set $T_k = T_+$, and define the T_j by $H_j(T_j) = H_k(T_+)$. Since all H_j are increasing functions of T, it follows that all $T_j \leq T_+$, so there exists such a sequence. This argument also shows there is only one such sequence.

We now reorder the reactions in order of increasing T_j, so that $T_1 \leq T_2 \leq \ldots \leq T_m = T_+$. It can happen that some of the T_j coincide. In fact, Example 2 at the end of this section indicates that to be "feasible," an allocation may well have to be adjusted so that two of them do coincide. (However, it seems unlikely that the case when three are the same is important.) Let $\{T_i^*\}$, $i = 1, \ldots, m^* \leq m$ be the (possibly shorter) sequence of distinct temperatures among the T_j; we denote the number of them by m^*.

It also may happen that two of the T_j almost coincide, that is, $0 < |T_i - T_j| = O(\varepsilon_i)$. This case is considered non generic, and is hereby excluded from consideration. Thus, we assume that two temperatures T_i, T_j either coincide (this may be required by the feasibility criterion given later) or differ by a quantity of larger order than $\varepsilon_i + \varepsilon_j$. We also assume that at each T_i^*, all but perhaps two of the quantities $H_k(T_i^*)$ are distinct.

Our attempt will be to construct a combustion front in which, roughly speaking, m^* sets of reactions go to completion at the temperatures T_i^* in that same order as one proceeds across the front's profile in the (upstream) direction of increasing temperatures. For that purpose, we use the notion of suballocation.

Def: *For each $\ell \leq m^*$, suballocations $\alpha^{(\ell)}$ are defined to be allocations for the subnetwork consisting only of the R_j for $T_j \leq T_\ell^*$. Let k_ℓ be the number of such reactions.*

Thus, $\alpha^{(\ell)} = (\alpha_1^{(\ell)}, \ldots, \alpha_{k_\ell}^{(\ell)})$, but for comparison purposes, we extend them to be sequences of length m by defining $\alpha_j^{(\ell)} = 0$ for $j > k_\ell$. We say $\alpha^{(\ell)} \leq \alpha^{(\ell')}$ if $\alpha_j^{(\ell)} \leq \alpha_j^{(\ell')}$ for all j.

The situation we envisage is when there exist suballocations $\alpha^{(\ell)}$ for all $\ell \leq m^*$ satisfying

$$\alpha^{(\ell)} \leq \alpha^{(\ell+1)} \quad , \ \ell = 1, \ldots, m^*-1 \quad ; \ \alpha^{(m^*)} = \alpha \quad . \tag{3.13}$$

There is a further condition which we shall require the $\alpha^{(\ell)}$ to satisfy. Let us fix $\ell \leq m^*$ and $\alpha^{(\ell-1)}$. There may be many suballocations $\alpha^{(\ell)} \geq \alpha^{(\ell-1)}$. We say that $\alpha^{(\ell)}$ is "maximal" if it allocates a maximal amount of the reactants to those reactions with the largest values of H at the temperature T_ℓ. This, simply put, is the principle that among competing reactions, the one with the largest rate (as defined by the functions $H_i(T)$) wins out. The precise definition is as follows. First suppose all the quantities $H_k(T_\ell^*)$ (T_ℓ^* fixed) are distinct. Let k_1 be the index that yields the largest of them; then k_2 the next largest, etc. Out of all possible suballocations $\alpha^{(\ell)} \geq \alpha^{(\ell-1)}$, $\alpha^{(\ell)}_{k_1}$ is required to be maximal. With it known, $\alpha^{(\ell)}_{k_2}$ is then maximal, etc. In this case, $\alpha^{(\ell)}$ is uniquely determined. Next, suppose two of the $H_k(T^*_\ell)$ coincide, say $H_{k_1}(T^*_\ell) = H_{k_2}(T^*_\ell)$. Then instead of maximizing $\alpha^{(\ell)}_{k_1}$, we maximize $\alpha^{(\ell)}_{k_1} + \alpha^{(\ell)}_{k_2}$. (This, of course, does not immediately yield unique separate values of $\alpha^{(\ell)}_{k_i}$.

If such suballocations exist, we define a new set of reaction vectors V'_j as follows:

For each $k \leq m^*$,

$$\alpha'_k \equiv \sum_j (\alpha_j^{(k)} - \alpha_j^{(k-1)})$$

(of course taking $\alpha_j^{(0)} = 0$). If $\alpha'_k \neq 0$, then

$$V'_k \equiv \frac{\sum_j (\alpha_j^{(k)} - \alpha_j^{(k-1)})\, V_j}{\alpha'_k} \tag{3.14}$$

If $\alpha'_k = 0$ for any values of k, we delete those temperatures T^*_k from the sequence and accordingly renumber the temperatures, as well as the V'_j, so there will be a fewer number of each. For convenience we also call the new number m^*.

In summary, we have constructed an alternative set of vectors V'_j, and number α'_j, with the properties that for all $\ell \leq m^*$,

$$Y_- + \sum_{j=1}^{\ell} \alpha'_j V'_j = Y_- + \sum_{j=1}^{k_\ell} \alpha_j^{(\ell)} V_j \in \bigcap_{j=1}^{k_\ell} \mathscr{E}_j \quad ,$$

and

$$\sum_{j=1}^{m^*} \alpha'_j V'_j = \sum_{j=1}^{m} \alpha_j V_j \quad ,$$

so that

$$Y_- + \sum_{j=1}^{m^*} \alpha'_j V'_j = Y_+(\alpha) \in \mathscr{E} \quad .$$

In some cases it may be enlightening to think of the V_j' as representing the reaction vector of an alternate reaction R_j' which is the same linear combination of the R_i, $i \leq j$, as is V_j' of the V_i. For instance, in Case 2 of Example 3 below, R_2' and R_3' would be $\frac{1}{2} A_2 \to P$ and $1/3\ A_1 \to P$, respectively.

Extended vectors K_k' are defined to be the same linear combinations of the K_j as the V_k' are of the V_j (3.14). In particular, the first component Q_k' will be a linear combination of the Q_j's. We can now determine the temperature at which the suballocation $\alpha^{(\ell)}$ would bring the corresponding subnetwork to equilibrium, if no reaction outside that subnetwork were present. We call it

$$\hat{T}_\ell = \sum_{j=1}^{\ell} \alpha_j' Q_j' + T_- \quad .$$

Def: The allocation α is feasible if

(1) there exist, for each $\ell \leq m^*$, suballocations $\alpha^{(\ell)}$ which are maximal and satisfy $\alpha^{(m^*)} = \alpha$, and

(2) $T_j^* > \hat{T}_j$ for $j < m^*$.

Example 1.

$$\begin{aligned} R_1&: A_1 \to A_2, & V_1 &= (-1,1)\\ R_2&: A_1 + A_2 \to P_1, & V_2 &= (-1,-1)\\ R_3&: A_2 \to P_2, & V_3 &= (0,-1)\\ R_4&: A_1 \to P_3, & V_4 &= (-1,0). \end{aligned}$$

These V_j satisfy (3.2). $\mathscr{E} = \{0\}$. We take $Y = (1,1)$; then all possible allocations are of the form $\alpha = (\theta, \phi, 1 + \theta - \phi, 1 - \theta - \phi)$, where $\theta, \phi \geq 0$ and $\theta + \phi \leq 1$. Suppose that for each α, the new ordering is the same as the old: $T_1 \leq T_2 \leq T_3 \leq T_4$. Then the only allocation with suballocations satisfying (3.13) is the one with $\theta = 1$, $\phi = 0$; $\alpha = (1, 0, 2, 0)$. The suballocations are $\alpha^{(1)} = (1)$, $\alpha^{(2)} = (1,0)$, $\alpha^{(3)} = (1,0,2)$, $\alpha^{(4)} = \alpha$. We then delete R_2 and R_4, so $m^* = 2$ and the new $\alpha = (1,2)$. Criterion (2) for feasibility still needs to be checked. If it is verified, we can expect a combustion front to exist in which only R_1 and R_3 occur, at two different temperatures. Since there is at most one feasible allocation, we expect there is no other such combustion front.

Example 2.

$$\begin{aligned} R_1&: A \to P_1 & V_1 &= -1\\ R_2&: A \to P_2 & V_2 &= -1\\ R_3&: A \to P_3 & V_3 &= -1. \end{aligned}$$

Again, $\mathscr{E} = \{0\}$. For $Y_- = 1$, all possible allocations are $\alpha = (\alpha_1, \alpha_2, \alpha_3)$, $\alpha_i \geq 0$, $\Sigma \alpha_i = 1$, and $T_+(\alpha) = \Sigma \alpha_j Q_j$. To discuss the feasible ones, we use the symbols $\alpha_*^{(i)}$ to denote the simple allocations: $\alpha_*^{(1)}$: (1,0,0), etc., and let $T^{(i)}$ be the corresponding burning temperatures: $T^{(i)} = T_- + Q_i$. Suppose $T^{(1)} < T^{(2)} < T^{(3)}$. Define $\bar{H}(T) = \text{Max } H_i(T)$. If, for some j = 1, 2, or 3, $\bar{H}(T^{(j)}) = H_j(T^{(j)})$, then $\alpha_*^{(j)}$ is a feasible allocation. If this is not true for any j, then it is a matter of checking that one of the following three possibilities must occur:

(a) there exists $T_{12} \in (T^{(1)}, T^{(2)})$ such that $H_1(T_{12}) = H_2(T_{12}) = \bar{H}(T_{12})$;

(b) similarly for a temperature T_{13};

(c) similarly for a temperature T_{23}.

In case (a), there is a feasible allocation $(\alpha_1, \alpha_2, 0)$ where α_i are chosen so that $T_{12} = \alpha_1 T^{(1)} + \alpha_2 T^{(2)}$. Similarly, for cases (b) and (c). Hence there is always at least one feasible allocation. The same conclusion holds for any number of competing reactions of this type.

Example 3. $A_1 \xrightarrow{R_1} A_2 \xrightarrow{R_2} A_3 \xrightarrow{R_3} P$,

$V_1 = (-1,1,0)$, $V_2 = (0,-1,1)$, $V_3 = (0,0,-1)$. Let $Y_- = (1,1,1)$. There is only one possible allocation: $\alpha = (1,2,3)$, and so the temperatures T_i are independent of α. We examine two of the possible cases for their ordering.

Case 1: $T_1 < T_2 < T_3$. There exist suballocations $\alpha^{(1)} = (1)$, $\alpha^{(2)} = (1,2)$, and $\alpha^{(3)} = \alpha$. The question of feasibility rests only on criterion (2).

Case 2: $T_3 < T_2 < T_1$. We reorder the reactions so that R_3 becomes R_1 and vice versa. With the new ordering, $\alpha = (3,2,1,)$, and there are suballocations $\alpha^{(1)} = (1)$, $\alpha^{(2)} = (2,1)$, $\alpha^{(3)} = (3,2,1)$, satisfying (3.13). The following table holds:

j	V_j	V'_j	Q'_j	$\hat{T}_j$
1	(0,0,-1)	(0,0,-1)	Q_1	$T_- + Q_1$
2	(0,-1,1)	(0,-½,0)	$(Q_1 + Q_2)/2$	$T_- + 2Q_1 + Q_2$
3	(-1,1,0)	(-1/3,0,0)	$(Q_1 + Q_2 + Q_3)/3$	$T_- + 3Q_1 + 2Q_2 + Q_3$

Moreover, $\alpha' = (1,2,3)$, so that indeed $\Sigma \alpha_j V_j = \Sigma \alpha'_j V'_j$. If condition (2) holds, we expect a front with reactions taking place as indicated by the V'_j: First, $A_3 \to P$ goes to completion at $T_1 (= T_1^*)$; then $A_2 \to P$ goes at T_2, with A_3 an intermediate which appears only momentarily (the reaction $R_1 : A_3 \to P$ is much faster than R_2 at this temperature); and finally $A_1 \to P$, passing through two quick intermediates.

Example 4. $A_1 \xrightarrow{R_1} A_2 \xrightarrow{R_2} P_1$, $A_1 \xrightarrow{R_3} P_2$; $Y_- = (1,1)$:

$\mathscr{E} = \{0\}$, $V_1 = (-1,1)$, $V_2 = (0,-1)$, $V_3 = (-1,0)$, and all allocations are of the form $\alpha = (\theta, 1+\theta, 1-\theta)$, $\theta \in [0,1]$. The details are tedious, but it can be shown that allocations always exist which satisfy the feasibility criterion (1). If a feasible allocation exists with $\theta \neq 0,1$, then necessarily two of the three T_i's coincide. This may be the only feasible allocation in some cases, and so is important. On the other hand, it would only accidentally happen that all three coincide.

4. THE LIMIT PROBLEM, LEVEL 1.

The fact is that the existence of a feasible allocation is by no means ensured; Example 3 above demonstrates this. If one does exist, the formulation of the appropriate limit problem will, under certain conditions, be described below. It is, in some sense, a formal approximation to the front problem and has a unique solution. We conjecture that if an allocation exists at all, then there always exists one satisfying criterion (1). If this allocation does not satisfy (2), then it may be possible to find two (or more) fronts traveling at different speeds such that the combined effect of the faster followed by the slower is to attain the burned state $U_+(\alpha)$, the various reactions implied by the suballocations going to completion in the prescribed order. This is certainly true in Example 3 at the end of Sec. 3. In general, suppose that criterion (1) holds, and also (2) for all j except one, say $j = \ell$. Then it is expected that there will exist a fast front, with burning governed by the "reactions" corresponding to $\alpha^{(\ell)}$ and final state $\hat{U}_\ell = U_- + \sum^{k_\ell} \alpha_j' K_j'$. There will also be a slower front with input $\hat{U}_\ell$, reactions described by the V_j' for $j > \ell$, and final state U_+.

Given a feasible allocation, we have the problem of determining the front's approximate speed and profile. The description of the procedure will be given below for the case when all the T_j are distinct, and D is diagonal; i.e.,

$$T_j = T_j^* \quad ; \quad D \text{ is diagonal} \quad . \tag{4.1}$$

The point of departure is the following boundary value problem, the natural extension of (2.1) to the case of several reactions:

$$DU'' - \bar{M}U' = -\sum_{j=1}^{m} \omega_j(U)\, K_j \quad , \; U(-\infty) = U_- \quad , \; U(\infty) = U_+(\alpha) \quad . \tag{4.2}$$

Here the vectors K_j are extended reaction vectors defined in Sec. 3.

The limit procedure proceeds under the approximation that all reaction rates vanish except where T is near one of the T_j^*'s; and that in the neighborhood of T_j^*, the requisite reactions which ensure that all R_k, for $k \le k_j$, reach equilibrium, all go to completion. The principle is that the effect of those reactions at T_j^* can be accounted for by prescribing a jump discontinuity in $\dot{U}$ at that point.

First, we need to rescale the problem as in Sec. 2. For each $j = 1, \ldots, m$, we define

$$\varepsilon_j = T_+^2/\theta_j \quad , \tag{4.3}$$

$$\eta_j = B_j\, \varepsilon_j^{1+\gamma(j)} \exp(-\theta_j/T_j) \quad , \tag{4.4}$$

where $\gamma(j)$ is the sum of left stoichiometric coefficients in reaction j of those species which go to zero in that reaction. We have

$$\omega_j(U) \cong \frac{\eta_j}{\varepsilon_j}\, \psi_j(u^{(j)}) \quad ,$$

where $u^{(j)} = \dfrac{U-U_j}{\varepsilon_j}$, and U_j is the value of U (not completely known beforehand, except when $j = m$) at the temperature T_j. Here ψ_j will be functions analogous to (2.8). Let $\eta = \eta_m$, and rescale as (2.6). The result is the equation

$$D\ddot{U} - M\dot{U} = -\sum_{j=1}^{m} \frac{r_j}{\varepsilon_j}\, \psi_j\!\left(\frac{U-U_j}{\varepsilon_j}\right) K_j \quad , \tag{4.5}$$

where

$$r_j = \eta_j/\eta \quad . \tag{4.6}$$

If x_j are the positions where $T = T_j^*$, the equations then become

$$D\ddot{U} - M\dot{U} = 0 \quad , \; x \neq x_j \quad , \; j = 1, \ldots, m^* \quad . \tag{4.7}$$

$$[D\dot{U}]_{x_j} = -\beta_j K_j' \quad , \; T(x_j) = T_j^* \quad , \tag{4.8}$$

$$U(-\infty) = U_- \quad , \; U(\infty) = U_+(\alpha) \quad , \tag{4.9}$$

where $[D\dot{U}]_{x_j} = \lim_{x \downarrow x_j} D\dot{U}(x) - \lim_{x \uparrow x_j} D\dot{U}(x)$, and the K_j' are the vectors defined in Sec. 3.

There are two conditions which we need to impose on the β_j in order to ensure a well posed problem. According to our definition of the suballocations $\alpha^{(\ell)}$, for each ℓ, the concentration vector

$$\hat{Y}_\ell \equiv Y_- + \sum_{j=1}^{k_\ell} \alpha_j^{(\ell)} V_j = Y_- + \sum_{j=1}^{\ell} \alpha_j' V_j'$$

is in equilibrium with respect to the set of reactions R_j, $j \leq k_\ell$. The vector $\hat{Y}_{\ell-1}$, however, is not in this same equilibrium cone; otherwise by Assumption 2, $\alpha_\ell' = 0$ and we would have discarded the temperature T_ℓ^* and renumbered the others. This means there is some index $\kappa(\ell)$ such that the κ-th component of $\hat{Y}_\ell$ vanishes, but not the κ-th component of $\hat{Y}_{\ell-1}$. Since $\hat{Y}_\ell - \hat{Y}_{\ell-1} = \alpha_\ell' V_\ell'$, this implies that the κ-th component of V_ℓ' is negative, and that the coefficient α_ℓ' is uniquely determined by the knowledge of $\kappa(\ell)$. For notational convenience, we set $\mathscr{E}_\kappa^0 = \{Y \in \mathscr{Y} : Y_\kappa = 0\}$. Thus, necessarily there is a function $\kappa(\ell)$ such that $\hat{Y}_\ell \in \mathscr{E}^0_{\kappa(\ell)}$, but $\hat{Y}_{\ell-1} \notin \mathscr{E}^0_{\kappa(\ell)}$. We shall require that the solution $U(x)$ of the reduced problem has the same property:

$$Y(x) \in \mathscr{E}^0_{\kappa(\ell)} \text{ for } x > x_\ell \; .$$

Since D is diagonal, this implies $D\,\dot{Y} \in \mathscr{E}^0_{\kappa(\ell)}$, $x > x_\ell$. In particular,

$$Y(x_\ell) \in \mathscr{E}^0_{\kappa(\ell)} \quad , \; D\,\dot{Y}(x_\ell +) \in \mathscr{E}^0_{\kappa(\ell)} \quad . \tag{4.10}$$

Finally, there is a condition on β_m. First, suppose that $K_m = K'_m$; i.e., only reaction R_m goes forward at $T_m = T_+$. Then the flame layer analysis given in Sec. 2 for the determination of β then can also be done at T_m; the reason essentially is that $U \equiv U_+ = \text{const}$ for $x > 0$. As a result, we have

$$\beta_m = \left(2\int_{-\infty}^{0} \psi_m(s\ D^{-1}\ K_m)\ ds\right)^{\frac{1}{2}} \ . \tag{4.11}$$

On the other hand if $K_m \neq K'_m$, then we have to account for other reactions occurring. Their rate functions will be of larger order of magnitude than that of R_m, and it turns out that a certain pseudo-steady-state approximation is valid near T_m, and reduces the analysis to the previous case. This results in a condition (4.11) with a slightly altered ψ_m, and K_m replaced by K'_m.

We need to state precisely what the problem consists of.

Limit Problem, Level 1: *Let U_+, K'_j, and T_j be determined from a feasible allocation α, and assume (4.1). Solve (4.7) - (4.9) under the conditions (4.10), (4.11) (possibly modifying (4.11) as indicated). Then the (approximate) actual mass flux and profile are given by*

$$\bar{M} = M\,\eta^{\frac{1}{2}} \quad , \ \bar{U}(\zeta) = U(\eta^{\frac{1}{2}}\zeta) \ .$$

The limit problem is easily solved. In fact, (4.7) shows that $D\dot{U} - MU$ is constant between the points x_j. In other words,

$$[D\dot{U} - MU]_{x_{j-1}+}^{x_j-} = 0, \quad \text{and (4.8) can be written } [D\dot{U} - MU]_{x_j-}^{x_j+} = -\beta_j\ K'_j .$$

Summing both over j, from 1 to ℓ, we find

$$[D\dot{U} - MU]_{-\infty}^{x_\ell+} = -\sum_{j=1}^{\ell} \beta_j\ K'_j \ .$$

But the left side is $D\dot{U}(x_\ell+) - MU(x_\ell) + MU_-$. We divide by M and disregard the 0-th component of this equation, to obtain

$$Y_- + \sum_{j=1}^{\ell} \frac{\beta_j}{M}\ V'_j = Y(x_\ell) - \frac{1}{M}\ D\dot{Y}(x_\ell+) \quad ,$$

which by requirement (4.10) must lie in $\mathscr{E}^0_{\kappa(\ell)}$. But we also know that $\hat{Y}_\ell = Y_- + \sum_{j=1}^{\ell} \alpha'_j V'_j \in \mathscr{E}_{\kappa(\ell)}$ and $\hat{Y}_{\ell-1}$ is not. By the previous argument, the coefficients α'_j are uniquely determined by the $\kappa(\ell)$, so from a simple recurrence argument:

$$\frac{\beta j}{M} = \alpha'_j \quad . \tag{4.12}$$

We determine M by setting j = m and using (4.11). Knowing M, we can find the β_j for j < m from (4.12) again.

We arbitrarily set $x_m = 0$. For x > 0, $D\ddot{U} - M\dot{U} = 0$, so $D\dot{U} - MU = \text{const} = -MU_+$. Thus $U = U_+ + e^{MD^{-1}x}C$ for some C. But since D^{-1} is positive definite, the only bounded solution is with C = 0. Therefore $U(x) \equiv U_+$ for x > 0. From (4.8), we have $D\dot{U}(0-) = \beta_m K'_m$. Solving (4.7) now, with initial data $U = U_+$, $D\dot{U}$ prescribed at x = 0, gives

$$U(x) = \hat{U}_{m-1} + \alpha'_m K'_m e^{MD^{-1}x} \quad , x_{m-1} < x < x_m \quad , \tag{4.13}$$

where

$$\hat{U}_{m-1} = U_+ - \alpha'_m K'_m = U_- + \sum_{j=1}^{m-1} \alpha'_j K'_j$$

is the end state given by the suballocation $\alpha^{(m-1)}$. Its first component is $\hat{T}_{m-1}$, defined earlier. The function on the right of (4.13) approaches $\hat{U}_{m-1}$ as $x \to -\infty$, so the condition $T_{m-1} > \hat{T}_{m-1}$ in the definition of feasibility ensures the existence of an $x_{m-1} < 0$ such that the first component of the right side of (4.13) equals T_{m-1} at $x = x_{m-1}$. This determines x_{m-1}, as well as $U(x_{m-1})$ and $D\dot{U}(x_{m-1}^+)$. Hence $D\dot{U}(x_{m-1}^-) = D\dot{U}(x_{m-1}^+) + \beta_{m-1}K'_{m-1}$ is found. Integrating backwards again, we may determine successively all the x_j and the complete profile.

A comment is in order regarding assumption (4.1). As shown in Example 3 of Sec. 3, for networks with competing reactions, it may happen that α must be chosen so that two of the T_j coincide; but it seems unlikely that there would be more than just one coincident pair. If $T_\ell = T_{\ell+1}$, say, then the limit problem (4.7-9) looks the same except that for $j = \ell$, the right side of (4.8) is replaced by a linear combination of vectors K_k for $k \leq k_\ell$, whose coefficients are not known a priori. They must be found from a flame layer analysis. The simplest example is that of the network of two competing reactions $A \to P_1$, $A \to P_2$. In this case, the details are given in [3].

The reason for assuming D to be diagonal is to be able to conclude (4.10) from the fact that $Y(x) \in \mathscr{E}^0_{\kappa(\ell)}$ for $x > x_\ell$. But this is only a mathematical requirement resulting from the assumption that D is constant. Physically, the $\kappa(\ell)$-th component of $D\dot{Y}$ represents the flux of $Y_{\kappa(\ell)}$, and that flux will certainly be zero, since $Y_{\kappa(\ell)} = 0$ for $x > x_\ell$. Also physically, D will generally not be diagonal, but there is no contradiction because it is also nonconstant, and depends on Y in such a manner that $(D\dot{Y})_\kappa = 0$ when $Y_\kappa = \dot{Y}_\kappa = 0$.

5. THE LIMIT PROBLEM, LEVEL 2.

On level 1, the limit problem contains the following apparent inconsistency: the jump conditions (4.8) replace the action of the rate functions $\omega_j(U)$ in (4.2). Therefore the β_j should be related to the ω_j. However, the only such relation used is that for $j = m$; indeed the other β's are determined uniquely as part of the solution process, with no reference to the ω's or ψ's. This, of course, is a great advantage in solving actual problems, because <u>only one flame layer analysis</u> (namely, that with $j = m$) <u>need be performed</u>.

It is not a true inconsistency at that level of approximation, however. To explain it, we proceed to level 2, where the relation between the ψ's and β's is brought in.

We assume (4.1) as before, and begin by defining the "flame locations" x_j more precisely as the positions where $Y_{\kappa(j)} = \varepsilon_j$, the function $\kappa(j)$ being that in Sec. 4. We then define $U_i \equiv U(x_j)$; and in particular, $T_j = T(x_j)$. This differs from the definition of T_j used in level 1, where these were obtained from (3.12). The difference turns out to be small, and this justifies using the former prescription for the T_j at the lowest level of approximation. Anticipating the small difference, we use the symbol T^0_j for the sequence defined by (3.12), so

$$H_j(T^0_j) = H_m(T_+) \text{ for all } j \quad , \tag{5.1}$$

and express

$$T_j = T_j^0 + \varepsilon_j T_j' \quad .$$

Limit Problem, Level 2. *Let the U_+, K_j', and T_j^0 be determined from a feasible allocation α, and assume (4.1). Solve (4.7) - (4.9) for M, $U(x)$, x_j, and β_j under conditions (4.10), (4.11). Also determine the T_j' for $j < m$ in accordance with the flame layer analyses to be described below. Then $\bar{M}$ and $\bar{U}(z)$ are found as before.*

In the case $K_j = K_j'$, the flame layer analysis proceeds by using stretched variables

$$\xi = \frac{x-x_j}{\varepsilon_j} \ , \ u^{(j)} = \frac{U-U_j}{\varepsilon_j} \quad .$$

The flame equation analogous to (2.9) is then

$$D\frac{\partial^2 u^{(j)}}{\partial \xi^2} + r_j \, \psi_j(u^{(j)}) \, K_j = 0 \quad . \tag{5.2}$$

It turns out that for any given vectors p_+ (with 0-th component ≥ 0) and $p_- = p_+ + \beta_j K_j$, there exists a positive number r_j such that (5.2) has a solution satisfying

$$D \frac{\partial u(j)}{\partial \xi} (\pm\infty) = p_\pm \quad .$$

This problem of determining r_j can be reduced to a standard problem of Liñan [5], [7]. The $p_\pm$ have already been determined from (4.7) - (4.9) as before; this now gives us r_j. Finally, taking the log of (4.6), using (4.4) and (3.10), we find

$$\varepsilon \,\ell n \, r_j = (H_j(T_j^0 + \varepsilon_j T_j') + (1+\gamma(j))\varepsilon \,\ell n \,\varepsilon_j)$$

$$- (H_m(T_+) + (1+\gamma(m))\varepsilon \,\ell n \,\varepsilon_m) \quad .$$

We write

$$H_j(T_j^0 + \varepsilon_j T_j') \cong H_j(T_j^0) + \varepsilon_j \, H_j'(T_j^0) \, T_j' = H_j(T_j^0) + \left(\frac{T_+}{T_j^0}\right)^2 \varepsilon_j \, T_j' \quad .$$

Hence from (5.1),

$$\ell n\ r_j = \left(\frac{T_+}{T_j^0}\right)^2 T_j' + (1+\gamma(j))\ \ell n\ \varepsilon_j - (1+\gamma(m))\ \ell n\ \varepsilon_m \quad .$$

This determines T_j'. Is is not necessarily true that $T_j' = O(1)$, but at least the temperature corrections $\varepsilon T_j' \ll 1$.

In the case $K_j' \neq K_j$, the comments in Sec. 4 regarding pseudo-steady-state analysis apply here as well.

6. FUTURE DIRECTIONS

Many aspects of the present treatment remain to be fully explored, for example, procedures for finding all possible feasible allocations, a systematic flame layer analysis in cases when $K_j \neq K_j'$, more than formal justification, etc. More importantly, the theory needs to be extended to cases ([6] being an example] where some reactions have zero or small activation energy. This is all in the context of plane laminar flames. But other crucial problems in flame theory abound; the significance of multiple reactions in them needs to be explored. Some of these directions are under current active investigation.

BIBLIOGRAPHY

1. H. Berestycki, B. Nicolaenko, and B. Scheurer, "Traveling Wave Solutions to Reaction Diffusion Systems Modeling Combustion," to appear in Proceedings of Special Summer Conference on Partial Differential Equations, Durham, NH, June 1982; American Mathematical Society.

2. J. Buckmaster and G. S. S. Ludford, "Theory of Laminar Flames," Cambridge University Press, 1982.

3. P. C. Fife and B. Nicolaenko, "The singular perturbation approach to flame theory with chain and competing reactions," to appear in Proceedings Conference on Differential Equations, Dundee, (1982), Springer-Verlag Lecture Notes in Mathematics.

4. P. C. Fife and B. Nicolaenko, "Asymptotic flame theory with complex chemistry," in preparation.

5. S. Hastings and A. Poore, "On a nonlinear differential equation from combustion theory," to appear.

6. A. Liñan, "A theoretical analysis of premixed flame propagation with an isothermal chain reaction. Instituto Nacional de Tecnica Aeroespacial "Esteban Terradas" (Madrid), USAFOSR Contract No. EOOAR 68-0031, Technical Report No. 1.

7. A. Liñan, "The asymptotic structure of counter flow diffusion flames for large activation energies," Acta Astronautica 1, 1007-1039 (1974).

PAUL C. FIFE
DEPARTMENT OF MATHEMATICS
UNIVERSITY OF ARIZONA
TUCSON, AZ 87521

BASIL NICOLAENKO
CENTER FOR NONLINEAR STUDIES, MS-B258
LOS ALAMOS NATIONAL LABORATORY
LOS ALAMOS, NM 87545

Contemporary Mathematics
Volume 17, 1983

STRONGLY NONLINEAR DETONATIONS

Robert Gardner[1]

ABSTRACT. The existence of one-dimensional steady detonations at high Mach numbers is proved for a model of reactive gas flow. The proof employs topological methods, including Conley's index of isolated invariant sets.

1. INTRODUCTION. This paper is concerned with the existence of one-dimensional steady detonation waves in a viscous, heat conducting, combustible gas. The results hold at arbitrarily large Mach numbers, i.e., the shock layer which initiates the combustion process can be of arbitrarily large strength. Thus we produce a continuum of waves which can be parametrized by the wave velocity, σ, for each σ exceeding a limiting velocity, σ_{CJ}, called the Chapman-Jouget velocity. Such waves describe the transient behavior of an overcompressed detonation, wherein the shock is initiated by a violent external source such as an explosive charge or the compressive action of a piston. The chemical reaction is assumed to be of the simplest form, $A \to B$, i.e. there is a single reactant and a single product. The governing equations consist of a reaction-diffusion equation, which describes the evolution of the nondimensionalized reactant concentration, coupled with the compressible Navier-Stokes equations (see (1)).

In related work, Majda [7] has proposed a simplified qualitative model wherein the fluid variables are "lumped" into a single variable. This model can then be used to describe the interior transition layer in a matched asymptotic expansion of Mach $1 + \varepsilon$ detonations; this approach requires that the shock layer be of sufficiently weak strength, (see Majda and Rosales [8], and Fife [4]). Detonation profiles at high Mach numbers have also recently been obtained by Holmes and Stewart [6]; their methods are different than those employed here. The results of [6] are somewhat less general than those obtained here in that they require zero species diffusion and high activation

---

1980 Mathematics Subject Classification 35K55.
[1]Research supported by NSF Grant #MCS8101644.

0271-4132/83/0000-1384/$03.25

energy asymptotics. In the absence of these hypotheses, the constructions are more delicate.

The details of the proof of the main result can be found in [5].

2. FORMULATION OF THE PROBLEM.

A. THE EQUATIONS OF REACTIVE GAS FLOW. The equations of a reacting gas in Lagrangian coordinates take the form

$$\tag{1} \begin{aligned} \tau_t - u_x &= 0 \\ u_t + p_x &= (\mu\tau^{-1}u_x)_x \\ (e + u^2/2)_t + (pu)_x &= (\mu\tau^{-1}uu_x)_x + (\lambda\tau^{-1}T_x)_x \\ z_t &= Dz_{xx} - \phi(T)z \ , \end{aligned}$$

where

τ = specific volume	u = fluid velocity
p = pressure	T = temperature
z = mass fraction of unburned gas	e = specific internal energy/unit mass of gas mixture
ϕ = reaction rate	λ,μ,D = heat conduction, viscosity and species diffusion coefficients,

(see Majda [7]). The gas mixture is assumed to be polytropic; thus

$$\tag{2} \begin{aligned} \tau P &= RT \\ e &= z(cT + g_u) + (1 - z)(cT + g_0) \\ &= cT + q_0 z + g_b , \end{aligned}$$

where R and c are positive constants, g_u and g_b are the energies of formation of the unburned and burned gases, respectively, and $q_0 = g_u - g_b$ is the liberated energy/unit mass. The reaction is assumed to be exothermic; thus $q_0 > 0$.

In order to avoid the cold boundary difficulty we assume ignition temperature kinetics. Thus ϕ is assumed to be a bounded nonnegative, and non-decreasing function of T such that $\phi(T) = 0$ for $T \le T_i$ and $\phi(T) > 0$ for $T > T_i$; T_i is the ignition temperature. This appears to provide a good approximation to the transient behavior of combustion waves on reasonable time and length scales (see Williams [9]).

B. TRAVELLING WAVES. A detonation profile is a travelling wave solution of (1), i.e., a solution which depends on the single quantity $\xi = x - \sigma t$; σ is the wave velocity. The wave should tend to distinct limits at $\xi = \pm\infty$; in particular, we require that $z(+\infty) = 1$ and $z(-\infty) = 0$, so that the gas is unburned (resp. burned) at $+\infty$ (resp. $-\infty$). It follows that σ must be positive. Such solutions satisfy an associated system of o.d.e.'s

$$\begin{aligned} -\sigma\tau' - u' &= 0 \\ -\sigma u' + p' &= (\mu\tau^{-1}u')' \\ -\sigma(e + u^2/2)' + (pu)' &= (\mu\tau^{-1}u'u') + (\lambda\tau^{-1}T')' \\ -\sigma z' &= Dz'' - \phi(T)z\ . \end{aligned}$$

If it is assumed that the gas at $\xi = +\infty$ is in a given state $(\tau_R, T_R, u_R, 1)$, the first three equations can be integrated from ξ to $+\infty$ to obtain

$$\begin{aligned} -\sigma[\tau] - [u] &= 0 \\ -\sigma[u] + [p] &= \mu\tau^{-1}u' \\ -\sigma[e + u^2/2] + [pu] &= \mu\tau^{-1}uu' + \lambda\tau^{-1}T', \end{aligned} \tag{3}$$

where $[X] = X - X_R$; the relations (2) have been used to determine p_R and e_R.

The first equation in (3) is an algebraic relation between τ and u. This observation, together with (2) can be used to eliminate u and u_R from the remaining equations. The final travelling wave system then takes the form

$$\begin{aligned} \mu\tau' &= -\sigma^{-1}(\sigma^2\tau^2 - A\tau + RT) \equiv F_1(\tau,T) \\ \lambda T' &= \sigma\tau(\sigma^2\tau^2/2 - A\tau + B - q_o(z - 1) - cT) \equiv F_2(\tau,T,z) \\ z' &= w \\ Dw' &= -\sigma w + \phi(T)z\ , \end{aligned} \tag{4}$$

where $A = \sigma^2\tau_R + p_R$ and $B = \sigma^2\tau_R^2/2 + \tau_R p_R + cT_R$; the details can be found in [5].

In the sequel, the vector of dependent variables will be denoted by $X = (\tau,T,z,w)$.

C. THE CRITICAL POINTS OF (4). Clearly $X_R = (\tau_R, T_R, 1, 0)$ is a critical point of (4). We now determine the other critical points of this system.

LEMMA 1. Suppose that (σ, q_o) lies in the shaded region in Figure 1a. Then (4) admits two isolated critical points

$$X_{L^*} = (\tau_{L^*}, 0, 0)$$

$$X_L^* = (\tau_L^*, T_L^*, 0, 0),$$

and a curve $C(z)$ of critical points parametrized by their z-component whose projection on the (T,z) plane is as indicated in Figure 1b. Furthermore, the inequalities

$$T \leq T_i < T_{L^*} < T_L^* \tag{5}$$

hold, where T is the temperature at any critical point along $C(z)$.

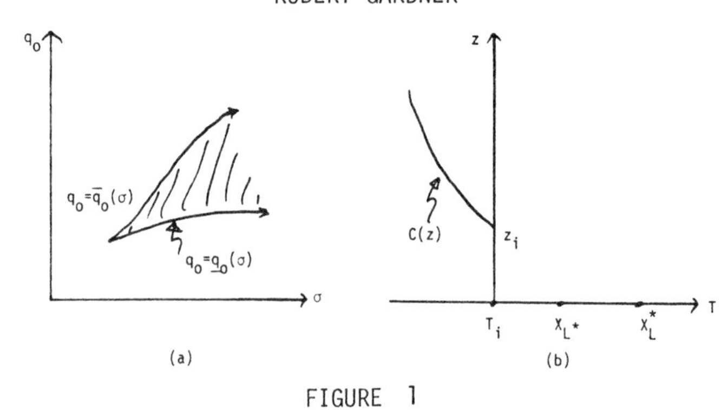

FIGURE 1

The lemma is proved by comparing the properties of the Hugoniot function associated with the jump conditions for the conservation laws and the additional condition that $w = \phi(T)z = 0$ at critical points. As a consequence of this analysis, it follows that the null clines of F_1 and F_2 have the aspect indicated in Figure 2 for various (fixed) values of z.

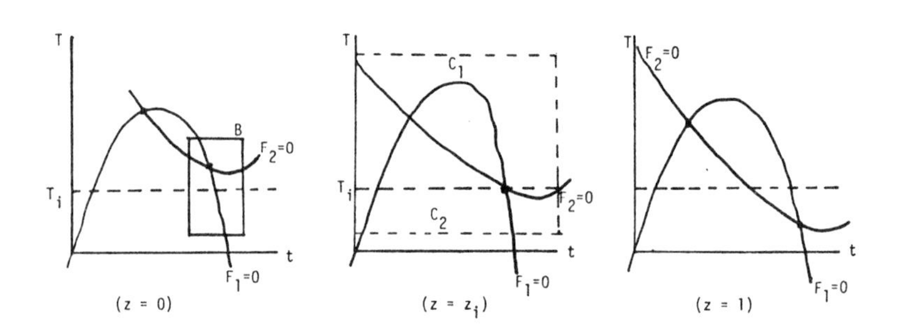

FIGURE 2

D. THE MAIN RESULT.

THEOREM 1. Suppose that (σ, q_0) lies in the shaded region of Figure 1a. There exist positive constants K_i, $1 \leq i \leq 4$, such that if

$$\begin{aligned} \lambda + \mu &< K_1 \\ \mu &< K_2 \\ D &< \min(K_3\lambda, K_4), \end{aligned} \tag{6}$$

then there exist complete solutions $X_w(\xi)$ and $X_s(\xi)$ of (4) such that

$$\lim_{\xi \to -\infty} X_s(\xi) = X_L^* \qquad \lim_{\xi \to +\infty} X_s(\xi) = X_R = C(1)$$

$$\lim_{\xi \to -\infty} X_w(\xi) = X_{L^*} \qquad \lim_{\xi \to +\infty} X_w(\xi) = C(z_*) ,$$

for some $z_* \in (z_i, 1)$.

Since the gas was assumed to be in the state X_R at $\xi = +\infty$, the solution $X_s(\xi)$ is the desired detonation profile; $X_s(\xi)$ is called a *strong detonation*. The solution $X_w(\xi)$, as we shall see in the next section, plays an essential role in determining the existence of $X_s(\xi)$; in particular, it is essential that $z_* < 1$. If the viscosity coefficients are too large, this will not be the case, and no detonation profile will exist. In the limiting case of vanishing viscosity and an infinitely fast reaction rate, $X_w(\xi)$ is called a *weak detonation*. Weak detonations are unstable solutions of the conservation laws since they violate the Lax shock conditions. The "correct" solution of the associated Riemann problem in this case is a Chapman-Jouget detonation followed by a centered rarefaction wave (see Chorin and Marsden [1]). This illustrates that care must be exercised in employing the viscosity method to pick out the stable solutions of the underlying conservation laws. It seems likely that $X_w(\xi)$ is an unstable solution of (1).

3. METHOD OF PROOF. The main tool in the proof of Theorem 1 is Conley's index of isolated invariant sets. We shall only give an outline of the main ideas; the details can be found in [2].

A. THE CONLEY INDEX. Given a flow on $\mathbb{R}^n$, an invariant set S is called an *isolated invariant set* if S is the maximal invariant set in some compact neighborhood N of itself; N is called an *isolating neighborhood*. A neighborhood N is therefore isolating if every solution curve which hits a point in ∂N eventually leaves N in either forward or backward time. The (possibly empty) set of solution curves which stay in N for all time, denoted $S(N)$, is therefore an isolated invariant set. A homotopy index, denoted by $h(S)$, can then be defined; (h is a homotopy type of pointed topological spaces). The index depends only on the isolated invariant set S and is invariant under deformations of the flow which preserve the isolation property (see Conley [2]). If S is a hyperbolic critical point with k positive eigenvalues, $h(S)$ is a pointed k-sphere, denoted by Σ^k. It can therefore be viewed as a generalization of the Morse index.

This invariant is useful in the location of connecting orbits, as can be seen from the following lemma.

LEMMA 2. (See Conley and Smoller [3]). Suppose N is an isolating neighborhood which contains exactly two hyperbolic critical points, X_1, X_2. If $h(S(N)) \neq h(X_1) \vee h(X_2)$ (where the latter is the space obtained by gluing $h(X_1)$ and $h(X_2)$ together at their distinguished point), then $S(N)$ contains nonconstant solutions.

In many situations, the flow admits a Liapunov function, so that the nonconstant solution in $S(N)$ is a connecting orbit. The index $h(S(N))$

therefore plays the role of an "intersection number" of stable and unstable manifolds, at least on some crude level. In the present context, a globally defined Liapunov function does not exist. However, the "nontriviality" of the index still furnishes much useful information.

B. MODIFICATION OF THE FLOW. The problem at hand is to connect X_L^* to X_R with a complete solution of (4). However, X_R is embedded in the curve $C(z)$ of critical points, and is therefore degenerate. The index theory outlined above cannot be applied directly.

The technique is to modify the kinetics $\phi(T)$ to a new function $\Phi(T,z)$ in such a manner as to eliminate this degeneracy. Thus we consider the system

$$(4)_m \qquad \begin{aligned} \mu\tau' &= F_1(\tau,T) \qquad & z' &= w \\ \lambda T' &= F_2(\tau,T,z) \qquad & Dw' &= -\sigma w + \Phi(T,z)z \ , \end{aligned}$$

where $\Phi = \phi$ for $T \geq T_i$ and where Φ changes sign in the manner indicated in Figure 3. The system $(4)_m$ admits precisely four hyperbolic critical points, X_R, X_i, X_{L^*}, and X_L^*, where $X_i = C(z_i)$. It can be seen by linearization that

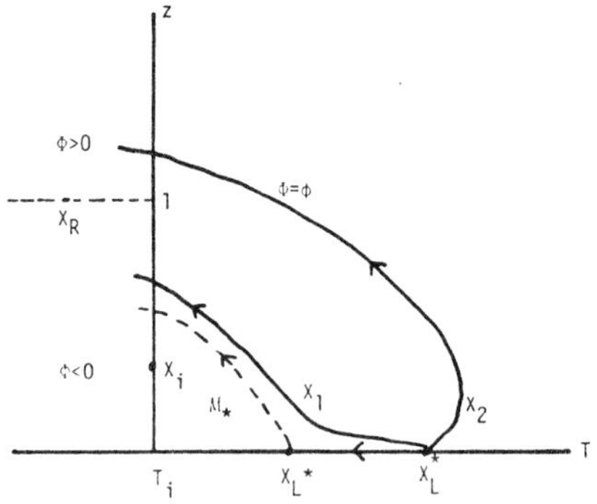

FIGURE 3

$$h(X_R) = \Sigma^1 \qquad h(X_i) = \Sigma^0$$
$$h(X_{L^*}) = \Sigma^1 \qquad h(X_L^*) = \Sigma^2 \ .$$

We remark that the (τ,T) components of (4) decouple from the (z,w) components whenever $T \leq T_i$; thus the nontrivial interaction between the fluid flow and the chemistry occurs when $T > T_i$. Since (4) and $(4)_m$ coincide in this region, it is reasonable to expect the latter system to contain most of the essential information concerning solutions of the former.

C. ISOLATING NEIGHBORHOODS FOR $(4)_m$. The goal is to construct isolating neighborhoods N_w and N_s of $(4)_m$ each of which contain exactly two critical points, X_{L^*}, X_i, and X_L^*, X_R, respectively, and to show that the indices of $S(N_w)$ and $S(N_s)$ are nontrivial in the sense of Lemma 2. This will enable us to recover substantial information about the solutions of (4).

The neighborhood N_w is constructed as follows. Let $\bar{z} \in (z_i,1)$ be given, let $L > \sigma^{-1}\max|\Phi|$, and let B be the rectangle indicated in the first and third diagrams in Figure 2. The region N_w is defined to be

$$N_w = B \times \{(z,w): 0 \le z \le \bar{z}, |w| \le L\} \cup A_*,$$

where A_* is a small neighborhood of X_*. Solution curves of $(4)_m$ which are transverse to ∂N_w leave N_w immediately in either forward or backward time; such solutions are clearly disjoint from $S(N_w)$. Other solution curves can be tangent to ∂N_w (e.g., if $z = \bar{z}$, $w = 0$, and $T < T_i$), and the possibility arises that the curve stays in N_w in both time directions. Such solutions must be followed in one time direction to points where they eventually leave N_w. In order to obtain control over the global behavior of such solutions, conditions on the viscosity coefficients must be imposed. We remark that the conditions(6) depend on the choice of $\bar{z}$; however, they are independent of the (nonphysical) modification Φ introduced into the equations.

The index $h(S(N_w))$ is computed by exploiting its homotopy invariance. In particular, we first allow D and then λ to approach zero. Since the equations are singular at the limiting parameter values, we introduce the change of variables

$$p(\xi) = Dw'(\xi) = -\sigma w + \Phi z$$

$$q(\xi) = \lambda T'(\xi) = F_2(\tau,T,z).$$

Under the change of variables $w \to p$, $(4)_m$ transforms to

$$(7) \qquad \begin{aligned} \mu\tau' &= F_1 & z' &= \sigma^{-1}\Phi(T,z)z + O(p) \\ \lambda T' &= F_2 & p' &= -\sigma D^{-1}p + O(1)\ . \end{aligned}$$

Clearly, $|p|$ must remain $O(D)$ as D tends to zero along solutions in $S(N_w)$. Moreover, N_w is mapped into a neighborhood $R_w \times I_p$, where R_w is the projection of N_w on the (τ,T,z) plane and $I_p = \{p: |p| \le \ell\}$, for some $\ell > 0$. For sufficiently small D, $R_w \times I_p$ is isolating for (7), and the indices of $S(N_w)$ and $S(R_w \times I_p)$ with respect to $(4)_m$ and (7) are the same. After a further homotopy, the equations can be continued to

$$\begin{aligned} \mu\tau' &= F_1 & z' &= \sigma^{-1}\Phi(T,z)z \\ \lambda T' &= F_2 & p' &= -\sigma p \end{aligned}$$

in the region $R_w \times I_p$.

A similar argument applies as $\lambda \to 0$; in particular, we make the change of variables $T \to q$. Noting that $F_2(\tau,T,z) = 0$ defines T as a function T_0 of (τ,z), the final equations take the form

$$\begin{aligned} \mu\tau' &= F_1(\tau,T_0(\tau,z)) = f(\tau,z) & p' &= -\sigma p \\ z' &= \Phi(T_0(\tau,z),z)z = g(\tau,z) & q' &= -\sigma q\ ; \end{aligned}$$

the isolating neighborhood is a region of the form $H_w \times I_p \times I_q$, where $I_q = \{q: |q| \le \ell\}$, and H_w is the projection of R_w on the (τ,z) plane.

(The flow in the (τ,z) plane is similar to Majda's qualitative model). It is now easily seen that $h(S(H_w \times I_p \times I_q))$ is the homotopy type of a point, which differs from $h(X_{L^*}) \vee h(X_i) = \Sigma^1 \vee \Sigma^0$. Thus by Lemma 1, $S(N_w)$ contains non-constant solutions. Moreover, our conditions on the viscosity coefficients imply that X_{L^*} is a repeller relative to the flow in $S(N_w)$; it follows that the (one-dimensional) unstable manifold of X_{L^*} (denoted by M_*) lies in $S(N_w)$. Thus $\bar{z}$ is an upper bound for z along M_*.

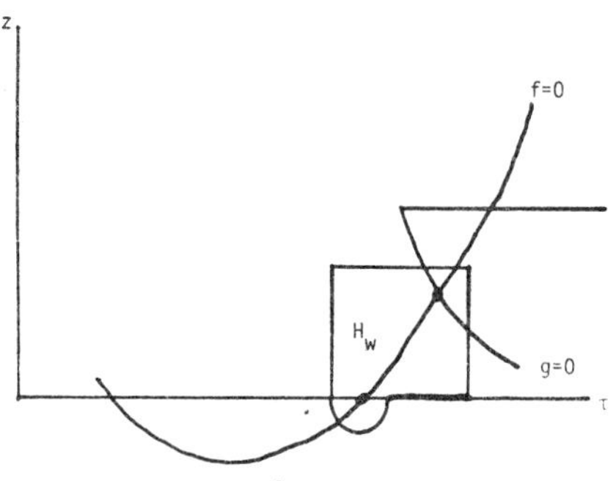

FIGURE 4

We next construct the isolating neighborhood N_s. To this end, let C be the union of the rectangles C_1 and C_2 in Figure 2, and let δ be small and positive. As a first approximation to N_s, define

$$N = C_1 \times \{(z,w): 0 \leq z \leq 1 + \delta,\ 0 \leq w \leq L\} \cup$$
$$C \times \{(z,w): |z - 1| \leq \delta,\ |w| \leq L\} \cup A^*,$$

where A^* is a small neighborhood of X_L^*. The difficulty is that X_i and X_{L^*} also lie in N. The final neighborhood is defined to be

$$N_S = N \setminus (A_* \cup A_i \cup T)$$

where A_* and A_i are small neighborhoods of X_{L^*} and X_i, and T is a thin neighborhood of the orbit segment M_* in the region $T \geq T_i$. In order to show that N_s is isolating, essential use is made of the fact that $z \leq \bar{z} < 1 - \delta$ along M_*.

The index of $S(N_s)$ is computed in a manner similar to that of $S(N_w)$. It too is "nontrivial", which provides us with a solution in the unstable manifold M^* of X_L^* which lies in $S(N_s)$.

D. PROOF OF THEOREM 1. The proof is completed as follows. It can be seen that the solutions of $(4)_m$ in the unstable manifolds of X_{L^*} and X_L^* obtained in C., above, eventually hit the plane $T = T_i$. Up till this point, they are also solutions of (4), since $\phi = \Phi$ for $T \geq T_i$. The points where these solutions first hit $T = T_i$ are then used as data for the original system, (4). It follows that these solutions tend to critical points along

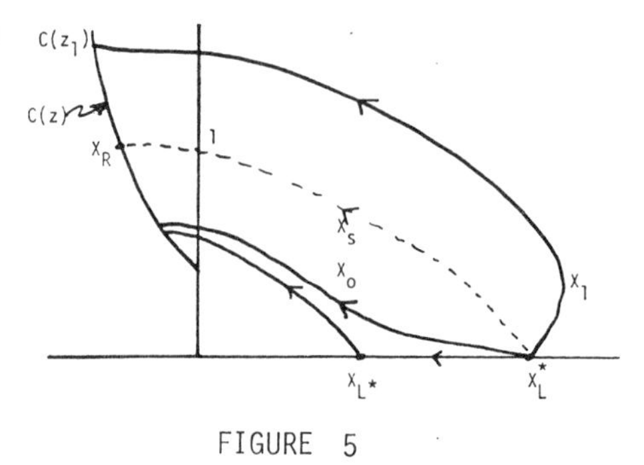

FIGURE 5

$C(z)$ in forward time, as indicated in Figure 5. In particular, X_1 tends to a critical point $C(z_1)$ with $z_1 > 1$.

Next, it is easily seen that there is an orbit connecting X_L^* to X_L* along which $z = w = 0$. Since the unstable manifold of X_L^* is two-dimensional, it follows that there is an orbit $X_0(\xi)$ in this manifold which closely approximates the connecting orbit until it reaches a small neighborhood of X_L*, at which point it closely approximates the (one-dimensional) unstable manifold of X_L*. It follows (from (7)) that X_0 tends to a critical point $C(z_0)$ with $0 < z_0 < 1$ in positive time.

An orbit $X_1(\xi)$ of (4) can be obtained from the nonconstant solution in $S(N_s)$ (with respect to $(4)_m$) which connects X_L^* to a critical point $C(z_1)$ with $z_1 > 1$. The argument is similar to the construction of $X_0(\xi)$.

The proof is concluded with a shooting argument in the unstable manifold of X_L^* (with respect to (4)).

BIBLIOGRAPHY

1. A. Chorin and J. Marsden, A Mathematical Introduction to Fluid Mechanics, Springer, New York, 1979.

2. C. Conley, Isolated Invariant Sets and the Morse Index, C.B.M.S. number 38, 1978.

3. C. Conley and J. Smoller, Existence of Heteroclinic Orbits and Applications, in "Dynamical Systems, Theory and Applications" (J. Moser, ed.), Lecture Notes in Physics, vol. 38, Springer, 511-524.

4. P. Fife, Propagating Fronts in Reactive Media, preprint.

5. R. Gardner, On the Detonation of a Combustible Gas, to appear: Trans. AMS.

6. Holmes and Stewart, Existence of One Dimensional Steady Detonations in a Simple Model Problem, Studies in Appl. Math., 1982.

7. A. Majda, A Qualitative Model for Dynamic Combustion, SIAM J. Appl. Math., 41, 70-93.

8. A. Majda and R. Rosales, Weakly Nonlinear Detonation Waves, preprint.

9. F. Williams, Combustion Theory, Addison Wesley Pub. Co., 1965.

DEPARTMENT OF MATHEMATICS
UNIVERSITY OF MASSACHUSETTS
AMHERST, MA 01003

Contemporary Mathematics
Volume 17, 1983

DIFFERENTIAL EQUATIONS AND CONVERGENCE ALMOST EVERYWHERE IN STRONGLY MONOTONE SEMIFLOWS

Morris W. Hirsch[1]

ABSTRACT. This paper is concerned with the asymptotic behavior of certain systems of autonomous differential equations. These include cooperative irreducible systems in $\mathbb{R}^n$, semilinear parabolic equations in smooth bounded domains, and certain parabolic semilinear evolution equations in Banach spaces. The unifying feature of the systems considered is that their solutions determine local semiflows which are strongly order-preserving in an ordered Banach space. The main results are general theorems that for such local semiflows, most solutions which have compact orbit closure are convergent or quasiconvergent. Several illustrative examples are given.

INTRODUCTION. Many kinds of differential equations lead to a local semiflow $\phi = \{\phi_t\}_{t \geqslant 0}$ in a suitable Banach space X -- usually $\mathbb{R}^n$ or a function space. The solution (perhaps "mild") through x is the map $t \longrightarrow \phi_t(x)$, denoted also by $x(t)$. The basic problem of dynamical systems is to deduce the asymptotic behavior of $x(t)$ as $t \longrightarrow +\infty$.

For some dynamical systems we have learned to expect sustained oscillation, turbulence, or chaos: in predator-prey systems, many Hamiltonian systems, wave equations, Navier-Stokes equations, and quadratic maps of the interval. For others we expect convergence to equilibria: two species in competition, gradient systems, heat equations, conservation laws, chemical reactions.

For many specific systems of the second class convergence results have been proved, often by constructing Liapunov functions. But there are few general results. It is the purpose of this article to identify a wide class of systems wherein most trajectories can be expected to converge, provided they lie in compact sets, even where there is no way to get Liapunov functions. These are the systems for which there is a strong comparison principle. For systems of ordinary differential equations of the kind we call cooperative such a principle was proved Kamke [Ka] . For parabolic partial differential equations standard maximum principles imply a comparison principle, as is well known. For abstract evolution equations special positivity properties of the

[1]This research was supported in part by NSF Grant MCS-80-02858.

0271-4132/83/0000-1386/$05.75

linear part are needed.

Such comparison principles can be used to prove that the local semiflow defined by solutions is *strongly monotone* in a suitable ordered Banach space X: this means that if $x > y$ then $x(t) \gg y(t)$ for $t > 0$ (that is, $x(t) - y(t)$ belongs to the interior of the nonnegative cone X_+). For most of our results we also need orbit closures to be compact. To meet these conditions for *PDE*s requires X to be carefully chosen.

Our main result is the following. Let $J \subset X$ be a simply ordered arc of initial values, all of whose points have compact orbit closures. Then there is a countable subset $J_0 \subset J$ whose complement in J consists entirely of quasiconvergent points, that is, points whose trajectories are asymptotic to the set E of stationary points. Under various further assumptions we prove discreteness of J_0, convergence for most or all points of J, existence of asymptotically stable equilibria, and other results.

An immediate corollary of the main result is that a strongly monotone local semiflow cannot have an orbitally stable nonconstant periodic trajectory.

Monotone maps have long been used to prove convergence. H. Amman introduced the use of abstract ordered Banach spaces in proving convergence for parabolic *PDE*s, although the basic methods in concrete function spaces go back earlier. See [Am-s], pp. 652-3 for historical discussion and references, as well as for an extensive review of the subject. The present work owes much to this and other papers by Amman. The books [Ma], [He] by R. Martin and D. Henry have also been very useful.

Monotonicity in cooperative *ODE*s has been applied in special situations by many people. Articles by J. Selgrade, J. Conlon, A. Lajmanovich and J. Yorke, S. Smale and S. Grossberg greatly influenced the development of the present work.

I am indebted to P. Chernoff, J. Marsden, A. Majda, X. Mora, J. Hale, and T. Kato for inspiring converstations and lectures.

This paper is organized as follows. Basic definitions are made in §0. In §1 we give examples of strongly monotone local semiflows. The main general results are given in §2 and §3. In §4 examples from differential equations are used to illustrate applications of the main results. Most proofs will appear elsewhere.

§0 TERMINOLOGY. The following terminology will be used throughout.

An *ordered Banach space* is a Banach space X together with a partial ordering $\leqslant$ on X, such that the set

$$X_+ = \{x \in X : x \geqslant 0\}$$

is a closed convex cone. We usually assume the interior of X_+, denoted by

Int X_+, is nonempty -- a severe restriction.

The Banach space of continuous (real) functions on a compact topological space has this property, as does the Banach space of C^k functions ($k \geqslant 0$) on a compact C^k manifold. But the space of continuous functions which vanish on a given nonempty subset usually does not have this property, nor do most function spaces on noncompact domains, nor most L^p spaces.

Let $a, b \in X$. The *order interval* $[a, b]$ is $\{x \in X : a \leqslant x \leqslant b\}$. The *open order-interval* $[[a, b]]$ is $Int\, [a, b] = \{x \in X : a \ll x \ll b\}$.

A subset $B \subset X$ is *order-bounded* if it is contained in some order interval. It is easy to see that B is order-bounded provided $A \leqslant B \leqslant C$ for some compact sets A, C in X.

A set $\Gamma \subset X$ is *p-convex* if whenever $a < b$ in Γ then the line segment from a to b is contained in Γ.

We write

$$x \leqslant y \quad \text{if} \quad y - x \in X_+ \ ,$$

$$x < y \quad \text{if} \quad x \leqslant y \text{ and } x \neq y \ ,$$

$$x \ll y \quad \text{if} \quad y - x \in Int\ X_+ \ .$$

The relations $y \geqslant x$, etc., have the obvious meanings. If P, Q are subsets of X then $P < Q$ means $p < q$ for all $p \in P$, $q \in Q$; similarly one defines $P \ll Q$, etc.

We define $\mathbb{R}_+ = \{t \in \mathbb{R} : t \geqslant 0\}$.

A map f between subsets of ordered Banach spaces is *monotone* if $x \leqslant y$ implies $f(x) \leqslant f(y)$, and *strongly monotone* if $x < y$ implies $f(x) \ll f(y)$.

If $T : X \longrightarrow Y$ is a linear map between ordered Banach spaces then T is monotone (resp. strongly monotone) if and only if $T(X_+) \subset Y_+$ (resp. $T(X_+\backslash\{0\}) \subset Int\ Y_+$). In this case we write $T \geqslant 0$ (resp. $T \gg 0$).

Let $W \subset X$ be a subset. A *local semiflow* in W is a continuous map $\phi : W_\# \longrightarrow W$ having the following properties:

(1) $W_\#$ is an open subset of $\mathbb{R}_+ \times W$ containing $\{0\} \times W$.

(2) For each $x \in W$ the set

$$I_x = \{t \in \mathbb{R}_+ : (t, x) \in W_\#\}$$

is a half-open interval

$$I_x = [0, \tau_x),\ 0 < \tau_x \leqslant \infty \ .$$

(3) For each $t \in \mathbb{R}_+$ define

$$W_t = \{x \in W : (t, x) \in W_\#\} \ ,$$

and define

$$\phi_t : W_t \longrightarrow W, \quad x \longmapsto \phi(t, x) \quad .$$

Then

$$\phi_0 = id_X = \text{identity map of } X,$$

and for all $s, t \geqslant 0$

$$W_{s+t} = \phi_t^{-1}(W_s) \quad ,$$

$$\phi_{s+t} = \phi_s \circ \phi_t \quad .$$

If $\tau_x = \infty$ for all x then ϕ is called a *semiflow*. If ϕ is a semiflow and each ϕ_t is a homemorphism of W onto W then ϕ is called a *flow*.

For $x \in W$ the map

$$I_x \longrightarrow W, \quad t \longmapsto \phi(t, x)$$

is the *trajectory* of x, its image is the *orbit* $O(x)$. The closure $clos_W\, O(x)$ of $O(x)$ in W is the *orbit closure* of x, denoted by $\overline{O}(x)$.

A subset $Y \subset W$ is *positively invariant* if $O(x) \subset Y$ for all $x \in W$. It is *invariant* if $Y \subset W_t$ and $\phi_t(Y) = Y$ for all t.

The ω-*limit set* of x is

$$\omega(x) = \cap_{t \in I_x} \overline{O}(x(t)) \quad .$$

Thus $y \in \omega(x)$ if $y = lim_{k \to \infty} x(t_k)$ for some sequence $t_k \longrightarrow \tau_x$ in I_x.

It is well know that $\tau_x = \infty$ if $\overline{O}(x)$ is compact. In this case $\omega(x)$ is a compact connected nonempty invariant set.

If $\phi_t(x) = x$ for all t then x is an *equilibrium* or *stationary point*. The set of equilibria is denoted by E.

If $lim_{t \to \infty} x(t)$ exists in W then it is an equilibrium, and we call x a *convergent point*. If $\overline{O}(x)$ is compact and $\omega(x) \subset E$ then x is a *quasiconvergent point*.

It is difficult, if not impossible, to distinguish convergent points from quasiconvergent ones by observing their trajectories (e.g. on a computer).

Notice that if every connected component of E is a single point, then a quasiconvergent point is necessarily convergent. In particular this happens whenever E is countable.

We say x has *period* $T \in \mathbb{R}$ if $T \neq 0$ and $x(T) = x$. In this case $x(s + T) = x(s)$ for all $s \geqslant 0$. The orbit of x is then called a *closed orbit*; it is *nondegenerate* if x is not stationary.

The local semiflow ϕ is *monotone* (respectively *strongly* monotone) if ϕ_t is monotone (resp. strongly monotone) for all $t > 0$. It is *compact* if $clos_W\, \phi_t(S)$ is compact whenever $t > 0$ and $S \subset W_t$ is bounded. And ϕ is *order-compact* if $clos_W\, \phi_t(S)$ is precompact whenever $t > 0$ and $S \subset W_t$ is *order-bounded in* W_t, which means $A \leqslant S \leqslant B$ for nonempty compact subsets $A, B \subset W_t$.

§1 EXAMPLES OF MONOTONE LOCAL SEMIFLOWS. We shall use the following simple result.

1.1 LEMMA. Let X, Y be ordered Banach spaces, $W \subset X$ a p-convex set, and $f : W \longrightarrow Y$ a map whose restriction to each line segment in W is continuously differentiable. Then f is (strongly) monotone provided each linear map $DF(x)$ is (strongly) monotone.

Our first example comes from ordinary differential equations. Later we shall see that it is closely related to partial differential equations.

Consider a C^1 local semiflow ϕ on a p-convex open set $W \subset \mathbb{R}^n$ generated by a C^1 vector field F on W. From 1.1 we see that if the matrices $D\phi_t(x)$ are all nonnegative (resp. positive) for $t > 0$ then ϕ is monotone (resp. strongly monotone).

We shall see that a condition guaranteeing $D\phi_t(x) \geqslant 0$ is that F be a *cooperative* vector field. This means that $\partial F_i/\partial x_j \geqslant 0$ everywhere, for all $i \neq j$. Equivalently, for every x there exists $\lambda > 0$ such that $DF(x) + \lambda I \geqslant 0$.

The vector field F is *irreducible* provided each matrix $DF(x)$ is irreducible. Recall that a matrix B is irreducible if it maps into itself no proper nontrivial subspace of the form $x_{i_1} = \ldots = x_{i_k} = 0$.

1.2 THEOREM. Let F be a cooperative vector field in the p-convex open set $W \subset \mathbb{R}^n$, generating the local semiflow ϕ. Then

(a) $D\phi_t(x) \geqslant 0$ for all $x \in W$, $t \geqslant 0$; and ϕ is monotone.

(b) If F is irreducible then $D\phi_t(x) \gg 0$ for all $x \in W$, $t > 0$; and ϕ is strongly monotone.

Proof of (a). (Compare Henry [He], pp. 60-62.) Fix x and set $B(t) = DF(x(t))$, $A(t) = D\phi_t(x)$. Then the matrix valued function $A(t)$ satisfies the variational equation

$$\frac{dA}{dt} = B(t)A; \qquad A(0) = I \tag{1}$$

where I is the identity matrix. Fix $\lambda > 0$, $T > 0$ so that for all $t \in [0, T]$,

$$B(t) + \lambda I \equiv C(t) \geqslant 0 .$$

Then (1) can be rewritten

$$\frac{dA}{dt} = -\lambda A + C(t)A; \qquad A(0) = I . \tag{2}$$

The solution to this is $A(t) = \lim_{k\to\infty} A_k(t)$ where $A_0(t) \equiv I$ and

$$A_{k+1}(t) = e^{-\lambda t}I + \int_0^t e^{-\lambda(t-s)}C(s)A_k(s)\,ds . \tag{3}$$

It follows by induction on k that $A_k(t) \geqslant 0$ for all $t \in [0, T]$. This proves $A(t) \geqslant 0$, and (a) is proved

Proof of (b). If F is irreducible then $C(s)$ is a nonnegative irreducible matrix for $0 \leq s \leq T$. Taking limits in (3) as $k \longrightarrow \infty$ we find $A(t)$ satisfies the integral equation

$$A(t) = e^{-\lambda t} I + \int_0^t e^{-\lambda(t-s)} C(s)A(s)\, ds \quad , \tag{4}$$

and $C(s) \geq 0$, $B(s) \geq 0$. Fix $i, k \in \{1, \ldots, n\}$ and $t \in (0, T]$; we show $A_{ik}(t) > 0$. Set

$$L = \{j : A_{jk}(t) : A_{jk}(t) = 0\} \quad .$$

Then the set $M = \{1, \ldots, n\} \backslash L$ is nonempty because $A(t)$ is nonsingular. It follows from irreducibility that there exists $j \in M$ with $C(s)_{ij} > 0$. This implies

$$[C(s)A(t)]_{ik} > 0 \quad .$$

Therefore from (4) we get $A_{ik}(t) > 0$. The proof of 1.2 is complete. Q.E.D.

Remarks. (i) Monotonicity of ϕ is an immediate consequence of a well-known result of Kamke [Ka]; see also Coppel [Co] for a proof.

(ii) In Hirsch [Hi] it is shown that compact limit sets of cooperative vector fields have rather special geometrical properties in $\mathbb{R}^n$ -- e.g. they cannot be linked or knotted, and they are nowhere dense.

(iii) In Smale [Sm] it is shown that *competitive* systems ($\partial F_i/\partial x_j \leq 0$ for $i \neq j$) can be constructed in $\mathbb{R}^n$ containing *any* $(n-1)$-dimensional system as an attractor. Reversing time shows that cooperative systems can have very complicated invariant sets -- but not as attractors! In particular they can have nondegenerate closed orbits.

For simplicity a map or manifold of class C^4 is termed *smooth*, although in most cases this can be relaxed.

EXAMPLE. Let $\overline{\Omega} \subset \mathbb{R}^n$ be a smooth compact n-dimensional submanifold with boundary $\partial\Omega$ and interior Ω. Set $D = \partial/\partial x_i$ and let

$$A = \sum_{i,j=1}^{n} a_{ij}(x) D_i D_j + \sum_{i=1}^{n} b_i(x) D_i$$

be a second order differential operator with smooth coefficients $a_{ij}, b_i : \overline{\Omega} \longrightarrow \mathbb{R}$ such that the matrices $[a_{ij}(x)]$ are positive definite. Let $\nabla = (D_1, \ldots, D_n)$ denote the gradient operator.

Suppose $g : \overline{\Omega} \times \mathbb{R} \longrightarrow \mathbb{R}$ is smooth, and consider a parabolic initial-boundary value problem for an unknown function

$$u(x, t), \qquad (x, t) \in \overline{\Omega} \times \mathbb{R}_+ \quad ,$$

of the type

(5a) $\frac{\partial u}{\partial t} = Au + g(x, u), \quad (x \in \overline{\Omega},\ t > 0),$

(5b) $Bu = 0,$

(5c) $u(x, 0) = v(x), \quad (x \in \overline{\Omega}),$

where the boundary operator B has the form

$$Bu : (x, t) \longmapsto (\alpha u + \beta \frac{\partial u}{\partial \nu})(x, t), \qquad (x \in \partial\Omega,\ t \geqslant 0) \quad .$$

Here $\beta = 0$ or 1, $\alpha : \partial\Omega \longrightarrow \mathbb{R}$ is smooth, and $\nu : \partial\Omega \longrightarrow \mathbb{R}^n$ is a smooth, outward-pointing vector field transverse to $\partial\Omega$. We assume the boundary conditions are of one of the following types:

$$\alpha = 1, \quad \beta = 0 \qquad (Dirichlet);$$

$$\alpha = 0, \quad \beta = 1 \qquad (Neumann);$$

$$\alpha \geqslant 0, \quad \beta = 1 \qquad (oblique).$$

We need the following compatibility condition: under Dirichlet boundary conditions we assume $g(x, 0) \equiv 0$ for all $x \in \partial\Omega$.

It is known (see e.g. Mora [Mo]) that there is a local semiflow ψ in $L^2(\Omega)$ such that the solution to (5) is given by

$$u(x, t) = \psi_t(v)(x)$$

with u continuous in t for each x. Moreover ψ restricts to a local semiflow ϕ on the Banach space

$$X = \{u \in C^1(\overline{\Omega}) : Bu = 0\} \equiv C^1_B(\overline{\Omega}) \quad ,$$

with $\psi_t(u) \in C^2(\overline{\Omega})$ for $t > 0$.

We order X by defining

$$X_+ = \{u \in X : u(x) \geqslant 0 \text{ for all } x \in \overline{\Omega}\} \quad .$$

In all cases Int X_+ *is nonempty:* for Neumann or oblique boundary conditions $u \in Int\ X_+ \Longleftrightarrow u > 0$ on $\overline{\Omega}_j$. For the Dirichlet case $u \in Int\ X_+ \Longleftrightarrow u > 0$ and $\partial u/\partial \mu < 0$ on $\partial\Omega$ where μ is the unit vector field outwardly normal to $\partial\Omega$.

A well-known comparison theorem says that ψ and ϕ are monotone. It follows from the maximum principle [P-W, p. 187, theorem 12] and the boundary point lemma [P-W, p. 190, theorem 14] that ϕ is strongly monotone. Moreover ϕ and ψ are compact, and in fact we have:

1.4 THEOREM. The local semiflow ϕ in $X = C^1_B(\overline{\Omega})$, defined by solutions to (4), is order-compact.

Proof. Let $S \subset X$ be order-bounded. Then S is L^2-bounded. Assume $\phi_t(S)$ is defined with $r + s = t$, $0 < r \leqslant s < t$. Then

$$\phi_t(S) = \phi_r(\phi_s(S)) = \psi_r(\psi_s(S)) \quad .$$

Now $\psi_s(S)$ has compact closure M in $L^2(\Omega)$ and $\psi_r : M \longrightarrow X$ is continuous. Therefore $\phi_t(S)$ lies in the compact set $\psi_r(M)$ in X. *Q.E.D.*

1.5 EXAMPLE. Let X be any ordered Banach space and A a densely defined linear operator in X which generates a continuous semigroup denoted by $\{e^{tA}\}_{t \geqslant 0}$. Let $f : X_0 \longrightarrow X$ be a map defined in a Banach space X_0 which is a linear subspace of X by a continuous injection map, and consider the initial value problem in X:

$$\text{(6a)} \qquad \frac{dx}{dt} = Ax + f(x), \qquad t > 0.$$

$$\text{(6b)} \qquad x(0) = y \ .$$

By a *mild solution* is meant a continous map $x : [0, T) \longrightarrow X$ such that

$$\text{(7a)} \qquad x(t) = e^{tA}y + \int_0^t e^{(t-s)A} f(x(x))\, ds, \qquad (0 < t < T),$$

$$\text{(7b)} \qquad x(0) = y \in X_0 \ .$$

Under various assumptions about f and X_0, mild solutions exist and have sufficient uniqueness and continuity properties to generate a local semiflow ϕ on X_0, where $\phi_t(y) = x(t)$ from (7). Usually the domain of A, $D(A)$, is contained in X_0 and ϕ_t takes values in $D(A)$ for $t > 0$. This often makes ϕ a compact local semiflow.

An important example is the case where $-A$ is sectorial (equivalently, $\{e^{tA}\}$ is an analytic semigroup), $X_0 = X^\alpha$ is the domain of a fractional power $(aI - A)^\alpha$ for some $a > 0$, $0 \leqslant \alpha < 1$, and f is locally Lipschitz for the graph norm on X^α (see [He]). In this case mild solutions constitute a local semiflow in X^α. If e^{tA} is compact for all $t > 0$ then the local semiflow is compact.

Martin [Ma] constructs local semiflows on $X = X_0$ when $f : X \longrightarrow X$ is locally Lipschitz. He also treats the case where X_0 is a closed convex subspace.

Suppose mild solutions to (6) constitute a local semiflow in X_0; when is it strongly monotone? Adapting an argument sketched in [He], pp. 60-61, similar to the proof of 1.2(b), one can prove the following result:

1.6 THEOREM. Suppose e^{tA} is strongly monotone for all $t > 0$, and every point of X_0 has a neighborhood U in X such that $f + \lambda I : X_0 \cap U \longrightarrow X$ is monotone for some $\lambda > 0$. Then the local semiflow ϕ on X_0 corresponding to (7) is strongly monotone.

Remark. It is well known that $e^{tA} \gg 0$ for $t > 0$ provided $(cI - A)^{-1} \gg 0$ for some $c > 0$. This follows from the formula

$$\text{(8)} \qquad e^{tA}x = \lim_{k \to \infty}(I - \tfrac{t}{k} A)^{-k}x \ .$$

X. Mora [Mo] has a very general theory of equation (6), which applies to initial-boundary value problems of the type

(7a) $$\frac{\partial u}{\partial t} = Au + f(x, u, \nabla u), \qquad (x \in \overline{\Omega},\ t > 0),$$

(7b) $$Bu = 0 ,$$

where A, Ω are as in Example 1.3, $f : \overline{\Omega} \times \mathbb{R} \times \mathbb{R}^n \longrightarrow \mathbb{R}$ is smooth, and B is either a Dirichlet or Neumann boundary operator.

Define

$$C_B^0(\overline{\Omega}) = \begin{cases} C^0(\overline{\Omega}) \text{ for Neumann b.c.,} \\ \\ \{u \in C^0(\overline{\Omega}) : u|\Omega = 0\} \text{ for Dirichlet b.c.} \end{cases} .$$

For Dirichlet boundary conditions assume the compatibility condition:

$$f(x, 0, v) \equiv 0 \quad \text{for} \quad x \in \partial\Omega,\ v \in \mathbb{R}^n .$$

1.7 EXAMPLE ([Mo]). Mild solutions to (7) constitue a strongly monotone order-compact local semiflow ϕ on $C_B^1(\overline{\Omega})$. When f is independent of ∇u then ϕ extends to a local semiflow ψ in $C_B^0(\overline{\Omega})$ which is monotone in the Dirichlet case and strongly monotone in the Neumann case.

Moreover ϕ and ψ are C^1, provided we assume the further compatibility conditions:

$$\frac{\partial f}{\partial v}(x, 0, v) \equiv 0 \quad \text{for} \quad x \in \partial\Omega,\ v \in \mathbb{R}^n .$$

Mora also considers equations in unbounded domains; similar results hold. In particular we have the following example:

1.8 EXAMPLE. Let $f : \mathbb{R}^n \times \mathbb{R} \times \mathbb{R}^n \longrightarrow \mathbb{R}$ be smooth; let A be as in Example 1.3. Mild solutions to

(8) $$\frac{\partial u}{\partial t} = Au + f(x, u, \nabla u), \qquad (t > 0,\ x \in \mathbb{R}^n)$$

constitute a local semiflow in the (nonseparable) Banach space $C^1\overline{(\mathbb{R}^n)}$ of bounded functions $\mathbb{R}^n \longrightarrow \mathbb{R}$ whose derivatives are bounded and uniformly continuous. When f is independent of ∇u then ϕ extends to a local semiflow ψ on the Banach space of bounded uniformly continuous functions. Both ϕ and ψ are strongly monotone and C^1. They are usually not compact.

Theorem 1.6 can be applied to weakly linked systems of parabolic equations where the reaction term is given by a cooperative vector field. Consider a system

(9a) $$\frac{\partial u_i}{\partial t} = A_i u_i + F_i(x, u_1, \ldots , u_m), \qquad i = 1, \ldots , m ,$$

(9b) $$B_i u_i = 0 ,$$

where each A_i is an elliptic operator like A in Example 1.3, in a smooth bounded domain $\overline{\Omega}_i \subset \mathbb{R}^{m_i}$, and for each $x \in \overline{\Omega}_1 \times \ldots \times \overline{\Omega}_m$, the map

$$\mathbb{R}^m \longrightarrow \mathbb{R}^m,$$

$$(y_1, \ldots, y_m) \longmapsto F_i(x, y_1, \ldots, y_m)$$

is a cooperative vector field (in other words, $\partial F_i/\partial y_j \geq 0$ for $i \neq j$). Each boundary operator B_i is like B in 1.3. Let $X_i = \{u \in C^2(\overline{\Omega}_i) : B_i u = 0\}$. Applying Theorem 1.6 to $A = A_i \times \ldots \times A_m$ operating in $X = X_1 \times \ldots \times X_m$, we obtain:

1.9 THEOREM. Assume that solutions to (9) define a local semiflow ϕ in X. Then ϕ is strongly monotone.

1.10 EXAMPLE. G. Auchmuty, upon request, has kindly supplied the following example of Equation (7) with Dirichlet boundary conditions, having a nondegenerate hyperbolic closed oribt. Let $\overline{\Omega}$ be the closed unit disk in $\mathbb{R}^2$. Take

$$Au = \Delta u + \frac{\partial u}{\partial \theta} - cu$$

where θ is the angular coordinate and the constant $c \in \mathbb{R}$ will be chosen later. Fix an integer $m \neq 0$. Then Δ has a (complex) eigenfunction in $\mathbb{C} \otimes C^1_B(\overline{\Omega})$ of the form (in polar coordinates)

$$\omega(r, \theta) = h(r)e^{im\theta} .$$

Take $c \in \mathbb{R}$ to be the eigenvalue for ω. Notice that ω is also an eigenfunction for $\partial/\partial\theta$ with eigenvalue im. If follows that the complexification of A has im as an eigenvalue. The imaginary and real parts of ω span an A-invariant 2-dimensional subspace E of $X = C^1_B(\overline{\Omega})$. The semiflow in X is defined by $\partial u/\partial t = Au$ operates in E by rotations: all orbits are closed; all except the origin are nondegenerate (in the sense of §0).

These closed orbits are not hyperbolic; to make them so perform a Hopf bifurcation: with μ a real parameter consider the family of equations

$$\text{(10a)} \qquad \frac{\partial u}{\partial t} = \Delta u + \frac{\partial u}{\partial \theta} + (\mu - c)u - u^3,$$

$$\text{(10b)} \qquad u|\partial\Omega = 0,$$

$$\text{(10c)} \qquad u(x, 0) = v(x) .$$

The conditions for a Hopf bifurcation at $u = 0$, $\mu = 0$ are fulfilled, and (10) has nonconstant time periodic solutions for small $\mu > 0$. It can be shown that the resulting periodic orbits are hyperbolic (for example by adapting the methods of Chapter 5 of Hassard-Kazarinoff-Wan [H-K-W]), but they cannot be attracting (by 2.3 below).

Although (10) has closed orbits, it can be shown that for any value of μ, any solution with initial datum $v \geq 0$ is convergent. The proof is like the one given in Example 4.2 below, which differs only in the boundary conditions. Moreover it follows from 2.5 below that for any $v_0 \in X$, $h \in X_+ - \{0\}$, there is countable discrete set $S \subset \mathbb{R}$ such that for any $\alpha \in \mathbb{R}\backslash S$, the solution to (10) with initial datum $v - v_0 + \alpha h$ is quasiconvergent.

1.11 EXAMPLE. A single hyperbolic conservation law leads to a monotone semiflow in a rather different context (see Crandall-Majda [C-M]), for which the results in the present paper do not appear to be useful, however. (But see Conlon [Co] for a connection between cooperative systems of *ODE* and shock waves.)

§2 CONVERGENCE AND LIMIT SETS. Throughout this section X denotes an ordered Banach space. (In fact the norm is irrelevant: X could be any ordered topological vector space.) We always assume X_+ has nonempty interior. This is a severe restriction, but in applications the main results can often be extended to more general spaces, owing to well-known smoothing properties of solution operators.

Let $W \subset X$ be a nonempty open set and ϕ a monotone local semiflow in W. As before we set

$$E = \textit{set of equilibira,}$$

$$C = \textit{set of convergent points,}$$

$$Q = \textit{set of quasiconvergent points,}$$

so that $E \subset C \subset Q \subset W$.

The two main technical tools are the following :

2.1 THEOREM. If $\overline{O}(x)$ is compact and $x \ll x(t)$ or $x \gg x(t)$ for some $t > 0$, then $x(t)$ converges to an equilibirum.

2.2 THEOREM. Suppose ϕ is strongly monotone, $x < y$ and $\overline{O}(x)$, $\overline{O}(y)$ are compact. Then either

(a) $\omega(x) \ll \omega(y)$ or

(b) $\omega(x) = \omega(y) \subset E$.

An immediate consequence of 2.1 is:

2.3 COROLLARY. There are no nondegenerate attracting closed orbits in a monotone local semiflow in W.

Proof. Let K be a closed orbit attracting a neighborhood N of K. Then there exists $x \ll y$ with $x \in K$, $y \in N$. Then $K = \omega(y)$, so every neighborhood of x contains some $y(t)$. In particular there exists $t > 0$ such that $y(t) \ll y$. It follows from 2.1 that y converges, thus $\omega(y)$ reduces to a single point x. Therefore $K = \{x\}$. *Q.E.D.*

More generally one can prove that every attractor K contains an equilibrium p, and if p is unique then p attracts a neighborhood of K. More powerful results are stated below for attractors in strongly monotone local semiflows.

In the rest of this section we make the following assumptions: ϕ is *strongly* monotone; and $J \subset W$ is a simply ordered open arc such that $\overline{O}(x)$ is compact (in W) for all $x \in J$. One of the following additional hypotheses will sometimes be needed:

2.4 ADDITIONAL HYPOTHESES.

(H1). X is separable.

(H2). ϕ is order compact.

(H3). The closure in W of $\cup_{x \in J}\omega(x)$ is compact.

(H4). X is continuously and monotonely embedded in an ordered Banach space $\hat{X}$, and ϕ extends to a monotone local semiflow $\hat{\phi}$ in a lattice $\hat{W} \subset \hat{X}$ such that $W \subset \hat{W}$ and $\hat{\phi}_t$ takes its values in W for all $t > 0$.

Notice that (H2) implies (H3). In each of the examples of §1, and in most applications, at least one of these additional hypttheses holds. One often has (H4) with $W = X = C^1_B(\overline{\Omega})$ and $\hat{W} = \hat{X} = C^0_B(\overline{\Omega})$ or $L^2(\Omega)$, depending on the boundary operator B. When Ω is bounded then $C^1_B(\overline{\Omega})$ is separable so (H1) holds.

Let $|S|$ denote the cardinality of a set S.

2.5 THEOREM. Assume that one of the hypotheses (H1) - (H4) of 2.4 holds. Then the following are true:

(a) $J\backslash Q$ is countable.

(b) If (H2), (H3), or (H4) holds then $J\backslash Q$ is discrete, and in fact every x in $J\backslash Q$ has a neighborhood N in J such that $N\backslash\{x\} \subset C$.

(c) If X is finite-dimensional then $X\backslash Q$ has Lebesgue measure zero.

(d) If E is finite then $|J\backslash C| < |E|$.

§3 ATTRACTORS AND STABILITY. In this section X is an ordered Banach space and ϕ is a strongly monotone local semiflow in an open set $W \subset X$. *We assume ϕ is order-compact and all orbit closures are compact.*

Let $K \subset W$ be a nonempty compact positively invariant set.We call K an *attractor* if it has a neighborhood N in W such that every point $x \in N$ is attracted to K (i.e. $\overline{O}(x)$ is compact and $\omega(x) \subset K$). We call K *orbitally stable* if every neighborhood of K contains a positively invariant neighborhood of K.

It follows easily from 2.1 that if K is an attractor or if K is orbitally stable, then $K \cap E \neq \phi$. This implies:

3.1 THEOREM. A nondegenerate closed orbit cannot be orbitally stable.

In contrast to 2.3 this result does not necessarily hold for flows that are merely monotone. A counter-example is the flow in $\mathbb{R}^3$ consisting of rotations about the z-axis. This is monotone for the ordering defined by: $(x, y, z) \geqslant 0$ if $z^2 \geqslant x^2 + y^2$ and $z \geqslant 0$.

An equilibirum p is called *orbitally order-stable* if every order-interval neighborhood of p contains a positively invariant one; the set of such p is denoted by E_0. If in addition p attracts some such neighborhood then we call p *asymptotically order-stable*; E_a denotes the set of such equilibria.

Suppose ϕ is C^1. Then $p \in E$ is called *simple* if for all sufficiently small $t > 0$, 1 is not in the spectrum Σ of $D\phi_t(p)$. If Σ is disjoint from the unit circle for some (hence any) $t > 0$ then p is *hyperbolic*. If Σ is in the closed unit disk then p is called *linearly contractive*; when Σ is in the open unit disk p is a *sink*. Under current assumptions a sink is asymptotically order-stable, while an orbitally order-stable equilibrium is orbitally stable and hence linearly contractive. A simple linearly contractive equilibrium is a sink. The equilibria which come from Theorem 2.1 are linearly contractive.

The following is a basic existence theorem for order-stable equilibria:

3.2 THEOREM. Let $E' \subset E$ be a compact simply ordered set of equilibria. Suppose E' is locally maximal in the sense that E' has a neighborhood in W which contains no larger simply ordered subset of E. Then $E' \cap E_0$ is nonempty and contains its supremum and infimum. Any point of $E' \cap E_0$ isolated in E' is asymptotically order stable.

Combining 3.2 with 2.2 we have:

3.3 COROLLARY. E_0 is not empty, and if E is finite then E_a is not empty.

It can be proved that an attractor or orbitally stable set K must contain an orbitally order-stable equilibirum. This has the interesting consequence that *the internal dynamics of K cannot be very complicated*: there cannot be a dense orbit (unless K is a singelton), nor can the union of nondegenerate closed orbits be dense in K.

Now let $B = [a, b]$ be a closed order-interval, $a \ll b$, contained in W and positively invariant: $\phi_t(B) \subset B$ for all $t > 0$. An equilibrium $p \in B$ is *orbitally order-stable rel B* if every order-interval neighborhood of p in W contains an order-interval neighborhood N of p such that $N \cap B$ is positively invariant. Let $E_0(B)$ denote the set of such equilibria.

3.4 THEOREM. $E_0(B)$ is nonempty. If $E_0(B) = \{p\}$ then p attracts all of B. If $K \subset Int\ B$ is a nonvoid compact invariant set, not a point, then there are three equilibria $p \ll c \ll q$ in B with p, q in $E_0(B)$ and $p \ll K \ll q$.

3.5 COROLLARY. If $|E \cap B| \leq 2$ or $|E_0(B)| \leq 1$ then every trajectory in B is convergent.

§4 APPLICATIONS TO DIFFERENTIAL EQUATIONS.

4.1 EXAMPLE. Consider equations (5) of §1 with Neumann boundary conditions:

$$\text{(1a)} \qquad \frac{\partial u}{\partial t} = Au + g(x, u), \qquad (x \in \overline{\Omega},\ t > 0),$$

$$\text{(1b)} \qquad Bu \equiv \frac{\partial u}{\partial \nu} = 0,$$

$$\text{(1c)} \qquad u(x, 0) = v(x).$$

Let ϕ be the resulting local semiflow in $X = C^1_B(\overline{\Omega})$. Then Ω is strongly monotone, C^1, and order-compact;(hence compact); ϕ extends to a C^1 strongly monotone local semiflow in $C^0(\overline{\Omega})$ and to a monotone one in $L^2(\Omega)$. These extensions are compact and order-compact, and take values in $C^2(\overline{\Omega})$ for $t > 0$.

Let c_1 and $c_2 > c_1$ be real constants such that $g(x, c_1) > 0 > g(x, c_2)$ for all $x \in \overline{\Omega}$. Then the order interval $B = \{u \in X : c_1 \leqslant u \leqslant c_2\}$ can be shown to be positively invariant; in fact $\phi_t(B) \subset Int\ B$ for all $t > 0$.

Theorem 2.2 can be interpreted as saying that *for almost any initial datum* v, *as* $t \longrightarrow \infty$ *the solution to* (1) *converges in* $C^1(\overline{\Omega})$ *to the solution set of the corresponding elliptic problem*

$$Au + g(x, u) = 0, \quad (x \in \overline{\Omega}); \qquad \frac{\partial u}{\partial \nu} = 0 .$$

It is interesting to find further assumptions guaranteeing convergence for all initial data in B.

Let $\lambda \in \mathbb{R}$ be the maximal eigenvalue of A and assume

$$\frac{\partial g}{\partial y}(x, y) + \lambda_A \leqslant 0 \tag{2}$$

for all $x \in \overline{\Omega}$, $y \in [c_1, c_2]$. An easy calculation then implies that ϕ is L^2-*contractive* in B; that is, for any $u, v \in B$ we have:

$$\frac{d}{dt}\|\phi_t(u) - \phi_t(v)\|_{L^2(\Omega)} \leqslant 0$$

Thus ϕ_t cannot increase L^2-distance in B. A well-known consequence of contractiveness is that $|\omega(x) \cap E| \leqslant 1$; thus quasiconvergence implies convergence.

We now show that *all trajectories in* B *are convergent*. If not let $J \subset B$ be a simply ordered arc as in 2.3, with $v \in J\backslash C = J\backslash Q$. It follows from 2.3 that for every $\varepsilon > 0$ there exist $u, w \in J \cap C$ such that $u < v < w$. Let

$$\delta = inf\ \{\|a - b\|_{L^2(\Omega)} : a \ll \omega(v) \ll b\}.$$

Then $\delta > 0$. Choose u, w so that $\|u - w\|_{L^2(\Omega)} < \delta$. Let $\phi_t(u), \phi_t(w)$ converge in X (hence in $L^2(\Omega)$) to p, q. Then by 2.2(a), $p \ll \omega(v) \ll q$. Hence $\|p - q\|_{L^2(\Omega)} \geqslant \delta$, contradicting L^2-contractivity.

4.2 EXAMPLE. Now drop assumption (2) and assume instead either:

$$\frac{\partial^2 f}{\partial y^2}(x, y) < 0 \ \text{for all} \ x \in \Omega, \quad c_1 < y < c_2; \quad \text{or} \tag{3a}$$

$$\frac{\partial^2 f}{\partial y^2}(x, y) > 0 \ \text{for all} \ x \in \Omega, \quad c_1 < y < c_2 \quad . \tag{3b}$$

If (3a) *or* (3b) *holds then every trajectory in* B *converges*. We assume (3a), the proof for (3b) being similar.

Suppose on the contrary that there exists $x \in E\backslash C$. By 3.4 with $K = \omega(x)$ there are equilibria $p \ll c \ll q$ in B with p, q linearly contractive:

$$\rho(D\phi_t(p)) \leqslant 1 \quad \textit{for all} \quad t > 0$$

where ρ indicates spectral radius.

Let r be any equilibrium in B such that $r > p$. I claim r is a sink, i.e. $\rho(D\phi_t(r)) < 1$ for all $t > 0$. To prove this observe that the linear semigroups $\{D\phi_t(p)\}$, $\{D\phi_t(r)\}$ are generated by the linear operators $DF(p)$, $DF(r)$ where $F : C^2_B(\overline{\Omega}) \longrightarrow C^0(\overline{\Omega})$ is the map $\omega \longmapsto \Delta\omega + g(\cdot, \omega)$. Hypothesis (3a) implies that DF is strictly decreasing (i.e. $DF(w_1)\xi < DF(w_2)\xi$ if $w_2 > w_1$ in B and $\xi > 0$ in $C^2_B(\overline{\Omega})$. From this one proves that $D\phi_t$ is strictly decreasing in W for each $t > 0$. Using strong monotonicity and compactness of $D\phi_t(w)$ and the Krein-Rutman theorem one proves that $\rho(D\phi_t(w))$ is a strictly decreasing function of w for each $t > 0$. It follows that r is a sink.

In particular all equilibria in $[c, b]$ are sinks. The fixed-point index of a sink is $+1$, and the sum of the indices of fixed points (of any $D\phi_t$, $t > 0$) in $[c, b]$ is also $+1$. Therefore $|E \cap [c, b]| = 1$, contradicting the existence of q. This proves that every trajectory in B is convergent.

Further analysis reveals that $|B \cap E| \leqslant 2$ and $B \cap E$ contains exactly one sink. There is a striking analogy with 1-dimensional equations $\dot{x} = f(x)$ with $f'' < 0$ [or $f'' > 0$] and $f(c\ ) > 0 > f(c\ )$. The graph of f crosses the x-axis at most twice for $c_1 \leqslant x \leqslant c_2$, and $f' < 0$ at exactly one root of f.

4.3 EXAMPLE. An example from population dynamics concerns the population densities $u(x, t)$, $v(x, t)$ of two species competing in a smooth bounded domain $\overline{\Omega} \subset \mathbb{R}^3$. Let Δ denote Laplace's operator, interpreted as diffusion. Suppose the densitites are governed by a system

$$\text{(4a)} \qquad \frac{\partial u}{\partial t} = \alpha_1 \Delta u + f(x, u, v) \ , \qquad \frac{\partial u}{\partial t} = \alpha_1 \Delta v + g(x, u, v)$$

with no-flux Neumann boundary conditions

$$\text{(4b)} \qquad \frac{\partial u}{\partial \nu} = \frac{\partial v}{\partial \nu} = 0$$

where α_1, α_2 are positive constants and ν is the outward unit normal field to $\partial\Omega$. Competition is expressed by the assumption

$$\frac{\partial f}{\partial v} \leqslant 0, \qquad \frac{\partial g}{\partial u} \leqslant 0 \ .$$

For each fixed x, the vector field $(f(x, \cdot, \cdot), g(x, \cdot, \cdot))$ in $\mathbb{R}^2$ is *not* cooperative, but the simple change of variable $v = -w$ converts (3), (4) into a system to which Theorem 1.9 and the results in §2 and §3 apply. We conclude, for example, that attracting closed orbits are not possible and, under reasonable technical assumptions, "most" trajectories will converge.

4.4 EXAMPLE. The preceding ecological example can also be modeled by a finite dimensional local flow. Let two species compete at n locations, $L_1, \ldots, L_n$. Let x_i, y_i be the two populations at L_i. Assume the two species migrate, *in any manner*, between locations. Suppose x_i and y_i compete, but not x_i, y_j if $j \neq i$ (because they live in different places). Also x_i and x_j cooperate (some x_j's migrate to L_i), and likewise y_k, y_j. Postulating a dynamical system, we are led to consider

$$(5) \qquad \frac{dx_i}{dt} = F_i(x, y); \qquad \frac{dy_i}{dt} = G_i(x, y)$$

with

$$\frac{\partial F_i}{\partial y_i} \leqslant 0 \text{ and } \frac{\partial G_i}{\partial x_i} \leqslant 0 \text{ for all } i ,$$

$$\frac{\partial F_i}{\partial x_j} \geqslant 0 \text{ and } \frac{\partial G_i}{\partial y_j} \geqslant 0 \text{ for } i \neq j .$$

Under the technical assumption that the community matrix

$$\begin{bmatrix} \frac{\partial F}{\partial x} & \frac{\partial F}{\partial y} \\ \frac{\partial G}{\partial x} & \frac{\partial G}{\partial y} \end{bmatrix}$$

is irreducible, we obtain an irreducible cooperative system by the change of variables $y_i = -z_i$, $i = 1, \ldots, n$. Applying Theorem 1.2 we obtain a strongly monotone local semiflow; the results of §2 can be applied and interpreted. *One thus expects the populations at each location to converge.*

For other examples in biology and chemistry see the papers by Grossberg, Lajmanovich-Yorke, Othmer, and Selgrade.

4.5 EXAMPLE. Consider the problem of numerically solving a totally nonlinear elliptic boundary value problem of the form

$$(6) \qquad \begin{aligned} H(\Delta u, \nabla u, u, x) &= 0, \qquad x \in \Omega \\ u(x, 0) &= 0, \qquad x \in \partial\Omega \end{aligned}$$

in a bounded domain $\Omega \subset \mathbb{R}^m$, where $H : \mathbb{R} \times \mathbb{R}^m \times \mathbb{R} \times \overline{\Omega} \longrightarrow \mathbb{R}$ is C^1.

One idea is to discretize x and solve numerically the resulting system of equations, say by Newton's method. This raises the problem of where to begin Newton's method.

A different idea, which might be used to find a starting place for Newton's method, is to numerically solve the initial boundary value problem

$$(7) \qquad \begin{aligned} \frac{\partial u}{\partial t} &= H(\Delta u, \nabla u, u, x), \qquad (t > 0,\ x \in \overline{\Omega}) , \\ u(x, t) &= 0, \qquad (t > 0,\ x \in \partial\Omega) ; \end{aligned}$$

then let $t \longrightarrow \infty$ and hope that $\lim_{t\to\infty} u(x, t)$ approaches a solution to (6). If we discretize x then (7) becomes a cooperative system of ordinary differential equations.

The results of §2 applied to cooperative vector fields suggest that under certain conditions this second approach has a good chance of success. The main requirement is that $\partial_1 H \geqslant K \geqslant 0$ everywhere. We also assume $|\partial_2 H| \leqslant L$, although this can be weakened. For simplicity we take $m = 1$, $\Omega = [a, b] \in \mathbb{R}$. Then (7) becomes

$$\text{(8)} \qquad \begin{aligned} u_t &= H(u_{xx}, u_x, u, x), \qquad (t > 0,\ a \leqslant x \leqslant b), \\ u(a, t) &= u(b, t) = 0 \ . \end{aligned}$$

Fix a large integer $k > 0$, set $h = (b - a)/(k + 1)$ and subdivide $[a, b]$ at the points

$$x_i = a + ih, \qquad i = 0, \ldots, k + 1 \ .$$

Define $H_i : \mathbb{R}^3 \longrightarrow \mathbb{R}^3$ by

$$H_i(z, w, y) = H(z, w, y, x_i) \ .$$

Think of $u_i(t) \in \mathbb{R}$ as $u(x_i, t)$. Using standard approximation to $\partial/\partial x$ and $(\partial/\partial x)^2$ we consider , in place of (8), the system of ordinary differential equations

$$\text{(9)} \qquad \frac{du_i}{dt} = H_i\left(\frac{u_{i+1} + u_{i-1} - 2u_i}{h^2}, \frac{u_{i+1} - u_i}{h}, u_i\right) \equiv F_i(u)$$

for $i = 1, \ldots, k$, where $u_0 = u_{k+1} = 0$, and $u = (u_1, \ldots, u_k)$. We want (9) to be a cooperative system, i.e. $\partial F_i/\partial u_j \geqslant 0$ for $i \neq j$. Evidently $\partial F_i/\partial u_j = 0$ if $j \neq i - 1, i, i \neq 1$. Since $\partial_1 H > 0$ we have

$$\frac{\partial F_i}{\partial u_{i-1}} > 0 \ .$$

Also

$$\frac{\partial F_i}{\partial u_{i+1}} = \frac{\partial_1 H}{h^2} + \frac{\partial_2 H}{h} \geqslant \frac{K}{h^2} - \frac{L}{h} \ .$$

It follows that the system is cooperative and irreducible provided $h < K/L$.

We can now conclude from Theorems 1.2 and 2.5 that *for almost every* $v \in \mathbb{R}^k$, *the solution to* (9) *with* $u(0) = v$ *either is unbounded or is quasiconvergent.* Assuming the latter, we find that $F(u(t)) \longrightarrow 0$ as $t \longrightarrow \infty$, while $u(t)$ stays bounded; and every convergent sequence $u(t_k)$, $t_k \longrightarrow \infty$ approaches an equilibrium $w \in \mathbb{R}^k$ for (9). Thus for any $\varepsilon > 0$ we can find $w \in \mathbb{R}^k$ such that $|F_i(w_1, \ldots, w_k)| < \varepsilon$, $i = 1, \ldots, k$. This w is an approximate solution to the original *BVP* (6) if we replace Δ and ∇ by their approximations.

We have ignored all problems of numerical stability and rounding and truncation error, which are of course crucial for practical implementation of any numerical scheme. Nevertheless the method seems reasonable. Since almot all bounded trajectories of (9) are guaranteed to quasiconverge, perhaps the numerical method of solving (9) need not be highly accurate. This may make (9) a cheaper approach to the *BVP* (6) thatn Newton's method. At the least, the stationary point of (9) ought to be a good place to begin Newton's method.

So far we have taken $m = 1$ merely for notational convenience. But for this case John Smillie has proved an interesting result which implies that *every* bounded solution to (9) converges. His general result (unpublished) is that this holds for any irreducible cooperative or competitive vector field F in $\mathbb{R}^n$ which is "tridiagonal" in the sense that $\partial F_i/\partial x_j = 0$ if $|i - j| > 1$.

An interesting problem is to demonstrate the existence of a solution to the *BVP* (6) by proving suitable convergence, as $k \longrightarrow \infty$, of stationary points of the *ODE* (9).

BIBLIOGRAPHY

[Am-S] Amman, H., "Fixed point equations and nonlinear eigenvalue problems in ordered Banach spaces", SIAM Rev. 18 (1976), 620-709.

[Am-R] Amman, H., "Periodic solutions of semilinear parabolic equations", in Nonlinear Analysis (ed. by L. Cesari, R. Kannan, H. Weinberger), Academic Press, New York, 1978.

[Am-Z] Amman, H., "Existence and multiplicity theorems for semi-linear elliptic boundary value problems", Math. Z. 150 (1976), 281-295.

[Con] Conlon, J.G., "A theorem in ordinary differential equations with an application to hyperbolic conservation laws", Adv. Math. 35 (1980), 1-18.

[Cop] Coppel, W.A., Stability and Asymptotic Behavior of Differential Equations, D.C. Heath, Boston, 1965.

[C-M] Crandall, M.G. and Majda, A., "Monotone difference aporoximations for scalar conservation laws", Math. Comp. 34 (1980), 1-21.

[Gr] Grossberg, S., "Competition, decision and consensus", J. Math. Anal. Appl. 66 (1978), 470-493.

[He] Henry, D., Geometric Theory of Semilinear Parabolic Equations, Lecture Notes in Mathematics 840, Springer-Verlag, New York, 1981.

[Hi] Hirsch, M.W., "Systems of differential equations which are competitive or cooperative. I: Limit sets", SIAM J. Math. Anal. 13 (1982), 167-179.

[H-K-W] Hassard, B.D., Kazarinoff, N.D. and Wan, Y.-H., Theory and Application of Hopf Bifurcation, London Mathematical Society Lecture Note Series 41, Cambridge University Press, Cambridge, England, 1981.

[Ka] Kamke, E., "Zur thoerie der systeme gewöhnlicher differentialgleichungen, II", Acta Math. 58 (1932), 57-85.

[L-Y] Lajmanovich, A. and Yorke, J., "A deterministic model for gonorrhea in a nonhomogeneous population", Math. Biosci. 28 (19676), 221-236.

[Ma] Martin, R.H., Nonlinear Operators and Differential Equations in Banach Spaces, John Wiley & Sons, New York, 1976.

[Mo] Mora, X., Semilinear problems define semiflows on C^k spaces, Math. Dept., U. Michigan, 1982 (Preprint).

[Oth] Othmer, H.G., "The qualitative dynamics of a class of biochemical control circuits", J. Math. Biol. 3 (1976), 53-78.

[P-W] Protter, M.H. and Weinberger, H., Maximum Principles in Differential Equations, Prentice-Hall, Englewood Cliffs, New Jersey, 1967.

[Se-S] Selgrade, J., "Mathematical analysis of a cellular control process with positive feedback", SIAM J. Appl. Math. 36 (1979), 219-229.

[Se-J] Selgrade, J., "Asymptotic behavior of solutions to single loop positive feedback systems", J. Diff. Eq. 38 (1960), 80-103.

[Sm] Smale, S., "On the differential equations of species in competition", J. Math. Biol. 3 (1976), 5-7.

[Wa] Walker, J.A., Dynamical Systems and Evolution Equations, Plenum Press, New York, 1980.

[WWa] Walter, W., Differential and Integral Inequalities, Springer-Verlag, New York, 1970.

Department of Mathematics
University of California
Berkeley, CA 94720

Contemporary Mathematics
Volume 17, 1983

SOME IDEAS IN THE PROOF THAT THE FITZHUGH-NAGUMO PULSE IS STABLE

Christopher K. R. T. Jones[1]

1. INTRODUCTION

The Fitzhugh-Nagumo equations are the system of reaction-diffusion equations

$$\begin{aligned} u_t &= u_{xx} + f(u) - w \\ w_t &= \varepsilon(u - \gamma w), \end{aligned} \tag{1}$$

where $f(u) = u(u - a)(1 - u)$, some a between 0 and $1/2$, $\gamma << 1$. Equation (1) originally arose as a simplification to the Hodgkin-Huxley nerve impulse equations. Many authors have proved the existence of a travelling pulse solution to (1) when $\varepsilon << 1$ that is a solution that is a function of the single variable $\xi = x - ct$ and approaches $(0,0)$ at $\xi \to \pm\infty$. For such results see Conley [2], Carpenter [1], Hastings [8] and Langer [10]. Call this wave $U_\varepsilon(\xi) = (u_\varepsilon(\xi), w_\varepsilon(\xi))$.

To discuss the stability of such a wave, one chooses a function space. The space of bounded uniformly functions with the sup norm is a convenient choice. $U_\varepsilon(\xi)$ is then said to be stable if there is a $\delta > 0$, so that for any solution $U(\xi,t) = (u(\xi,t), w(\xi,t))$ of (1) which satisfies

$$||U(\xi,0) - U_\varepsilon(\xi + k_1)|| < \delta$$

for some k_1, there is a k_2 so that

$$||U(\xi,t) - U_\varepsilon(\xi + k_2,t)|| \to 0$$

as $t \to +\infty$.

In this report I shall sketch some of the ideas in the following theorem.

THEOREM. If ε is small enough, $U_\varepsilon(\xi)$ is stable.

A standard technique to prove such a theorem is the linearised stability criterion. Reset (1) in a moving co-ordinate frame and linearise the right hand side. One obtains the operator

1980 Mathematics Subject Classification 35B40.

[1]Supported by the National Science Foundation under grant MCS 8200392

0271-4132/82/0000-1165/$02.00

$$L\begin{pmatrix}p\\r\end{pmatrix} = \begin{pmatrix}p'' + cp' + f'(u_\varepsilon(\xi))\,p - r\\ cr' + \varepsilon(p - \gamma r)\end{pmatrix} \tag{2}$$

Evans [3-6] has laid the foundation for studying the stability of travelling waves in this fashion. In [3] he showed that the following properties of L, as an operator on the space mentioned above, are sufficient for stability.

(I) $\sigma(L)\setminus\{0\} \subset \{\lambda: \mathrm{Re}\,\lambda < a\}$ for some $a < 0$

(II) 0 is a simple eigenvalue.

One therefore has to prove (I) and (II).

2. STEPS IN THE PROOF

The first step in the proof of (I) is to locate $\sigma_e(L)$ = essential spectrum of L = spectrum other than isolated eigenvalues of finite multiplicity. Consider the linearisation about the rest state $(0,0)$, call this L_0. L_0 is L with $u_\varepsilon(\xi)$ replaced by 0 in (2). $\sigma(L_0)$ is easily seen to lie in a set of the form $N = \{\lambda: \mathrm{Re}\,\lambda < a\}$. A standard result, together with an estimate of Evans, that shows for $\lambda > 0$ and large $\lambda \notin \sigma(L)$, entails that $(\mathbb{C}\setminus N) \cap \sigma_e(L) = \emptyset$.

Evans [6] defined a function $D(\lambda)$ with domain $G = \mathbb{C}\setminus N$ which is analytic and whose zeroes are eigenvalues of L. Moreover the order of the zero is the algebraic multiplicity of the eigenvalue.

I shall explain how $D(\lambda)$ is derived, it is motivated by looking for eigenvalues. Writing $(L - \lambda I)\binom{p}{r} = 0$ as a system:

$$\begin{aligned} p' &= q\\ q' &= -cq + (\lambda - f'(u_\varepsilon(\xi))p + r\\ r' &= -(\varepsilon/c)p + ((\lambda + \varepsilon\gamma)/c)r. \end{aligned} \tag{3}$$

Abbreviate this as $z' = Az$ where $z = (p,q,r) \in \mathbb{C}^3$.

There is a unique solution to (3), up to normalization, which is bounded as $\xi \to -\infty$. This follows from an analysis of the asymptotic system, which is (3) with $u_\varepsilon(\xi)$ replaced by 0. Call this solution $\zeta(\lambda,\xi)$. $\zeta(\lambda,\xi)$ is a candidate, in fact the only one up to a scalar multiple, for an eigenfunction; $\lambda \in \mathbb{C}\setminus N$ is an eigenvalue if and only if $\zeta(\lambda,\xi)$ is bounded as $\xi \to +\infty$.

To test the boundedness of $\zeta(\lambda,\xi)$ as $\xi \to +\infty$, consider the adjoint of (3)

$$z^{*\prime} = Bz^* \tag{4}$$

where $B = -A^*$. It is trivial to check that if $z^*(\xi)$ satisfies (4) and $z(\xi)$ satisfies (3), then $z^* \cdot z$ (dot product in $\mathbb{C}^3$) is constant in ξ. There is a unique solution to (4) that is bounded as $\xi \to +\infty$, call this $\eta(\lambda,\xi)$. This has a specific meaning relative to (3) because of the duality of 2 and 1 dimensional spaces in $\mathbb{C}^3$. As $\xi \to +\infty$, $\eta(\lambda,\xi)$ is normal to the stable subspace for the asymptotic system of (3). Therefore $\zeta(\lambda,\xi)$ is bounded as $\xi \to +\infty$ if and only if

$$\lim_{\xi\to+\infty} \zeta(\lambda,\xi) \cdot \eta(\lambda,\xi) = 0.$$

But $\zeta \cdot \eta$ is independent of ξ. So λ is an eigenvalue iff $D(\lambda) = \zeta(\lambda,\xi) \cdot \eta(\lambda,\xi) = 0$. ζ and η can both be chosen to be analytic in λ, for fixed ξ. $D(\lambda)$ is therefore an analytic function and its zeroes in $\mathbb{C} \setminus N$ are the eigenvalues there.

The heart of the proof then is to show that in $\mathbb{C} \setminus N$, $D(\lambda)$ has only one zero, at zero, and it is simple.

The travelling wave is constructed, recall for ε small, as a singular perturbation. If one writes the equation for a travelling wave as a system, one obtains

$$\begin{aligned} u' &= v \\ v' &= -cv - f(u) + w \\ w' &= -(\varepsilon/c)(u - \gamma w). \end{aligned} \tag{5}$$

Set $\varepsilon = 0$ in (5). Each plane $w = \text{constant}$ is then invariant. The $w = 0$ plane contains a heteroclinic orbit (travelling front) running from $(0,0,0)$ to $(1,0,0)$, for some $c = c^*$. For this c^*, there is a w^* for which another heteroclinic orbit exists, i.e. it lies in the $w = w^*$ plane. The singular solution consists of these two connections together with curves of critical points (called the slow manifolds) joining their ends, see Fig. 1.

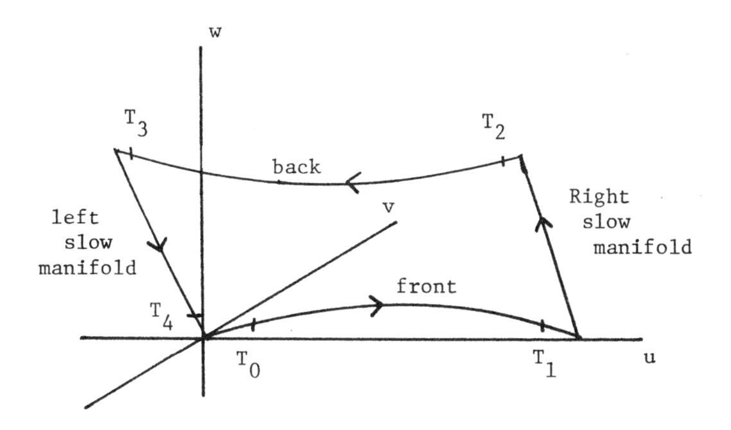

Figure 1

The existence theorem states that if $\varepsilon << 1$ then there is an orbit of (5), which tends to $(0,0,0)$ as $\xi \to \pm\infty$, that is close to the singular solution. See Conley [2], Carpenter [1], Hastings [8] and Langer [10] for details.

The two heteroclinic orbits represent the front and the back of the pulse respectively. They are each travelling waves of a reduced PDE. With respect to these equations their stability is well understood, see Fife and McLeod [7]. Call the spectrum of the linearised reduced operators about the front and the back, σ_F and σ_B respectively

The next step in the proof is to show that $\sigma(L) \cap G$ lies close to $\sigma_F \cup \sigma_B$. To be more precise, given $\delta > 0$, if V_δ is the union of balls of radius δ about each point of $\sigma_F \cup \sigma_B$, then there is an ε_0 so that if $\varepsilon < \varepsilon_0$, $\sigma(L) \cap G \subset G \cap V_\delta$.

Because the front and the back are stable, the point of largest real part in both σ_F and σ_B is zero. Therefore any eigenvalue that could cause instability must lie close to zero. One must count how many eigenvalues are near zero.

Since D is analytic, the winding number of $D(K) \subset \mathbb{C} \setminus \{0\}$, counts the number of zeroes of D inside a circle K. Let K be a small circle about 0, the next step is to show that the winding number of $D(K)$ is 2. There is a technical complication here. K may not lie in the domain of D but D has a natural analytic continuation that can be used.

It is a standard fact that 0 is an eigenvalue, due to translations of the wave also being waves. Therefore there is only one eigenvalue that could have positive real part, and it obviously must be real. Evans showed that $D(\lambda) > 0$ for $\lambda > 0$ and large for a class of problems that covers the one here. The proof will therefore be finished if it can be proved that $D'(0) > 0$.

Evans devised a beautiful technique for computing the sign of $D'(0)$. The traveling wave exists for a value of c at which the unstable manifold of $(0,0,0)$ intersects its stable manifold. As c varies through this value the unstable manifold crosses the stable manifold. The direction in which it crosses determines the sign of $D'(0)$! This direction is implicit in Langer's construction of the wave. The proofs of the middle steps require approximating by the reduced systems.

In the next section, I shall outline how this is achieved.

3. APPROXIMATION BY THE REDUCED SYSTEMS

The eigenvalue equations (3) can be coupled with the travelling wave equations (5) to form an autonomous system on $\mathbb{R}^3 \times \mathbb{C}^3$, for each λ. This space can be thought of as the complexified tangent bundle to $\mathbb{R}^3$. If the $\mathbb{R}^3$ part is restricted to the travelling wave, c must be set correctly, one derives a flow on a space that is the homoclinic orbit with a copy of $\mathbb{C}^3$ attached at each point. $\zeta(\lambda,\xi)$ lies in this space, for each ξ it sits above the point $U_\varepsilon(\xi)$ on the homoclinic orbit.

The key to the approximation then is to follow certain information about $\zeta(\lambda,\xi)$ around the orbit. $\zeta(\lambda,\xi)$ is determined near $-\infty$, information is encoded in ζ as ξ increases to $+\infty$. At $+\infty$, $\eta(\lambda,\xi)$ is determined and $D(\lambda)$ can be evaluated there with all the information coming from ζ.

To show that $D(\lambda) \neq 0$, only the "direction" of $\zeta(\lambda,\xi)$ is important. Since (5) is linear, it induces a flow on 2-d complex projective space, $\mathbb{C}P^2$. Let $\pi: \mathbb{C}^3 \to \mathbb{C}P^2$ be the natural map and set $\hat{z} = \pi(z)$. The proof then proceeds by estimating $\hat{\zeta}(\lambda,\xi)$ for large ξ and $\lambda \in G \setminus V_\delta$.

To obtain this information, $\hat{\zeta}(\lambda,\xi)$ is followed around the orbit. Control is afforded in different pieces using the approximations by the reduced systems. The homoclinic orbit splits up naturally into four pieces, the front, right hand slow manifold, back and the left hand slow manifold. Estimates on $\hat{\zeta}(\lambda,\xi)$ are proved at the edges of each of the pieces, i.e., at times T_i, $i = 0,1,2,3,4$ at which $U_\varepsilon(T_i)$ is at (or near) each of the corners, see Fig. 1. These estimates are proved succesively, each one dependeing on the previous estimate and the approximation given by the underlying orbit in between. In this fashion enough control is obtained on $\hat{\zeta}(\lambda,T_4)$, with $T_4 >> 0$ to show that $D(\lambda) \neq 0$ for $\lambda \in G \setminus V_\delta$.

To show the winding number is 2 for $D(K)$, K as described in the previous section, the projectivized flows do not contain enough information. They are however useful. Let $\zeta(\lambda,\xi) = (p,q,r) \in \mathbb{C}^3$, then $(q/p, r/p)$ forms local coordinates on $\mathbb{C}P^2$, which require $p \neq 0$. It can be shown that for each T_i, $i = 0,1,2,3,4$, the first component of $\zeta(\lambda,T_i)$ is non-zero. Define $\gamma_i(\lambda)$ by

$$\zeta(\lambda,T_i) = \gamma_i(\lambda)\tilde{\zeta}(\lambda,T_i)$$

where $\tilde{\zeta}(\lambda,T_i) = (1, q/p, r/p)$ and so $\gamma_i(\lambda) = p(\lambda,T_i)$. It is easy to see that $W(D(K))$ (winding number of $D(K)$ equals $W(\gamma_4(K))$. The argument that $W(\gamma_4(K)) = 2$ is again established iteratively. In fact it is shown that $W(\gamma_i(K))$ are 0,1,1,2,2 for $i = 0,1,2,3,4$. The jumps from 0 to 1 and 1 to 2 occur as the homoclinic orbit passes over the front and back respectively. This reflects the fact that these orbits are stable relative to their own reduced systems. The fact that the winding number does not change in the other regions is due to the fact that not much is happening on the slow manifolds.

BIBLIOGRAPHY

1. Carpenter, G., A geometric approach to singular perturbation problems with applications to nerve impulse equations, J. Diff. Eq. 23 (1977), 335-367.

2. Conley, C., On travelling wave solutions on nonlinear diffusion equations, in Dynamical Systems Theory and Applications (J. Moser Ed.), Lecture Notes in Physics 38, Springer-Verlag, Berlin.

3. Evans, J. W., Nerve axon equations I: linear approximations, Indiana Univ. Math. J. 21 (1972), 877-855.

4. Evans, J. W., Nerve axon equations II: stability at rest, Indiana Univ. Math. J. 22 (1972), 75-90.

5. Evans, J. W., Nerve axon equations III: stability of the nerve impulse, Indiana Univ. Math. J. 22 (1972), 577-594.

6. Evans, J. W., Nerve axon equations IV: the stable and the unstable impulse, Indiana Univ. Math. J. 24 (1975), 1169-1190.

7. Fife, P. and McLeod, J. B., The approach of solutions of nonlinear diffusion equations to travelling front solutions, Arch. Rat. Mech. Anal. 65 (1977), 335-361.

8. Hastings, S. P., On the existence of homoclinic and periodic orbits for the Fitzhugh-Nagumo equations, Quart. J. Math., Oxford, Ser. 27 (1976), 123-134.

9. Jones, C., Stability of travelling wave solutions for the Fitzhigh-Nagumo system, to appear.

10. Langer, R., Existence of homoclinic travelling wave solutions to the Fitzhugh-Nagumo equations, Ph.D. Thesis, Northeastern University (1980).

DEPARTMENT OF MATHEMATICS
UNIVERSITY OF ARIZONA
TUCSON, ARIZONA 85721

Contemporary Mathematics
Volume 17, 1983

DIFFUSION INDUCED CHAOS

James P. Keener

ABSTRACT. There are systems of autonomous differential equations with linearly stable rest points for which the addition of diffusion renders the system oscillatory. These same systems, if forced by small, slowly varying periodic parameter variations, have chaotic trajectories in the presence of diffusion. Although these oscillations do not occur in all systems, for any linearization with diffusive instability, there are open sets of nonlinear terms which yield oscillations. Similarly for almost all choices of quadratic nonlinearity, there are corresponding open sets of diffusively unstable linearizations for which oscillations occur.

1. INTRODUCTION. One of the fascinating consequences of the recent study of diffusion reaction equations is that the addition of diffusion to certain reaction equations can destabilize a stable uniform steady state and yield stable spatially nonuniform solutions [18], [19]. However, a stable uniform steady state for a system in $\mathbb{R}^2$ cannot lose its stability directly to temporal oscillations by the addition of diffusion. The effect of diffusion is to decrease the trace of the linearization, which because of linear stability, is already negative without diffusion.

The goal of this paper is to demonstrate that the addition of diffusion to systems of nonlinear differential equations can produce temporal oscillations through multiple bifurcations. Specifically, there are systems of differential equations

$$\mathring{u} = f(u) \qquad u \in \mathbb{R}^2 \tag{1.1}$$

with $f(u_0) = 0$, u_0 a linearly stable rest point, such that the partial differential equation

$$\frac{\partial u}{\partial t} = D\nabla^2 u + f(u), \quad x \in [0,\pi], \quad u \in \mathbb{R}^2, \tag{1.2}$$

with boundary data

$$u(0,t) = u(\pi,t) = u_0, \tag{1.3}$$

or

0271-4132/83/0000-1388/$08.00

$$(1.4) \qquad u_x(0,t) = u_x(\pi,t) = 0,$$

has temporal and spatial oscillations for certain parameter values. We will further show that for the nonautonomous system

$$(1.5) \qquad \mathring{u} = f(u) + \varepsilon^2 g(u,\varepsilon t)\ , \ 0 < \varepsilon << 1,\ \ g(u_0,\varepsilon t) = 0,$$

the diffusive system

$$(1.6) \qquad \frac{\partial u}{\partial t} = D\nabla^2 u + f(u) + \varepsilon^2 g(u,\varepsilon t)$$

with boundary data (1.3) or (1.4) has space-time trajectories which are chaotic, that is, trajectories which depend sensitively on initial data for $\varepsilon << 1$.

The oscillations of (1.2) cannot appear as a direct Hopf bifurcation from the steady solution u_0. They appear, instead, through a sequence of three bifurcations, the third of which is a Hopf bifurcation from a spatially nonuniform steady solution. This branch of periodic solutions then terminates at a solution of infinite period (a heteroclinic orbit). For parameter values near this heteroclinic orbit, the nonautonomous system (1.4) is chaotic.

The results of this paper are comparable to results of Smale [16] where it was shown that a system of differential equations in $\mathbb{R}^4$ with a stable rest point begins to oscillate when it is coupled to a second identical system by linear cross terms. The oscillations occur because of the presence of a van der Pol oscillator in a subsystem of the coupled $\mathbb{R}^8$ system. In the present work, we start with a system in $\mathbb{R}^2$ coupled by continuous, rather than discrete, diffusion and show multiple bifurcations which lead to the induced oscillation.

The outline of this paper is as follows. In the next section we review known results concerning the bifurcation of nonuniform steady solutions for the case of Dirichlet boundary data. In sections 3 and 4, we discuss the relevant center manifold equations in more detail, first to show the multiple bifurcation to periodic oscillations and the heteroclinic orbit and then to determine parameter regions in which periodic forcing renders the motion chaotic. In section 5 we show how similar results are obtained with Neumann data or when the equation (1.1) is invariant under the transformation $u \to -u$. Finally in section 6, we reach the main goal of this paper, to show that these bifurcation phenomena do indeed occur in the neighborhood of linearly stable rest points if the nonlinear terms are chosen carefully. A summary of the explicit results is also given in section 6 for the convenience of the reader not interested in plowing through the bifurcation theory upon which the results depend.

2. PRELIMINARIES.

Consider the system of diffusion reaction equations

$$\frac{\partial U}{\partial t} = D\frac{\partial^2 U}{\partial x^2} + f(U),\ x \in [0,\pi],\ U \in \mathbb{R}^2 \tag{2.1}$$
$$U(0,t) = U(\pi,t) = U_0$$

where $f(U_0) = 0$. By setting $U = u + U_0$ we write (2.1) in the form

$$\frac{\partial u}{\partial t} = D\frac{\partial^2 u}{\partial x^2} + Au + Q(u,u) + R(u) \tag{2.2}$$
$$u(0,t) = u(\pi,t) = 0$$

Here $D = (\delta_{ij}D_i)$ is a diagonal matrix of positive diffusion coefficients, $A = (a_{ij})$ is a 2×2 matrix, $Q(u,u)$ is a vector valued, symmetric, bilinear form, and all higher order remainder terms are collected in $R(u)$. We choose to study Dirichlet boundary data in this section, but will discuss Neumann data in section 5.

The first step in the analysis of equation (2.2) is a linear stability analysis. Setting $u = e^{\mu t}\phi(x)$ and keeping only linear terms, we are led to the eigenvalue problem

$$D\nabla^2\phi + (A-\mu I)\phi = 0,\quad \phi \in \mathbb{R}^2, \tag{2.3}$$
$$\phi(0) = \phi(\pi) = 0.$$

The solutions of this eigenvalue problem are the functions

$$\phi_n(x) = \phi_0 \sin nx \qquad \phi_0 \in \mathbb{R}^2$$

provided

$$(A - n^2D - \mu I)\phi_0 = 0. \tag{2.4}$$

The eigenvalues μ_n are

$$\mu_n^{\pm} = \frac{1}{2}\operatorname{tr} A_n \pm \frac{1}{2}\sqrt{(\operatorname{tr} A_n)^2 - 4\operatorname{Det}(A_n)} \tag{2.5}$$

where $A_n = A - n^2D$, $\operatorname{tr} A = \operatorname{trace}(A)$.

We suppose that the diffusion free system $(D=0)$ is stable. That is, we require

$$\operatorname{tr} A = a_{22} + a_{11} < 0 \tag{2.6}$$
$$\operatorname{Det} A = a_{11}a_{22} - a_{12}a_{21} > 0$$

where $A = (a_{ij})$. When (2.6) holds, the addition of diffusion can destabilize the system in only one way. Since $\operatorname{Tr} A_n < \operatorname{Tr} A < 0$ for all n with positive D_i, Hopf bifurcations from the uniform steady state are impossible, and the only possible instability appears if

$$(2.7)\qquad \operatorname{Det}(A_n) = (a_{11} - n^2D_1)(a_{22} - n^2D_2) - a_{12}a_{21} = 0.$$

In Figure 1 are sketched the neutral stability curves $\det(A_n) = 0$ in the

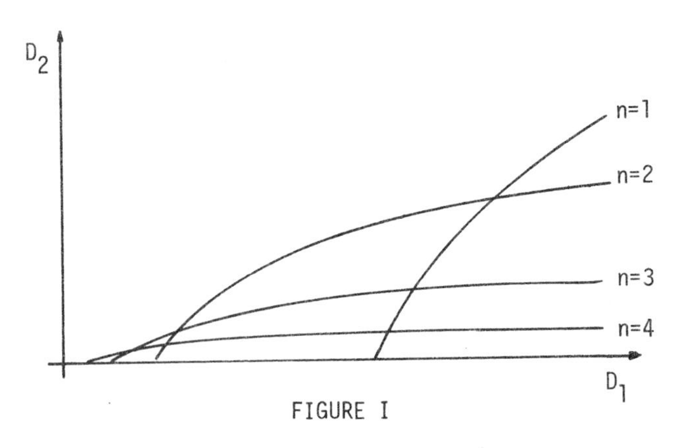

FIGURE I

D_1, D_2 parameter plane for different values of n. For this plot we take $a_{11} < 0$ (without loss of generality) and $0 < a_{22} < -a_{11}$. If $a_{22} < 0$ as well, the neutral stability curves are never in the first quadrant. By choosing D_1, D_2 appropriately the principal instability can have spatial shape $\sin nx$ for any n. Bifurcation analysis for transversal crossings of a single neutral stability curve has been given by various authors [2], [15].

Of interest to us is that each neutral stability curve intersects every other curve, and the shape of the principal instability changes at intersections with the adjacent curve. It is near such intersections that we will do our bifurcation analysis.

Suppose the numbers D_1, D_2, a_{ij} are chosen so that (2.6) and (2.7) hold for consecutive integers $n = m_1$ and $n = m_2$ where m_1 is even, $|m_1-m_2| = 1$. The requirement that m_1 and m_2 are consecutive guarantees that all other modes are linearly stable. Now we let

$$(2.8)\qquad A = A_0 + \varepsilon B, \quad A_0 = (a_{ij}), \quad B = (b_{ij}), \quad 0 < \varepsilon \ll 1.$$

The functions $\{\sin nx\}_{n=1}^{\infty}$ are complete in $\mathcal{L}^2(0,\pi)$ and therefore we write

$$(2.9)\qquad u = \varepsilon \sum_{k=1}^{\infty} u_k(t)\sin nx, \qquad u_k \in \mathbb{R}^2.$$

If we denote $\underset{\sim}{u} = (u_1, u_2, \ldots)$, then the parabolic equations (2.2) can be written as an infinite set of ordinary differential equations

$$(2.10)\qquad \frac{du_k}{dt} = (A_0-k^2D+\varepsilon B)u_k + \varepsilon Q_k(\underset{\sim}{u},\underset{\sim}{u}) \qquad k = 0, 1, 2, \ldots$$

where

$$Q(u,u) = \sum_{k=1}^{\infty} Q_k(\underset{\sim}{u},\underset{\sim}{u})\sin kx.$$

The assumptions on m_1, m_2 assure that the matrices A_0-k^2D have two eigenvalues with negative real part if $k \neq m_1, m_2$. However, by construction, $A_0-m_i^2D$ has one negative eigenvalue and one zero eigenvalue. When ε is small, we expect the evolution of (2.10) to be a rapid (exponential) decay onto a two dimensional manifold (the center manifold) corresponding to the zero eigenvalues of system (2.10). It is the motion on this two dimensional manifold which we wish to study.

The equations which govern the flow on this two dimensional manifold are known as the center manifold equations, the outer expansion (in singular perturbation theory), or the quasi-steady-state approximation. The calculations can be done using perturbation methods or normal forms, but in any case, the calculation is tedious. We relegate the details of this calculation to the appendix for the interested reader. Quoting from the appendix, we have that (2.10) reduces to

$$\text{(2.11)}\quad \begin{aligned} \frac{dx_1}{d\tau} &= a_1x_1 + b_1x_1x_2 + \varepsilon x_1(c_1x_1^2 + d_1x_2^2) + O(\varepsilon^2) \\ \frac{dx_2}{d\tau} &= a_2x_2 + b_2x_1^2 + b_3x_2^2 + \varepsilon x_2(c_2x_1^2 + d_2x_2^2) + O(\varepsilon^2) \end{aligned}$$

where x_i, $i = 1, 2$ are amplitudes corresponding to spatial variations $\sin m_ix$, $i = 1, 2$. The coefficients are listed in the appendix. From the derivation it is apparent that the first of equations (2.11) is odd in x_1 to all orders of ε, so that when $x_1 = 0$ then $\frac{dx_1}{d\tau} = 0$ as well. This special symmetry results since m_1 is even and m_2 odd.

3. ANALYSIS OF THE CENTER MANIFOLD EQUATIONS. In this section we review the bifurcation phenomena of (2.11) associated with independent variations of the parameters a_1, a_2. When higher order terms are ignored $(\varepsilon = 0)$ the equations (2.11) reduce to

$$\text{(3.1)}\quad \begin{aligned} \frac{dx_1}{d\tau} &= a_1x_1 + b_1x_1x_2 \\ \frac{dx_2}{d\tau} &= a_2x_2 + b_2x_1^2 + b_3x_2^2 \end{aligned}$$

This system of equations occurs in a different context in [5], [8], [10]. The steady solutions of (3.1) are easily found to be

$$
(3.2)\quad
\begin{aligned}
&\text{a)} \quad x_1 = x_2 = 0\\
&\text{b)} \quad x_1 = 0, \ \ x_2 = -a_2/b_3\\
&\text{c)} \quad x_2 = -a_1/b_1, \ x_1^2 = \frac{a_1}{b_1^2 b_2}(a_2 b_1 - a_1 b_3)\\
&\qquad\qquad \text{provided } \ x_1^2 > 0.
\end{aligned}
$$

From (3.2) one can easily summarize the bifurcation of steady state solutions, as follows. The trivial solution branch 3.2-a exists for all parameter values. There is a primary bifurcation to $x_2 \neq 0$ from the trivial branch at $a_2 = 0$. There is another primary bifurcation to $x_1 \neq 0$ and $x_2 \neq 0$ at $a_1 = 0$. Finally, there is a secondary bifurcation from the primary branch (3.2-b) at the parameter values $a_1 b_2 - a_1 b_3 = 0$. For certain parameter values the secondary branch is the same as the second primary branch (3.2-b). But an important observation is that, because of the unbroken symmetry of the full equations (2.11), the secondary bifurcation always occurs (unless $b_1 = b_3 = 0$) and the stability of the steady solutions is determined by the behavior of (3.1). Higher order terms may change the specific values of the bifurcation points, but cannot change their qualitative features. More detail of the qualitative features of these steady solutions is shown in [5], [7].

For most parameter values, the study of the steady solutions gives the full picture; there are no additional dynamics of interest, and the higher order corrections in equation (2.11) yield no interesting information. However, in the case $b_1 b_2 < 0$, $b_2 b_3 > 0$, a full understanding of the dynamics comes only when one retains more terms.

To illustrate why this is true, phase portraits for (3.1) with $b_1 b_2 < 0$, $b_2 b_3 > 0$, are shown in Figures II-a through II-d and II-h, and are summarized in Figure III. Figures II and III are drawn for the representative case $b_1 < 0$.

In each of the phase portraits Figure II a, b, c, d there are only hyperbolic fixed points, so that higher order terms do not change the structure. However, the phase portrait Figure II-h, which occurs only on the line H in Figure III, is not structurally stable. On H, the line $2a_1 b_3 - b_1 a_2 = 0$, the system (3.1) is integrable, and there is a family of closed orbits bounded by a heteroclinic trajectory and the x_2-axis.

The phase portrait Figure II-h portrays another interesting bifurcation. Crossing the line H transversely, there is a critical Hopf bifurcation. That is, on the line H, there is a family of periodic solutions which originate as small amplitude oscillations and terminate at the heteroclinic

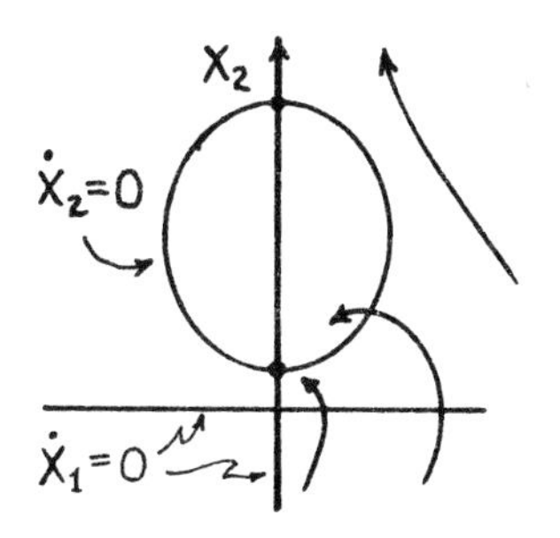

FIGURE II-a

FIGURE II-b

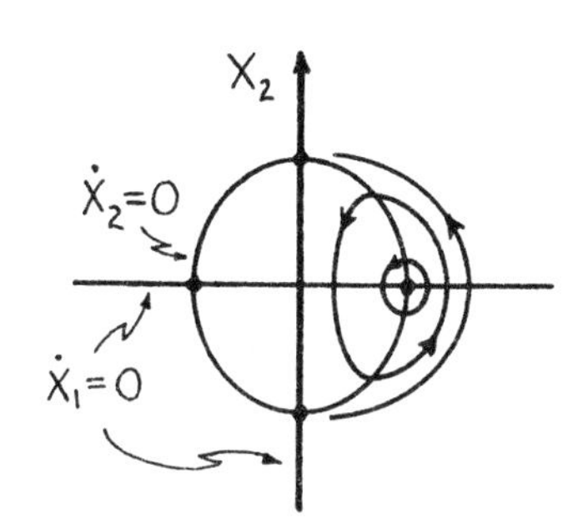

FIGURE II-h

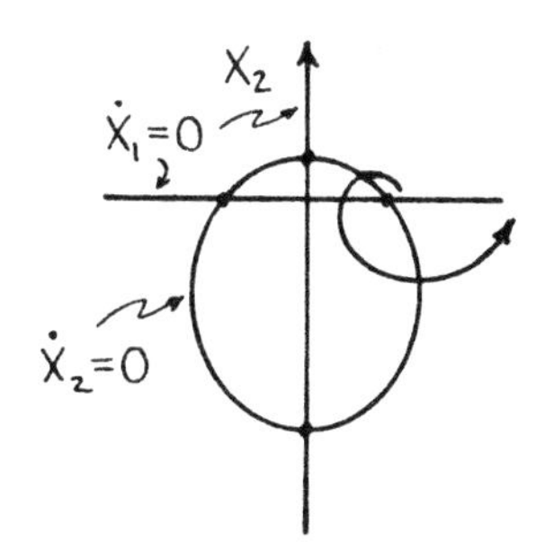

FIGURE II-c

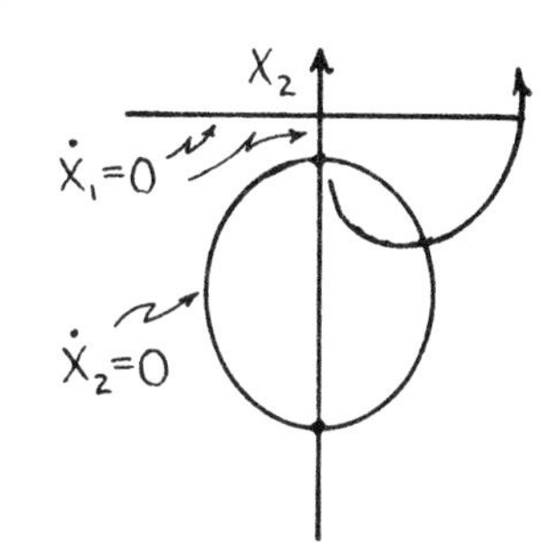

FIGURE II-d

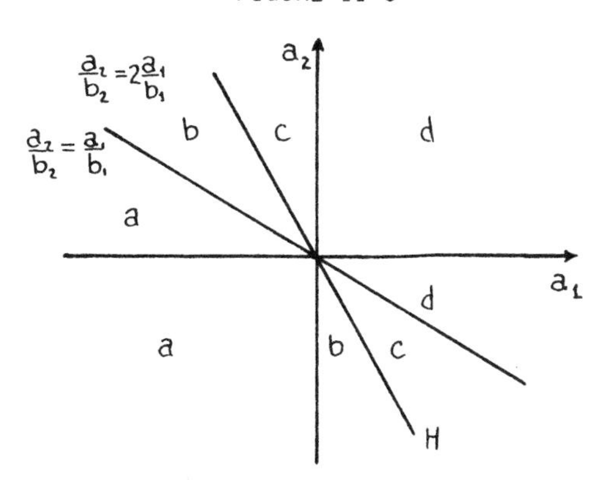

FIGURE III-a

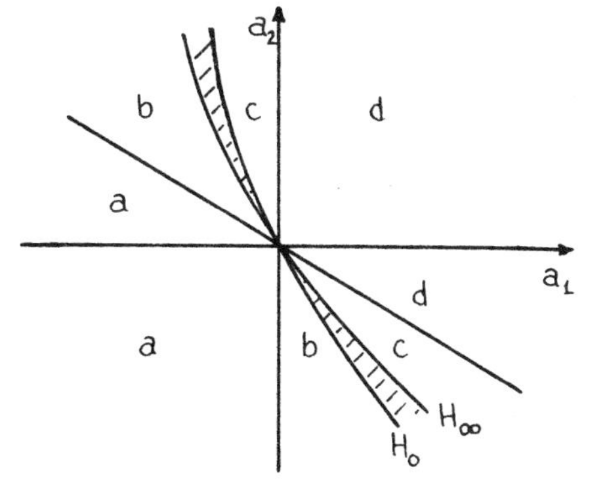

FIGURE III-b

trajectory, so that on the line H we see the entire "global" branch of periodic solutions. This special structure will not, in general, be preserved under perturbation. However, the Hopf bifurcation, the branch of periodic solutions and the heteroclinic trajectory will all occur on parameter curves different than the curve H. The critical Hopf bifurcation will become either subcritical or supercritical. We expect higher order terms to split the curve H into two bounding curves H_0 and H_∞, on which a Hopf bifurcation and a heteroclinic trajectory occur, respectively. Between H_0 and H_∞ there are periodic solutions of (2.11) [8].

To carry out the analysis necessary to find the curves H_0 and H_∞ explicitly, it is convenient to rewrite (2.11) as

$$\frac{dy}{d\tau} = 2y(b_1 z+\varepsilon c_1 y+\varepsilon d_1(z-\alpha_1)^2) + O(\varepsilon^2)$$

(3.3)

$$\frac{dz}{d\tau} = b_2 y + b_3(z^2-\alpha_1^2) + \varepsilon(z-\alpha_1)(2b_3\alpha_2+c_2 y+d_2(z-\alpha_1)^2) + O(\varepsilon^2)$$

where $y = x_1^2$, $z = x_2 + \frac{a_1}{b_1}$, $\frac{a_1}{b_1} = \alpha_1$, $\frac{a_2}{2b_3} = \alpha_1 + \varepsilon\alpha_2$. A modest calculation reveals that the steady solutions of (3.3) are

(3.4)

a) $$y = 0, \quad z = \pm\alpha_1 + O(\varepsilon)$$

b) $$y = \frac{b_3\alpha_1^2}{b_2} + O(\varepsilon), \quad z = \frac{-\varepsilon\alpha_1^2}{b_1 b_2}(c_1 b_3 + d_1 b_2)$$

and that for the solution (3.4-b), the trace of the linearization of (3.3) is

$$\text{(3.5)} \qquad \mathrm{Tr} = \varepsilon\left\{2b_3\alpha_2 + \frac{\alpha_1^2}{b_1 b_3}[2b_1 b_3 c_1+3b_1 b_2 d_2+b_1 b_3 c_2-2c_1 b_3^2-2d_1 b_2 b_3]\right\}.$$

In terms of original variables, the curve $\mathrm{Tr} = 0$ is given by

$$\frac{a_2}{2b_3} = \frac{a_1}{b_1} - \varepsilon\left(\frac{a_1}{b_1}\right)^2 \frac{K_1}{2b_1 b_2 b_3} + O(\varepsilon^2)$$

where

$$K_1 = 2c_1 b_3(b_1-b_3) + c_2 b_1 b_3 - 2d_1 b_2 b_3 + 3d_2 b_1 b_2.$$

This curve is depicted in Figure III-b as H_0.

At this point we could also calculate the local direction of bifurcation from H_0 for (3.3) by standard techniques. However, a more interesting calculation is to determine the location in parameter space of the heteroclinic trajectory.

With $\varepsilon = 0$, the system (3.3) becomes

$$\frac{dy}{d\tau} = 2b_1 yz \tag{3.6}$$
$$\frac{dz}{d\tau} = b_2 y + b_3(z^2-\alpha_1^2)$$

This system is integrable and its first integral is

$$H(y,z) = -b_1 y^\gamma(\alpha_1^2 - z^2 + \frac{b_2}{b_1-b_3} y), \; \gamma = -b_3/b_1 \tag{3.7}$$

and the equations (3.6) can be written as

$$\frac{dy}{d\tau} = y^{1-\gamma} \frac{\partial H}{\partial z} \tag{3.8}$$
$$\frac{dz}{d\tau} = -y^{1-\gamma} \frac{\partial H}{\partial y}.$$

Integral curves for this system are given by $H(y,z) = H_0$ = constant. These orbits are closed for all values of H_0 satisfying

$$\hat{H}_0 = -\frac{b_1\alpha_1^2}{b_1-b_3} \left(\frac{\alpha_1^2 b_3}{b_2}\right)^\gamma \leqslant H_0 \leqslant 0. \tag{3.9}$$

For $H_0 = \hat{H}_0$, the orbit is the one point $y = \frac{b_3\alpha_1^2}{b_2}$, $z = 0$ and for $H_0 = 0$, the orbit is heteroclinic, connecting the saddle points $y = 0$, $z = \pm\alpha_1$. There are, in fact, two heteroclinic orbits; one given by $y = 0$ and the other by

$$\frac{b_2}{b_2-b_3} y + \alpha_1^2 - z^2 = 0. \tag{3.10}$$

The trajectory $y = 0$ is of no interest here since it is trivially heteroclinic and remains so when $\varepsilon \neq 0$ because of the symmetry of (3.3). For us, the trajectory of importance is (3.10) for which

$$y_0(t) = \frac{b_3-b_1}{b_2} \alpha_1^2 \operatorname{sech}^2 b_1 \alpha_1 \tau \tag{3.11}$$
$$z_0(t) = -\alpha_1 \tanh b_1 \alpha_1 \tau.$$

To find a heteroclinic trajectory for (3.3) we seek a bounded solution of (3.3) of the form

$$y = y_0(t) + \varepsilon y_1(t) + \dots, \; z = z_0(t) + \varepsilon z_1(t) + \dots$$

and determine by standard arguments that $y_1(t)$, $z_1(t)$ must satisfy

$$\begin{pmatrix} \frac{dy_1}{d\tau} \\ \frac{dz_1}{d\tau} \end{pmatrix} = A(\tau)\begin{pmatrix} y_1 \\ z_1 \end{pmatrix} + \begin{pmatrix} g_1(y_0,z_0) \\ g_2(y_0,z_0) \end{pmatrix}$$

(3.12) where

$$A(\tau) = \begin{pmatrix} 2b_1 z_0(t) & 2b_1 y_0(\tau) \\ b_2 & 2b_3 z_0(\tau) \end{pmatrix}$$

and

$$g_1(y,z) = 2y(c_1 y + d_1(z-\alpha_1)^2)$$

$$g_2(y,z) = (z-\alpha_1)(2b_3\alpha_2 + c_2 y + d_2(z-\alpha_1)^2).$$

It follows from the Fredholm alternative that bounded solutions of (3.10) exist if and only if

$$\int_{-\infty}^{\infty} (Yg_1 + Zg_2)dt = 0 \tag{3.13}$$

where Y, Z are bounded integrable solutions of the homogeneous adjoint equation

$$\begin{pmatrix} \frac{dY}{d\tau} \\ \frac{dZ}{d\tau} \end{pmatrix} + A^T(\tau)\begin{pmatrix} Y \\ Z \end{pmatrix} = \begin{pmatrix} 0 \\ 0 \end{pmatrix}. \tag{3.14}$$

It follows from (3.8), (3.12) that the solutions of (3.14) are

$$Y = y_0^{\gamma-1}\frac{dz_0}{d\tau}, \quad Z = -y_0^{\gamma-1}\frac{dy_0}{d\tau}$$

so that the condition (3.13) becomes

$$I_1 = \int_{-\infty}^{\infty} y_0^{\gamma-1}\left(\frac{dz_0}{d\tau}g_1(y_0,z_0) - \frac{dy_0}{d\tau}g_2(y_0,z_0)\right)d\tau = 0. \tag{3.15}$$

It is a straightforward calculation to show that (3.15) is equivalent to the requirement

$$\frac{a_2}{2b_3} = \frac{a_1}{b_1} - \varepsilon\left(\frac{a_1}{b_1}\right)^2 \frac{K_2}{b_1 b_2 b_3(2b_3-3b_1)} + O(\varepsilon^2) \tag{3.16}$$

where

$$K_2 = -2b_3c_1(b_3-b_1^2) + c_2b_1b_3(b_3-b_1) + 2d_1b_2b_3(2b_1-b_3) + 3d_2b_1b_2(b_3-2b_1).$$

This last curve is depicted in Figure III-b as H_∞. It is this curve (to this order of approximation) on which a heteroclinic trajectory occurs.

Similarly, the curve H_0, given by (3.5) is the curve of Hopf bifurcation points. Between these two curves (the hatched region in Fig. III-b.) there is a family of periodic orbits of (3.3). The stability of the periodic orbits between these two curves depends entirely on the order in which they are crossed when moving from region b to region c. If the curves are oriented as shown in Figure III-b, the periodic orbits are stable, whereas if the orientation is opposite that shown, the orbits are unstable. This orientation is determined by the number

$$\Delta = \left(\frac{K_2}{2b_3-3b_1} - \frac{K_1}{2}\right)\frac{1}{b_1b_2b_3} \tag{3.17}$$

In particular, if $\Delta < 0$ then the curve H_0 is everywhere below the curve H_∞ in Figure III-b, so that the periodic orbits are supercritical and hence stable. On the other hand, if $\Delta < 0$, the curve H_0 is everywhere above the curve H_∞ in the a_1-a_2 parameter plane so that the periodic orbits are unstable.

Finally it should be noted that although the regions shown in Figure III-b are known only approximately, the dynamics associated with each region shown is robust; higher order terms cannot change the qualitative features depicted therein for $\varepsilon > 0$ sufficiently small.

4. CHAOTIC BEHAVIOR. In this section we show that if the coefficients of equation (3.1) are allowed to vary slowly, then in certain regions of parameter space (which we determine explicitly), trajectories of (2.1) are chaotic. There are, of course, many ways to allow parameters to vary, and the message is that essentially any parametric variation renders the motion chaotic. However, for definiteness and simplicity, we take the matrix A in (2.2) (see also (2.8)) to be of the form

$$A = A_0 + \varepsilon B + \varepsilon^2 C(\varepsilon t) \tag{4.1}$$

where $C(t)$ is a harmonic matrix with zero mean. With this slight change, we calculate (in the Appendix) that the center manifold equations are now

$$\begin{aligned}\frac{dx_1}{d\tau} &= (a_1 + \varepsilon\delta_1(\tau))x_1 + b_1x_1x_2 + \varepsilon x_1(c_1x_1^2 + d_1x_2^2) + O(\varepsilon^2)\\ \frac{dx_2}{d\tau} &= (a_2 + \varepsilon\delta_2(\tau))x_2 + b_2x_1^2 + b_3x_2^2 + \varepsilon x_2(c_2x_1^2 + d_2x_2^2) + O(\varepsilon^2)\end{aligned} \tag{4.2}$$

where the functions δ_1, δ_2 are sinusoidal with period $T = \frac{2\pi}{\omega}$. To show that the system (4.2) has chaotic trajectories it suffices to show [17] that the period T Poincaré map has transverse heteroclinic points, i.e., transverse intersections of a stable and unstable manifold of hyperbolic fixed points. In fact, Smale has shown [17] that if the Poincaré map has transverse heteroclinic points, then

a) The set of nonwandering points contains an invariant Cantor set

b) Some iterate of the Poincaré map restricted to this Cantor set is equivalent to the shift automorphism on doubly infinite sequences [13].

We take this as our definition of chaos.

To actually find transverse heteroclinic points, we use a slight modification of the perturbation calculation of Section 3. In particular, we seek solutions of (4.2) which are bounded for all time, and approach the heteroclinic trajectory of the autonomous system as $\varepsilon \to 0$. These solutions, if they exist in open sets of parameter space are solution trajectories, which, when viewed in the Poincare map, are exactly the heteroclinic points we seek. This idea to find heteroclinic points for the Poincare map from heteroclinic (or homoclinic) trajectories of the reduced autonomous equation has been used by a number of authors [3], [6], [9], [14].

We proceed directly to the calculation. We look for open regions of parameter space in which certain solutions of (4.2) are bounded and near to the heteroclinic trajectory when ε is small. Using the variables of (3.3) we seek solutions

$$\begin{aligned} y &= y_0(\tau) + \varepsilon y_1(\tau) + \dots \\ z &= z_0(\tau) + \varepsilon z_1(\tau) + \dots \end{aligned} \tag{4.3}$$

for the equation

$$\begin{aligned} \frac{dy}{d\tau} &= 2y(b_1 z + \varepsilon c_1 y + \varepsilon d_1(z-\alpha_1)^2 + \delta_1(\tau)) + O(\varepsilon^2) \\ \frac{dz}{d\tau} &= b_2 y + b_3(z^2-\alpha_1^2) + \varepsilon(z-\alpha_1)(2b_3\alpha_2 + c_2 y + d_2(z-\alpha_1)^2 + \delta_2(\tau)) + O(\varepsilon^2). \end{aligned} \tag{4.4}$$

The functions $y_0(\tau)$, $z_0(\tau)$ are given by (3.11) and the functions $y_1(\tau)$, $z_1(\tau)$ must satisfy

$$\begin{pmatrix} \dfrac{dy_1}{d\tau} \\ \dfrac{dy_2}{d\tau} \end{pmatrix} = A(\tau)\begin{pmatrix} y_1 \\ z_1 \end{pmatrix} + \begin{pmatrix} g_1(y_0,z_0) \\ g_2(y_0,z_0) \end{pmatrix} + \begin{pmatrix} 2y_0\delta_1(\tau) \\ (z_0-\alpha_1)\delta_2(\tau) \end{pmatrix} \tag{4.5}$$

where $A(\tau)$, g_1, g_2 are the same as in (3.12). It follows that bounded solutions of (4.5) exist if and only if

$$I_1 + I_2 = I_1 + \int_{-\infty}^{\infty} y_0^{\gamma-1}(\tau)[2\,\frac{dz_0}{d\tau}\,y_0\delta_1(\tau+\phi) - \frac{dy_0}{d\tau}(z_0-\alpha_1)\delta_2(\tau+\phi)]d\tau = 0 \tag{4.6}$$

where I_1 is the integral expression (3.15) corresponding to the Fredholm condition for the autonomous perturbation to (4.5). It is important to realize that the phase ϕ in (4.6) is a parameter which can be varied freely. This follows since the autonomous solutions $y_0(\tau)$, $z_0(\tau)$ have been unnecessarily restricted so that $z_0(0) = 0$. Instead of allowing arbitrary shifts of the origin for y_0 and z_0, we introduce a phase shift in the time periodic functions $\delta_1(\tau)$, $\delta_2(\tau)$.

Suppose $\delta_1(\tau) = \delta_1 \cos \omega\tau$, $\delta_2(\tau) = \delta_2 \cos(\omega\tau+\eta)$. Then the integral condition (4.6) is equivalent to

$$\frac{a_2}{2b_3} = \frac{a_1}{b_1} - \varepsilon\left(\frac{a_1}{b_1}\right)^2 \frac{\Gamma_2}{(2b_3-3b_1)(b_1b_2b_3)} - \varepsilon P(\phi) \tag{4.7}$$

where

$$P(\phi) = \frac{J_2\left(-\frac{2b_3}{b_1}, \frac{\omega}{a_1}\right)}{J_1\left(-\frac{b_3}{b_1} - 1\right)} \frac{(2b_3-b_1)}{b_3b_1} \left\{ \frac{b_1^2\omega^2 + 4b_3^2a_1^2}{2a_1^2b_3(2b_3-b_1)} (\delta_2 \cos(\phi+\eta) - \delta_1 \cos \phi) - \frac{\delta_2}{2} \frac{\omega b_1}{a_1b_3} (\sin(\phi+\eta) + 2 \cos(\phi+\eta)) \right\}, \tag{4.8}$$

$$J_1(m) = \sqrt{\pi} \frac{\Gamma(m+1)}{\Gamma(m+3/2)}, \quad m > -1, \tag{4.9}$$

and (from [4])

$$J_2(\nu,\omega) = \int_0^\infty \operatorname{sech}^\nu\tau \cos \omega\tau \, d\tau = 2^{\nu-2} \frac{\Gamma(\frac{\nu}{2} + \frac{i\omega}{2})\Gamma(\frac{\nu}{2} - \frac{i\omega}{2})}{\Gamma(\nu)} . \tag{4.10}$$

For all but exceptional values of γ_1, γ_2, η, the functions $P(\phi)$ is sinusoidal with nonzero amplitude, zero mean. Thus, the solvability condition (4.6) can be satisfied everywhere on the interior of the open domain

$$\left| \frac{a_2}{2b_3} - \frac{a_1}{b_1} + \varepsilon\left(\frac{a_1}{b_1}\right)^2 \frac{\Gamma_2}{(2b_3-3b_1)(b_1b_2b_3)} \right| \leq \varepsilon\hat{P}(\gamma_1,\gamma_2,\eta) + O(\varepsilon^2) \tag{4.11}$$

$$\hat{P} = \max_\phi P(\phi)$$

Furthermore, on the interior of this domain, the homoclinic points are transverse. Thus, for each fixed γ_1, γ_2, η for which $P \neq 0$, there is, for ε sufficiently small, a narrow band about the curve H_∞ in Figure III-b inside which trajectories of (4.2) are chaotic in the sense described earlier.

In summary, there are regions of oscillation (section 3) and regions of chaotic behavior (section 4) provided the coefficients b_1, b_2, b_3 satisfy $b_1b_2 < 0$, $b_2b_3 > 0$. The goal of section 6 is to show that these constraints can indeed be satisfied for linearly stable systems.

5. FURTHER EXTENSIONS. The results of sections 3, 4 do not hold if Neumann boundary data (1.4) are imposed, since $b_1 = b_2 = b_3 \equiv 0$ [1]. Similarly, if for some unlikely reason $Q(u,u) \equiv 0$, then $b_1 = b_2 = b_3 \equiv 0$. In any case, if the coefficients b_i, $i = 1, 2, 3$ are zero, one must consider higher order corrections of the center manifold equations. It follows that these equations take the form

$$\frac{dx_1}{d\tau} = x_1[a_1 + c_1x_1^2 + d_1x_2^2] + O(\varepsilon)$$
$$\frac{dx_2}{d\tau} = x_2[a_2 + c_2x_1^2 + d_2x_2^2] + O(\varepsilon) \tag{5.1}$$

where we have used the different scaling $\tau = \varepsilon^2 t$, $A = A_0 + \varepsilon^2 B$. The coefficients can be found in the appendix.

The equations (5.1) have been discussed elsewhere [1], [7], [11], so we content ourselves with a discussion of how oscillations, heteroclinic trajectories, and chaos arise.

The system (5.1) can be made slightly more tractable by setting $v = x_1^2$, $w = x_2^2$ so that

$$\frac{dv}{d\tau} = 2v[a_1 + c_1v + d_1w]$$
$$\frac{dw}{d\tau} = 2w[a_2 + c_2v + d_2w] \tag{5.2}$$

where v, w are restricted to the first quadrant.

The equations (5.2) are readily studied by phase plane techniques, and there is a wide variety of bifurcation phenomena [1], [7]. However, a Hopf bifurcation only occurs if

$$c_1d_2 < 0, \quad c_1d_2 - d_1c_2 > 0. \tag{5.3}$$

In addition if

$$c_1d_1 > 0 \tag{5.4}$$

there is a nontrivial heteroclinic trajectory for certain values of a_1, a_2.

In Figures IV and V we have displayed the relevant phase portraits for equations (5.2) in the representative case $c_1 > 0$, with (5.3), (5.4) satisfied. In the first and fourth quadrants of the a_1-a_2 plane of Figure V there are no stable rest points (with v, w nonnegative). In the third quadrant of the a_1-a_2 plane the origin is the only stable rest point (Figure IV-a). By passing from the third quadrant into the second quadrant the interesting bifurcations take place. For example in region b of Figure V there is a nontrivial stable steady state on the w-axis (Figure IV-b). Crossing the critical line $a_1d_2 = a_2d_1$, there is in Figure IV-c a stable spiral point with $v > 0$, $w > 0$. Crossing the line H_0 (given by $a_1d_2(c_1-c_2) = a_2c_1(d_2-d_1)$) there is a Hopf bifurcation through which the spiral point loses stability.

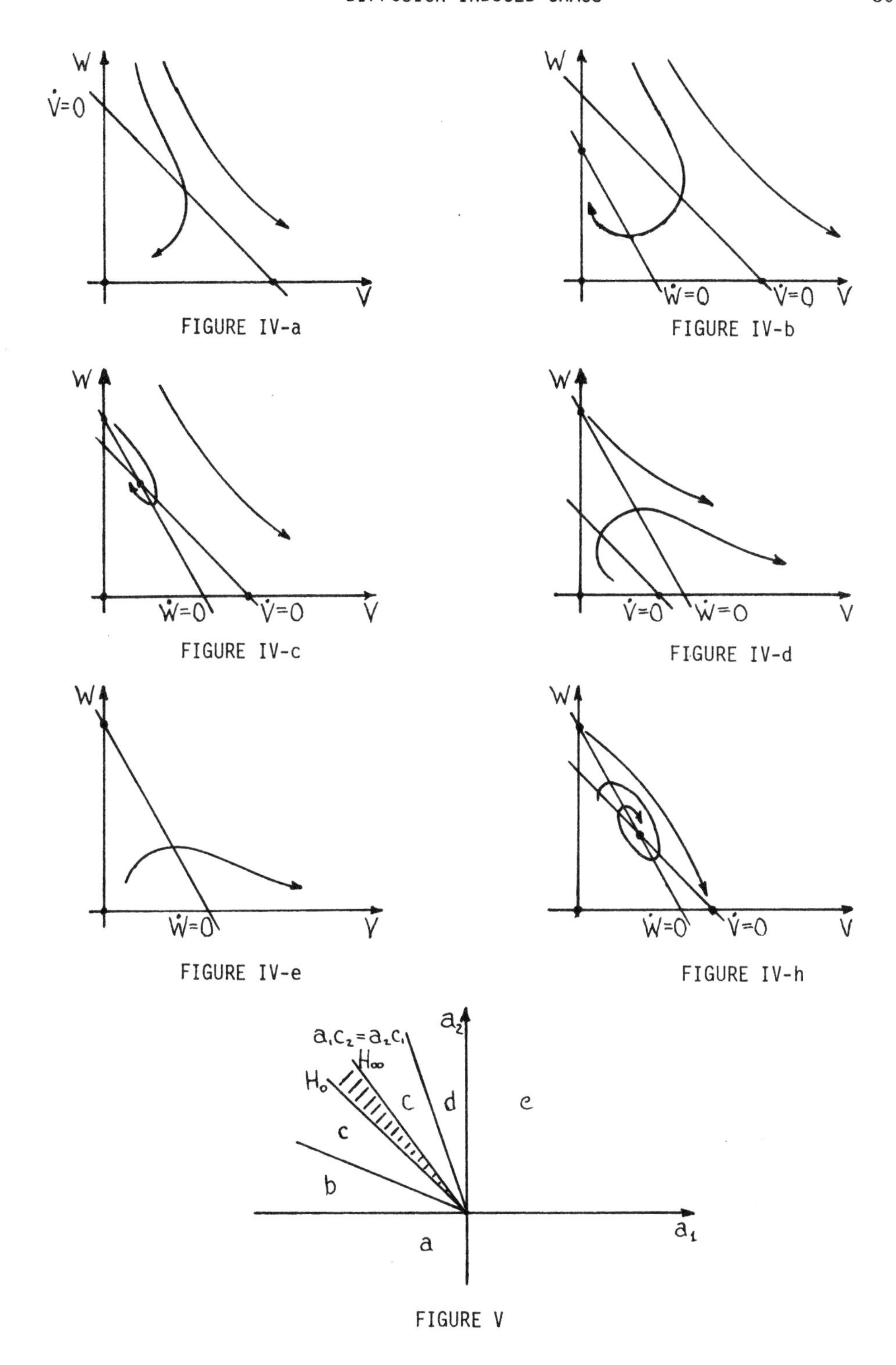

FIGURE IV-a

FIGURE IV-b

FIGURE IV-c

FIGURE IV-d

FIGURE IV-e

FIGURE IV-h

FIGURE V

Crossing the line $a_1c_2 = a_2c_1$ the unstable spiral point collapses into an unstable critical point on the v-axis, with resulting phase portrait shown in Figure IV-d. This unstable critical point collapses into the unstable origin as a_1 becomes positive (Figure IV-e).

In addition to the critical curves mentioned there is another critical curve H_∞ on which the phase portrait (Figure IV-h) has a heteroclinic orbit. The curve H_∞ can be either above or below H_0 in Figure V depending on parameter values. The two curves coincide identically if $c_1 = d_2$ and $d_1 = c_2$ when $a_1 = -a_2$ in which case the system (5.2) is integrable where

$$(5.5)\qquad (vw)^k(a_1 + c_1u + c_1v) = \text{constant} \qquad k = \frac{c_1}{d_1-c_1},$$

is a constant of the motion.

In general, however, the Hopf bifurcation and heteroclinic orbit occur for different parameter values. Between the curves H_0 and H_∞ (the hatched region) periodic orbits exist, and these periodic solutions are stable whenever H_∞ lies above H_0 in Figure V, but unstable otherwise.

Notice that the oscillations here appear as a tertiary bifurcation just as occured in Section 3. In a similar way we expect chaotic trajectories for the slightly nonautonomous problem in a neighborhood of the curve H_∞. However, in this problem we cannot find this region explicitly since we do not have an expression for the heteroclinic orbit (except in the very restrictive case $c_1 = d_2$, $c_2 = d_1$).

In summary, we find a sequence of bifurcations leading to a tertiary Hopf bifurcation of periodic solutions in the case $c_1d_2 < 0$, $c_1d_2 - d_1c_2 > 0$. It is the goal of the next section to show that these conditions can be realized with linearly stable dynamics.

6. CONDITIONS FOR DIFFUSION INDUCED OSCILLATIONS. The results of section 3 can be summarized as follows:

PROPOSITION 6.1. Suppose the differential equation

$$(6.1)\qquad \mathring{u} = f(u,\varepsilon) \qquad u \in \mathbb{R}^2$$

has the form

$$(6.2)\qquad f(u,\varepsilon) = (A_0 + \varepsilon B)u + Q(u,u) + R(u,\varepsilon)$$

where R is cubic or higher order in u. Suppose the system (6.1) is linearly stable but admits diffusive instability, i.e.,

$$(6.3)\qquad \begin{array}{ll} a_{11}a_{22} < 0 & a_{12}a_{21} < 0 \\ a_{11} + a_{22} < 0 & a_{11}a_{22} - a_{21}a_{12} > 0. \end{array}$$

Then for any consecutive integers m_1, m_2 with m_1 even, there are positive diffusion coefficients D_1, D_2 so that

$$(6.4) \qquad (a_{11}-m_i^2D_1)(a_{22}-m_i^2D_2) - a_{12}a_{21} = 0, \quad i = 1, 2.$$

For m_1, m_2 fixed, define the vectors

$$(6.5) \qquad p_i = \begin{pmatrix} a_{22} - m_i^2D_2 \\ -a_{21} \end{pmatrix}, \quad q_i = \begin{pmatrix} a_{21} \\ m_i^2D_1 - a_{11} \end{pmatrix}, \quad i = 1, 2.$$

If

$$(6.6) \qquad b_1b_2 < 0, \quad b_2b_3 > 0$$

where

$$b_1 = q_1^T Q(p_1,p_2)$$
$$b_2 = q_2^T Q(p_1,p_1)$$
$$b_3 = q_2^T Q(p_2,p_2)$$

then there are open sets of parameter values B so that for $\varepsilon > 0$ sufficiently small the equation

$$(6.7) \qquad \begin{aligned} \frac{\partial u}{\partial t} &= D\nabla^2 u + f(u,\varepsilon), \quad D = (\delta_{ij}D_i) \\ u(0,t) &= u(\pi,t) = 0 \end{aligned}$$

has time periodic solutions. Furthermore, there are stable periodic solutions if the number Δ in (3.20) is negative.

A similar summary of section 5 is as follows:

<u>PROPOSITION 6.2</u>. Suppose the assumptions (6.1) through (6.5) of Proposition 6.1 hold.

a) Suppose that $Q(u,u) \equiv 0$ and that

$$(6.8) \qquad R(u,\varepsilon) = C(u,u,u) + O(\varepsilon)$$

where C is a trilinear form in u.

If

$$(6.9) \qquad c_1d_2 - c_2d_1 > 0, \quad c_1d_2 < 0$$

where

$$(6.10) \qquad \begin{aligned} c_1 &= 3/4\, q_1^T C(p_1,p_1,p_1) \\ c_2 &= 3/2\, q_2^T C(p_1,p_1,p_2) \\ d_1 &= 3/2\, q_1^T C(p_1,p_2,p_2) \\ d_2 &= 3/4\, q_2^T C(p_2,p_2,p_2). \end{aligned}$$

Then there are open sets of parameter values B so that for $\varepsilon > 0$ sufficiently small the equation (6.7) has time periodic solutions. If $c_1 d_1 > 0$ then there is a heteroclinic trajectory for certain parameter values.

b) Suppose (6.8) holds and that

$$c_1 d_2 - c_2 d_1 > 0, \quad c_1 d_2 < 0$$

where

$$c_1 = 3/4\, q_1^T C(p_1,p_1,p_1) - \tfrac{1}{2} q_1^T Q(p_1, A_{2m_1}^{-1} Q(p_1,p_1)) - q_1^T Q(p_1, A_0^{-1} Q(p_1,p_1))$$

$$c_2 = 3/2\, q_2^T C(p_1,p_1,p_2) - q_2^T Q(p_1, A_{m_1+m_2}^{-1} Q(p_1,p_2)) - q_2^T Q(p_1, A_1^{-1} Q(p_1,p_2)$$

$$\text{(6.11)} \qquad - q_2^T Q(p_2, A_0^{-1} Q(p_1,p_1)).$$

$$d_1 = 3/2\, q_1^T C(p_1,p_2,p_2) - q_1^T Q(p_1, A_0^{-1} Q(p_2,p_2)) - q_1^T Q(p_2, A_{m_1+m_2}^{-1} Q(p_1,p_2))$$

$$- q_1^T Q(p_2, A_1^{-1} Q(p_1,p_2))$$

$$d_2 = 3/4\, q_2^T C(p_2,p_2,p_2) - 1/2\, q_2^T Q(p_2, A_{2m_2}^{-1} Q(p_2,p_2)) - q_2^T Q(p_2, A_0^{-1} Q(p_2,p_2)).$$

Then there are open sets of parameter values of B so that for $\varepsilon > 0$ sufficiently small the equation

$$\text{(6.12)} \qquad \frac{\partial u}{\partial t} = D\nabla^2 u + f(u,\varepsilon), \quad D = (\delta_{ij} D_i)$$
$$u_x(0,t) = u_x(\pi,t) = 0$$

has time periodic solutions. If $c_1 d_1 > 0$, then there is a heteroclinic trajectory for certain parameter values.

Here the main difficulty of our endeavor comes into focus. Although the aforementioned bifurcations to periodic solutions are generic, we have no guarantee as yet that either of (6.6) or (6.9) hold under the specific assumptions we have placed on A_0.

In this section we want to explore the above conditions more thoroughly to see what types of nonlinearities yield diffusive oscillations. Specifically, we will first show that for any fixed matrix A there are quadratic forms Q which yield oscillations in (6.7), and conversely, that for almost all quadratic forms Q there are matrices A so that diffusive oscillations occur in (6.7). Finally we show that there are always trilinear forms C for which all conditions of Proposition 6.2 are satisfied.

We first show that for any A satisfying (6.3) there are quadratic forms Q for which (6.6) hold. We write the bilinear form Q as

$$\text{(6.13)} \qquad Q(u,v) = \underset{\sim}{\beta}_1 u_1 v_1 + \underset{\sim}{\beta}_2 (u_1 v_2 + v_1 u_2) + \underset{\sim}{\beta}_3 (u_2 v_2)$$

where

$$\underset{\sim}{\beta}_i \in \mathbb{R}^2 \qquad i = 1, 2, 3, \qquad \underset{\sim}{u} = \begin{pmatrix} u_1 \\ u_2 \end{pmatrix}, \quad \underset{\sim}{v} = \begin{pmatrix} v_1 \\ v_2 \end{pmatrix}, \quad \beta_i = \begin{pmatrix} \beta_{i1} \\ \beta_{i2} \end{pmatrix} .$$

We can view the definition of b_i, $i = 1, 2, 3$ as a linear mapping of $\underset{\sim}{\beta}^T \in \mathbb{R}^6$ into $\mathbb{R}^3$, where $\underset{\sim}{\beta}^T = (\beta_{11}, \beta_{21}, \beta_{31}, \beta_{12}, \beta_{22}, \beta_{32})$.

<u>LEMMA 6.3</u>. The mapping $\underset{\sim}{\beta}^T \to (b_1, b_2, b_3) = \underset{\sim}{b}$, has rank 3.

<u>PROOF</u>: From (6.13), we can write equations (6.6) as a linear transformation

$$\underset{\sim}{b} = [Q_1 M,\ Q_2 M] \underset{\sim}{\beta} \tag{6.14}$$

$$Q_i = \begin{pmatrix} q_{1i} & 0 & 0 \\ 0 & q_{2i} & 0 \\ 0 & 0 & q_{2i} \end{pmatrix}, \qquad i = 1, 2$$

$$M = \begin{bmatrix} p_{11}p_{21} & p_{12}p_{21} + p_{11}p_{21} & p_{12}p_{22} \\ p_{11}^2 & 2p_{11}p_{12} & p_{12}^2 \\ p_{21}^2 & 2p_{21}p_{22} & p_{22}^2 \end{bmatrix}$$

and $p_i = \begin{pmatrix} p_{i1} \\ p_{i2} \end{pmatrix}$, $q_i = \begin{pmatrix} q_{i1} \\ q_{i2} \end{pmatrix}$, $i = 1, 2$. It follows that $\det M = (p_1 \times p_2)^3$. But $m_1 \neq m_2$ implies $p_1 \times p_2 \neq 0$ so that $\det M \neq 0$ and $\mathrm{Rank}\,[Q_1 M, Q_2 M] = 3$. Therefore, there are open sets of coefficients $\underset{\sim}{\beta}_i$ for which (6.6) is satisfied.

It is a more difficult matter to show that for a fixed quadratic form Q, a matrix A can be found so that Proposition 6.1 holds. Toward this end a preliminary lemma is useful.

<u>LEMMA 6.4</u>. Consider the scalar quadratic function

$$q(u,v) = \beta_1 uv + \beta_2(u+v) + \beta_3$$

a) Suppose $\beta_2^2 - \beta_1\beta_3 > 0$.

If $q(u,u) \neq 0$ for all u in some range $u_0 \leq u \leq u_1$, then $q(u,v) \neq 0$ for all u, v in the same range $u_0 \leq u, v \leq u_1$.

b) Suppose $\beta_2^2 - \beta_1\beta_3 > 0$. There is a nonzero number $\mu > 1$ so that $q(u,v) \cdot q(u,u) > 0$ for all u, v satisfying

$$\frac{1}{\mu} < \frac{v}{u} < \mu.$$

This lemma follows directly from considerations of the graph of the curve $q(u,v) = 0$, and is left to the reader.

We are now able to state the main result of this section.

LEMMA 6.5. Suppose the quadratic form $Q(u,v)$ in (6.13) is such that

$$\underset{\sim}{\beta}_i \times \begin{pmatrix} 1 \\ 0 \end{pmatrix} \neq 0 \qquad \text{and} \qquad \underset{\sim}{\beta}_i \times \begin{pmatrix} 0 \\ 1 \end{pmatrix} \neq 0$$

for $i = 1$ or 3. Then for all pairs of consecutive integers m_1, m_2 with m_1 even, there are open sets of matrices A_0 and diffusion coefficients D_1, D_2 for which Proposition 6.1 holds.

PROOF. The proof is in two parts. We first show that for a fixed quadratic form Q, vectors p_i, q_i can be chosen so that (6.6) holds. We then show that these vectors p_i, q_i can be restricted in a way that a matrix A_0 and diffusion coefficients satisfying (6.3), (6.4) can be found.

For convenience and without loss of generality we will consider only those vectors p_i, q_i whose second components are positive. For any vector $p \in \mathbb{R}^2$ we use the notation $r(p)$ to denote the ratio of components of p, $r(p) = p_1/p_2$. Then the signs of the components of Q are the same as the signs of the components of the expression

$$\underset{\sim}{\beta}_1 r(u)r(v) + \underset{\sim}{\beta}_2(r(u) + r(v)) + \underset{\sim}{\beta}_3.$$

By assumption $0 < |r(\underset{\sim}{\beta}_i)| < \infty$ for $i = 1$ or $i = 3$. We suppose $0 < |r(\underset{\sim}{\beta}_1)| < \infty$ without loss of generality.

Let q_0 be a vector orthogonal to $\underset{\sim}{\beta}_1$, $q_0 \cdot \underset{\sim}{\beta}_1 = 0$. Let q_1, q_2 be vectors close to q_0 so that $(q_1 \cdot \underset{\sim}{\beta}_1)(q_2 \cdot \underset{\sim}{\beta}_2) < 0$, and let $N > 1$ be the largest integer for which $(\frac{N-1}{N})^2 \leqslant \frac{r(q_1)}{r(q_2)} \leqslant (\frac{N}{N-1})^2$. N can be made as large as we wish by choosing q_1, q_2 sufficiently close to q_0.

Define the scalar quadratic forms

$$Q_1(u,v) = q_1^T\, Q(u,v)$$
$$Q_2(u,v) = q_2^T\, Q(u,v).$$

Observe that since $(q_1 \cdot \underset{\sim}{\beta}_1)(q_2 \cdot \underset{\sim}{\beta}_1) < 0$, for all vectors p_i with $|r(p_i)| > r_0$, $i = 1, 2$, r_0 sufficiently large, then

$$Q_2(p_1,p_1)Q_2(p_2,p_2) > 0$$
$$Q_1(p_i,p_i)Q_2(p_i,p_i) < 0 \qquad i = 1, 2.$$

Now from Lemma 6.4, if, in addition, $r(p_1) \cdot r(p_2) > 0$, and $\frac{1}{\mu} < \frac{r(p_1)}{r(p_2)} < \mu$, for some fixed number $\mu > 1$, then

$$Q_1(p_1,p_2) \cdot Q_1(p_1,p_1) > 0$$

as well, so that the conditions (6.6) are satisfied. For reasons that will become apparent shortly, we pick q_1, q_2 so that $(\frac{N}{2N-1})^2 < \mu$ and pick p_1, p_2 so that $r(p_i) = -Kr(q_i)$, $i = 1, 2$, $K > [r(q_0)]^2$. There are nonempty, open sets of such vectors with $K > 0$ sufficiently large. Notice also that the pair of vectors p_1, q_1 is interchangeable with the pair p_2, q_2.

Now that p_i, q_i, $i = 1, 2$ have been chosen we must determine the matrix A_0 and the diffusion coefficients D_1, D_2. To find A_0 and D_1, D_2, we write (6.4) and (6.5) as a set of linear equations

$$\begin{aligned} &a_{22} - m_i^2 D_2 + a_{21} r(p_i) = 0 \\ &a_{22} - m_i^2 D_2 + a_{12} r(q_i) = 0 \qquad i = 1, 2 \\ &a_{11} - m_i^2 D_1 + a_{21}/r(q_i) = 0. \end{aligned} \tag{6.15}$$

This system of six equations in six unknowns has only the trivial solution unless

$$\frac{r(p_1)}{r(q_1)} = \frac{r(p_2)}{r(q_2)}$$

which was required above. Now the family of nontrivial solutions of (6.14) can be written as

$$D_1 = \frac{a_{21}}{m_1^2-m_2^2}\left(\frac{1}{r(q_1)} - \frac{1}{r(q_2)}\right)$$

$$D_2 = \frac{a_{21}}{m_1^2-m_2^2}\left(r(p_1) - r(p_2)\right)$$

$$a_{12} = a_{21}\frac{r(p_i)}{r(q_i)}$$

$$a_{11} = \frac{a_{21}}{m_1^2-m_2^2}\left(\frac{m_2^2}{r(q_1)} - \frac{m_1^2}{r(q_2)}\right)$$

$$a_{22} = \frac{a_{21}}{m_1^2-m_2^2}\left(m_2^2 r(p_1) - m_1^2 r(p_2)\right)$$

where a_{21} is arbitrary.

It suffices to show that the conditions (6.3) hold.

Note first that since $\frac{r(p_1)}{r(q_1)} = \frac{r(p_2)}{r(q_2)} < 0$, then $a_{12}a_{21} < 0$ and $D_1 D_2 > 0$

as required. We now calculate

$$a_{11}a_{22} = \frac{a_{12}a_{21}(m_2^2r(p_2)-m_1^2r(p_1))(m_2^2r(p_1)-m_1^2r(p_2))}{(m_1^2-m_2^2)^2r(p_1)r(p_2)} \tag{6.16}$$

which is negative whenever $r(p_1)$, $r(p_2)$ lie between the curves

$$r_1 = \frac{m_1^2}{m_2^2}\, r_2 \qquad \text{and} \qquad r_1 = \frac{m_2^2}{m_1^2}\, r_2.$$

Since $\frac{r(p_1)}{r(p_2)} = \frac{r(q_1)}{r(q_2)}$, $a_{11}a_{22}$ is negative whenever $\max(m_1,m_2) \leq N$. It follows that

$$a_{11}a_{22} - a_{12}a_{21} = -\frac{a_{12}a_{21}m_1^2m_2^2(r(p_1)-r(p_2))^2}{(m_1^2-m_2^2)^2r(p_1)r(p_2)} > 0. \tag{6.17}$$

Finally we have that

$$a_{11} + a_{22} = \frac{a_{21}}{(m_1^2-m_2^2)}\left[m_2^2(r(p_1) + \frac{1}{r(q_1)}) - m_1^2(r(p_2) + \frac{1}{r(q_2)})\right]. \tag{6.18}$$

All the requirements (6.3) can be satisfied by choosing the sign of a_{12} appropriately if, in addition

$$(r(q_2) - r(q_1))[m_2^2(\frac{1}{r(q_1)} - Kr(q_1)) - m_1^2(\frac{1}{r(q_2)} - Kr(q_2)] > 0 \tag{6.19}$$

where $K = -\frac{a_{21}}{a_{12}} > 0$. The point to make here is that for fixed q_1, q_2, the inequality (6.19) prescribes a unique ordering for the numbers m_1, m_2. For example, if $\sqrt{K} > r(q_2) > r(q_1) > 0$ then (6.19) requires $m_2 > m_1$. Furthermore, for each fixed q_1, q_2 there are at most a finite number of pairs m_1, m_2 which satisfy (6.19), but the number of pairs can be increased without bound by letting $\frac{r(q_2)}{r(q_1)}$ approach 1, provided $K \neq (r(q_0))^2$. Recall that m_1 must always be even. For a given pair of consecutive integers, the order prescribed by(6.19) may be inconsistent with choosing m_1 even. To satisfy (6.19) for all consecutive pairs with m_1 even, it may be necessary to interchange q_1 and q_2, but as noted earlier, this can always be done. Thus, all of the conditions of Proposition 6.1 can be satisfied for any consecutive pair of integers m_1, m_2.

If the quadratic form Q is identically zero, or if we wish to apply Neumann boundary conditions, then Proposition 6.2 contains the relevant requirements. We now show that

LEMMA 6.6. There are open sets of trilinear forms $C(u,u,u)$ for which the conditions (6.9) hold.

PROOF. The trilinear form $C(u,v,w)$ can be written in the form

$$C(u,v,w) = \sum c_{ijk} u_i v_j w_k \tag{6.20}$$

where the coefficients c_{ijk} are each vectors in $\mathbb{R}^2$. It is not difficult to show that the equations (6.10) can be written in the form

$$\begin{pmatrix} c_1 \\ d_1 \\ c_2 \\ d_2 \end{pmatrix} = [Q_1 M,\ Q_2 M] \begin{pmatrix} \Gamma_1 \\ \\ \Gamma_2 \end{pmatrix}$$

where $\Gamma_i = (\gamma_{1i}, \gamma_{2i}, \gamma_{3i}, \gamma_{4i})^T \quad i = 1, 2.$

$$\begin{aligned} \underset{\sim}{\gamma}_1 &= c_{111} \\ \underset{\sim}{\gamma}_2 &= c_{112} + c_{121} + c_{211} \\ \underset{\sim}{\gamma}_3 &= c_{122} + c_{212} + c_{221} \\ \underset{\sim}{\gamma}_4 &= c_{222}. \end{aligned}$$

$$Q = \begin{bmatrix} \frac{3}{4} q_{1i} & & & 0 \\ & \frac{1}{2} q_{1i} & & \\ & & \frac{1}{2} q_{2i} & \\ 0 & & & \frac{3}{4} q_{2i} \end{bmatrix} \qquad i = 1, 2.$$

and

$$M = \begin{bmatrix} p_{11}^3 & p_{11}^2 p_{12} & p_{11} p_{12}^2 & p_{12}^3 \\ 3p_{11}^2 p_{21} & p_{11}^2 p_{22} + 2p_{11} p_{12} p_{21} & p_{12}^2 p_{21} + 2p_{11} p_{12} p_{22} & 3p_{12}^2 p_{22} \\ 3p_{11} p_{21}^2 & 2p_{11} p_{21} p_{22} + p_{12} p_{21}^2 & p_{11} p_{22}^2 + 2p_{12} p_{21} p_{22} & 3p_{12} p_{22}^2 \\ p_{21}^3 & p_{21}^2 p_{22} & p_{21} p_{22}^2 & p_{22}^3 \end{bmatrix}$$

It follows that $\det M = (\underset{\sim}{p}_1 \times \underset{\sim}{p}_2)^6$ which is nonzero since $p_1 \times p_2 \neq 0$. Thus, there are open sets of coefficients c_{ijk} for which (6.9) is satisfied. In fact, for any fixed quadratic form Q coefficients c_{ijk} can be found so that the more complicated conditions (6.9), (6.11) are satisfied as well. We suspect, but have not yet proven, that a converse statement is also true.

7. CONCLUSION. In the foregoing sections we have shown that certain systems of differential equations in $\mathbb{R}^2$ can have diffusion induced oscillations and chaos even though the diffusion free system is linearly stable. This can occur provided the systems are subject to diffusive instability. These oscillations arise as a tertiary Hopf bifurcation from a spatially nonuniform steady profile, and they terminate with infinite period at a heteroclinic trajectory.

The diffusion induced oscillations do not occur for all diffusively unstable systems. However, for every diffusively unstable linearization A there are open sets of nonlinear terms which yield oscillation. With Dirichlet data and for nearly all choices of quadratic terms, there are linearizations A for which the diffusive system is oscillatory. For Neumann data, the oscillations occur for any choice of matrix A and quadratic terms Q provided cubic terms are chosen from the appropriate open set. With Dirichlet data oscillations occur if the quadratic term is chosen appropriately.

It remains to determine what these restrictions on nonlinear terms imply for the global dynamics of the linearly stable diffusion free system. For example, we conjecture that one consequence of Lemma 6.5 is that there are globally stable dynamics for which there are diffusion induced oscillations, although we have not explored this question in detail.

APPENDIX

In this section we give a brief derivation of the center manifold equations for the equation

$$\text{(A-1)}\qquad \begin{aligned} \frac{\partial u}{\partial t} &= D\nabla^2 u + (A_0 + \varepsilon B + \varepsilon^2 C(\varepsilon t))u + Q(u,u) + C(u,u,u) + (\text{h.o.t.}) \\ u(0,t) &= u(\pi,t) = 0 \end{aligned}$$

where $0 < \varepsilon << 1$, Q and C are bilinear and trilinear forms respectively. Setting $u = \varepsilon \sum_{k=1}^{\infty} u_k(t)\sin nx$, $u_k \in \mathbb{R}^2$, we can write (A-1) as the system of ordinary differential equations

$$\text{(A-2)}\qquad \frac{du_k}{dt} = (A_k + \varepsilon B + \varepsilon^2 C(\varepsilon t))u_k + \varepsilon Q_k(\underset{\sim}{u},\underset{\sim}{u}) + \varepsilon^2 C_k(\underset{\sim}{u},\underset{\sim}{u},\underset{\sim}{u})$$

where

$$A_k = A_0 - k^2 D, \quad \underset{\sim}{u} = (u_1, u_2, u_3, \ldots).$$

We define

$$P_i = \begin{pmatrix} a_{22} - m_i^2 D & a_{11} - m_i^2 D_1 \\ -a_{21} & a_{21} \end{pmatrix} = (p_i, \pi_i) \qquad i = 1, 2$$

$$P_i^{-1} = \frac{1}{a_{21}T_r(A_{m_i})}\begin{pmatrix} a_{21} & -a_{11} + m_i^2D_1 \\ a_{21} & a_{22} - m_i^2D_2 \end{pmatrix} = (q_i, \sigma_i)^T.$$

Then

$$P_i^{-1}A_{m_i}P_i = \begin{pmatrix} 0 & 0 \\ 0 & Tr(A_{m_i}) \end{pmatrix}.$$

We set $\begin{pmatrix} x_k \\ y_{m_k} \end{pmatrix} = P_k^{-1}u_{m_k}$, $k = 1, 2$ and make the change of time scale $\tau = \varepsilon t$, and find

$$\varepsilon \frac{d}{d\tau}\begin{pmatrix} x_k \\ y_{m_k} \end{pmatrix} = \begin{pmatrix} 0 & 0 \\ 0 & -\lambda_{m_k} \end{pmatrix}\begin{pmatrix} x_k \\ y_{m_k} \end{pmatrix}$$

$$\text{(A-4)} \qquad + \varepsilon P_k^{-1}\left((B+\varepsilon C(\varepsilon t))P_k\begin{pmatrix} x_k \\ y_{m_k} \end{pmatrix} + Q_{m_k}(\underset{\sim}{u},\underset{\sim}{u}) + \varepsilon C_{m_k}(\underset{\sim}{u},\underset{\sim}{u},\underset{\sim}{u})\right) \qquad k = 1, 2,$$

$$\varepsilon \frac{du_j}{d\tau} = (A_1+\varepsilon B+\varepsilon^2C(\tau))u_j + \varepsilon Q_j(\underset{\sim}{u},\underset{\sim}{u}) + \varepsilon^2C_j(\underset{\sim}{u},\underset{\sim}{u},\underset{\sim}{u}) \qquad j \neq m_1, m_2.$$

where

$$\lambda_{m_k} = -Tr(A_{m_k}) \qquad k = 1, 2.$$

We want to find y_{m_1}, y_{m_2} and u_j, $j \neq m_1, m_2$ as functions of x_1, x_2 and ε. We can do this by ignoring initial transients and finding the long time (quasisteady state) behavior

$$\text{(A-5)} \qquad \begin{aligned} y_{m_k} &= \frac{\varepsilon}{\lambda_k}\sigma_k^T[Bp_kx_k + Q_{m_k}(u_0,u_0)] + O(\varepsilon^2), \quad && k = 1, 2, \\ u_j &= -\varepsilon A_j^{-1}Q_j(\underset{\sim}{u}_0,\underset{\sim}{u}_0) + O(\varepsilon^2), && j \neq m_1, m_2, \end{aligned}$$

where $u_0 = p_1x_1 \sin m_1x + p_2x_2 \sin m_2x$ and

$$\text{(A-6)} \qquad \frac{dx_k}{d\tau} = q_k^T[(B+\varepsilon C(\tau))P_k\begin{pmatrix} x_k \\ y_{m_k} \end{pmatrix} + Q_{m_k}(\underset{\sim}{u},\underset{\sim}{u}) + \varepsilon C_{m_k}(\underset{\sim}{u},\underset{\sim}{u},\underset{\sim}{u})].$$

When we substitute (A-5) into (A-6), we find a closed system of differential equations for x_1, x_2. The main difficulty is that we must now calculate $Q_k(\underset{\sim}{u},\underset{\sim}{u})$.

An identity we use is

$$\sin jx \sin kx = \sum_{p=1}^{\infty} \gamma_{jkp} \sin px$$

(A-7)

$$\text{where } \gamma_{jkp} = \frac{1}{2\pi}\left\{\frac{1-(-1)^{j-k+p}}{j-k+p} + \frac{1-(-1)^{k-j+p}}{k-j+p} + \frac{1-(-1)^{k+j-p}}{k+j-p} - \frac{1-(-1)^{j+k+p}}{j+k+p}\right\}$$

Notice that $\gamma_{jkp} = 0$ whenever $j + k + p$ is even. Using this identity and $u = \sum_{j=1}^{\infty} u_j \sin jx$, we find that

$$Q(u,u) = \sum_{j,k} Q(u_j,u_k) \sin jx \sin kx$$

so that

$$Q_p(\underset{\sim}{u},\underset{\sim}{u}) = \sum_{j,k} Q(u_j,u_k)\gamma_{jkp}. \tag{A-8}$$

It follows that

(A-9)

$$Q_j(u_0,u_0) = Q(p_1,p_1)x_1^2\gamma_{m_1m_1j} + Q(p_2,p_2)x_2^2\,\gamma_{m_2m_2j},\ j \text{ odd}$$

$$Q_j(u_0,u_0) = 2Q(p_1,p_2)x_1x_2\gamma_{m_1m_2j},\ \ j \text{ even.}$$

It also follows that

$$\begin{aligned} Q_{m_1}(\underset{\sim}{u},\underset{\sim}{u}) = {} & 2Q(u_{m_1},u_{m_2})\gamma_{m_1m_1m_2} \\ & + 4\sum_{\substack{k \text{ odd} \\ k\neq m_2}} Q(u_{m_1},u_k)\gamma_{m_1m_1k} + 4\sum_{\substack{k \text{ even} \\ k\neq m_1}} Q(u_{m_2},u_k)\gamma_{m_1m_2k} + O(\varepsilon^2) \end{aligned}$$

and

$$\begin{aligned} Q_{m_2}(\underset{\sim}{u},\underset{\sim}{u}) = {} & Q(u_{m_1},u_{m_1})\gamma_{m_1m_1m_2} + Q(u_{m_2},u_{m_2})\gamma_{m_2m_2m_2} \\ & + 4\sum_{\substack{k \text{ odd} \\ k\neq m_2}} Q(u_{m_2},u_k)\gamma_{m_2m_2k} + 4\sum_{\substack{k \text{ even} \\ k\neq m_1}} Q(u_{m_1},u_k)\gamma_{m_1m_1k} + O(\varepsilon^2) \end{aligned}$$

It is much easier to calculate the contributions from the cubic terms,

$$C_{m_1}(u_0,u_0,u_0) = \tfrac{3}{4}\, C(p_1,p_1,p_1)x_1^3 + \tfrac{3}{2}\, C(p_1,p_2,p_2)x_1x_2^2,$$

$$C_{m_2}(u_0,u_0,u_0) = \tfrac{3}{2}\, C(p_1,p_1,p_2)x_1^2x_2 + \tfrac{3}{4}\, C(p_2,p_2,p_2)x_2^3,$$

where by $3C(p_1,p_1,p_2)$ we actually mean $C(p_1,p_1,p_2) + C(p_1,p_2,p_1) + C(p_2,p_1,p_1)$, etc.

On combining (A-5), (A-9), (A-10) and (A-11) we find that

(A-12)
$$\frac{dx_1}{d\tau} = x_1[a_1 + b_1x_2 + \varepsilon x_1(c_1x_1^2 + d_1x_2^2) + \varepsilon\delta_1(\tau)] + O(\varepsilon^2)$$
$$\frac{dx_2}{d\tau} = a_2x_2 + b_2x_1^2 + b_3x_2^2 + \varepsilon x_2(c_2x_1^2 + d_2x_2^2) + \varepsilon\delta_2(\tau)x_2 + O(\varepsilon^2)$$

where

$$a_k = q_k^T\left(I + \varepsilon\frac{B\pi_k\sigma_k^T}{\lambda_k}\right)Bp_k, \qquad k = 1,\ 2$$

$$\gamma_k(\tau) = q_k^T C(\tau)p_k, \qquad k = 1,\ 2$$

$$b_1 = 2\Gamma_1 q_1^T\left\{\left(I + \varepsilon\frac{B\pi_1\sigma_1^T}{\lambda_1}\right)Q(p_1,p_2) + \frac{\varepsilon}{\lambda_1}Q(p_2,\pi_1)\sigma_1^T Bp_1 + \frac{\varepsilon}{\lambda_2}Q(p_1,\pi_2)\sigma_2^T Bp_2\right\}$$

$$b_2 = \Gamma_1 q_2^T\left\{\left(I + \varepsilon\frac{B\pi_1\sigma_2^T}{\lambda_2}\right)Q(p_1,p_1) + \frac{2\varepsilon}{\lambda_1}Q(p_1,p_2)\sigma_1^T Bp_1\right\}$$

$$b_3 = \Gamma_2 q_2^T\left\{\left(I + \varepsilon\frac{B\pi_2\sigma_2^T}{\lambda_2}\right)Q(p_2,p_2) + \frac{2\varepsilon}{\lambda_2}Q(p_2,\pi_2)\sigma_2^T Bp_2\right\}$$

$$c_1 = \frac{2\Gamma_1^2}{\lambda_2}q_1^T Q(p_1,\pi_2)\sigma_2^T Q(p_1,p_1) - 4\sum_{\substack{k\ \mathrm{odd}\\ k\neq m_2}}\gamma_{m_1m_1k}^2 q_1^T Q(p_1,A_k^{-1}Q(p_1,p_1)) + \frac{3}{4}q_1^T C(p_1,p_1,p_1)$$

$$c_2 = \frac{2\Gamma_2\Gamma_1}{\lambda_2}q_2^T Q(p_2,\pi_2)\sigma_2^T Q(p_1,p_1) + \frac{4\Gamma_1^2}{\lambda_1}q_2^T Q(p_1,\pi_1)\sigma_1^T Q(p_1,p_2)$$
$$- 4\sum_{\substack{k\ \mathrm{odd}\\ k\neq m_2}}\gamma_{m_2m_2k}\gamma_{m_1m_1k}\, q_2^T Q(p_2,A_k^{-1}Q(p_1,p_1))$$
$$- 8\sum_{\substack{k\ \mathrm{even}\\ k\neq m_1}}\gamma_{m_1m_2k}^2 q_2^T Q(p_1,A_k^{-1}Q(p_1,p_2)) + \frac{3}{2}q_2^T C(p_1,p_1,p_2)$$

$$d_1 = \frac{2\Gamma_1\Gamma_2}{\lambda_2}q_1^T Q(p_1,\pi_2)\sigma_2^T Q(p_2,p_2) + 4\Gamma_1^2 q_1^T Q(p_2,\pi_1)\sigma_1^T Q(p_1,p_2)$$
$$- 4\sum_{\substack{k\ \mathrm{odd}\\ k\neq m_2}}\gamma_{m_2m_2k}\, q_1^T Q(p_1,A_k^{-1}Q(p_2,p_2))$$
$$- 8\sum_{\substack{k\ \mathrm{even}\\ k\neq m_1}}\gamma_{m_1m_2k}\, q_1^T Q(p_2,A_k^{-1}Q(p_1,p_2)) + \frac{3}{2}q_1^T C(p_1,p_2,p_2)$$

$$d_2 = \frac{2\Gamma_2^2}{\lambda_2} q_2^T Q(p_2,\pi_2)\sigma_2^T Q(p_2,p_2) - 4 \sum_{\substack{k \text{ odd} \\ k\neq m_2}} \gamma^2_{m_2 m_2 k} q_2^T Q(p_2, A_k^{-1} Q(p_2,p_2)$$

$$+ \frac{3}{4} q_2^T C(p_2,p_2,p_2)$$

$$\Gamma_1 = \gamma_{m_1 m_1 m_2} = \frac{8}{\pi} \frac{m_1^2}{m_2(4m_1^2 - m_2^2)}$$

$$\Gamma_2 = \gamma_{m_2 m_2 m_2} = \frac{8}{3\pi m_2} .$$

The derivation is similar when the Neumann boundary conditions (1.4) are imposed. We set $u = \varepsilon \sum_{k=0}^{\infty} u_k(t) \cos kx$ and let $A = A_0 + \varepsilon^2 B$, $\tau = \varepsilon^2 t$. Following the same steps as before one finds that the resulting equations are (provided $m_2 \neq 1$)

$$\text{(A-13)} \qquad \begin{aligned} \frac{dx_1}{d\tau} &= x_1[a_1 + x_1(c_1 x_1^2 + d_1 x_2^2)] + O(\varepsilon) \\ \frac{dx_2}{d\tau} &= x_2[a_2 + x_2(c_2 x_1^2 + d_2 x_2^2)] + O(\varepsilon) \end{aligned}$$

where

$$a_k = q_k^T B p_k \qquad k = 1, 2$$

$$c_1 = \frac{3}{4} q_1^T C(p_1,p_1,p_1) - q_1^T Q(p_1, A_0^{-1} Q(p_1,p_1)) - \frac{1}{2} q_1^T Q(p_1, A_{2m_1}^{-1} Q(p_1,p_1))$$

$$d_1 = \frac{3}{2} q_1^T C(p_1,p_2,p_2) - q_1^T Q(p_1, A_0^{-1} Q(p_2,p_2))$$

$$- q_1^T Q(p_2, A_1^{-1} Q(p_1,p_2)) - q_1^T Q(p_2, A_{m_1+m_2}^{-1} Q(p_1,p_2))$$

$$c_2 = \frac{3}{2} q_1^T C(p_1,p_1,p_2) - q_2^T Q(p_2, A_0^{-1} Q(p_1,p_1))$$

$$- q_2^T Q(p_1, A_1^{-1} Q(p_1,p_2)) - q_2^T Q(p_1, A_{m_1+m_2}^{-1} Q(p_1,p_2))$$

$$d_2 = \frac{3}{4} q_2^T C(p_2,p_2,p_2) - q_2^T Q(p_2, A_0^{-1} Q(p_2,p_2)) - \frac{1}{2} q_2^T Q(p_2, A_{2m_2}^{-1} Q(p_2,p_2)).$$

BIBLIOGRAPHY

1. M. Ashkenazi and H. G. Othmer, Spatial patterns in coupled biochemical oscillators, J. Math. Biol., 5 (1978), 305-350.

2. J. A. Boa and D. S. Cohen, Bifurcation of localized disturbances in a model biochemical reaction, SIAM J. Appl. Math., 30 (1976), 123-135.

3. S. H. Chow, J. K. Hale and J. Mallet-Paret, An example of bifurcation to homoclinic orbits, J. Diff. Eqns., 37 (1980), 351-373.

4. V. A. Ditkin and A. P. Prudnikov, Integral transforms and operational calculus, Pergamon Press, 1965.

5. J. Guckenheimer, On a codimension two bifurcation, Lecture Notes in Mathematics, 898, ed. D. A. Rand and L.-S. Young, Springer, 1981.

6. P. J. Holmes, Averaging and chaotic motions in forced oscillations, SIAM J. Appl. Math., 38 (1980), 65-80.

7. J. P. Keener, Secondary bifurcation in nonlinear diffusion reaction equations, Stud. Appl. Math., 55 (1976), 187-211.

8. J. P. Keener, Infinite period bifurcation and global bifurcation branches, SIAM J. Appl. Math., 41 (1981), 127-144.

9. J. P. Keener, Chaotic behavior in slowly varying systems of nonlinear differential equations, Stud. Appl. Math.,67 (1982), 25-44.

10. W. F. Langford, Periodic and steady-state mode interactions lead to Tori, SIAM J. Appl. Math., 37 (1979), 22-48.

11. W. F. Langford, A. Arneodo, P. Coullet, C. Tresser and J. Coste, A mechanism for a soft mode instability, Phys. Letters, 78A (1980), 11-14.

12. T. J. Mahar and B. J. Matkowsky, A model biochemical reaction exhibiting secondary bifurcation, SIAM J. Appl. Math., 32 (1977), 394-404.

13. J. Moser, Stable and random motions in dynamical systems, Annals of Mathematical Studies, 77, Princeton University Press, 1973.

14. F. K. Melnikov, On the stability of the center for time periodic perturbations, Trans. Moscow Math. Soc., 12 (1967), 1-57.

15. G. Nicolis and J. V. B. Auchmuty, Dissipative structures, catastrophes and pattern formation: A bifurcation analysis, Proc. Nat. Acad. Sci., 71 (1974), 2748-2751.

16. S. Smale, A mathematical model of two cells via Turing's equations, in Lectures on Mathematics in the Life Sciences, 6 (1974) (J. Cowan, ed.), Am. Math. Soc.

17. S. Smale, Differentiable dynamical systems, Bull. Amer. Math. Soc., 73 (1967), 747-817.

18. L. A. Segal and J. L. Jackson, Dissipative structure: An explanation an an ecological example, J. Theor. Biol., 37 (1972), 545-559.

19. A Turing, The chemical basis of morphogenisis, Phil. Trans. Roy. Soc. London, Ser. B, 237 (1952), 32-72.

DEPARTMENT OF MATHEMATICS
UNIVERSITY OF UTAH
SALT LAKE CITY, UT 84112

Contemporary Mathematics
Volume 17, 1983

MONOTONE AND OSCILLATORY EQUILIBRIUM SOLUTIONS OF A PROBLEM ARISING IN POPULATION GENETICS

Henry L. Kurland[1]

ABSTRACT. In this article we give an expository account of existence for a Fisher equation for small enough positive ε of four strictly monotone equilibrium solutions on a finite interval satisfying zero Neumann boundary conditions and two finite families of oscillatory equilibrium solutions on the same interval satisfying the same boundary conditions where the cardinalities of the families increases with decreasing epsilon. The monotone solutions have boundary and transition layers, and existence of the monotone solutions follows from a general topological existence theorem for two-point boundary value problems of the form $\varepsilon v' = F(v,x,\varepsilon,\theta)$, $v(a) \epsilon E(a,\varepsilon,\theta)$, $v(b) \epsilon E(b,\varepsilon,\theta)$ where v is n-dimensional, x is the independent variable, θ is a parameter, and $E(a,\varepsilon,\theta)$ and $E(b,\varepsilon,\theta)$ are the locus of points satisfying the left and right boundary conditions respectively. Existence of the oscillatory solutions follows from the adiabatic invariance of the action for a Hamiltonian system.

1. INTRODUCTION.

In this article we announce some existence results for equilibrium solutions of the Fisher equation

$$u_t = \varepsilon^2 u_{xx} + u(1-u)(u-a(x)) \tag{1}$$

over the finite interval $-1 \le x \le 1$ satisfying zero Neumann boundary conditions for all small enough positive ε. The solutions exhibited consist of a class of monotone solutions and a class of oscillating solutions.

The class of monotone solutions found extends results of Peletier and Fife [P1,P2]. The existence of solutions in this class follows from a general existence theorem, Theorem B below, for boundary value problems over a finite interval $a \le x \le b$ of the form

$$\begin{aligned} &\varepsilon v' = F(u,x,\varepsilon,\theta) \\ &v(a) \in E(a,\varepsilon,\theta) \ , \quad v(b) \in E(b,\varepsilon,\theta) \end{aligned} \tag{2}$$

where $v \in R^n$ and $E(a,\varepsilon,\theta)$ and $E(b,\varepsilon,\theta)$ are subsets in R^n which are the locus of points that satisfy the left and right boundary conditions

[1] Supported by NSF Grant MCS-8103459.

0271-4132/83/0000-1392/$06.00

respectively and which may vary with ε and an external parameter θ as indicated. The exact statement is given in Theorem B below. Details of the proof of this theorem and this application will appear in [K4]. Because the application given here does not require it, we suppress the external parameter θ from our notation except for the statement of the theorem.

The existence of the class of oscillating solutions is joint work with M. Levi [K-L] and follows from an application of the adiabatic invariance of the action [A] and the results of J. Smoller [S] on period vs. amplitude maps.

As is well known [A-W,P1], Equation (1) serves as a simple model for the genetic evolution of a diploid population in a non-homogeneous finite one-dimensional habitat, scaled by the x-variable, corresponding to the heterozygote inferior case; i.e., the heterozygote genotype Aa is considered inferior to either of the homozygote genotypes AA or aa where the variable u is the proportion of alleles a relative to the total number of alleles a + A present in the population at a given point in space and time. The assumption of zero Neumann boundary conditions corresponds to the assumption that the population is genetically isolated; hence there is no influx of genetic material into the environment, say through migration. The function $a(x)$ in the selection term of (1) satisfies $0 < a(x) < 1$ over $[-1,1]$, and if $a(x) > 1/2$ at a point x in the habitat, then the genotype aa is superior to the genotype AA at x with the reverse ordering holding when $a(x) < 1/2$.

Written as a first order system, the boundary value problem we are explicitly interested in is

$$\begin{aligned} \varepsilon u_x &= w \\ \varepsilon w_x &= -u(1-u)(u-a(x)) \\ u_x(\pm 1) &= 0,\ 0 \le u \le 1 . \end{aligned} \tag{3}$$

Thus we will apply the general theorem where $v = (u,w) \in R^2$, $F(v,x) = (w,-u(1-u)(u-a(x))$, and $E(\pm 1,\varepsilon) = E(\pm 1) = \{(u,w): w = 0\}$. Our results for this boundary value problem are the following.

THEOREM A. If $a(x)$ is continuous on $[-1,1]$, then for all small enough positive ε,

(1) if $a(-1) > 1/2$, then there exists a unique strictly increasing equilibrium solution $u_{\varepsilon,-1}$ of (1) which has a boundary layer at $x = -1$ and satisfies $\lim_{\varepsilon \downarrow 0} u_{\varepsilon,-1}(x) = 1$ for $-1 < x \le 1$;

(2) if $a(+1) < 1/2$, then there exists a unique strictly increasing equilibrium solution $u_{\varepsilon,+1}$ of (1) which has a boundary layer at $x = +1$ and satisfies $\lim_{\varepsilon \downarrow 0} u_{\varepsilon,+1}(x) = 0$ for $-1 \le x < 1$.

(3) if $-1 < \bar{x} < 1$ and $(a(x)-1/2)$ changes sign at $\bar{x}$, there

exists a unique pair of strictly monotone solutions, $u_{\varepsilon,\bar{x}\uparrow}$ and $u_{\varepsilon,\bar{x}\downarrow}$, to (2), the former monotone increasing, the latter monotone decreasing, and each member of the pair exhibits an interior transition layer at $\bar{x}$; i.e.,

$$\lim_{\varepsilon\downarrow 0} u_{\varepsilon,\bar{x}\uparrow}(x) = \begin{cases} 0 & \text{if } -1 \le x < \bar{x} \\ 1 & \text{if } \bar{x} < x \le 1 \end{cases}$$

$$\lim_{\varepsilon\downarrow 0} u_{\varepsilon,\bar{x}\downarrow}(x) = \begin{cases} 1 & \text{if } -1 \le x < \bar{x} \\ 0 & \text{if } \bar{x} < x \le 1 \end{cases} ;$$

(4) if $a(-1) > 1/2 > a(+1)$ and $(a(x) - 1/2)$ change sign at $\bar{x}$, $-1 < \bar{x} < 1$, then $u_{\varepsilon,-1} > u_{\varepsilon,\bar{x}\uparrow} > u_{\varepsilon,+1}$, and if $-1 < x_1 < x_2 < 1$ and $(a(x) - 1/2)$ change sign at both x_1 and x_2, then $u_{\varepsilon,x_1\uparrow} > u_{\varepsilon,x_2\uparrow}$ and $u_{\varepsilon,x_1\downarrow} < u_{\varepsilon,x_2\downarrow}$.

(5) If in addition $a(x)$ is at least twice continuously differentiable, there exist four discrete families of oscillating solutions. Such a family can be ennumerated $u^{\varepsilon,1}, u^{\varepsilon,2}, \ldots, u^{\varepsilon k(\varepsilon)}$ so that where $|u^{\varepsilon,j}|$ is the number of oscillations of $u^{\varepsilon,j}$, we have $|u^{\varepsilon,j+1}| = |u^{\varepsilon,j}| + 1$ and $\lim_{\varepsilon\downarrow 0}|u^{\varepsilon,1}| = \infty$ and $\lim_{\varepsilon\downarrow 0} k(\varepsilon) = \infty$. These families need not account for all the oscillatory equilibrium solutions of (1).

In [P1,P2] existence of the solution which we call $u_{\varepsilon,\bar{x}\uparrow}$ is shown via a sub- and super-solution method derived from a non-linear maximum principle and a monotone iteration scheme, but requires that $a(x)$ be differentiable with $a'(x) < 0$ and $a(-1) > 1/2 > a(+1)$. Also it is not shown that $a_{\varepsilon,\bar{x}\uparrow}$ has an internal transition layer. The solutions $u_{\varepsilon,\pm 1}$ are also claimed to exist via a shooting method, i.e., varying $u(+1)$ while holding $u'(+1) = 0$ and shooting backwards, but no details are given. The boundary layer behavior of $u_{\varepsilon,\pm 1}$ is not mentioned, but the relation $u_{\varepsilon,-1} > u_{\varepsilon,\bar{x}\uparrow} > u_{\varepsilon,+1}$ is claimed. Also in the case that $a(x) - 1/2$ is an odd function, the stability of $u_{\varepsilon,\bar{x}\uparrow}$ as a solution of the partial differential equation is shown via energy methods.

2. BACKGROUND FOR THE EXISTENCE THEOREM.

To state the general existence theorem for boundary value problems of the form (2) and then to apply it to (3) requires a fair number of definitions and results from C.C. Conley's theory of isolated invariant sets [C] and some additional results from work of the author [K1,K2,K3]. To help motivate these notions let us first manipulate (2), hence also (3), into a form more amenable to our purpose by rescaling the independent variable. That is, if we reparameterize integral curves of (2) with independent variable τ defined by

$x = \varepsilon\tau$, then straightforward application of the chain rule shows that the reparameterized curves satisfy the equation

$$\begin{aligned} \dot{v} &= F(v,x,\varepsilon) \\ \dot{x} &= \varepsilon \end{aligned} \tag{4}$$

where the dot above a variable denotes differentiation with respect to τ. Clearly there is a bijection between the integral curves of (4) and those of (2) for $\varepsilon > 0$; however, (4) has the advantage of still being a differential equation when $\varepsilon = 0$ which we call the fast system associated to (2),

$$\begin{aligned} \dot{v} &= F(v,x,0) \\ \dot{x} &= 0 \ . \end{aligned} \tag{5}$$

We regard (5) as having phase space $R^n \times [a,b]$ with coordinates (v,x). However, we can also obviously regard (5) as defining a one-parameter family of equations with parameter x, and for fixed $x \in [a,b]$

$$\dot{v} = F(v,x,0) \tag{6}$$

the fast system at x associated to (2) which we regard as having phase space R^n which we identify with the slice $R^n \times \{x\}$ of $R^n \times [a,b]$. We also use this identification to regard $E(a,\varepsilon)$ as a subset of $R^n \times \{a\}$ and $E(b,\varepsilon)$ as a subset of $R^n \times \{b\}$.

The general theorem, whose underlying idea is conceptually similar to the Wazewski retract method gives conditions which guarantee that the flow of the differential equation in (4), hence (2), will carry $E(a,\varepsilon)$ into a subset of $R^n \times \{b\}$ in such a way that it necessarily intersects $E(b,\varepsilon)$. The integral curve through any such point of intersection when appropriately parameterized is a solution of the boundary value problem (3). Because we use the fast system (6) to approximate the behavior of (4) on specially selected subsets of $R^n \times [a,b]$, our theorem only yields existence for small enough positive ε.

Before giving the relevant definitions from the theory of isolated invariant sets let us examine the phase portraits of the fast systems associated to (3),

$$\begin{aligned} \dot{u} &= w \\ \dot{w} &= -u(1-u)(u-a(x)) \\ \dot{x} &= 0 \ , \end{aligned} \tag{7}$$

so that we may have some concrete examples readily at hand. There are only three possible qualitatively distinct phase portraits associated to (7) for a fixed x, the three possibilities arising from the three possibilities for the value of $a(x)$; (i) $1 > a(x) > 1/2$, (ii) $a(x) = 1/2$, (iii) $1/2 > a(x) > 0$. The phase portraits are easily sketched by noting that for a given x (7)

defines a Hamiltonian system with one degree of freedom where the Hamiltonian function is

$$H(u,w,x) = w^2/2 + \int_0^u s(1-s)(s-a(x))ds$$

and then drawing the level curves of H to obtain the phase portraits. The three possible phase portraits are sketched in Figure 1 below.

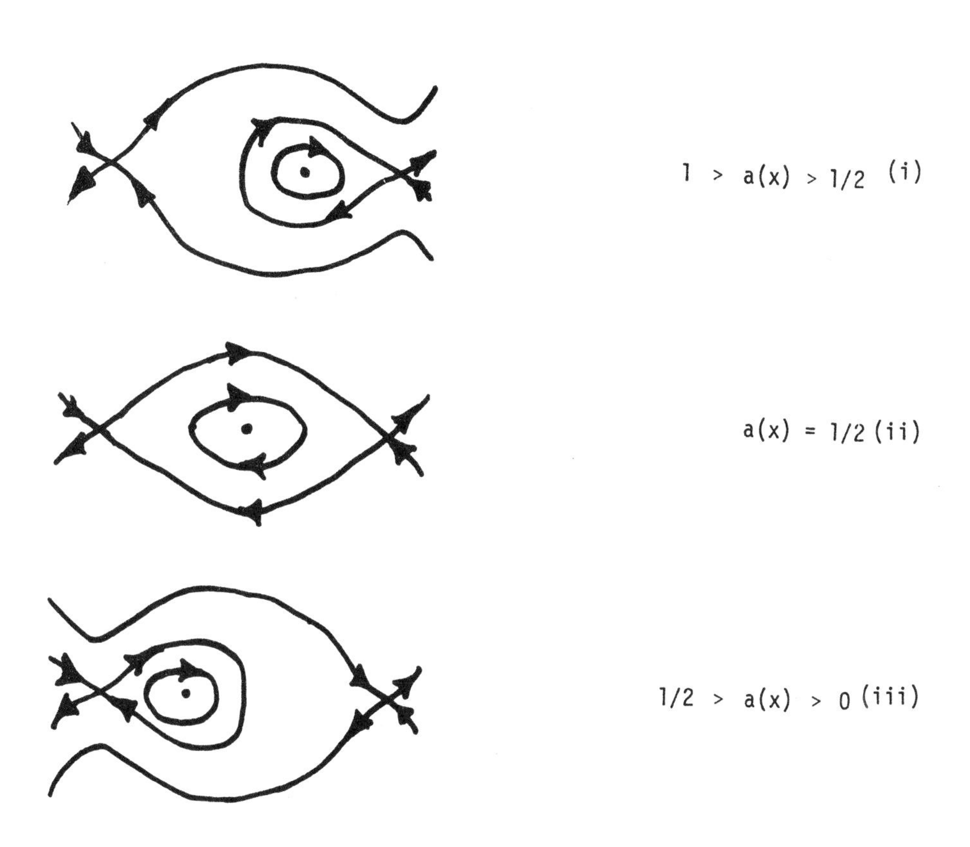

Figure 1

In particular because for each $x \in [-1,1]$, $u = 0$ and $u = 1$ correspond to local maxima of the potential energy function

$$U(u,x) = \int_0^u s(1-s)(s-a(x))ds,$$

for each x, the points $(0,0)$ and $(1,0)$ are hyperbolic critical points of the fast system at x. Also for each x, the point $(a(x),0)$ is a center for the fast system at x since $u = a(x)$ corresponds to a local minimum of

$U(u,x)$. Because when $1 > a(x) > 1/2$ the inequality $U(1,x) < U(0,x)$ holds, it follows that the center $(a(x),0)$ is in this case contained in a two-cell bounded by an orbit homoclinic to $(1,0)$, whereas when $1/2 > a(x) > 0$, the reverse inequality holds between the potential energies; whence it follows that the center $(a(x),0)$ is in this case contained in a two cell bounded by an orbit homoclinic to $(0,0)$. Finally, when $a(x) = 1/2$, the center $(a(x),0)$ is contained in a two-cell bounded by two heteroclinic orbits, the upper portion of the boundary a heteroclinic orbit from $(0,0)$ to $(1,0)$ and the lower portion of the boundary a heteroclinic orbit from $(1,0)$ to $(0,0)$. For each $x \in [-1,1]$, let $C(x)$ be the closure of the two-cell containing the center $(a(x),0)$ as just described. The hyperbolic critical points and the two-cells $C(x)$ are examples of isolated invariant sets of a flow which we will now formally define.

An autonomous differential equation defined by a continuous vectorfield and which has a unique solution to each initial value problem defines a continuous flow on its phase space P which we denote as a right-action of the real line R on P. Thus if $p \in P$ and $t \in R$ the result of t acting on p is $p \cdot t$ obtained by following the integral curve to the equation with initial condition p for time t, and for $t_1, t_2, \in R$, $(p \cdot t_1) \cdot t_2 = p \cdot (t_1 + t_2)$. Of course in general the flow of a differential equation is not complete; i.e., there may be $p \in P$ for which $p \cdot t$ is defined only for t in a bounded interval. However, differential equations defined by vectorfields with compact support have complete flows. Because the solutions we seek must lie in a bounded region of phase space known a priori, by multiplying the vectorfield by a bump function which is identically one on this region and zero outside some large ball, we can assume without loss of generality that the vectorfield has compact support. Then for $p \in P$, the orbit through p is $p \cdot R$.

An invariant subset of P is one which is the union of orbits. If $S \subset P$ is invariant, then S is an isolated invariant set if it is the largest invariant set contained in some compact neighborhood N of itself. Such a neighborhood is called an isolating neighborhood for S. Note that a compact neighborhood N is an isolating neighborhood if, and only if, for each point $p \in \partial N$, $p \cdot R \not\subset N$. This forces the largest invariant set contained in N to lie in the interior of N. As examples of isolating neighborhoods, take any sufficiently small closed neighborhood of either of the hyperbolic critical points $(0,0)$ or $(1,0)$ or of the cells $C(x)$ relative to any of the fast equations at x for any $x \in [-1,1]$.

Associated to each isolated invariant set S of a flow is an index $h(S)$, called the homotopy index of S. $h(S)$ is the homotopy type of any member of a certain class of quotient spaces arising from certain special isolating neighborhoods which we now describe. Any isolating neighborhood N of S

contains an isolating neighborhood B of S with the property that the exit time map $\sigma|B$ and the entrance time map $\sigma^*|B$ are continuous maps on B into $[0,\infty]$ where

$$\sigma|B(p) = \sup\{t \geq 0: p\cdot[0,t] \subset B\} ,$$

$$\sigma^*|B(p) = \sup \{t \geq 0: p\cdot[-t,0] \subset B\} .$$

As isolating neighborhood B with this property is called an <u>isolating block</u>. Geometrically, if we take an isolating neighborhood B which is a manifold with boundary and possibly corners, then the continuity of $\sigma|B$ and $\sigma^*|B$ is assured if no flow line tangent to the boundary of B is internally tangent; i.e., it must not enter the interior of B near the point of tangency. Thus continuity is assured if the flow is transverse to the boundary of B. Locally in such a case $\partial B = f^{-1}(0)$, $\text{int } B = f^{-1}(\infty,0)$ and if $f(p) = 0$ we require $(\partial f/\partial t)(p\cdot t)\big|_{t=0} \neq 0$.

The <u>exit set</u> of a block B, denoted B^-, is defined by $B^- = (\sigma|B)^{-1}(0)$ and the <u>entrance set</u> of a block B is defined by $B^+ = (\sigma^*|B)^{-1}(0)$. Clearly the exit and entrance sets are closed subsets of ∂B. Examples of isolating blocks will be given shortly after defining the Morse index.

The index $h(S)$ is the homotopy type of the pointed space B/B^- with $[B^-]$ as base point. B/B^- is called an <u>index space</u> for S. The definition of $h(S)$ is independent of the block used to define it. In fact if B_1 and B_2 are both isolating blocks for S, then there are maps defined by inclusions and the flow

$$m : B_1/B_1^- \to B_2/B_2^-$$

$$\overline{m} : B_2/B_2^- \to B_1/B_1^-$$

which are homotopy equivalences; i.e, $m \circ \overline{m}$ and $\overline{m} \circ m$ are base point homotopic to the appropriate identity map. The homotopy classes of these maps are canonical in the sense that certain arbitrary choices are made in the construction of m and $\overline{m}$, but the homotopy class of m is independent of these choices and similarly for $\overline{m}$. Hence there is a unique equivalence up to homotopy between any two index spaces. The composition of two such equivalences is again an equivalence of the appropriate type.

The <u>Morse index</u> of S, $\mathcal{I}(S)$, consists of the collection of index spaces for S and of the collection of the canonical homotopy equivalences between the index spaces, that is $\mathcal{I}(S)$ is a category, and because between any two spaces there is a unique canonical equivalence, $\mathcal{I}(S)$ is called a <u>connected simple system</u>.

To illustrate these ideas consider the hyperbolic critical point (0,0) for any of the fast systems (7) at $x \in [-1,1]$. Any square in the u-w plane with vertices at $(\pm r,0)$, $(0,\pm r)$ for r small enough, say $0 < r < \min(a(x),1-a(x))$, is an isolating block B_r for the critical point (0,0). The exit set consists of the two line segments in the boundary with negative slope, and the entrance set those two segments in the boundary with positive slope. To see this consider for example the segment connecting (0,r) to (r,0) points of which satisfy the equation $w + u - 1 = 0$. The derivative of the left-hand side along a flow line of (7) is

$$-u(1-u)(u-a(x)) + w \tag{8}$$

which is strictly positive for all points on the segment. The positivity of this derivative forces points on this boundary segment to exit the square immediately in positive time and to enter the interior of the square immediately in negative time (except at the corner points where the flow line is extrenally "tangent"). An analogous analysis prevails on the other segments to give the exit and entrance sets as claimed.

More relevant to the solution of the boundary value problem (3) is a family of isolating blocks for (0,0) which is obtained by modifying slightly the choice of two of the diametrically opposite corner points. Namely consider the parallelogram $\overline{B}_r$ with vertices $(0,\pm r)$, $(r + \delta_r, -\delta_r)$, $(-r-\delta_r,\delta_r)$ where δ_r is chosen small enough (independent of x) via considerations of the following sort. Note that because the expression (8) is positive at $(u,w) = (r,0)$, by continuity it remains positive for nearby points, in particular for all points on the line segment from $(r,0)$ to $(r + \delta_r, -\delta_r)$; (the uniform continuity of $a(x)$ over $[-1,1]$ allows the choice of δ_r to be independent of x). Thus for an appropriate choice of δ_r, the exit set of the block will again be the segments of the boundary with negative slope and the entrance set the segments with positive slope. Note that for $x = -1$, the intersection of $E(-1)$, the points satisfying the left-end boundary condition, and $\overline{B}_r$ is precisely the line segment

$$\sigma_r = \{(u,0) : -r \le u \le r\}$$

and which lies in the interior of $\overline{B}_r$ except for the endpoints which lie in distinct components of $\overline{B}_r^- \backslash \overline{B}_r^+$ (see Figure 2 below).

Note that by identifying the two components of the exit set to a single point we see that the index space $\overline{B}_r/\overline{B}_r^-$ is a "cigar band" pinched at the point $[\overline{B}_r^-] \equiv *$; hence the homotopy index $h(S)$ of $S = \{(0,0)\}$ is the homotopy type of a circle with base point, Σ^1. In particular, note that after identifying $\overline{B}_r^-$ to a point that the image of σ_r is a circle which is a strong deformation retract of $\overline{B}_r/\overline{B}_r^-$. By taking the singular homology with integer coefficients Z of the pair $(\overline{B}_r/\overline{B}_r^-, *)$, which we denote $\tilde{H}_*(\overline{B}_r/\overline{B}_r^-)$, we associate a family of algebraic objects to S. Now the k-th homology group is trivial for $k \neq 1$, and $\tilde{H}_1(\overline{B}_r/\overline{B}_r^-) \simeq Z$. The non-zero generating class of $\tilde{H}_1(\overline{B}_r/\overline{B}_r^-)$ is precisely the homology class of the circle σ_r viewed as a singular chain.

It is convenient for our purposes to be able to associate a single algebraic object to an isolated invariant S rather than a family of isomorphic objects, and this is easily done by defining the homology of the Morse index, $\tilde{H}_*(\mathcal{I}(S))$, as a collection of equivalence classes. Specifically, if α and $\overline{\alpha}$ are homology classes in $\tilde{H}_*(B/B^-)$ and $\tilde{H}_*(\overline{B}/\overline{B}^-)$ respectively where B and $\overline{B}$ are both blocks for S, we say α and $\overline{\alpha}$ are equivalent if they correspond under the isomorphism of homology groups induced by the canonical equivalence in $\mathcal{I}(S)$ between the index spaces B/B^- and $\overline{B}/\overline{B}^-$. Define $\tilde{H}_*(\mathcal{I}(S))$ to be the collection of equivalence classes under this equivalence relation. $\tilde{H}_*(\mathcal{I}(S))$ inherits the algebraic structure of $\tilde{H}_*(B/B^-)$ for B any block for S by making the canonical map of sending a homology class to its equivalence class in $\tilde{H}_*(\mathcal{I}(S))$ an isomorphism. Because there is a unique canonical homotopy equivalence between any two index spaces for S, the algebraic structure defined on $\tilde{H}_*(\mathcal{I}(S))$ is in this way independent of the choice of block. Thus each of the segments σ_r in our example above thought of as a singular simplex represents the same element of $\tilde{H}_*(\mathcal{I}(S))$ for $S = \{(0,0)\}$ for the fast flow at $x = -1$. More generally, for any isolated invariant set S with block B, if c is a singular chain with support in B and if ∂c has support in B^-, then we say c <u>represents homology in</u> $\tilde{H}_*(\mathcal{I}(S))$.

Also required for the statement of Theorem B is the definition of a continuous curve of isolated invariant sets associated to a continuous one-parameter family of differential equations. For our purposes here we need only consider the case of a one-parameter family of fast equations (6). For each $x \in [a,b]$, let $S(x)$ be an isolated invariant set for (6), the fast system at x. We say that the correspondence $x \to S(x)$ is a <u>continuous curve of isolated invariant sets</u> for the fast system if for each $x_0 \in [a,b]$, if N is an isolating neighborhood for $S(x_0)$, then for some neighborhood J of x_0 in $[a,b]$, if $x \in J$ then N is an isolating neighborhood for $S(x)$.

For the fast system (7), as noted above for each $x \in [-1,1]$, $\{(0,0)\}$ and $\{(1,0)\}$ are isolated invariant sets. The curves $x \to \{(0,0)\}$ and $x \to \{(1,0)\}$ are examples of continuous curves of isolated invariant invariant sets. Note that the correspondence $x \to C(x)$ where $C(x)$ is the two-cell defined above in general fails to be continuous at each value of x at which $a(x) = 1/2$ because if x_0 is such a value, then $C(x_0)$ contains both critical points

(0,0) and (1,0), but if x is near x_0 and $a(x) \neq 1/2$, then $C(x)$ does not contain one of these critical points, (0,0) if $a(x) > 1/2$ but (1,0) if $a(x) < 1/2$, and any isolating neighborhood of $C(x)$ cannot contain the critical point not in $C(x)$; however, any isolating neighborhood of $C(x_0)$ necessarily contains both of these critical points. Thus $x \to C(x)$ is not continuous at x_0.

Given a one-parameter family of equations such as (b) above with parameter interval $[a,b]$, it is shown in [C,K3] that to each continuous curve of isolated invariant sets, $\eta : x \to S(x)$, can be associated a canonical equivalence between the Morse indices $I(S(a))$ and $I(S(b))$. We denote this canonical equivalence by $I(\eta)$ and regard it as a map between Morse indices, $I(\eta) : I(S(a)) \to I(S(b))$. In reality $I(\eta)$ is a collection of square diagrams of index spaces which commute up to homotopy. Let $B(\lambda)$ and $\overline{B}(\lambda)$ be isolating blocks for $S(\lambda)$, $\lambda = a,b$. Then a typical square of $I(\eta)$ is

$$\begin{array}{ccc} B(a)/B(a)^- & \xrightarrow{I(\eta)(B(a),)B(b))} & B(b)/B(b)^- \\ m_a \downarrow & & \downarrow m_b \\ \overline{B}(a)/\overline{B}(a)^- & \xrightarrow{I(\eta)(\overline{B}(a),\overline{B}(b))} & \overline{B}(b)/\overline{B}(b)^- \end{array}$$

where m_a and m_b are the unique canonical homotopy equivalences in $I(S(a))$ and $I(S(b))$ respectively between the indicated index spaces and where $I(\eta)(B(a),B(b))$ and $I(\eta)(\overline{B}(a),\overline{B}(b))$ are canonically defined homotopy equivalences depending only on the (homotopy class of the) curve η and the blocks at the endpoints. An outline of how $I(\eta)$ is constructed goes as follows. Let $a = x_0 < x_1 < \dots < x_n = b$ be a partition of $[a,b]$ so fine that on each sub-interval $[x_{i-1},x_i]$ of the partition a single subset of phase space B_i serves as an isolating block for $S(x)$ for all x in the sub-interval and so that the exit set does not change with x. In particular, on the end sub-intervals the given blocks, say $B(a)$ and $B(b)$, should be chosen. Then B_i and B_{i+1} are both blocks for $S(x_i)$ for $i = 1,\dots,n-1$, hence there is a unique canonical homotopy equivalence in $I(S_i)$, $m_i : B_i/B_i^- \to B_{i+1}/B_{i+1}^-$. Then $I(\eta)(B(a),B(b))$ is the composition $m_{n-1} \circ m_{n-2} \circ \dots \circ m_1$. Of course it must be shown that the definition is independent of all the choices made and that the squares commute which is a fair amount of work.

All the definitions above were made in terms of an isolating block modulo its exit set. However, the analogous definitions can be made for an isolating block modulo its entrance set. For example let $h^*(S)$ denote the homotopy type of B/B^+ where B is a block for S, and let $I^*(S)$ and $\tilde{H}_*(I^*(S))$ denote the corresponding Morse index and its homology.

Finally, if α_1 and α_2 are two arcs which we regard as singular one-chains, both having support in R^2, and if the boundary of each arc is disjoint from the other arc (a technical requirement) then the intersection number of the two arcs, $\alpha_1 \circ \alpha_2$, can be defined homologically and gives the number of non-degenerate intersections of the two arcs. By a non-degenerate intersection we mean one that persists if either or both of the arcs is changed slightly. The precise definition is given in [D] and can of course be carried out for two singular chains c_1 and c_2. If c_1 and c_2 represent homology classes γ_1 and γ_2, then the intersection number of the homology classes, $\gamma_1 \circ \gamma_2$, is defined to be $c_1 \circ c_2$ and is independent of the choice of chains chosen to represent γ_1 and γ_2. We can now state our theorem.

THEOREM B. Assume

(i) for each $x \in [a,b]$, $S(x)$ is an isolated invariant set for (6), the fast system at x;

(ii) the map $\eta : x \to S(x)$ is a continuous curve of isolated invariant sets, $x \in [a,b]$;

(iii) $\mathcal{O}$ is an open neighborhood in $R^n \times R$ of the graph of η and $B(\lambda) \subset R^m \times \{\lambda\} \cap \mathcal{O}$ is an isolating block for $S(\lambda)$ for $\lambda = a,b$;

(iv) $\alpha \in \tilde{H}_*(I(S(a)))$ and $\gamma \in \tilde{H}_*(I^*(S(b)))$ are non-zero homology classes and the intersection number $I(\eta)_*\alpha \circ \gamma$ is non-zero;

(v) for each pair $(\varepsilon,\theta) \in (0,\bar{\varepsilon}\,] \times [\theta_0,\theta_1]$, there exist singular chains $\alpha_{\varepsilon\theta}$ and $\gamma_{\varepsilon\theta}$ which have support in $E(a,\varepsilon,\theta) \cap B(a)\backslash B(a)^+$ and $E(b,\varepsilon,\theta) \cap B(b)\backslash B(b)^-$ respectively and which represent α and γ respectively.

Then there exists ε^*, $0 < \varepsilon^* \le \bar{\varepsilon}$, so that if $0 < \varepsilon \le \varepsilon^*$, then (3) has a solution $v_{\varepsilon,\theta}$ whose graph lies in $\mathcal{O}$ for each $\theta \in [\theta_0,\theta_1]$. Moreover $v_{\varepsilon,\theta}(a) \in |\alpha_{\varepsilon,\theta}|$ and $v_{\varepsilon,\theta}(b) \in |\gamma_{\varepsilon,\theta}|$ where $|\alpha_{\varepsilon\theta}|$ and $|\gamma_{\varepsilon\theta}|$ are the supports of the singular chains.

In (iv), $I(\eta)_*$ is the map induced on homology by any homotopy equivalence $I(\eta)(B(a),B(b))$ where $B(a)$ and $B(b)$ are blocks for $\eta(a)$ and $\eta(b)$ respectively and is well-defined by virtue of the commutative squares defining $I(\eta)$.

3. MONOTONE SOLUTIONS.

We can apply Theorem B directly to prove existence of the solutions $u_{\varepsilon,\pm 1}$ and $u_{\varepsilon,\bar{x}\uparrow}$ mentioned in Theorem A. Let $r_0 = \inf\min\{a(x),1-a(x)\}$ where the infimum is over $x \in [-1,1]$, and choose any r_1, $0 < r_1 < r_0/2$.

PROPOSITION 1. If $a(1) < 1/2$, then (1) has an equilibrium solution $u_{\varepsilon,1}$ for all small enough ε. If $a(-1) > 1/2$, then (1) has an equilibrium solution $u_{\varepsilon,-1}$ for all small enough ε.

OUTLINE OF PROOF. Note $\eta : x \to \{(0,0)\}$ is a continuous curve of isolated invariant sets for (7). Let C be the cylinder in $R^2 \times [-1,1]$ which has for cross-section in $R^2 \times \{x\} \simeq R^2$ the open disc with center at $(0,0)$ and radius r_1. Let h_{1t} be the top half of the orbit homoclinic to $\eta(1)$. Note

$$h_{1t} = \{(u,w) : H(u,w,1) = 0 \text{ and } u,w \geq 0\}$$

where H is the Hamiltonian defined in section two above. Choose $\delta > 0$ small enough so that the closed δ-neighborhood of h_{1t} does not contain the center $(a(1),0)$ and so that its projection onto the u-axis lies in the interval $-r_1 < u < 1$. Let M be the open δ-neighborhood of h_{1t}. Because $a(x)$ is continuous, the phase portraits of the fast systems at x for x close enough to 1 are qualitatively the same as that for $x = 1$, and in fact for x close enough to 1 say $1 - c_0 \leq x < 1$, the top half of the orbit homoclinic to $\eta(x)$, call it h_{xt} will also be contained in M and M will not contain $(a(x),0)$. Let $\mathcal{M} = M \times [1 - c_0, 1]$ and let $\mathcal{O} = C \cup \mathcal{M}$. Note $\mathcal{O}$ is a neighborhood of the graph of $\eta : x \to \{(0,0)\}$.

Let $B(-1)$ be a parallelogram $\overline{B}_r$ as described above with vertices $(0,\pm r)$, $(r + \delta_r, -\delta_r)$, $(-r - \delta_r, \delta_r)$ where r is small enough so that $B(-1) \times \{-1\} \subset \mathcal{O}$. $\overline{B}_r$ is sketched in Figure 2(i). As noted above the arc σ_r considered as a singular chain generates the one-dimensional homology of $\overline{B}_r/\overline{B}_r^-$ and hence defines a non-zero class α in $\tilde{H}_1(I(\eta(-1)))$ which generates this homology group.

Let B_r be a square block for $\eta(+1)$ as described above. Note that the local unstable and stable manifolds of $\eta(+1)$ in B_r are arcs; call them respectively α_r and γ_r which have transverse intersection at $(0,0)$. Also, the endpoints of α_r (resp. γ_r) lie in different components of B_r^- (resp. B_r^+). It follows that α_r defines a non-zero generating class $\overline{\alpha}$ of $\tilde{H}_1(I(\eta(1)))$ and γ_r defines a non-zero generating class of γ of $\tilde{H}_1(I^*(\eta(1)))$ and the intersection product $\overline{\alpha} \circ \gamma$ is not zero. Because the map $I(\eta)$ is an equivalence, the induced map on homology is an isomorphism. Hence $I(\eta)_*\alpha = \overline{\alpha}$; whence $I(\eta)_*\alpha \circ \gamma$ is non-zero.

Let $B(+1)$ be a block as pictured in Figure 2(ii) which satisfies $B(+1) \times \{1\} \subset \mathcal{O}$. Then there is a segment $L \subset B(+1) \cap E(1)\backslash B(+1)^-$,

$$L = \{(u,0) : a(1) < u_1 \leq u \leq u_2\} ,$$

such that the endpoints of L lie in the distinct components of $B(+1)^+$. Since $B(+1)/B(+1)^+$ has the homotopy type of a circle and since the image of L is a strong deformation retract of $B(+1)/B(+1)^+$, when L is regarded as a one simplex it generates the one-dimensional homology of $B(+1)/B(+1)^+$;

hence must define the homology class γ of $\tilde{H}_1(I^*(\eta(1)))$.

Thus all the hypotheses of the theorem are satisfied. Hence for all small enough ε, there is a solution $v_{\varepsilon,1}(x) = (u_{\varepsilon,1}(x), w_{\varepsilon,1}(x))$ of (3) with $v_{\varepsilon,1}(-1) \in \sigma_r$ and $v_{\varepsilon,1}(+1) \in L$. Since $(0,0) \notin L$, this solution is not the trivial solution. Set $u_\varepsilon(t) = u_{\varepsilon,1}(-1 + \varepsilon t)$, $w_\varepsilon(t) = w_{\varepsilon,1}(-1 + \varepsilon t)$, and $x_\varepsilon(t) = -1 + \varepsilon t$. Then $c(t) = (u_\varepsilon(t), w_\varepsilon(t), x_\varepsilon(t))$ is an integral curve of (4) and its image for $0 \le t \le 2/\varepsilon$ lies in $\mathcal{O}$. Thus by definition of the vectorfield, w_ε is increasing so long as $u_\varepsilon(t) < a(x_\varepsilon(t))$. By choice of $\mathcal{O}$ (in particular C) this cannot fail unless $(2-c)/\varepsilon < t$, but must fail somewhere for $(2-c)/\varepsilon < t < 2/\varepsilon$ since $u_{\varepsilon,1}(1) \ge u_1 > a(1)$. However by choice of M, on any interval in which w_ε is decreasing it remains positive; otherwise $c(t)$ could not remain in M and have w_ε increase back to zero when $t = 2/\varepsilon$. Thus u_ε continues to increase for $(2-\varepsilon)/\varepsilon < t < 2/\varepsilon$ and shows that $u_{\varepsilon,1}$ is strictly monotone increasing for $-1 \le x \le 1$. By decreasing the parameters r, δ, and c_0 in the construction of $\mathcal{O}$ we force $\lim_{\varepsilon \downarrow 0} u_{\varepsilon,1}(x) = 0$ for $-1 \le x < 1$.

An analogous argument is used to obtain $u_{\varepsilon,-1}$ for ε small enough. Here of course the curve of invariant sets is $\eta : x \to \{(1,0)\}$. The blocks $B(\pm 1)$

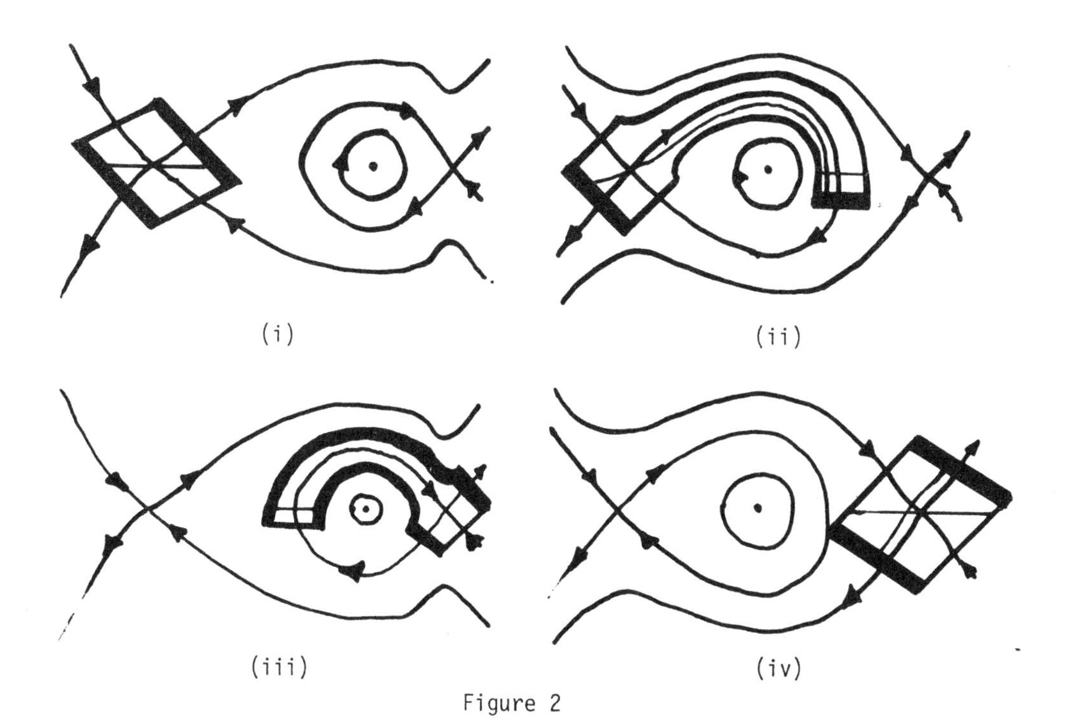

Figure 2

Exit sets are heavily drawn for all blocks.

(i) and (ii) are $B(-1)$ and $B(+1)$ respectively for $u_{\varepsilon,1}$.

(iii) and (iv) are $B(-1)$ and $B(+1)$ respectively for $u_{\varepsilon,-1}$.

are chosen as sketched in Figures 2(iii),(iv). $\mathcal{O}$ is chosen analogously to have cross-sectional diameter less than r_1 for x away from the boundary layer $x = -1$, but to contain a δ-neighborhood of the top-half of the orbit homoclinic to $\eta(x)$ for x close to -1. ///

PROPOSITION 2. If $-1 < \bar{x} < 1$, $a(-1) > 1/2 > a(+1)$, and $(a(x) - 1/2)$ changes sign at $\bar{x}$, then for all small enough ε, (1) has an equilibrium solution $u_{\varepsilon,\bar{x}\uparrow}$ and $u_{\varepsilon,-1} > u_{\varepsilon,\bar{x}\uparrow} > u_{\varepsilon,1}$.

OUTLINE OF PROOF. For each $x \in [-1,1]$ let $\eta(x) = C(x) \cup \{(0,0),(1,0)\}$. Then for each x we are adjoining at most one point to $C(x)$, namely $(0,0)$ if $a(x) > 1/2$, but $(1,0)$ if $a(x) < 1/2$. The points $(0,0)$ and $(1,0)$ are already in $C(x)$ if $a(x) = 1/2$. The curve η is a continuous curve of isolated invariant sets for (7) as is easily seen. (By adjoining the continuous curves of critical points to $C(x)$ we have eliminated the discontinuities arising in C when $a(x) = 1/2$). $\mathcal{O}$ is defined by assuming that its cross-section in $R^2 \times \{x\}$ is a δ-neighborhood of $\eta(x)$ which has two components near $x = \pm 1$, one containing $C(\pm 1)$ and the other containing the critical point $(0,0)$ or $(1,0)$ as appropriate. Blocks $B(\pm 1)$ are sketched in Figure 3. Note that the component of $B(\pm 1)$ containing $C(\pm 1)$ is a block, and modulo either its exit or entrance set has the homotopy type of a point; hence contributes no homology to its Morse indices. The component of $B(-1)$ containing $(0,0)$ is a block $\bar{B}_r$ as in section two. The component of $B(+1)$ containing $(1,0)$ is a parallelogram with vertices $(1, \pm r)$, $(1 + r + \delta_r, \delta_r), (1 - r - \delta_r, -\delta_r)$ and is a block $\tilde{B}_r$ for $(1,0)$. $E(-1) \cap \bar{B}_r$ is an arc α_r which defines a generating class α of the homology group $\tilde{H}_1(\mathcal{I}(\eta(-1)))$, and $E(+1) \cap \tilde{B}_r$ is an arc γ_r which defines a generating class γ of the homology group $\tilde{H}_1(\mathcal{I}^*(\eta(+1)))$. Similar to the proof of existence for $u_{\varepsilon,1}$ it follows that $\mathcal{I}(\eta)_*\alpha \circ \gamma$ is non-zero. The existence of a solution to the boundary value problem for small enough ε follows and has the properties desired of $u_{\varepsilon,\bar{x}\uparrow}$ including $u_{\varepsilon,-1} > u_{\varepsilon,\bar{x}\uparrow} > u_{\varepsilon,1}$ as follows by a somewhat harder analysis along the lines of the proof for $u_{\varepsilon,1}$. A different proof of existence for $u_{\varepsilon,\bar{x}\uparrow}$ using a more sophisticated version of Theorem B which makes the analysis easier is given in [K4]. ///

The existence of $u_{\varepsilon,\bar{x}\downarrow}$ under the hypotheses given cannot be shown just with Theorem B. A more sophisticated theorem is needed which analyzes the relation between $(0,0)$, $(1,0)$ and $C(x) \cup \{(1,0),(0,0)\}$ as x varies over $[-1,1]$. This requires the introduction of repeller-attractor pairs, the long coexact sequence of a repeller-attractor pair, and the connection and splitting maps for such a sequence. The reader is referred to [K4] for the proof.

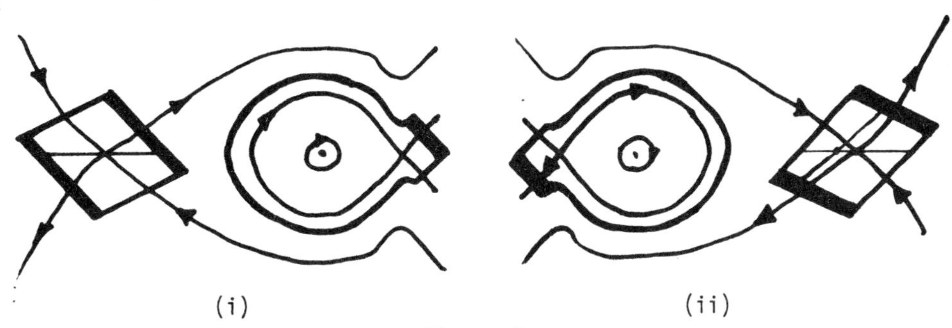

Figure 3

Figure 3(i) is the block $B(-1)$ for $C(-1) \cup \{(0,0)\}$. The left-hand component of $B(-1)$ is a block for $\{(0,0)\}$. The horizontal arc in this component is α_r.

Figure 3(ii) is the block $B(+1)$ for $C(+1) \cup \{(1,0)\}$. The right hand component of $B(+1)$ is a block for $\{(1,0)\}$. The horizontal arc in this component is γ_r. The arc with positive slope is the image under the flow of a subarc of α_r.

4. OSCILLATORY SOLUTIONS

The equilibrium solutions of (1) discussed in this section arise from the perturbation of the periodic solutions about $(a(x),0), x \in [-1,1]$, for the fast equation (7). In particular in the x-u plane, these solutions oscillate about the graph of $a(x)$ and are projections of phase curves in x-u-w space which spiral about the curve of centers $x \to (x,a(x),0)$.

The proof of existence of these oscillatory solutions depends upon a canonical change of coordinates to action-angle variables which preserve the Hamiltonian form of the system (7). Recall that for a point (u,w) on a closed orbit of a Hamiltonian system with one degree of freedom, the action coordinate I of the point is the area enclosed by the closed orbit divided by 2π, and recall that the angle coordinate φ is chosen so that it increases at a constant rate with time and increases by 2π for each circuit of the closed orbit.

Geometrically φ can be described as follows. Fix a point (u_0,w_0) on the closed orbit which we take to have angle coordinate $\varphi = 0$, and let (u,w) be another point on the same closed orbit which necessarily has the same action coordinate I as (u_0,w_0). Let Σ_0 be a local cross-section of the flow through (u_0,w_0) which is moved under the action of the flow to a local cross-section Σ through (u,w). Then for small increments in the action ΔI, Σ_0 (hence Σ) intersects a nearby closed orbit whose points have action coordinate $I + \Delta I$. The angle coordinate φ of the point (u,w)

is defined to be the limit as ΔI goes to zero of the ratio of an area to ΔI, where the area is the area between the two closed orbits of action I and $I + \Delta I$ swept out by Σ_0 as it moves to Σ. It is then immediate that φ increases by 2π for each circuit of the orbit. Also φ has a constant rate of increase with time due to the area preservation property of the Hamiltonian flow. A good physical analogue is the flow of an incompressible fluid in a pipe of varying diameter where the walls of the pipe correspond to the two closed orbits. In narrow sections of the pipe the fluid has a higher velocity than in thicker sections of the pipe, but the volume swept out in time Δt by a moving cross-section is constant throughout the pipe. Below we identify the angle variable φ with the point $e^{i\varphi}$ on the standard unit circle S^1. Hence φ is only determined up to an additive integral multiple of 2π. We are now ready to describe the existence proof.

PROPOSITION 3. If $a(x)$ is C^2, then part (iv) of Theorem A is true.

OUTLINE OF PROOF. There is a smallest area among the areas of all the two cells $C(x)$ defined above, and let A be this value. Let $0 < c \ll A$ be a small positive number, and let $I_M = (2\pi)^{-1}(A-c)$, $I_m = (2\pi)^{-1}c$. Then for each $x \in [-1,1]$ there is a closed annular region $A(x)$ in the interior of $C(x)$ whose boundary consists of two closed orbits of (7), one containing area c, the other area $A-c$. Each point in such a region has action coordinate I that satisfies $I_m \leq I \leq I_M$. Thus the region of phase space of interest in each x-slice is homeomorphic (via the symplectic transformation to action-angle variables) to the cylinder $K = S^1 \times [I_m, I_M]$, which we refer to as the phase cylinder. On the phase cylinder with coordinates (φ, I), the fast system at x has the form

$$\begin{aligned} \dot{\varphi} &= \omega(I,x) \\ \dot{I} &= 0 \\ \dot{x} &= 0 \quad . \end{aligned} \tag{9}$$

However the time-dependent Hamiltonian system (4), where $v = (u,w)$, $F(v,x,\varepsilon) = (w,-u(1-u)(u-a(x)))$, transforms via a time-dependent symplectic transformation to the system

$$\begin{aligned} \dot{\varphi} &= \omega(I,x) + \varepsilon f(I,\varphi,x) \\ \dot{I} &= \varepsilon g(I,\varphi,x) \\ \dot{x} &= \varepsilon \end{aligned} \tag{10}$$

for some functions f, g and ω which are C^2 on $K \times [-1,1]$ since $a(x)$ is C^2. Of course (10) reduces to (9) when $\varepsilon = 0$. The natural phase space for this system is $K \times [-1,1]$. The details of how the system (4) is transformed to (10) is carried out in [A, section 52F] where it is shown that in the present situation the action coordinate is an adiabatic invariant.

Explicitly, it is shown that there exists a constant k so that

$$| I_{\varepsilon}(t) - I_{\varepsilon}(0)| < k\varepsilon \quad \text{for} \quad 0 \le t \le 2/\varepsilon \tag{11}$$

where $I_{\varepsilon}(t)$ is the I-component of the integral curve for (10) with initial condition $(\varphi, I, -1) \epsilon S^1 \times (I_m, I_M) \times [-1,1]$.

It is clear from our description above that the angle coordinate φ can be chosen so that for $x = \pm 1$ the set of points in K corresponding to the subset of the left and right boundary conditions given by $\{(u,0) \epsilon A(x): u > a(x)\}$ is the line segment $L_0 \equiv \{0\} \times [I_m, I_M]$. Then because for each $x \epsilon [-1,1]$, the phase portrait of (7), the fast system at x, is symmetric with respect to the u-axis, it follows when $x = \pm 1$, that the subset of the left and right boundary conditions given by $\{(u,0) \epsilon A(x): u < a(x)\}$ corresponds to the line segment $L_{\pi} \equiv \{\pi\} \times [I_m, I_M]$. The segments L_0 and L_{π} each give rise to two families of oscillatory solutions in the same manner. Accordingly we only consider solutions starting in L_0 in the $x = -1$ slice. Set $L_0' = [I_m + k\varepsilon_0, I_M - k\varepsilon]$. In Figure 4 we have drawn two pictures of the phase cylinder K. The one on the left corresponds to $x = -1$ where the dark vertical line segment is L_0'. The

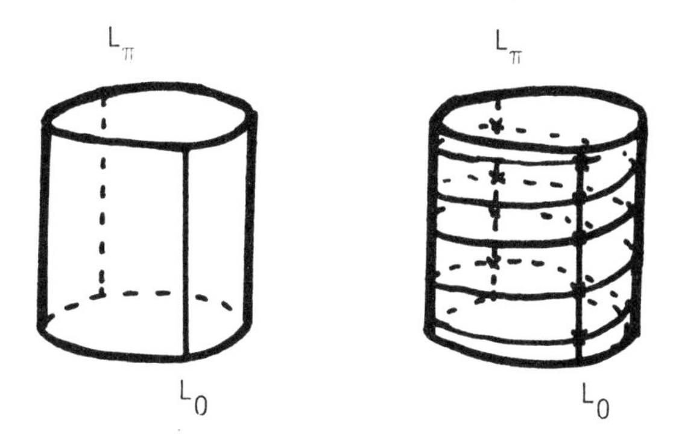

Figure 4

cylinder on the right corresponds to $x = +1$, where we claim that the "helix" is the image of L_0' under the flow of (10) after time $2/\varepsilon$ and the diametrically opposite vertical line segments are L_0 and L_{π}. Each intersection of the "helix" with L_0 or L_{π} is then clearly the right endpoint of a solution to the boundary value problem. We will see that the number of revolutions in the helix corresponds to a difference in the frequency of oscillation as the action coordinate of the initial points on L_0' is increased, where the frequency decreases as the action coordinate increases. Thus proceeding down L_0 or L_{π}, the solution corresponding to each intersection point has one more oscillation than the solution corresponding to the intersection point preceeding it. We will also see below that both the number of oscillations of the solution corresponding to the highest intersection point and the number of intersection points of L_0 with the "helix" increase to infinity as ε decreases towards zero.

To see that the image of L_0' is as claimed we note first that by (11) the action coordinate of each point in L_0' remains unchanged within order $O(\varepsilon)$ over the time interval $0 \le t \le 2/\varepsilon$. Thus the image of L_0' after time $2/\varepsilon$ will be an arc in the phase cylinder which within $O(\varepsilon)$ runs from the bottom of the cylinder to the top. Next we must examine the frequency of oscillation of an orbit at a given level of the action. To do this, we return to an examination of the periods of the closed orbits in the cells $C(x)$ in the phase planes of the fast systems (7).

An analysis of the periods of these orbits has been carried out in [S]. In particular for $x \in [-1,1]$, choose Cartesian coordinates (β,w) in $C(x)$ where w is as before and $\beta = u-a(x)$ if $a(x) \ge 1/2$, $\beta = a(x)-u$ if $a(x) < 1/2$. β is the distance along the u-axis measured positively from the center $(a(x),0)$ towards the homoclinic critical point in $C(x)$ (at least when $a(x) \ne 1/2$). Note that when $a(x) = 1/2$, the phase portraits are symmetric about $u = 0$. For $x \in [-1,1]$, each closed orbit in $C(x)$ corresponds monomorphically with a positive value of β determined by the intersection of the orbit with the positive β-axis, and let $T(\beta,x)$ denote the elementary period of the orbit in $C(x)$ corresponding to $\beta > 0$. Then the following has been proved.

LEMMA [S, Theorem 3.1]. $(\partial/\partial\beta)T(\beta,x) > 0$.

Clearly each positive value of β corresponds to exactly one value of the action I, namely the area enclosed by the orbit corresponding to β divided by 2π. Clearly $\beta'(I) > 0$; hence we can define $T(I,x) \equiv T(\beta(I),x)$. Then note that $(\partial/\partial I)T(I,x) > 0$. In particular in (9) and (10), $\omega(I,x) = 2\pi/T(I,x)$; whence $(\partial/\partial I)\omega(I,x) < 0$. Referring to (10) it is clear that the number of oscillations $\nu_\varepsilon(I_0)$ made by the solution to (10) with initial condition $(0,I_0,-1) \in L_0 \times \{-1\}$ is given by

$$\nu_\varepsilon(I_0) = \frac{1}{2\pi}\int_0^{2/\varepsilon} \{\omega(I_\varepsilon(t,I_0),x_\varepsilon(t)) + \varepsilon f(I_\varepsilon(t,I_0),\varphi_\varepsilon(t,I_0),x_\varepsilon(t))\}dt, \quad (12)$$

where the integral curve to (10) with initial condition $(0,I_0,-1)$ is given by $c_\varepsilon(t,I_0) \equiv (\varphi_\varepsilon(t,I_0),I_\varepsilon(t,I_0),x_\varepsilon(t))$. (Note $x_\varepsilon(t) \equiv -1 + \varepsilon t$, and is therefore independent of I_0). Because ω and f have bounded C^2 norms on $K \times [-1,1]$ and because the action is an adiabatic invariant it can be shown that

$$\nu_\varepsilon(I_0) = \frac{1}{2\pi\varepsilon}\int_{-1}^{1} \omega(I_0,x)dx + O(1) \quad (13)$$

uniformly for $I_m < I_0 < I_M$ as ε decreases to zero. It is then immediate that $\nu_\varepsilon(I_0)$ approaches infinity as ε goes to zero uniformly for $I_m < I_0 < I_M$.

Suppose $I_m < I_1 < I_2 < I_M$. Define $\Delta \nu_\varepsilon(I_1,I_2) = \nu_\varepsilon(I_1) - \nu_\varepsilon(I_2)$. From (13) it follows that

$$\Delta \nu_\varepsilon(I_1,I_2) = \frac{1}{2\pi\varepsilon} \int_{-1}^{1} \frac{\partial\omega}{\partial I}(\xi,x)(I_1 - I_2)dx + O(1)$$

where $I_1 \leq \xi \leq I_2$. As $(\partial/\partial I)\omega$ is continuous and strictly negative on $K \times [-1,1]$, it follows that there is $\Omega > 0$ so that

$$\Delta \nu_\varepsilon(I_1,I_2) \geq \frac{1}{2\pi\varepsilon} \Omega(I_2 - I_1) + O(1) \tag{14}$$

uniformly for $I_1,I_2 \in (I_m,I_M)$. Thus given $h > 0$, for small enough positive ε, if $0 < \Delta I = I_2 - I_1 \leq h$, then $\Delta \nu_\varepsilon(I_1,I_2)$ is positive and increases to infinity as ε goes to zero. It follows that if we partition $L_0' \simeq L_0' \times \{-1\}$ in increments of ΔI, then for small enough positive ε, the image of the partition segments under the flow of (10) after time $2/\varepsilon$ makes some positive number of revolutions around the phase cylinder K at $x = +1$. For small ΔI, this shows that the image of L_0' after time $2/\varepsilon$ is basically helical and the number of revolutions in the helix goes to infinity as ε goes to zero. Hence from our discussion of Figure 4 above, the number of solutions to the boundary value problem which corresponds to the number of intersections of the "helix" with L_0 or L_π also becomes infinite as ε goes to zero. Finally it is immediate from (13) that the number of oscillations of the solution with the highest final action coordinate on L_0 or L_π becomes infinite as ε goes to zero. This completes the proof. ///

BIBLIOGRAPHY

[A] V.I. Arnold, Mathematical Methods of Classical Mechanics, Springer-Verlag, New York, 1978.

[A-W] P.G. Aronson and H.F. Weinberger, "Nonlinear diffusion in population genetics, combustion, and nerve propagation", Proceedings of the Tulane Program in Partial Differential Equations, Lecture Notes in Mathematics, 446, Springer-Verlag, New York, 1975.

[C] Charles C. Conley, Isolated Invariant Sets and the Morse Index, CBMS-AMS, Providence, R.I., 1978.

[K1] H.L. Kurland, "The Morse index of an isolated invariant set is a connected simple system", J. Differential Equations, 42 (1981), 234-259.

[K2] H.L. Kurland, "Homotopy invariants of repeller-attractor pairs. I. The Püppe sequence of an R-A pair", J. Differential Equations, to appear October 1982.

[K3] H.L. Kurland, "Homotopy invariants of repeller-attractor pairs. II. Continuation of an R-A pair, J. Differential Equations, to appear April 1982.

[K4] H.L. Kurland, "Boundary value problems via homology in the Morse Index", J. Differential Equations, to appear.

[P1] L.A. Peletier, "On a non-linear diffusion equation arising in population genetics", Proceedings of the fourth conference held at Dundee, Scotland, March 30-April 2, 1976, Lecture Notes in Mathematics, 564, Springer-Verlag, New York 1976, 365-371.

[P2] L.A. Peletier, "A non-linear eigenvalue problem occurring in population genetics", Journel d'Analyse Non-Lineare, Proceedings Besancon, France, 1977, Lecture Notes in Mathematics, 655, Springer-Verlag, New York, 1978, 170-187.

[S] J. Smoller and A. Wasserman, "Global bifurcation of steady-state solutions", J. Differential Equations 39 (1981), 269-290.

[D] A. Dold, Lectures on Algebraic Topology, Springer-Verlag, New York, 1972.

[K-L] H.L. Kurland and M. Levi, "Adiabatic invariants and a problem in population genetics", to appear.

DEPARTMENT OF MATHEMATICS
BOSTON UNIVERSITY
BOSTON, MA 02215

Contemporary Mathematics
Volume 17, 1983

Somé convection-diffusion equations arising in population dynamics

Masayasu Mimura

ABSTRACT. This paper treats a nonlinear diffusion equation described by

$$u_t + ((K*u)u)_x = (u^m)_{xx}$$

which serves as a spatially aggregating population model in which the component are by (nonlinear) diffusion and (nonlocal) convection. For several specific kernels $K(x)$, the long time-behavior of solution to the initial-value problem is investigated.

1. INTRODUCTION. From an ecological viewpoint, it is interesting to understand the significance of grouping such as half-wild herds of cattle, fish schooling, bird flocking and insect swarming (for instance [9]). One of such grouping is gregarious behavior as a form of cover-seeking in which each animal tries to reduce its chance of being caught by a predator. In view of this explanation, Hamilton [5] presented a model consisting of a colony of frogs on the fringe of a circular pond and a water snake (a predator) in the pond. The model assumes that the frogs move along the fringe rather than escaping into the water, and that in each round of jumping a frog stays put only if the gap it occupies is smaller than both neighboring gaps, otherwise it jumps into the smaller of these gaps. Thus Hamilton [5] showed a computer simulation experiment in which one or more heaps of aggregated frogs occured (Fig.1). He concluded from this result that the selfish avoidance of a predator can lead to the aggregation.

The purpose of this paper is to present a continuous version of Hamilton's model in the framework of diffusion equaiton models. The model in one space dimension is described by

$$u_t + [(K*u)u]_x = (u^m)_{xx}, \tag{1}$$

in which two migrational processes, (nonlinear degenerate) diffusion and (nonlocal) convection, are involved. The kernel $K(x)$ is specified by Hamilton's idea. A special case of (1) when $m>1$ and $K\equiv 0$ occurs in the theory of flow through porous medium and has been investigated by authors (for instance, [1], [10]). In ecological terms, the convective term implies an aggregation of

0271-4132/83/0000-1394/$03.25

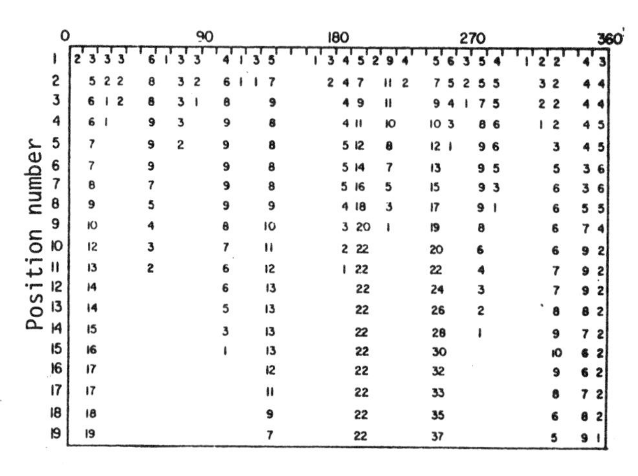

Fig.1.

Gregarious behavior of 100 frogs is shown in terms of the numbers found successively within 10° segments on the margin of the pool. The initial scatter (position 1) is random. Frogs jump simultaneously giving the series of positions shown. They pass neighbours' posotions by one-third of the width of the gap([5]).

biological individuals and the diffusive term indicates dispersal influenced by local population pressures ([3], [4]).

We expect that a balance between the dehomogenizing (or aggregative) process by nonlocal convection and the homogenizing one by diffusion is reached to produce a solitary wave pattern exhibiting phenomenologically an aggregation of individuals. This motivates us to study the properties of solutions to the initial-value problem for (1).

2. BASIC EQUATION. Let us present a diffusion equation model as a continuous version of Hamilton's model. We start with a continuity equation

$$u_t + J_x = 0,$$

where $u(t,x)$ is the population density at time t and position x and J is the population flux. Here the flux J is assumed to include two processes, diffusion and convection, described by

$$J = -Du_x + Vu,$$

where D is the diffusivity and V is the convection velocity. We assume D and V to satisfy the following conditions: The enviroment is spatially uniform, the diffusivity D depends on local population pressures, i.e., $D = D(u)$ with $D(u) \geq 0$ and $D'(u) \geq 0$ for $u \geq 0$, and the velocity V takes the form of convolution $V = K*u$. The kernel $K(x)$ satisfies the property to be caused an aggregative mechanism, motivated by Hamilton's idea. The basic assumption on $K(x)$ is

$$K(x) = \begin{cases} -k(x) & x > 0 \\ k(x) & x < 0, \end{cases}$$

where $k(x)$ is a non-negative and monotonically non-increasing function (Fig.2). Let us briefly interpret the property of $K(x)$. Suppose that an initial function $u_0(x)(\geq 0)$ satisfies the support of $u_0(x)$, say $\mathrm{supp}(u_0)$, is compact. Then it follows from the equation $u_t + [(K*u)u]_x = 0$ that the support of a solution $u(t,x)$, $\mathrm{supp}(u(t))$ becomes shorter as time t goes on. This indicates phenomenologically an aggregation of biological individuals. The kernel $K(x)$ will be assumed more specifically from an ecological point of view.

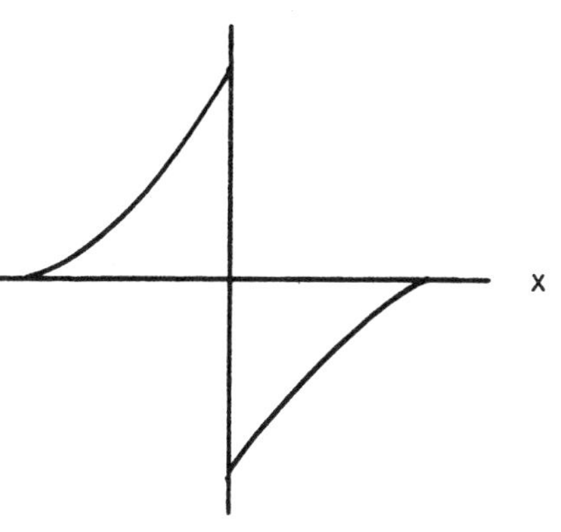

Fig.2. Specific form of K(x)

We now limit to the special case in which $D(u)$ has the form $D(u)=du^m$ where d is a posotive constant and $m(\geq 1)$ is constant, and study the large time behavior of solutions to the initial-value problem

$$u_t + [(K*u)u]_x = d(u^m)_{xx}, \quad t>0, \ x \in \mathbb{R} \tag{2}$$

and the initial condition

$$u(o,x) = u_0(x), \ x \in \mathbb{R}, \tag{3}$$

where $u_0(x)$ is non-negative, $\lim_{|x|\to\infty} u_0(x) = 0$ and $u_0(x)$ belongs to $L^1(\mathbb{R})$, i.e., $\| u_0 \|_1 = I(<+\infty)$, which means the total population.

3. PRINCIPAL VALUE INTEGRAL. Let us start with an integral

$$\int_{-\infty}^{\infty}\left[-\frac{1}{2\delta}\coth\frac{\pi(z-\xi)}{2\delta}\right]v(\xi) \equiv \tilde{v}(z),$$

for a real function $v(x)$, where $\mathrm{Im}\, z \neq 0$ and δ is an arbitrary real constant.

If $v(x)$ satisfies a Hölder condition on the x-real axis, then it follows from the Plemelj formulae that the boundary values of $\tilde{v}(z)$ on $z = x$ satisfy

$$\lim_{z\downarrow o} \tilde{v}(z) = Tv(x) + iv(x) \equiv v^+(x) \tag{4}$$

and

$$\lim_{z\uparrow o} \tilde{v}(z) = Tv(x) - iv(x) \equiv v^-(x), \tag{5}$$

respectively, where

$$Tv(x) = \mathrm{p.v.}\int_{-\infty}^{\infty}\left[-\frac{1}{2\delta}\coth\frac{\pi(z-\xi)}{2\delta}\right]v(\xi)d\xi. \tag{6}$$

We easily see from (4) and (5) that

$$v(x) = \frac{i}{2}[v^-(x) - v^+(x)] \tag{7}$$

and

$$Tv(x) = \frac{1}{2}[v^-(x) + v^+(x)]. \tag{8}$$

We consider the equation (2) when the convolution $K*u$ takes the form $K*u = Tu$ and $m = 1$, i.e.,

$$u_t + [Tu\cdot u]_x = du_{xx}. \tag{9}$$

We study the initial-value problem to (9) subject to (3) by considering the exact linearization of (9). Substituting (7) and (8) into (9), we have

$$(u^- - u^+)_t + \frac{1}{2}[(u^- - u^+)(u^- + u^+)]_x = d(u^- - u^+)_{xx}. \tag{10}$$

Here $u^\pm$ are defined in a similar way to $v^\pm$ in (4) and (5). Let us introduce new functions $f^\pm(t,x)$ defined by

$$u^\pm(t,x) = -2d[\log f^\pm(t,x)]_x + \frac{I}{2\delta}, \tag{11}$$

where $\| u_0 \|_1 = I$. Substituting $u^\pm$ into (10), then we obtain an equation for $f^\pm$

$$(f_t^- - df_{xx}^- + \frac{I}{2\delta} f_x^-)/f^- - (f_t^+ - df_{xx}^+ + \frac{I}{2\delta} f_x^+)/f^+ = 0.$$

Thus, if one could find functions $f^\pm$ satisfying the linear diffusion equation

$$f_t = df_{xx} - \frac{I}{2\delta} f_x \tag{12}$$

and the analyticity of $(\log f^\pm)_x$, then a solution $u(t,x)$ to (9) can be obtained through the relation

$$u = -id[\log f^-/f^+]_x. \tag{13}$$

By using this exact linearization, we can find steady state solutions to (9). Let us difine a function $u(x;p)$ by

$$u(x;p) = \frac{dp\sin p\delta}{\cosh px + \cos p\delta}, \tag{14}$$

for a real constant p satisfying $0 < p\delta < \pi$. $u(x;p)$ takes a single-peak distribution and the amplitude is zero as $p \downarrow 0$ and infinite as $p \uparrow \pi/\delta$. Since (14) is rewritten as

$$u(x;p) = -id[\log\{1 + \exp(p(x + \delta i))\} - \log\{1 + \exp(p(x - \delta i))\}]_x, \tag{15}$$

we may put $f^\pm$ as

$$f^\pm = 1 + \exp(p(x \mp \delta i)). \tag{16}$$

Noting that $\| u(\cdot;p) \|_1 = 2\delta dp$, we find that $f^\pm$ is t independant solutions of (12) when $2\delta dp = I$. Hence, it turns out that $u(x;p)$ is a t independent solution of (9), i.e., one parameter family of steady state solutions of (9).

Thus, we find that there exist steady state solutions of (9) when the total population I satisfies $0 < I < 2d\pi$. From this result, we next have the question of the stability of $u(x;p)$ and of the existance of steady state solution for $2d\pi \leqq I$. We let partially answer this question. Take an initial function as

$$u_0(x) = u(x-a;p) + u(x+a;p) \tag{17}$$

for a non-negative constant a, where $\|u_0\|_1 = 4\delta dp$. Noting that $u_0(x)$ is rewritten as

$$u_0(x) = -\, id[\log\{(1 + \exp(p(x-a+\delta i)))(1 + \exp(p(x+a+\delta i)))\}$$
$$-\log\{(1 + \exp(p(x-a-\delta i)))\}\{(1 + \exp(p(x+a+\delta i)))\}]_x,$$

we easily find that $f^{\pm}(0,x)$ are represented by

$$f^{\pm}(0,x) = \{1 + \exp(x-a \mp \delta i)\}\{1 + \exp(x+a \mp \delta i)\}. \tag{18}$$

Thus solutions of (12) subject to (18) can be explicitly represented by

$$f^{\pm}(t,x) = 1 + \{\exp(p(x-a\mp\delta i)) + \exp(p(x+a\mp\delta i))\}\exp(-p^2 t)$$
$$+ \exp(2p(x\mp\delta i)). \tag{19}$$

From the representation of $f^{\pm}(t,x)$, we conclude that

i) $\lim_{t\to\infty} u(t,x) = u(x;2p)$ for $0 < p\delta < \pi/2$,

ii) $u(t,x)$ grows up as $t\to\infty$ for $p\delta = \pi/2$, and

iii) $u(t,x)$ blows up at $t = t_0 = p^{-2}\log(-\cos p\delta/\cosh pa)$, for $\pi/2 < p\delta < \pi$.

Remark. Suppose that the total population I is an adjustable parameter. Then the results in the above are expressed by

i) $\lim_{t\to\infty} u(t,x) = u(x;p)$ for $0 < I < 2d\pi$

ii) $u(t,x)$ grows up as $t\to\infty$ for $I = 2d\pi$, and

iii) $u(t,x)$ blows up at $t = t_0$ for $2d\pi < I < 4d\pi$.

The precise discussion is in Satsuma and Mimura [11]. From this remark, it is understood that the convective term expressed by singular integrals could cause bolw up behavior. Furthermore we conjecture that there is a critical value I_0 determining whether $u(t,x)$ tends to the steady state $(0 < I < I_0)$ or it blows up (or grows up)$(I_0 \leqq I)$ for general initial functions $u_0(x)$ with $\|u_0\|_1 = I$. Unfortunately, this problem remains open.

4. NONLINEAR DEGENERATE DIFFUSION. We next study the problem (2), (3) in the case when $K(x)$ is the simplest kernel

$$K(x) = \begin{cases} -k, & x > 0 \\ k, & x < 0 \end{cases}$$

for a positive constant K. The assumption on $K(x)$ is imposed not to retain some realistic meaning from an ecological point of view, but to make the mathematical consideration easy. However, we assume $m > 1$, which means the dispersal depending on local population pressures. Then the equation (2) leads to

$$u_t + k[(\int_x^{\infty} u(t,\xi)d\xi - \int_{-\infty}^{x} u(t,\xi)d\xi)u]_x = d(u^m)_{xx}. \tag{20}$$

By the preservation of the total population, i.e., $\|u_0\|_1 = \|u(t,\cdot)\|_1 = I$, (20) is rewritten as

$$u_t + k[(I - 2\int_{-\infty}^{x} u(t,\xi)d\xi)u]_x = d(u^m)_{xx}. \tag{21}$$

In this section we treat a more general equation than (21),

$$u_t + [\phi(\int_{-\infty}^{x} u(t,\xi)d\xi)u]_x = d(u^m)_{xx}, \tag{22}$$

where $\phi(s)$ is a smooth function of s satisfying $\phi'(s) < 0$ for $s > 0$. When $\phi(s)$ is specified as $\phi(s) = k(I-2s)$, (22) reduces to (21). We consider the initial-value problem for (22) under the assumption that the initial function $u_0(x)$ is of compact support in $\mathbb{R}$. In the absence of the convection term, (22) leads to the porous medium equation which exhibits a variety of interesting phenomena. One of them is the occurrence of interfaces. Suppose that $u_0(x)$ satisfies

$$u_0(x) > 0 \qquad \text{if} \quad x \in (x_1, x_2) \quad (-\infty < x_1 < x_2 < -\infty)$$

and

$$u_0(x) = 0 \qquad \text{if} \quad x \notin (x_1, x_2).$$

Define $P(t)$ by $P(t) = \{x \in \mathbb{R} : u(t,x) > 0\}$. Then it is known that there exist curves $x = s_i(t)(i=1,2)$ such that $P(t) = \{x \in \mathbf{R} : s_1(t) < x < s_2(t)\}$. The curves $s_i(t)(i=1,2)$ are called the interface curves. Furthermore, it is shown that $u(t,x) = O(t^{-1/(m+1)})$ and $|s_i(t)| \to \infty$ $(i=1,2)$ as $t \to \infty$. From this behavior, we could expect that interfaces occur in our problem.

THEROEM 1. (INTERFACES). Let $u(t,x)$ be a solution of the problem (22), (3). There exist constants α and β such that $u(t,x) \equiv 0$ outside of $\alpha \leq x-ct \leq \beta$ with $cI = \Phi(I) - \Phi(0)$, where $\Phi' = \Phi$.

REMARK. When $\Phi(s)=k(I-s)s$ (from (21)), we have $c=0$. Thus it turns out that the interface curves $s_i(t)$ $(i=1,2)$ satisfy $\alpha<s_1(t)$ and $s_2(t)<\beta$. In ecological terms, the interval (α,β) may be called the largest habitat for individuals.

We next discuss the results about the asymptotic behavior of solutions $u(t,x)$ of the problem (22),(3).

THEOREM 2. (ASYMPTOTIC BEHAVIOR) The solution $u(t,x)$ of (22),(3) tends to the traveling solitary wave solution $w(s)$ $(s=x-ct)$ with the speed $c = (\Phi(I)- \Phi(0))/I$ satisfies the problem

$$\frac{d}{ds}\left[cw + \frac{d}{ds}(w^m) - \Phi\left(\int_{-\infty}^{s} w(\eta)d\eta\right)\right]=0, \quad s \in \mathbb{R}, \tag{23}$$

$$w(-\infty) = w(\infty) = 0 \tag{24}$$

$$\int_{-\infty}^{\infty} w(s)ds = I \tag{25}$$

and

$$\int_{-\infty}^{\infty}\int_{-\infty}^{x} [u_0(\xi) - w(\xi)]d\xi dx = 0. \tag{26}$$

REMARK. The traveling wave solution $w(s)$ has compact support, that is

$$w(s) > 0 \quad \text{if} \quad s \in (s_1,s_2) \ (-\infty < s_1 < s_2 < \infty)$$

and

$$w(s) \equiv 0 \quad \text{if} \quad s \notin (s_1,s_2),$$

where

$$s_2 - s_1 = \int_0^I [\Phi(z) - \Phi(0) - cz]^{-1/m} dz.$$

Since the equaiton (22) is invariant under a translation in x, $w(x+x_0-ct)$ is also a solution of (22) for an arbitrarily fixed x_0. However, the condition (26) uniquely determines one traveling wave solution. The proofs are in Nagai and Mimura [7].

For more ecologically realistic kernels $K(x)$, the equation (2) has been investigated. Recently Nagai [6] has shown the uniqueness, existence and some regularity results for the initial-value problem (2), (3) under the assumptions that $K(x)$ is differentiable on $\mathbb{R}$ expect for a finite number of discontinuity points of the first kind and belongs to $L^{\infty}(\mathbb{R})$, and dK/dx belongs to $L^{1}(\mathbb{R})$. Unfortunately the stationary problem has not been discussed.

5. SOME REMARKS. In our model, we did not take the population supply by births and deaths of individuals into account. If the model includes such an effect, it is described by

$$u_t + [K*u)u]_x = d(u^m)_{xx} + f(u), \tag{27}$$

where $f(u)$ means the rate at which individuals are suppled per unit interval. In the absence of the covective term, (27) has been studied by several authors ([3],[4]). For a specific form $f(u) = u(1-u)$, the existence of traveling wave solutions $u(t,x) = w(x - ct)$ satisfying $w(-\infty)=1$ and $w(\infty)=0$ are known ([2],[8]). It is interesting for the equation (27) to discuss such plane wave solutions. This is currently being studied.

BIBLIOGRAPHY

1. Aronson. D. G., "Regularity properties of flow through prous media", SIAM J. Appl. Math. 17 (1969), 461-467.

2. Aronson. D. G., "Density-dependent interaction-diffusion systems" in dynamics and modelling of reactive systems ed. by W. E. Stewart, W. H. Ray and C. C. Conley.(1980), 161-176.

3. Gurney W. S. C. and R. M. Nisbet, "The regulation of inhomogeneous populations", J. Theor. Biol., 52 (1975), 441-457.

4. Gurtin. M. and R. MacCamy, "On the diffusion of biological populations", Math. Biosci., 33 (1977), 35-49.

5. Hamilton. W. D., "Geometry for the selfish herd", J. Theor. Biol., 31 (1971), 295-311.

6. Nagai. T., "Some nonlinear degenerate diffusion equations with a nonlocally convective term in ecology", manuscript.

7. Nagai. T. and M. Mimura., "Asymptotic behavior for a nonlinear degenerate diffusion equation in population dynamics", SIAM J. Appl. Math. (in press).

8. Newman. W. I., "Some exact solutions to a non-linear diffusion problem in population genetics and combustion, J. Theor. Biol. 85 (1980), 325-334.

9. Okubo. A., Diffustion and ecological problems :mathematical models. Biomath. 10 , Springer-Verlag, New York, 1980.

10. Oleinik. O. A., S. A. Kalashnikov., and Chzou Yui-lin., "The Caudly problem and boundary value problems for equations of the type of nonstationary filtration", Izv. Akad. Nank, SSSR. 22 (1958), 667-704.(In Russian)

11. Satsuma. J. and M. Mimura., "Exact treatment of nonlinear diffusion equations with singular integral terms, manuscript (1982).

DEPARTMENT OF MATHEMATICS
HIROSHIMA UNIVERSITY
HIROSHIMA, JAPAN 730

Contemporary Mathematics
Volume 17, 1983

FINITE-DIMENSIONAL ATTRACTING MANIFOLDS IN REACTION-DIFUSION EQUATIONS

Xavier Mora

ABSTRACT. Reaction-difusion equations are considered as defining a semiflow on certain function spaces. A condition is given, which ensures that this semiflow is attracted to a finite-dimensional invariant manifold of a given dimension. In some sense, this condition can be viewed as a generalization of a well-known condition which ensures attraction to the manifold of homogeneous states.

1. When considering the dynamical structure of semilinear parabolic problems over bounded domains, it is interesting to note that in many cases one can ensure that the corresponding semiflow (defined in a function space conveniently chosen) is entirely attracted to a finite-dimensional invariant manifold. This implies that, asymptotically, the system evolves according to a finite-dimensional dynamics; in particular, one cannot have other kinds of attractors than those appearing in finite-dimensional flows.

We here consider this question for reaction-diffusion problems. Specifically, we look at problems of the form

$$D_t u = K \Delta u + f(u) \tag{1}$$

with boundary conditions of either Dirichlet type :

$$u|_{\partial\Omega} = 0 \tag{2}$$

or Neumann type :

$$\nu \cdot \nabla u|_{\partial\Omega} = 0 \tag{3}$$

Here, Ω is a bounded domain of $\mathbb{R}^n$, $u = (u^1,\dots,u^N)$ is a vector of functions of $x \in \Omega$ and $t \in \mathbb{R}$, ν denotes the unit normal to $\partial\Omega$, and K denotes a positive-definite symmetric $N \times N$ matrix (which, for simplicity, will be assumed to be independent of $x \in \Omega$). Finally, f denotes a function from $\mathbb{R}^N$ to $\mathbb{R}^N$.

We want to look at these equations as defining a dynamical system on a suitable function space. Given a Haussdorff space X , by a <u>dynamical system</u>

1980 Mathematics Subject Classification. 35B40, 35K60.

(or semiflow) on X we understand a mapping $\phi : (u,t) \in X \times \mathbb{R}_+ \longmapsto \phi_t u \in X$ with a domain of the form $X^\omega \equiv \{(u,t) \mid u \in X,\ 0 \leqslant t < \omega(u)\}$, where the $\omega(u)$ are real numbers in the interval $0 < \omega(u) \leqslant \infty$; in order to define a dynamical system on X, this mapping must satisfy the following properties : (D0) $\phi_0 u = u$ ($\forall u \in X$) ; (D1) $\phi_{t+s} u = \phi_s \phi_t u$ ($\forall u \in X$), this relation being understood in the sense that whenever one side is defined the other one is also defined and the equality holds; (D2) For every $t \in \mathbb{R}_+$, the set $X^t \equiv \{u \in X \mid t < \omega(u)\}$ is open in X, and the mapping $\phi_t : X^t \longrightarrow X,\ u \longmapsto \phi_t u$ is continuous; (D4) If $\omega(u) < \infty$, then the orbit of u, i. e. the set $\{\phi_t u \mid 0 \leqslant t < \omega(u)\}$, is not contained in any compact set of X. A dynamical system on a metric space is said to be <u>compact</u> if the evolution operators ϕ_t $(t > 0)$ are compact (i.e. the image of a bounded set is relatively compact). If the state space X is a Banach space and the operators $\phi_t : X^t \longrightarrow X$ are of class C^r, then we say that the dynamical system ϕ_t is <u>of class C^r</u>.

2. For our purposes it is interesting to reformulate problems (1, 2) or (1, 3) in the form

$$\dot{u} = Au + F(u) \tag{4}$$

where u represents now a variable taking values in a Banach space E, $\dot{u}$ denotes its time derivative, A is a linear operator on E, and F is a non-linear operator on E. To ensure that equation (4) defines a dynamical system on a certain Banach space X, it suffices to have the following conditions : (H0) The linear operator A is the generator of an analytic semigroup e^{At} on the Banach space E, and the non-linear operator F is continuous from X to E, where X is a Banach space embedded in E ; (H1) The family of operators e^{At} $(t \geqslant 0)$ restricts to a semigroup on the space X ; (H2) The semigroup e^{At} maps E to X with a bound of the form

$$\|e^{At}\|_{E \to X} \leqslant C\, t^{-\alpha} \qquad (0 < \forall t < \tau)$$

for some $\tau > 0$ and some α in the interval $0 \leqslant \alpha < 1$; (H3) The non-linear operator $F : X \quad E$ is Lipschitz on bounded sets. Specifically, one has the following result :

THEOREM 1. Under hypotheses (H0, H1, H2, H3), equation (4) defines a dynamical system on the space X. This dynamical system ϕ_t satisfies the following property :

$$\omega(u) < \infty \ \Rightarrow \ \lim_{t \nearrow \omega(u)} \|\phi_t u\|_X = \infty \tag{5}$$

If the non-linear operator $F : X \quad E$ is of class C^r, then the dynamical system ϕ_t is of the same class. In the case where A has compact resolvent, then one has also the following properties :

The dynamical system is compact. (6)

If the orbit $\{ \phi_t u \mid 0 \leq t < \omega(u) \}$ is bounded in X, then it is necessarily a relatively compact set (which implies that $\omega(u) = \infty$). (7) ∎

In the particular case where X is one of the fractional power spaces associated to the semigroup e^{At}, these results can be found in the book of Henry [1974 - 81]. The general case considered here is treated in Mora [1982] (see also Weissler [1979, 1980]).

In the following, $C^q(\bar{\Omega})$ denotes the Banach space of functions $u : \Omega \to \mathbb{R}^N$ whose derivatives of order $\leq q$ are uniformly continuous; its norm is defined as

$$\|u\|_{C^q} \equiv \max_{1 \leq i \leq N} \max_{|\alpha| \leq q} \sup_{x \in \Omega} |D^\alpha u(x)|$$

Furthermore, for a fixed set of boundary conditions, and for $q = 0, 1$, we define $\hat{C}^q(\bar{\Omega})$ as the closed subspace of $C^q(\bar{\Omega})$ consisting of those functions which satisfy the boundary conditions of order $\leq q$. In Mora [in print], problems like (1, 2) or (1, 3) are reformulated in the form (4) with E and X taken of the form $\hat{C}^q(\bar{\Omega})$, and it is shown that the resulting problem satisfies the hypotheses of Theorem 1. As a consequence, one obtains the following result:

THEOREM 2. Assume that the function $f : \mathbb{R}^N \to \mathbb{R}^N$ is locally Lipschitz, and in the case of Dirichlet boundary conditions assume also that $f(0) = 0$; fix q equal to 0 or 1, and assume that Ω is a bounded domain with boundary of class $C^{2+q+\alpha}$. Under these conditions, problems (1, 2) and (1, 3) define dynamical systems on the Banach spaces $\hat{C}^q(\bar{\Omega})$; these dynamical systems satisfy properties (5), (6), and (7) of Theorem 1. If the function f is of class $C^{1+\alpha}$, then these dynamical systems are of class C^1. ∎

3. Let us now enter into the question of finite-dimensional attracting invariant manifolds. For this, we shall consider problem (4) in the following particular situation: (a) E is a Hilbert space and A is a self-adjoint operator with compact resolvent; (b) F maps E into itself and it is globally Lipschitz and globally bounded, i.e. there exist finite constants, L and M, such that

$$\|F(u) - F(v)\|_E \leq L \|u - v\|_E \qquad (\forall u, v \in E) \tag{8}$$

$$\|F(u)\|_E \leq M \qquad (\forall u \in E) \tag{9}$$

These conditions imply the validity of (H0, H1, H2, H3) of § 2 with $X = E$ and $\alpha = 0$; therefore, they imply that equation (4) defines a dynamical system on the space E . Let λ_k ($k = 1, 2, \ldots$) denote the eigenvalues of $-A$ ordered so that $\lambda_{k+1} \geqslant \lambda_k$, and repeated according to their respective multiplicities. Our search for a finite-dimensional attracting manifold will be associated to a partition of $\Sigma(A)$, the spectrum of A , in the form $\Sigma(A) = \Sigma_1 \cup \Sigma_2$, $\Sigma_1 = \{-\lambda_1, \ldots, -\lambda_k\}$, $\Sigma_2 = \{-\lambda_{k+1}, \ldots\}$, where k is a natural number such that $\lambda_{k+1} > \lambda_k$ and $\lambda_{k+1} > 0$. In the following, E_1 and E_2 denote the subspaces of E spanned by the eigenfunctions corresponding respectively to Σ_1 and Σ_2 .

THEOREM 3 . If, in addition to these conditions, there exists some $\ell \in (0,1]$ satisfying

$$\lambda_{k+1} - \lambda_k > (2 + \ell + \ell^{-1}) L \tag{10}$$

then the dynamical system ϕ_t has an invariant manifold H given by the graph of a mapping $h : E_1 \to E_2$; this mapping satisfies a Lipschitz condition with constant ℓ and it is bounded by the constant M / λ_{k+1} . The invariant manifold H has the following properties: it is uniformly asymptotically stable; it is globally attracting with asymptotic phase (i.e., for every $u \in E$, there exist $\bar{u} \in H$ such that $\|\phi_t u - \phi_t \bar{u}\|_E \to 0$ as $t \to \infty$). ■

The proof of this result can be found in Mora [1982] . In fact, that work includes the more general situation of having conditions (H0, H1, H2, H3) of § 2 plus the analogous of (8) and (9) .

4. Let us now consider the application of this type of results to problems of the type (1, 2) or (1, 3) . Assume for the moment that we were using an analog of Theorem 3 but applicable to the problem resulting of (1, 2) or (1, 3) when taking $E = X = \hat{C}(\bar{\Omega})$. In order to have properties (8) and (9) , we would need analogous properties to hold for the function $f : \mathbb{R}^N \to \mathbb{R}^N$. This is certainly a very special situation. In many cases however, one knows a closed set $Q_0 \subset \mathbb{R}^N$ such that $Q \equiv \{ u \in \hat{C}(\bar{\Omega}) \mid u(\Omega) \subset Q_0 \}$ is a positively invariant set in the dynamical system on $\hat{C}(\bar{\Omega})$; in fact, one can often find positively invariant sets of this form which are also globally attracting (see Chueh , Conley , Smoller [1977]). Since such sets Q_0 must be necessarily convex (Chueh , Conley , Smoller [1977]) , it is easy to see that, given a region Q_0 of this type where f is bounded and satisfies a Lipschitz condition, we can modify f outside Q_0 so that the new f satisfies these properties over all of $\mathbb{R}^N$ and with exactly the same constants. Obviously, this modification of f outside Q_0 does not change the flow inside $Q \subset \hat{C}(\bar{\Omega})$, and Q is still positively invariant for the new global flow on $\hat{C}(\bar{\Omega})$. Therefore, if H is a globally attracting invariant manifold obtained for the modified f , it follows that $H \cap Q$ is necessarily non-empty, and it still attracts all the flow inside Q for the original f . Furthermore, after having modified f in this way,

it is easily realised that the corresponding functional operator F maps $L_2(\Omega) \equiv L_2(\Omega)^N$ into itself, and this mapping is bounded and satisfies a Lipschitz condition. Specifically, if we consider $\mathbb{R}^N$ with the euclidean norm, and $L_2(\Omega) \equiv L_2(\Omega)^N$ with the norm $(\sum_i \|u^i\|^2_{L_2})^{1/2}$, then F satisfies properties (8) and (9) with constants $L = L_0$ and $M = |\Omega|^{1/2} M_0$, where L_0 and M_0 are the corresponding constants for the original function f restricted to Q_0 . By applying Theorem 3 , we therefore obtain the following result: If there exist $k \in \mathbb{N}$ and $\ell \in (0,1]$ such that condition (10) holds, then, in the dynamical system that the modified problem defines on $L_2(\Omega)$, the flow inside the positively invariant set $\overline{Q}$ (the L_2-closure of Q) is attracted to a k-dimensional invariant manifold with all the properties described in Theorem 3 .

If the set Q_0 is bounded, then we can easily see that the attractivity of H remains true in the space $\hat{C}(\overline{\Omega})$ (the fact that $H \subset \hat{C}(\overline{\Omega})$ can be derived from the invariance of H together with the regularizing properties of parabolic equations). This property can be easily obtained by combining the asymptotic phase property of Theorem 3 (i.e., that for every $u \in L_2(\Omega)$ there exist $\bar{u} \in H \subset \hat{C}(\overline{\Omega})$ such that $\|\phi_t u - \phi_t \bar{u}\|_{L_2} \to 0$ as $t \to \infty$) together with the fact that the orbit of a point $u \in Q$ must be relatively compact in $\hat{C}(\overline{\Omega})$ (Theorems 1 and 2).

Note also that if we are not particularly interested in the value of ℓ , then condition (10) reduces to

$$\lambda_{k+1} - \lambda_k > 4 L \tag{11}$$

Summarizing these observations, we can state the following result :

THEOREM 4 . Together with the hypotheses of Theorem 2 , assume that we know a compact set $Q_0 \subset \mathbb{R}^N$ such that $Q \equiv \{ u \in \hat{C}(\overline{\Omega}) \mid u(\Omega) \subset Q_0 \}$ is positively invariant and the function f satisfies

$$\|f(u) - f(v)\| \leq L \|u - v\| \qquad (\forall u,v \in Q_0)$$

$$\|f(u)\| \leq M_0 \qquad (\forall u \in Q_0)$$

($\|\cdot\|$ denotes the euclidean norm in $\mathbb{R}^N$). If there are $k \in \mathbb{N}$ such that condition (11) holds, then the flow inside $Q \subset \hat{C}(\overline{\Omega})$ is attracted to a k-dimensional invariant manifold H ; this manifold is uniformly asymptotically stable and the attractivity has the property of asymptotic phase. ■

According to condition (11) , it is clear that if Q satisfies the hypotheses of Theorem 4 and the sequence λ_k ($k \to \infty$) satisfies $\lambda_{k+1} - \lambda_k \to \infty$, or at least $\limsup_{k\to\infty} (\lambda_{k+1} - \lambda_k) = \infty$, then we always have a finite-dimensional attracting invariant manifold. In the case of reaction-diffusion problems, the first of these properties holds in the case of Ω being a bounded interval of

$\mathbb{R}^1$, and the second one holds for at least some domains of $\mathbb{R}^2$.

It is interesting to note that these results are of the same nature that a well-known simpler result which gives a condition of strong diffusion under which the system is necessarily attracted to the manifold of homogeneous states (i.e., $\mathbb{R}^N$ in the case of Neumann boundary conditions, and $\{0\}$ in the case of Dirichlet boundary conditions). According to Conway, Hoff, Smoller [1978], this property holds whenever one has the following condition

$$d_1 \alpha_1 > L \qquad (12)$$

where d_1 is the minimal eigenvalue of K , and α_1 is the first non-zero eigenvalue of $-\Delta$. Taking into account that the eigenvalues and eigenfunctions of $-A = -K\Delta$ are given by $(d_i\alpha_q, \theta_i\psi_q)$ where (α_q, ψ_q) are the eigenvalues and eigenfunctions of $-\Delta$, and (d_i, θ_i) are the eigenvalues and eigenvectors of K , we see that condition (12) can be rewritten as follows:

$$\lambda_1 > L \quad \text{(for Dirichlet boundary conditions)}$$

$$\lambda_{N+1} - \lambda_N > L \quad \text{(for Neumann boundary conditions)}$$

which shows the similarity with (11) . Let us note also that our way of seeing that L_2-attractivity implies C-attractivity is simpler than the one used in Conway, Hoff, Smoller [1978] .

5. To obtain the attracting invariant manifold of Theorem 4 , we use essentially the method of Krylov and Bogoljubov (see Hale [1969: Ch. VII] and Henry [1974 - 81: Ch. 6]). This is not however the only possible method to obtain these results. For instance, in the work of Kurzweil [1967, 1968] one finds a different principle which probably can be used to obtain similar results* . A third method can be found in Mañé [1977] and Kamaev [1981] ; in Mañé's paper, the conditions that determine the minimal value of k for which one obtains the invariant manifold are a bit involved; however, it is worth noting that, for the particular situation considered in § 3 , these conditions reduce exactly to (11) .

In connection with these subjects, we cannot leave without comment the interesting investigations of Foiaş, Prodi, Temam and Ladyženskaja about the attractor of Navier-Stokes equations. Motivated by the theories of Hopf and Landau on turbulence, Foiaş, Prodi [1967] and Ladyženskaja [1972] showed that, in some sense, the asymptotic behavior of the solutions is determined by their projection on a finite-dimensional space. Recently, Foiaş, Temam [1979] and Ladyženskaja [1982] have completed these investigations by showing that

* We have known of the work of Kurzweil by indication of Jack Hale in this conference.

the Haussdorff dimension of the total attractor is finite. Another interesting work in the same direction is that of Babin, Višik [1982a, b].

In order to obtain attracting invariant manifolds of a dimension as small as possible, the results here obtained have the inconvenient of being restricted to manifolds representable as graphs of mappings $h: E_1 \longrightarrow E_2$. In the direction of eliminating this restriction we have the interesting works of Kurzweil [1970] and Kamaev [1980a, b]. In particular, this last author considers a situation similar to that of § 3 (but with α not necessarily equal to zero) and obtains the following result: If problem (4) has a hyperbolic invariant set Θ, then there exists a $k_0 \in \mathbb{N}$ such that, for every $k \geq k_0$, the k-th Galerkin approximation of (4), i.e. the equation

$$\dot{u}_1 = P_1 A u_1 + P_1 F(u_1)$$

(u_1: function with values in E_1, P_1: orthogonal projection on E_1), has a hyperbolic invariant set homeomorphic to Θ.

BIBLIOGRAPHY

1. Babin, A. V., Višik, M. I., 1982a. (Бабин, А. В., Вишик, М. И.) Существование и оценка размерности аттракторов квазилинейных параболических уравнений и системы Навье-Стокса. Успехи Мат. Наук 37(3): 173-174.

2. Babin, A. V., Višik, M. I., 1982b. (Бабин, А. В., Вишик, М. И.) Аттракторы системы Навые-Стокса и параболических уравнений и оценка их размерности. Зап. Научн. Сем. Ленинград. Отдел. Мат. Инст. Стеклов. 115: 3-15.

3. Chueh, K. N., Conley, C. C., Smoller, J. A., 1977. Positively invariant regions for systems of nonlinear diffusion equations. Indiana Univ. Math. J. 26: 373-392.

4. Conway, E., Hoff, D., Smoller, J., 1978. Large time behavior of solutions of systems of nonlinear reaction diffusion equations. SIAM J. Appl. Math. 35: 1-16.

5. Foiaş, C., Prodi, G., 1967. Sur le comportement global des solutions non-stationnaires des équations de Navier-Stokes en dimension 2. Rend. Sem. Mat. Univ. Padova 39: 1-34.

6. Foiaş, C., Temam, R., 1979. Some analytic and geometric properties of the solutions of the evolution Navier-Stokes equations. J. Math. Pures Appl. 58: 339-368.

7. Hale, J. K., 1969. Ordinary Differential Equations. Wiley-Interscience, New York.

8. Henry, D., 1974, 1981. Geometric Theory of Semilinear Parabolic Equations. Univ. Kentucky Lecture Notes (1974). Lecture Notes in Math. 840 (1981).

9. Kamaev, D. A., 1980a. (Камаев, Д. А.) Гиперболические предельные множества одного класса параболических уравнений и метод Галеркина. Докл. Акад. Наук СССР 254: 282-286. / Trad.: Hyperbolic limit sets of a class of parabolic equations and Galerkin's method. Soviet Math. Dokl. 22: 344-348.

10. Kamaev, D. A., 1980b. (Камаев, Д. А.) Гипержолические предельные мнобества эволюционныхуравнений и метод Галеркина. Успехи Мат. Наук 35(3): 188-192.

11. Kamaev, D. A., 1981. (Камаев, Д. А.) О гипотезе Хопфа одного класса уравнений химической кинетики. Зап. Научн. Сем. Ленинград. Отдел. Мат. Инст. Стеклов. 110: 57-73.

12. Kurzweil, J., 1967. Invariant manifolds for flows. Differential Equations and Dynamical Systems (ed. J. K. Hale, J. P. LaSalle; Academic Press, New York): 431-468.

13. Kurzweil, J., 1968. (Курцвейль, Я.) Инвариантные мнобества дифференциальных систем. Дифференциальные Уравнения 4: 785-797. / Trad.: Invariant sets of differential systems. Differential Equations 4: 406-412.

14. Kurzweil, J., 1970. Invariant manifolds I . Comment. Math. Univ. Carolin. 11: 309-336.

15. Ladyženskaja, O. A., 1972. (Ладыбенская, О. А.) О динамической системе, поробдаемой уравнениями Навье-Стокса. Зап. Научн. Сем. Ленинград. Отдел. Мат. Инст. Стеклов. 27: 91-114. / Trad.: A dynamical system generated by the Navier-Stokes equations. J. Soviet Math. 3: 458-479 (1975).

16. Ladyženskaja, O. A., 1982. (Ладыбенская, О. А.) О конечномерности ограниченных инвариантных мнобесть для системы Навье-Стокса и других диссипативных систем. Зап. Научн. Сем. Ленинград. Отдел. Мат. Инст. Стеклов. 115: 137-155.

17. Mañé, R., 1977. Reduction of semilinear parabolic equations to finite dimensional C^1 flows. Lecture Notes in Math. 597: 361-378.

18. Mora, X., 1982. Sistemes Dinàmics Determinats per Equacions Diferencials Semilineals sobre Espais de Banach. Ph. D. Dissertation, Universitat Autònoma de Barcelona, Spain.

19. Mora, X., in print. Semilinear parabolic problems define semiflows on C^k spaces. Trans. Amer. Math. Soc.

20. Weissler, F. B., 1979. Semilinear evolution equations in Banach space. J. Funct. Anal. 32: 277-296.

21. Weissler, F. B., 1980. Local existence for semilinear parabolic equations in L^p . Indiana Univ. Math. J. 29: 79-102.

DEPARTMENT OF MATHEMATICS
UNIVERSITY OF MICHIGAN
ANN ARBOR, MICHIGAN 48109

Current Address:
Secció de Matemàtiques
Universitat Autònoma de Barcelona
Bellaterra, Barcelona, Spain.

Contemporary Mathematics
Volume 17, 1983

TRAVELING WAVE SOLUTIONS OF MULTISTABLE REACTION-DIFFUSION EQUATIONS

David Terman

ABSTRACT. A new method for proving the existence of traveling wave solutins for equations of the form

$$u_t = u_{xx} + F'(u)$$

is presented. It is assumed that F is sufficiently smooth, $\lim_{|u|\to\infty} F(u) = -\infty$, F has only nondegenerate critical points, and if A and B are distinct critical points of F then $F(A) \neq F(B)$. The results describe when, for a given function F, there must exist zero, exactly one, a finite number, or an infinite number of waves which connect two fixed, stable rest points.

1. INTRODUCTION.

Consider the equation

$$u_t = u_{xx} + F'(u) \tag{1.1}$$

which arises in various branches of mathematical biology including population genetics, ecology, and nerve conduction (see [1],[3]). We assume throughout that F satisfies:

(1.2)
- (a) $F \in C^1\mathbf{R})$,
- (b) $\lim_{|u|\to\infty} F(u) = -\infty$,
- (c) every critical point of F is nondegenerate,
- (d) if A and B are distinct critical points of F, then $F(A) \neq F(B)$.

We are interested in finding traveling wave solutions of (1.1). These are nonconstant, bounded solutions of the form $u(x,t) = U(z)$, $z = x + \theta t$.

1980 Mathematics Subject Classification. 35K55
Sponsored by the United States Army under Contract No. DAAG29-80-C-0041. This material is based upon work supported by the National Science Foundation under Grant No. MCS80-17158.

0271-4132/83/0000-1400/$05.50

If $U(z)$ is a traveling wave solution of (1.1) then U satisfies the first order system of ordinary differential equations:

$$U' = V \tag{1.3}$$
$$V' = \theta V - F'(U).$$

For boundary conditions we take

$$\lim_{z\to-\infty} U(z) = A \quad \text{and} \quad \lim_{z\to+\infty} U(z) = B \tag{1.4}$$

where A and B are stable rest points of equation (1.1). Note that the stable rest points of equation (1.1) correspond to those values of U for which F assumes a local maximum. If (U,U') is a solution of equations (1.3) and (1.4) we shall sometimes say that "U connects $A \to B$". We shall also sometimes call a traveling wave solution a "connection". In this paper we develop a technique to determine, for a given function F and two stable rest points A and B, how many different waves connect $A \to B$. We shall see that there may exist zero, a finite number, or a countably infinite numbers of such waves.

The case when $F(u)$ has exactly two local maxima has been considered by a number of authors (see [1] for references). Some work on the multistable case is in the paper of Fife and McLeod [3]. If F has just two local maxima then $F'(u)$ has the familiar cubic shape. In this case there exists a unique wave with positive speed which connects the stable rest points.

It follows from (1.3) that if $U(z)$ is a traveling wave solution, then

$$\frac{d}{dz}\{\tfrac{1}{2} V^2 + F(u)\} = \theta V^2. \tag{1.5}$$

We assume throughout that the speed, θ, is positive. An immediate consequence of this assumption and (1.5) is that if U connects $A \to B$, then $F(A) < F(B)$. Note that if $u(x,t)$ is a traveling wave solution moving to the left with positive speed θ, then $u(-x,t)$ is a traveling wave solution moving to the right with negative speed, $-\theta$.

The (U,V) phase plane is the natural place to study the solutions of system (1.3). In the phase plane, stable rest points of equation (1.1) correspond to saddles, while traveling wave solutions correspond to trajectories which "connect" the saddles. One can only expect saddle-saddle connections to exist for special values of the speed θ. The difficulty with phase plane analysis is that the phase planes become much too complicated for a general function F. For example, even when F has only three local maxima there may exist an infinite number of traveling wave solutions.

To illustrate the approach we take, let F be as shown in Figure 1.

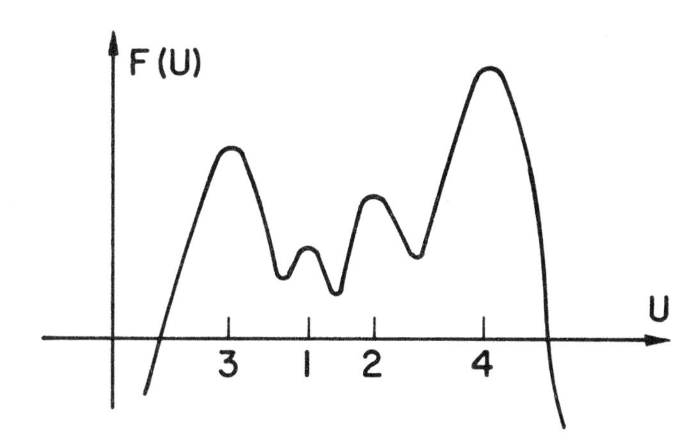

Figure 1. Notice that the local maxima of F are ordered according to the height determined by F.

We suppose that there are four values of U, say A, B, C, and D, where F assumes a local maximum. Assume that $A < B < C < D$ and $F(D) > F(A) > F(C) > F(B)$. Notice that in Figure 1 we ordered the four critical points according to the height determined by F. That is, we set $A = 3$, $B = 1$, $C = 2$, and $D = 4$. Unless stated otherwise we shall always order the stable rest points, or local maxima of F, in this manner. Our description of how many traveling waves exist shall be in terms of this ordering. Two functions which satisfy the conditions (1.2) are said to be in the same equivalence class if they have the same ordering of their local maxima. Hence, given a positive integer n, each permutation of the integers $1,2,\ldots,n$ determines a unique equivalence class of functions.

For a given function F there exists a speed θ_0 such that no saddle-saddle connections exist for $\theta > \theta_0$. It is not very hard to determine what the phase plane must look like for $\theta > \theta_0$. If F is as shown in Figure 1 then the phase plane for θ sufficiently large looks, qualitatively, like what is illustrated in Figure 2A. As θ approaches $+\infty$ the unstable manifolds (trajectories which approach the saddles in backwards time) become more and more vertical while the stable manifolds (trajectories which approach the saddles in forward time) become more horizontal. In Figure 2B we show the phase plane for $\theta = 0$.

The basic approach is to begin at $\theta = \theta_0$ and then start decreasing θ. By comparing the phase plane for $\theta = \theta_0$ with that for $\theta = 0$ one is able to determine all the possibilities for the fastest wave. For the function F

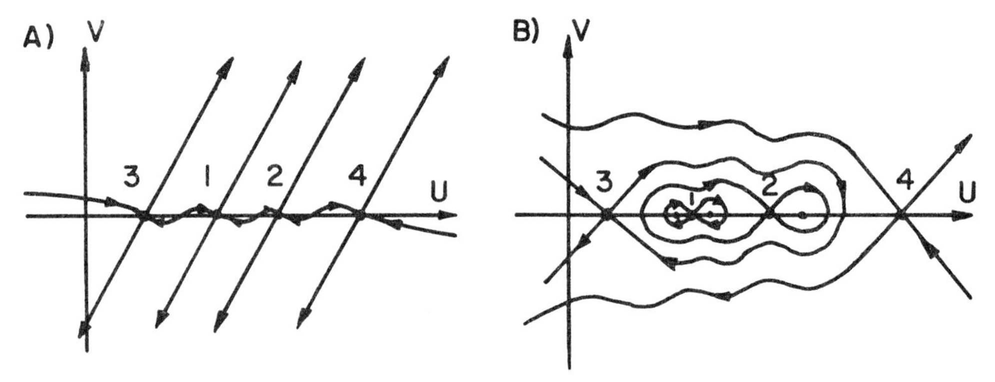

Figure 2. Phase planes for System 1.2 where F is as shown in Figure 1. In (A) the speed, θ, is very large, and in (B), $\theta = 0$.

shown in Figure 1 these possibilites are the 1 T 3, 1 T 2, and 2 T 4 connections. Of course, which one of these is the fastest wave depends on the specific function F. Let us suppose that for a particular F the fastest wave connects 1 T 3. Then the qualitative features of the phase planes change after the 1 T 3 connection (see Figure 3). We can then decrease r further and determine, by comparing the new phase plane with the phase plane at $r = 0$, all the possibilites for the next fastest connection. These possibilities are the 2 T 3, 1 T 2, and 2 T 4 connections. A similar analysis can be done if the fastest wave was either the 1 T 2 or the 2 T 4 connection. In this manner we can construct a directed graph as shown in Figure 3. This graph illustrates all the possible orderings of which connections can take place. To each equivalence class of functions there corresponds such a directed graph. To each specific function there corresponds a path in the directed graph determined by its equivalence class. This path first shows the fastest wave, then the second fastest wave, etc... Our immediate goal is to be able to construct, and understand, these graphs.

One approach to constructing these graphs would be to draw a lot of phase planes. This would be very tedious, if not impossible, since the phase planes become very complicated. What is needed is a way to quantify the essential information contained in each phase plane. That is, we need a way to translate each phase plane into an array of numbers. Then, by just looking at the array of numbers, we can hopefully determine what the possibilities are for the next fastest wave. After a connection takes place the qualitative, or topological, features of the phase planes change. This implies that the array of numbers also changes. Hence, we need an algorithm which tells us how to change the array of numbers after a particular connection has occurred.

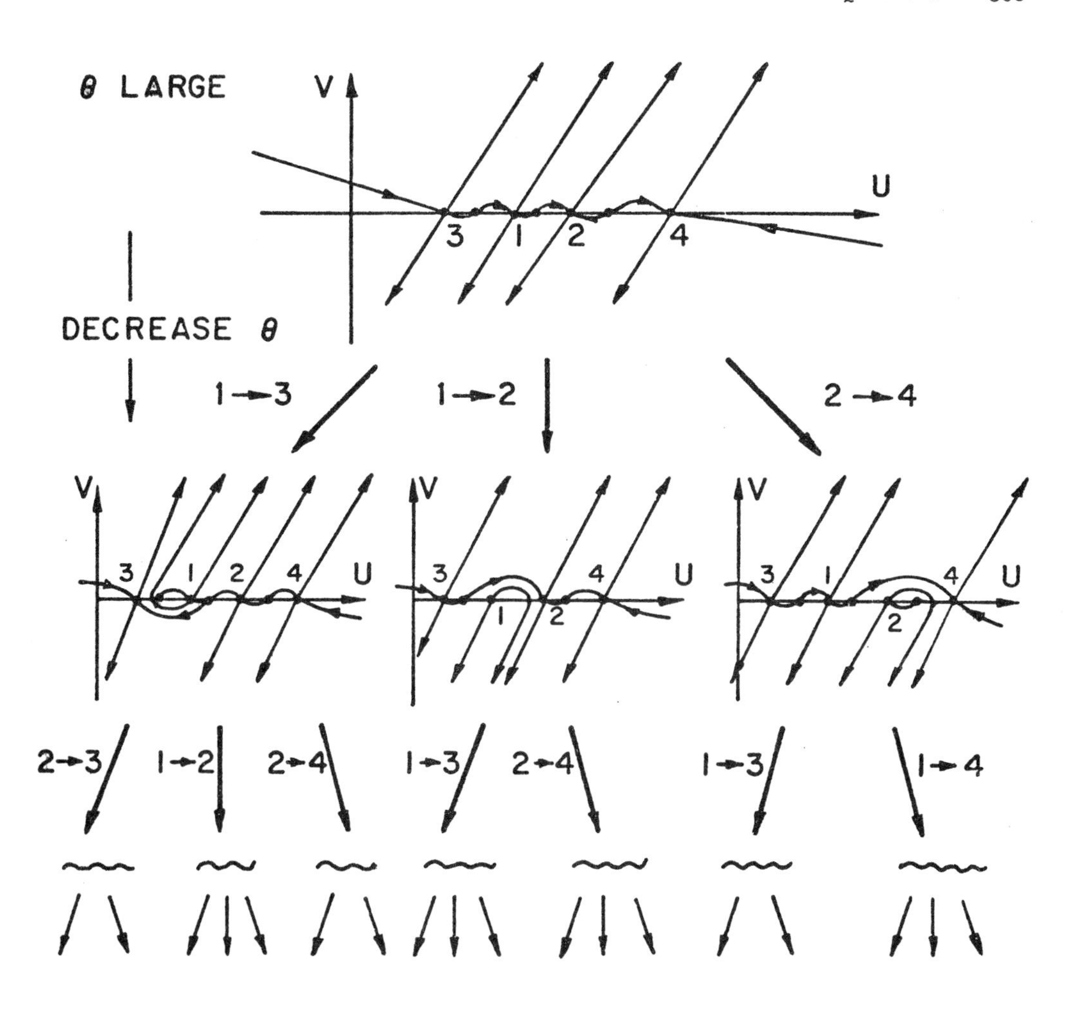

Figure 3. The directed graph for the equivalence class of functions shown in Figure 1.

In section 2 we describe how to

(a) assign an array of numbers to each phase plane,

(1.6) (b) determine what the possibilities are for the next fastest wave by just looking at the array,

(c) change the array after a connection has taken place.

With these three operations we will be able to construct all of the directed graphs. In section 3 we present some applications of the technique just described. In section 4 we present further applications and give a complete description of the tristable equation. A complete proof of all the results presented in this paper may be found in [4].

Acknowledgement: The author would like to thank Professor Charles Conley for many enlightening discussions of the problem.

2. DIRECTED GRAPHS. We illustrate how to do the three operations described in (1.6) with an example. Suppose we are given a function F which is in the equivalance class of functions illustrated in Figure 1. Furthermore, suppose we know that at $\theta = \theta_0$ the phase plane is as shown in Figure 4. We wish to determine what the possibilities are for the next fastest wave. Only the unstable manifolds are shown in Figure 4 because, in order to determine what the possibilities are for next fastest wave, that is all we need to know.

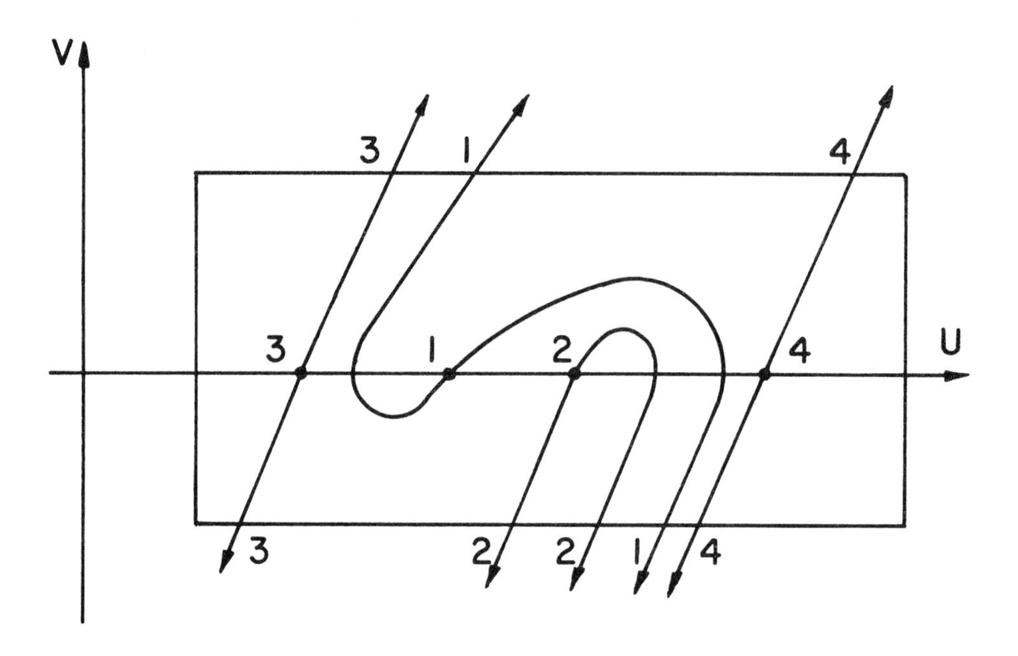

Figure 4. The array of numbers associated with this phase plane is $\frac{314}{32214}$.

The first step in assigning an array of numbers to the phase plane is to draw a big rectangle, R, around the rest points. Since the set of bounded trajectories is compact (see Conley [2]), R can be chosen so that all the connections, for all values of θ, lie inside of it.

The next step is to locate those points on the boundary of R which lie on one of the unstable manifolds. By analyzing system (1.3) one finds that R can be chosen so that these points either lie on the top side or the bottom side of R. To each one of these points there corresponds a number;

that number being the saddle on whose unstable manifold the point lies. We now have two lists of numbers; one corresponding to the top side of R, and the other to the bottom side. For the example shown in Figure 4 we have the two lists $\{3,1,4\}$ and $\{3,2,2,1,4\}$. Combining the two lists we obtain:

$$\frac{314}{32214}\,. \tag{2.1}$$

This is the desired array of numbers!

We claim that the array shown in (2.1) determines all the possibilities for the next fastest traveling wave solution. We explain how these possibilities are realized with the following proposition.

PROPOSITION 1: Suppose that for a given function F we have, at some speed $\theta = \theta_0$, the ordering $\frac{T_1\ T_2\ \cdots\ T_m}{B_1\ B_2\ \cdots\ B_n}$. If, for some k, $T_k < T_{k+1}$ then there exists a connection $T_k \to T_{k+1}$ for some $\theta \in (0,\theta_0)$. If, for some k, $B_k > B_{k+1}$ then there exists a connection $B_{k+1} \to B_k$ for some $\theta \in (0,\theta_0)$. These give all the possibilities for the next fastest wave. That is, if the next fastest wave corresponds to an $A \to B$ connection then there exists an integer k such that either $A = T_k$ and $B = T_{k+1}$, or $A = B_{k+1}$ and $B = B_k$.

Let us apply this proposition to the phase plane shown in Figure 4. Looking at the array (2.1) and applying Proposition 1 we find that the possibilities for the next fastest wave are the $1 \to 4$, $2 \to 3$, and $1 \to 2$ connections. Note that the $1 \to 3$ connection may or may not exist for some $\theta \in (0,\theta_0)$, but it cannot be the next fastest connection. Furthermore, the $1 \to 4$, $2 \to 3$, and $1 \to 2$ connections must all exist for some speeds less than θ_0.

We now need an algorithm which tells us how the array changes after a connection has taken place. This algorithm is described in the following proposition. For this proposition we assume that the array is known for some value of the speed, say $\theta = \theta_0$. We also assume that the next fastest connection is an $A \to B$ connection. Note that there must be two B's in the array since to each saddle there corresponds two unstable directions. Of course, there are two A's also, but the 'other A' will play no role. In the proposition we consider two cases depending on whether the 'other B' is on the top or the bottom of the array.

PROPOSITION 2. If the other B is on the top then after the $A \to B$ connection everything in the array remains exactly the same except the A is moved to the immediate right of the other B. If the other B is on the bottom then after the $A \to B$ connection everything in the array remains exactly the same except the A is moved to the immediate left of the other B.

Here are two examples of what may happen:

(a) $$\frac{..AB..B..}{....} \xrightarrow{A\to B} \frac{..B..BA..}{....}$$

(b) $$\frac{..AB...}{..B..} \xrightarrow{A\to B} \frac{..B...}{..AB..}$$

Applying this proposition to the phase plane shown in Figure 4 we have the following portion of the graph:

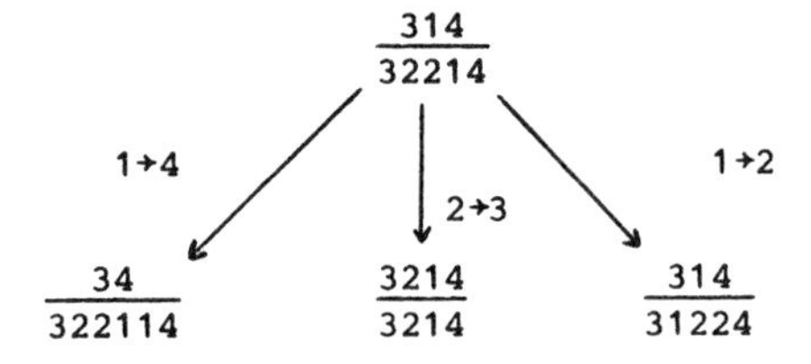

In order to complete the description of how to construct the directed graph it is necessary to explain how one starts the graph when the speed is very large. It is not hard to show that if the ordering of the saddles is $A_1, A_2, \ldots, A_n$ then the graph begins with the array.

$$\frac{A_1\ A_2\ \cdots\ A_n}{A_1\ A_2\ \cdots\ A_n} .$$

This is because when θ is very large, then in the phase plane the unstable trajectories are nearly verticle. For example, if F is in the equivalence class of functions shown in Figure 1 then the graph starts with the array $\frac{3124}{3124}$. The rest of the directed graph is shown in Figure 5.

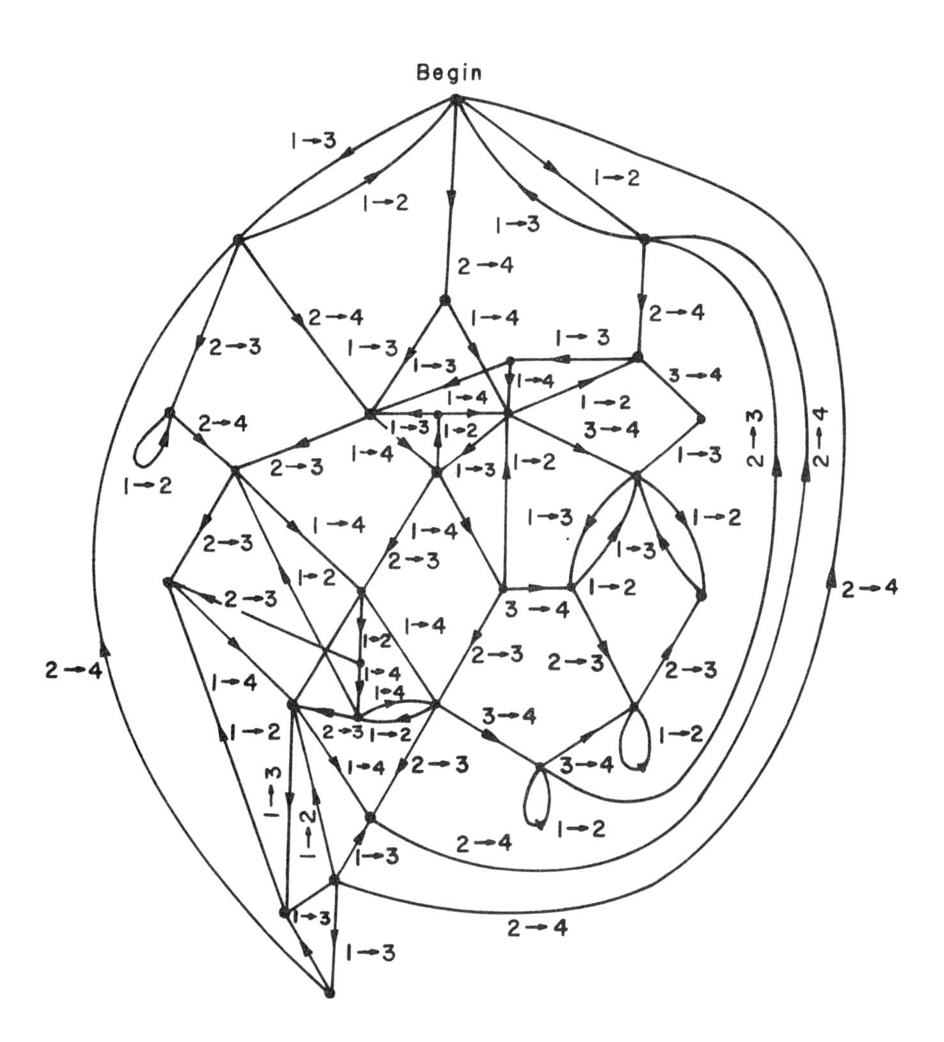

Figure 5. The complete directed graph associated with the equivalence class of functions shown in Figure 1.

3. APPLICATIONS

Suppose that F is in the equivalence class of functions illustrated by Figure 6.

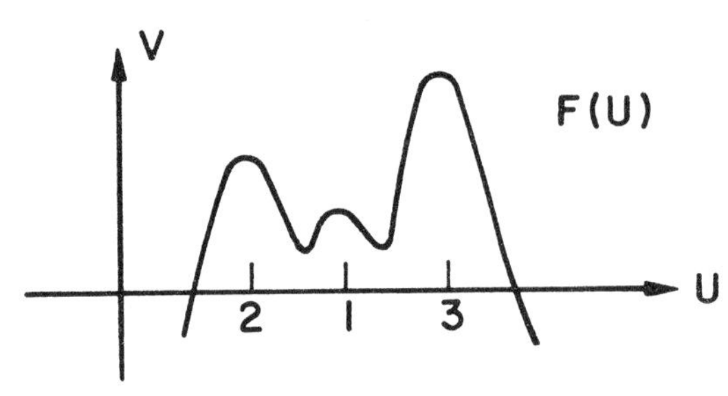

Figure 6. For the equivalence class of functions shown here there exists an infinite number of $1 \to 2$ connections, a finite number of $1 \to 3$ connections and precisely one, $2 \to 3$ connection.

It follows from a simple shooting argument that there exists a unique wave connecting $2 \to 3$. We use this fact and the ideas described in the previous section to demonstrate that there exists a finite number of waves connecting $1 \to 3$ and an infinite number of waves connecting $1 \to 2$.

To prove these results we consider the graph associated with this equivalence class of functions. The graph begins when θ is very large with the array $\frac{213}{213}$. Then, using Propositions 1 and 2 we find that the desired graph can be represented as simply:

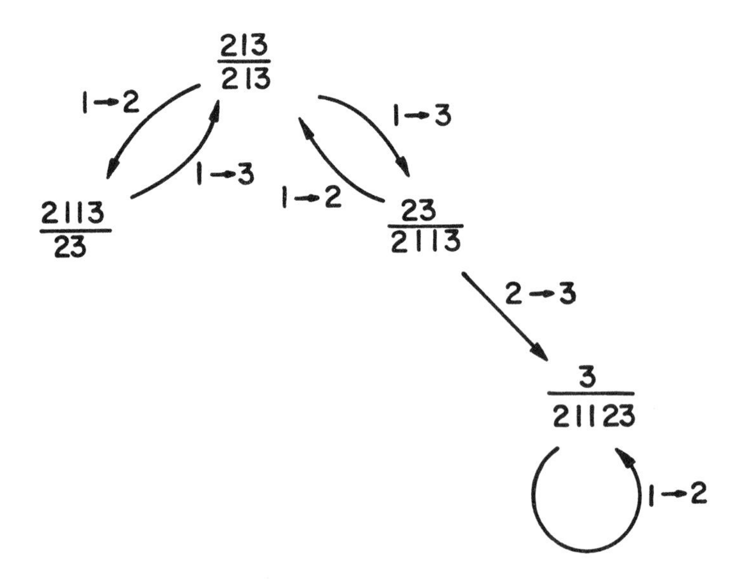

Figure 7. The directed graph associated with the equivalence class of functions shown in Figure 6.

Each function in the equivalence class shown in Figure 6 determines a path in the above directed graph. This path represents the order, starting with the fastest, in which the various connections take place. Until now all we know about this path is that it must obey the arrows in the directed graph, and eventually a $2 \to 3$ connection must take place. However, there is only one $2 \to 3$ connection in the graph. After the $2 \to 3$ connection we have the array $\frac{3}{21123}$. Proposition 1 implies that the next connection has to be a $1 \to 2$ connection. This, however, leaves the array unchanged, and is, therefore, represented by the loop in Figure 7. The phase plane, of course, changes after the $1 \to 2$ connection, but not the array. Therefore, after the $2 \to 3$ connection we are forced to go around this loop an infinite number of times, proving that there must exist an infinite number of $1 \to 2$ connections.

A separate argument shows that there can be at most a finite number of $1 \to 3$ connections with speeds greater than the speed of the $1 \to 2$ connection. There cannot be any $1 \to 3$ connections with slower speeds since then we're caught up in the $1 \to 2$ loop. The directed graph shows, however, that there must be at least one $1 \to 3$ connection, so this completes the proof.

Our second application is a generalization of the first one. Here it is assumed that $\bar{A} = (A,0)$ and $\bar{B} = (B,0)$ are saddles in the (U,V) phase plane with $A < B$ and $F(A) < F(B)$. Furthermore, if $\bar{C} = (C,0)$ is any other saddle such that $A < C < B$, then $F(C) < F(A)$. Let $\bar{D}_1, \bar{D}_2, \ldots, \bar{D}_n$ be the saddles which satisfy $A < D_k < B$ for each k. Here $\bar{D}_k \equiv (D_k,0)$. We assume that $F(D_1) > F(D_2) > \ldots > F(D_n)$. Then,

THEOREM 1: a) There exists an infinite number of waves connecting $D_1 \to A$. b) There exists an infinite number of waves connecting $D_2 \to A$.

It is natural to ask how many connections there must be from $D_k \to A$ for $k \geq 3$. We conjecture there may exist zero, a finite number, or an infinite number of such connections. While we do not know of a rigorous proof of this result, we shall indicate why we believe it is true after we outline the proof of Theorem 1.

Theorem 1a is proved using a shooting argument. The basic idea of the shooting argument was suggested to the author by Professor John Mallet-Paret. To set up the shooting argument we must first introduce some notation. Let P be the critical point of F immediatly to the right of A. Note that the F assumes a local minimum at $U = P$. Let ℓ be the ray, in the phase plane, $U = P$, $V \leq 0$. For a given value of θ, let $A^{\theta}_{SE}(z)$ be the trajectory which satisfies:

(a) $\lim_{z\to+\infty} A^{\theta}_{SE}(z) = \bar{A}$,

(b) $A^{\theta}_{SE}(z)$ 'approaches' $\bar{A}$ from the quadrant $U > A$, $V < 0$.

Let $D^{\theta}_{SW}(z)$ be the trajectory that satisfies

(a) $\lim_{z\to-\infty} D^{\theta}_{SW}(z) = \bar{D}_1$

(b) $D^{\theta}_{SW}(z)$ 'leaves' $\bar{D}_1$ into the quadrant $U < D_1$, $V < 0$.

Note that $A^{0}_{SE}(z)$ must intersect ℓ at some point. We denote the V coordinate of this point by γ_0. Hence, for θ sufficiently small, $A^{\theta}_{SE}(z)$ intersects ℓ at least once. Let $\gamma_A(\theta)$ be the V coordinate of the first place where $A^{\theta}_{SE}(z)$ intersects ℓ as $A^{\theta}_{SE}(z)$ is followed backwards starting at $\bar{A}$. Clearly, $\lim_{\theta\downarrow 0} \gamma_A(\theta) = \gamma_o$.

Let N be a fixed positive integer. If θ is sufficiently small then $D^{\theta}_{SW}(z)$ must intersect ℓ at least N times. Let $\gamma^N_D(\theta)$ be the V coordinate of the point where $D^{\theta}_{SW}(z)$ intersects ℓ for the Nth time. In [4] is proved that there exists positive constants θ_1 and θ_2 such that $\theta_1 < \theta_2$ and:

(3.1)

(a) $\gamma_A(\theta)$ and $\gamma^N_D(\theta)$ are continuous functions of θ for $\theta < \theta_2$,

(b) $\gamma_A(\theta_1) < \gamma^N_D(\theta_1)$,

(c) $\gamma_A(\theta_2) > \gamma^N_D(\theta_2)$.

The reason that (3.1b) is true is because $F(A) > F(D_1)$. To prove (3.1c) one uses the fact that $\lim_{\theta\to+\infty} \gamma_A(\theta) = 0$.

We conclude from (3.1) that there exists some speed, say $\theta = \theta_o$, for which $\gamma_A(\theta_0) = \gamma^N_D(\theta_0)$. This corresponds to a connection, $D_1 \to A$, which winds around the phase plane N times. Since N is arbitrary there must exist an infinite number of $D_1 \to A$ connections.

We indicate why Theorem 1b is true with a specific example. Suppose that F is in the equivalence class of functions shown in Figure 1. Theorem 1a implies that there exists an infinite number of waves which connect

$2 \to 3$. A simple shooting argument shows that there must exist a $3 \to 4$ connection. Using these facts we show that there exists an infinite number of $1 \to 3$ connections.

Consider the graph associated with this equivalence class of functions. It begins when θ is very large with the array $\frac{3124}{3124}$. Using Proposition 1 and the fact that there exists a $3 \to 4$ connection, we conclude that eventually a '3' must be next to the '4' on the top side of the array. For example, we may have the sequence of connections:

$$\frac{3124}{3124} \xrightarrow{2\to 4} \frac{314}{31224} \xrightarrow{1\to 4} \frac{34}{312214} \xrightarrow{3\to 4} \frac{4}{3122134}. \tag{3.2}$$

Of course, other sequences of connections are possible, but we do know that, since there exists a $3 \to 4$ connection, the array must eventually be of the form $\frac{4}{3....34}$. The four dots represent some permutation of the numbers 1, 1, 2, and 2.

We now wish to use the fact that there exists an infinite number of $2 \to 3$ connections. To illustrate what has to happen suppose we start where we left off in (3.2) with the array $\frac{4}{3122134}$. The only way a $2 \to 3$ connection can take place is if a two is eventually next to the left-handed three on bottom side of the array. This cannot happen immediately because of the one separating the two and the left handed three. Therefore, there must eventually be a $1 \to 3$ connection. For example, we may have the sequence of connections:

$$\frac{4}{3122134} \xrightarrow{1\to 3} \frac{4}{3221134} \xrightarrow{2\to 3} \frac{4}{3211234} \xrightarrow{2\to 3} \frac{4}{3112234}.$$

It is not hard to show that, because there exists an infinite number of $2 \to 3$ connections, the array must equal $\frac{4}{3112234}$ an infinite number of times. But we know that there always has to be another $2 \to 3$ connection. Hence, a two has to be next to the left-handed three again, and there must be another $1 \to 3$ connection. Repeating this argument we conclude that there must exist an infinite number of $1 \to 3$ connections.

Recall that after the statement of Theorem 1 we conjectured that there exist functions for which there do not exist any waves which connect $D_3 \to A$. To understand why we believe this to be true consider the equivalence class of functions illustrated in Figure 8.

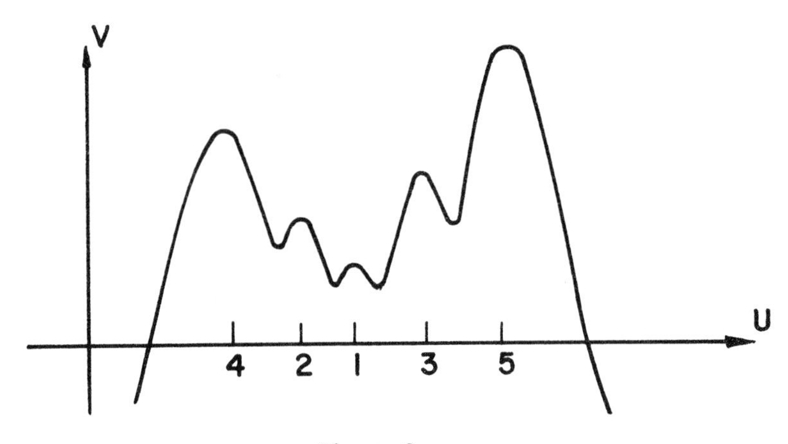

Figure 8

We claim that there does not have to exist any $1 \to 4$ connection. In (3.3) we show a path in the directed graph which does not contain any $1 \to 4$ connection. It does, however, contain an infinite number of $3 \to 4$, $2 \to 4$, $2 \to 3$, $1 \to 3$, and $1 \to 2$ connections which we know, from the previous results, must exist.

$$
\begin{aligned}
&\frac{42135}{42135} \xrightarrow{1\to 3} \frac{4235}{421135} \xrightarrow{\substack{3\to 5\\ 2\to 5\\ 4\to 5}} \frac{5}{421133245}^{\,*} \xrightarrow{\substack{1\to 2\\ 1\to 2}} \frac{5}{423311245}\\
&\xrightarrow{\substack{2\to 4\\ 3\to 4}} \frac{5}{431122345} \xrightarrow{\substack{1\to 3\\ 1\to 3}} \frac{5}{432211345} \xrightarrow{\substack{2\to 3\\ 2\to 3}} \frac{5}{431122345} \xrightarrow{\substack{1\to 3\\ 1\to 3}} \frac{5}{432211345}\\
&\xrightarrow{3\to 4} \frac{5}{422113345} \xrightarrow{2\to 4} \frac{5}{421133245}^{\,*} \longrightarrow \cdots\cdot
\end{aligned}
\tag{3.3}
$$

Note that the array $\frac{5}{421133245}$ appears twice in (3.3). We have labelled these two arrays with the symbol '*'. Hence, we can just keep repeating the connections between the two *'s to obtain the desired path.

This, of course, is not a rigorous proof that a $1 \to 4$ connection may not exist, since there may not exist a specific function $F(U)$ which realizes the path shown in (3.3).

It is also possible (in fact, easier) to construct paths for which there does exist an infinite number of $1 \to 4$ connections.

4. FURTHER APPLICATIONS AND THE TRISTABLE EQUATION IN DETAIL.

In this section we treat, in detail, the case when equation (1.1) has three stable rest states. We assume, throughout, that F assumes a local maximum precisely when U is equal to either A, B, or C, where $A < B < C$. In the previous section we showed that if $F(B) < F(A) < F(C)$, then there exists an infinite number of waves which connect $B \to A$. If $F(B) < F(C) < F(A)$ then, by symmetry, there exists an infinite number of waves which connect $B \to C$. There are essentually two other cases to consider. These are:

$$\text{(3.1)} \qquad \begin{array}{ll} \text{(a)} & F(A) < F(C) < F(B) \\ \text{(b)} & F(A) < F(B) < F(C) \end{array}$$

If (3.1a) is satisfied then there exists a unique wave connecting $A \to B$ and another connecting $C \to B$. There are no waves connecting $A \to C$ or $C \to A$. These facts are proved by considering the directed graph corresponding to the equivalence class of functions satisfying (3.1a). The directed graph is shown in Figure 9A. In Figure 9A we used the required ordering $A = 1$, $B = 3$, and $C = 2$.

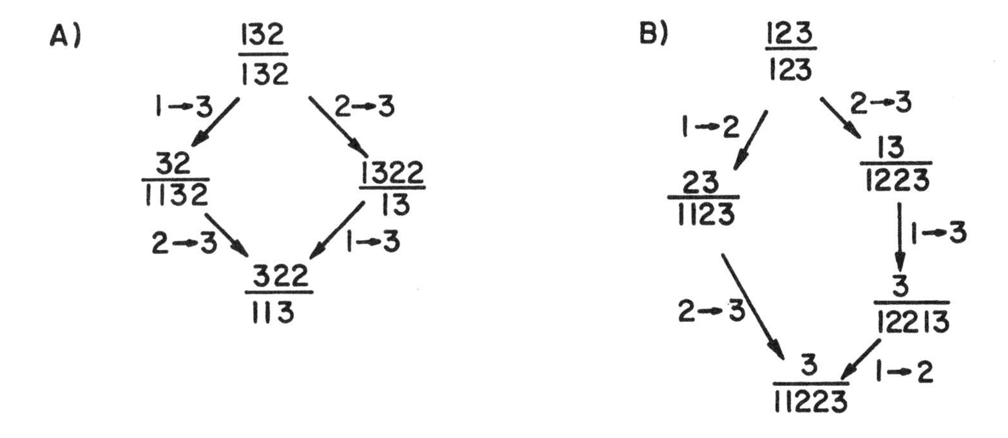

Figure 9. Directed graphs for functions with three local maxima. If these critical points are at U equal to A, B, and C, then in (A), $F(A) < F(C) < F(B)$. In (B), $F(A) < F(B) < F(C)$.

If (3.1b) is satisfied then there exists a unique wave connecting $A \to B$ and a unique wave connecting $B \to C$. There is a unique wave connecting $A \to C$ if and only if the wave connecting $A \to B$ is slower than the wave connecting $B \to C$. The directed graph for this equivalence class of functions is shown in Figure 9B. In Figure 9B we set $A = 1$, $B = 2$, and $C = 3$. This result is also proved in the paper of Fife and Mcleod [3].

Throughout the remainder of this section we assume that $F(B) < F(C)$. We wish to think of the number $F(A)$ as a bifurcation parameter. We have shown that if $F(A) > F(C)$ then there exists an infinite number of waves which connect $B \to C$, if $F(A) \in (F(B), F(C))$ then there exists an infinite number of waves which connect $B \to A$, and if $F(A) < F(B)$ then there exists only a finite number of waves. These waves are all illustrated in Figure 10. Figure 10 gives a qualitative description of which waves exist for a given value of the speed, θ, and another parameter, λ, which is related to the value of $F(A)$. Let us denote the function $F(U)$ at a particular value of λ by $F_\lambda(U)$. As λ is increased, $F_\lambda(A)$ decreases. If $\lambda < \lambda_1$ then $F_\lambda(A) > F_\lambda(C)$, if $\lambda \in (\lambda_1, \lambda_2)$ then $F_\lambda(A) \in \left(F_\lambda(B), F_\lambda(C)\right)$, and if $\lambda > \lambda_2$ then $F_\lambda(A) < F_\lambda(B)$. In Figure 10, the $B \to A$ connections are represented by solid curves, the $B \to C$ connections by dashed curves, the $C \to A$ connections by the solid curve with small circles, the $A \to C$ connections by the solid curve with small triangles, and the $A \to B$ connections by the solid curves with small squares. Note that there should be an infinite number of dashed curves, or $A \to B$ connections, and an infinite number of solid curves, or $B \to A$ connections. These two sets of curves are nested about the axis $\theta = 0$.

Figure 10 was drawn by considering how various phase planes change as the parameter $F(A)$ changes. Of course, the precise quantitative features of each curve in Figure 10 depends on exactly how the functions $F_\lambda(U)$ are chosen to vary with λ. Figure 10 does, however, illustrate the qualitative relationships described below between the various curves.

For each set of curves ($B \to A$ connections, $B \to C$ connections, etc...), the top, or fastest, curve corresponds to monotone traveling waves. These waves are asymptotically stable with respect to the partial differential equation (1.1). This was proved by Fife and McLeod [3]. The other curves in Figure 10 correspond to nonmonotone waves. For the dashed and solid curves ($B \to C$ and $B \to A$ connections) the nth curve from the top, or the nth fastest wave, corresponds to a connection which winds around $n-1$ times in the phase plane.

The most interesting points in Figure 10 are when (λ, θ) is equal to $(\lambda_1, 0)$ and $(\lambda_2, 0)$. When $\lambda = \lambda_1$ then $F(A) = F(C)$, and there exists waves with zero speed which connect $A \to C$ and $C \to A$. Note that when $\lambda = \lambda_1$ there exists an infinite number of $B \to A$ and $B \to C$ connections. When $\lambda = \lambda_2$, $F(A) = F(B)$, and there exist waves with zero speed which

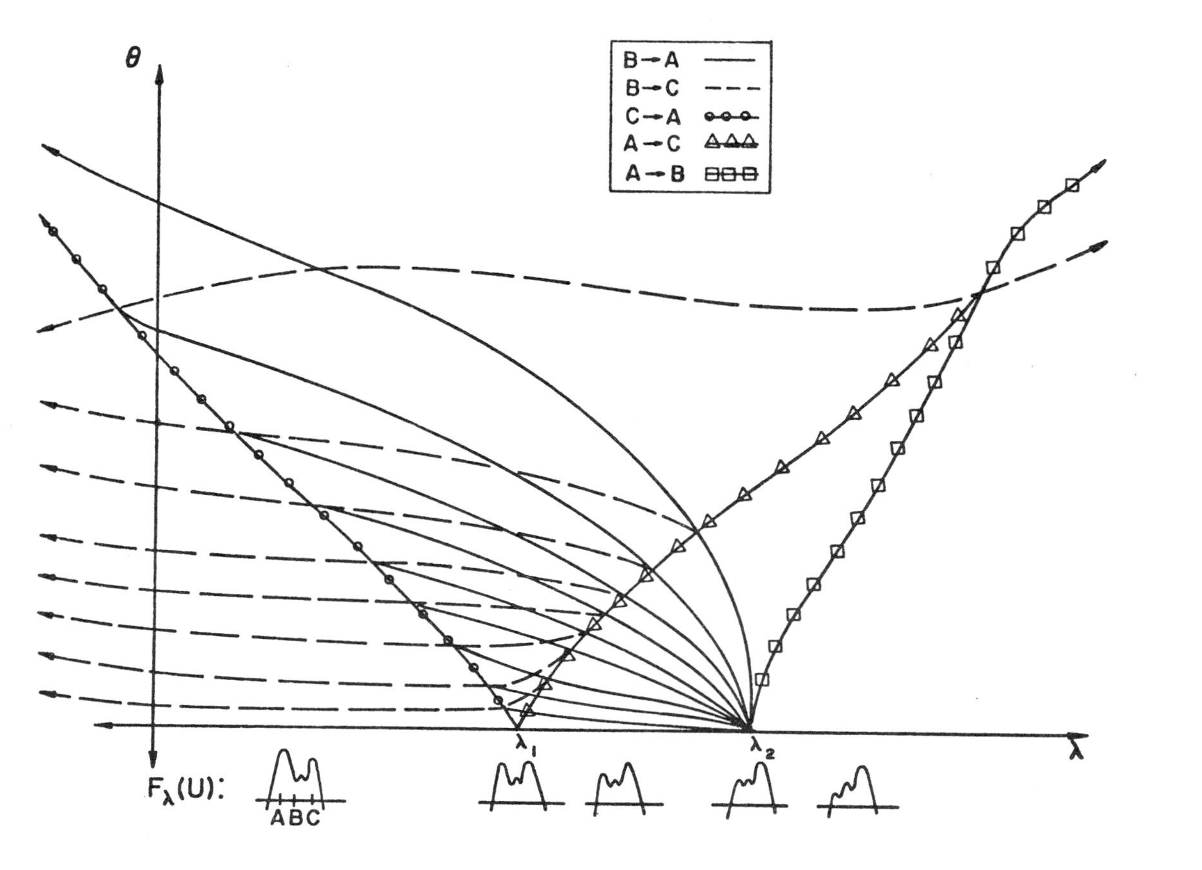

Figure 10. Global description of the tristable equation. It is assumed that for each value of λ, $F_\lambda(U)$ has a local maximum at $U = A, B,$ and C. For $\lambda < \lambda_1$, $F(A) > F(C) > F(B)$. For $\lambda \in (\lambda_1, \lambda_2)$, $F(C) > F(A) > F(B)$, and for $\lambda > \lambda_2$, $F(C) > F(B) > F(A)$. There should be an infinite number of dashed curves ($B \to C$ connections) and solid curves ($B \to A$ connections). These curves converge to the axis $\theta = 0$.

connect $A \to B$ and $B \to A$. As λ is decreased from $\lambda = \lambda_2$, so that $F(B) < F(A)$, the infinite branches of $B \to A$ connections bifurcate from the point $(\lambda, \theta) = (\lambda_2, 0)$.

REFERENCES

[1] Aronson, D. G. and Weinberger, H. F., Nonlinear diffusion in population genetics, combustion and nerve propagation, in Proceedings of the Tulane Program in Partial Differential Equations and Related Topics, Lecture Notes in Mathematics 446, Springer, Berlin, 1975, 5-49.

[2] Conley C., Isolated invariant sets and the generalized Morse index, C.B.M.S. Regional Conference Series in Mathematics, No. 38, American Mathematical Society, Providence, Rhode Island.

[3] Fife, P. C. and McLeod, J. B., The approach of solutions of nonlinear diffusion equations to traveling front solutions, Arch. Rat. Mech. Anal. 65, 335-361: Bull. Amer. Math. Soc. 81 (1975), 1075-1078.

[4] Terman, D. Traveling wave solutions of a multistable reation-diffusion equation II, to be issued as a MRC Technical Summary Report, University of Wisconsin-Madison.

Mathematics Research Center
University of Wisconsin-Madison
Madison, WI 53706

Contemporary Mathematics
Volume 17, 1983

CREATION AND BREAKING OF SELF-DUALITY SYMMETRY -A MODERN ASPECT OF CALCULUS OF VARIATIONS

Mel S. Berger[1]

ABSTRACT. Self-duality symmetry for certain nonlinear generalizations of Maxwell's equations arising from the calculus of variations is created by imposing special boundary conditions on the competing functions. We apply these ideas to both Ginzberg-Landau and Yang-Mills equations. Self-duality symmetry is broken in these cases by slightly altering the action functional in accord with the physical problems involved. In the case of Ginzberg-Landau equations an interesting nonlinear eigenvalue problem is studied. For the Yang-Mills equations a novel type of axi-symmetrical free boundary problem is obtained.

The material we discuss below comprises a novel method of studying global problems of solutions to nonlinear elliptic systems. The systems we shall discuss arise in modern theoretical physics by extending Maxwell's equations of electromagnetism to a nonlinear context. In general, we shall study absolute minima of an associated nonquadratic action functional that we shall arrive at by canonical means. The key idea in self-duality will be the utilization of boundary conditions at infinity for the action functional to generate topological invariants for the variational problem. However in studying the breaking of self-duality, the need to keep close to the physical problems of science will lead to deeper mathematical insights and this is my point in the subsequent development.

We intend to develop new directions for the study of nonlinear partial differential equations inspired from Maxwell's theory and thus only indirectly connected with continuum mechanics. In fact, great insights in modern theoretical physics and contempory differential geometry have resulted from the directions we shall discuss. The question that we pose in this article is the follow-

---

[1]Supported by the NSF.

0271-4132/83/0000-1380/$05.00

ing:

[Q]: What are the implications of these new advances for nonlinear elliptic partial differential equations?

SECTION 1 INTRODUCTION

Our article is organized as follows: in section 2 we consider the simplest example of the "creation" of self-duality symmetry and show that it leads to a great simplification in the solution of a nonlinear boundary value problem on the real line R^1. In section 3, we consider the second example of self-duality symmetry, namely we start with a two dimensional analogue of the example in section 2. We then invoke the gauge principle of H. Weyl to derive a variational formulation of the Ginzberg-Landau Theory of Superconductivity. After that, we show how Taubes [1] used the recent self-duality computations of Bogomolnyi [2] to find the complete resolution of the associated nonlinear elliptic Euler-Lagrange equations in the special case that the Ginzberg-Landau parameter $\lambda = 1$. The mathematical implication of these results leads to an extension of the notion of "natural constraints" that arose in my earlier work [3], [4] and [5]. The breaking of self-duality in the case of the Ginzberg-Landau functional is important in theoretical physics and occurs by simply changing the Ginzberg-Landau parameter from the specially assigned value $\lambda = 1$. The resulting problem then becomes a nonlinear eigenvalue problem and we shall discuss it briefly below. An interesting development in this connection is the special global linearization that occurs as the Ginzberg-Landau parameter $\lambda \to \infty$. In this case, all the topological invariants become focused within delta functions known as "vortex filaments." In fact, much of the theory of Type II Superconductivity can be understood on this basis. Mathematically this idea is formulated in terms of our ideas on "nonlinear desingularization" as we discussed in earlier connections with ideal fluids and vortices [6] and [7].

Our third example that comprises section 4 of the creation of self-duality is in some sense a general one and is connected with Maxwell's equations themselves written in terms of differential forms. We briefly consider the pure Maxwell's equations with no sources or currents in this case and relate this formulation to the calculus of variations via an action principle using arguments of Atiyah, Hitchin, Singer and Schwartz. A main step here is to note that the Maxwell's equations are independent of metric and can be formulated in terms of the covariant derivative of the potential A. The key mathematical aspect which we address is breaking of self-duality in this case. Here we are motivated by the notion of the "bag model" in contemporary high-energy physics. The mathematical questions involved are the following :

1. Can a free boundary problem result from a minimization process associated with a variational integral without a priori assumptions on the existance of a bounded confined domain for the solution?
2. What is the simplest nonlinear extension of Maxwell's equations that yields a strict free boundary problem consistant with the needs of specific physical problems?

In these cases it is of great value to compare our work with vortex motion of an ideal fluid. At one time the associated notion of vortices was totally neglected but is currently viewed as a basis of interesting fluid motions. The notion of vortices in superconductivity was used by Abrikosov in fundamental studies of 1957 based on the Ginzberg-Landau equations. In fact, this is how I personally became involved with attempting to extend this notion to more general physical contexts. The notion of confinement that I discuss in section 4 arose from an attempt to extend the notion of free boundary problems to high-energy physics based on recent work of the theoretical physicist S. Adler. Adler's ideas contain, I believe, new mathematical structures of great importance for nonlinear partial differential equations.

SECTION 2 A SIMPLE EXAMPLE EXHIBITING CREATION OF SELF-DUALITY SYMMETRY

We begin by considering a simple one-dimensional example of a nonlinear boundary value problem on $(+\infty, -\infty)$ that exhibits self-duality symmetry. Consider the following problem for real parameter $\lambda > 0$,

$$\frac{d^2u}{dx^2} + 2\lambda(u - u^3) = 0 \qquad x \in (+\infty, -\infty) \tag{1}$$

$$u(\infty) = 1 \qquad u(-\infty) = -1$$

This boundary value problem can be rewritten as the Euler-Lagrange equation of the following functional [2],

$$I(u) = \int_{-\infty}^{\infty} [\tfrac{1}{2}\left(\frac{du}{dx}\right)^2 + \frac{\lambda}{2}(u^2 - 1)^2] \tag{2}$$

The boundary value problem (1) is a standard one in nonlinear theory but is generally considered on a bounded domain. The modern physical aspect of the problem requires us to study the problem on the whole real line $(+\infty, -\infty)$. The boundary conditions at $+\infty$ and $-\infty$ have been especially chosen to create self-duality symmetry by which we mean the functional $I(u)$ can be written as the sum of squares of first order differential operators together with the divergence of a nonlinear expression. This divergence term then gives rise to a constant of integration or a conserved quantity utilizing the given boundary conditions.

In this particular case, the argument proceeds as follows

$$I(u) = \int_{-\infty}^{\infty} \tfrac{1}{2}[\frac{du}{dx} + \sqrt{\lambda}(u^2 - 1)]^2 - \int_{-\infty}^{\infty} \sqrt{\lambda}\frac{du}{dx}(u^2 - 1) \tag{3}$$

where the second integral becomes

$$= \sqrt{\lambda}\frac{d}{dx}\, u(\frac{u^2}{3} - 1)$$

$$= \sqrt{\lambda}u(\frac{u^2}{3} - 1) \Big|_{-\infty}^{\infty} = -\frac{4}{3}\sqrt{\lambda}$$

$$\text{thus } I(u) \geq \frac{4}{3}\sqrt{\lambda}$$

Now comes the key step in utilizing self-duality symmetry. Generally to determine the absolute minima of a given functional (2), one has to solve a second order nonlinear boundary value problem but self-duality enables us to simplify this expression since (3) shows $\inf_H I(u)$ over a reasonable Sobolev space H is attained by solving the following first order nonlinear boundary value problem

$$\frac{du}{dx} + \sqrt{\lambda}(u^2 - 1) = 0 \quad \text{with } u(\infty) = 1 \text{ and } u(-\infty) = -1 \tag{4}$$

The explicit solution of (4) is

$$u(x) = \frac{1 + k\exp(-2\sqrt{\lambda}x)}{1 - k\exp(-2\sqrt{\lambda}x)} \quad \text{for } k > 0 \tag{5}$$

However this solution has a singularity for finite x and in fact the determination of smooth minima seems to require systems more elaborate that (1). Another point to observe here is that this self-duality exists for all positive values of the nonlinear eigenvalue λ . To break the self-duality in this case is very easy and can be obtained without changing the boundary condition but by simply altering the cubic term u^3 in (1) to u^5 .

SECTION 3 A KEY EXAMPLE FOR CREATION AND BREAKING OF SELF-DUALITY SYMMETRY ON R^2

3(i) The Creation of Self-Duality. The second example of self-duality symmetry that we discuss illustrates a canonical procedure to determine gauge invariant systems of nonlinear partial differential equations that generalize Maxwell's equations to a nonlinear context. There are four steps in this procedure. First, we write a simple scalar equation extending equation (1) of section 2 to a higher dimension except that we allow the unknown Ψ to be complex valued. The equation we obtain is

$$\Delta\Psi + \lambda\Psi(1 - |\Psi|^2) = 0 \tag{6}$$

We then write down a functional (7) whose Euler-Lagrange equations yield (6),

$$I_0(\Psi) = \tfrac{1}{2}\int_{R^2}[|\nabla\Psi|^2 + \tfrac{\lambda}{4}(|\Psi|^2 - 1)^2] \tag{7}$$

which is analogous to the functional (2) of section 2.

The second step in this procedure, the so-called Gauge Principle of H. Weyl dates back to 1929 and consists in extending the functional (7) to be compatible with Maxwell's equations. This is achieved as follows : first we add a kinetic energy term to the functional $I_0(\Psi)$ expressed in terms of the electromagnetic potential A. In terms of differential forms, this expression can be written $||dA||^2_{L_2}$. The next point in Weyl's procedure is to change the Dirichlet integral of Ψ into the L_2norm of the covariant derivative of Ψ with respect to the potential A. In symbols the covariant derivative is

$$\nabla_A = (\partial_j - iA_j)\Psi \qquad (j = 1, 2) \tag{8}$$

The geometric reason for doing this is that in this case the new function I(Ψ, A) will be invariant under the gauge transformation

$$(\Psi, A) \to (\Psi e^{i\phi}, A + \nabla\phi) \tag{9}$$

This invariance is crucial for the determination of the more subtle boundary conditions at infinity in this case. The final requirement for the gauge principle of Weyl is that the nonlinear term involving λ in the functional is left unchanged. Thus the resulting functional $I(\Psi, A)$ can be written

$$I(\Psi, A) = \tfrac{1}{2}\int[|\nabla\Psi|^2 + \tfrac{1}{2}|dA|^2 + \tfrac{\lambda}{4}(|\Psi|^2 - 1)^2] \quad (10)$$

Step three in this procedure is to impose boundary conditions at infinity that allow us to invoke self-duality symmetry. This is attained very easily from the functional (10) by merely requiring that each term in the variational integral be in $L_2(R^2)$. In particular, this imposes restrictions on the behavior of the integrands of each of the three terms at infinity. Thus

$$\text{(i)}\ |\Psi|^2 \to 1 \qquad \text{(ii)}\ |\nabla_A\Psi| \to 0 \qquad \text{(iii)}\ |F| = |dA| \to 0$$

These conditions imply the following: first that

$$\Psi \to e^{i\eta(\theta)} \quad \text{with } \eta(\theta + 2\pi) = \eta(\theta) + 2\pi N \quad (11)$$

where N will be a (topologically invariant) integer, from (ii)

$$A \to -i\partial\ln\Psi \quad (12)$$

These last two restrictions imply from the principle of the argument that by taking a line integral around a large contour $C_R = \{|z| = R\}$ where R is very large that

$$N = (2\pi i)^{-1}\oint_{C_R} \Psi'/\Psi = (2\pi i)^{-1}\oint_{C_R} d\ln\Psi \quad (13)$$

thus as $R \to \infty$ via equation (12),

$$N = (2\pi)^{-1}\int_{R^2}\text{curl}A \quad \text{(by Green's theorem)} \quad (14)$$

Continuous variations of Ψ and A subject to the L_2 restrictions stated above do not change N. Thus N is a topological invariant and can in fact be related to the homotopy group $\pi_1(S^1)$, cf.[8].

This topological invariant was first termed "flux quantization" by Abrikosov in 1957 in the pioneering research article on Type II Superconductivity. In fact, the functional (10) had been used by the great Russian physicists Ginzberg and Landau in developing the theory of superconductivity from the point of view of nonlinear partial differential equations. However it was Abrikosov who first discovered that when λ increased beyond 1 (a normalized reference number) that new physical phenomena occur due to topological invariant N. We will return to this aspect below.

In 1976 Bogomolnyi observed that the Ginzberg-Landau functional (10) for fixed $\lambda=1$ exhibits self-duality symmetry defined in section 2. To achieve this he rewrote (10) for arbitrary λ as follows [denote equation below (*)]

$$I_\lambda = \tfrac{1}{2}\int_{R^2}[J_1^2(\Psi,A) + J_2^2(\Psi,A) + J_3^2(\Psi,A)] + \tfrac{1}{2}\int_{R^2}\mathrm{curl}A + \frac{(\lambda-1)}{8}\int_{R^2}(|\Psi|^2-1)^2$$

in terms of the invariant N. Here the different operators J_1, J_2, J_3 can be written as follows for $N > 0$, with $\Psi = \phi_1 + i\phi_2$

$$J_1 = (\partial_1\phi_1 + A_1\phi_2) - (\partial_2\phi_2 - A_2\phi_1) \tag{15}$$
$$J_2 = (\partial_2\phi_1 + A_2\phi_2) + (\partial_1\phi_2 - A_1\phi_1)$$
$$J_3 = \mathrm{curl}A + \tfrac{1}{2}(\phi_1^2 + \phi_2^2 - 1)$$

So to minimize I_λ with N fixed and $\lambda=1$, we set $J_1 = J_2 = J_3 = 0$. The resulting first order system of nonlinear partial differential equations is of great interest since (15) forms a nonlinear Fredholm operator of index 2N. Moreover, the system can be completely analyzed by using complex notation and the calculus of variations. In fact, in 1980 Taubes [1] proved the following two key results: (a) the smooth solution (A, Ψ) of "vortex number" N are parametrized by R^{2N} (the points in the complex plane where Ψ vanishes (with multiplicity included)) and (b) every smooth finite action critical point of the Ginzberg-Landau functional with $\lambda = 1$ is a self-dual solution. In order to see how these results were obtained we introduce complex notation as follows

$$A = A_1 + iA_2 \qquad \Psi = \phi_1 + i\phi_2 \tag{16}$$
$$\partial = \tfrac{1}{2}(\partial_1 - i\partial_2) \qquad \bar{\partial} = \tfrac{1}{2}(\partial_1 + i\partial_2)$$
$$\Psi = e^f \quad \text{with } f = f_1 + if_2$$

In this notation the equations $J_1 = J_2 = 0$ are equivalent to

$$2\bar{\partial}\Psi = iA\Psi \tag{17}$$

This equation can be integrated to yield the following result

$$A = -2i\log\Psi \tag{18}$$

Moreover, in this notation $\mathrm{curl}A = -\Delta f_1$ (19) and thus the equation $J_3 = 0$ can be written as the boundary value problem

$$\Delta f_1 - \tfrac{1}{2}(e^{2f_1} - 1) = 0 \quad \text{with } f_1 \to 0 \text{ at } \infty \tag{20}$$

An interesting fact about this system is that there are four unknowns namely the real and imaginary parts of A and Ψ and only three equations relating them. This is due to the gauge invariance of the system. This means that the action functional (*) is invariant under the gauge transformations (9) thus is A and Ψ are a solution of the self-dual equations so are $(A+\nabla\phi)$ and $\Psi e^{i\phi}$ for arbitrary smooth ϕ. In addition to this, equation (20) can be extended to measure the number of times that Ψ vanishes in the complex plane. This amended equation results in the appearence of delta functions on the right hand side of (20) to yield

$$\Delta f_1 - \tfrac{1}{2}(e^{2f_1} - 1) = -4\pi \sum_{i=1}^{N} \delta(x - a_i) \tag{21}$$

In fact the solution of (20) has topological invariant 0 whereas the solution of (21) has topological invariant N.

The result of all of this is that solutions of the self-dual equations can be completely analyzed and in fact when $\lambda=1$ these self-dual solutions yield a complete solution for the smooth critical points of the Ginzberg-Landau functional (10).

Now let us try to derive the implications of this result for general problems of the calculus of variations. To do this, let us recall the variational characterization of eigenvalues and eigenfunctions of the Laplacian Δ. These quantities can be characterized in two ways: first as abstract minimax critical points of appropriate quadratic functionals or as strict minima over various explicit sets defined successively by orthogonality. For nonlinear problems, the first characterization can be carried over via Morse Theory or the Ljusternik Schnirrelmann Theory. However the second characterization has not been well-developed up to now. In previous work, I discussed the notion of "natural constraints" for saddle points of nonquadratic functionals. By this I mean, a set of level surfaces A_c, $c \in C$ (an index set) and an appropriate Banach space with two properties (a) every critical point (admissible) of funtional $F(u)$ lies on some A_c, $c \in C$ and (b) <u>every</u> critical point of $F(u)$ can be represented as an absolute minimum restricted to some A_c, viz. $\min_{A_c} G(u)$. (22)
Now the problem of "natural constraints" for the present case can be phrased as follows

Can all critical points of $I_\lambda(\Psi, A)$ be obtained as $\min\limits_{H_N} \tilde{I}(\Psi, A)$ for some integer N ? (23)

The exciting results discussed above show that in the self-dual case $\lambda=1$ the problem of natural constraints has a solution $\tilde{I} = I_1$ and H_N defined by (14).

<u>3(ii) The Breaking of Self-Duality.</u> The self-duality of the Ginzberg-Landau functional (10) is easily destroyed by simply considering values of the parameter $\lambda \neq 1$. Note self-duality is not broken in the example of section 2 by this change. The breaking of self-duality in this case forces us to consider what results of

the self-dual arguments stated above can be carried over to this more general context. This is especially important for concrete physical problems since the entire field of Type II Superconductivity currently of great importance in high technology, involves $\lambda>1$ where the stable solutions have topological index N=1.

First a few remarks about this more general system. Once the gauge of the resulting system is fixed we can write the system in the form $\operatorname{div} A_0 = 0$, $\Psi = |\Psi| e^{i\phi}$ and $A = A_0 + \nabla\phi$

$$(i\nabla + A_0)^2 |\Psi| = \lambda(|\Psi|^2 - |\Psi|^3) \qquad (24)$$
$$\Delta A_0 = |\Psi|^2 A_0$$

Indeed, we work with the gauge invariant quantities $|\Psi|$ and H=curlA. Here for the case N=1 we seek solutions where $|\Psi|\to 1$ at $x\to\infty$ and at $x=0$, $|\Psi|=0$. Moreover we require that H=curlA vanishes at infinity and is finite at the origin. This is simply a nonlinear eigenvalue problem of the type that has been studied for the past twenty years. The case $\lambda=1$ has just been treated and thus affords a global starting point to work for values of λ different from 1. What is interesting in this case is that as $\lambda\to\infty$ the system (24) is known physically to have a global linearization defined by

$$H = dA \text{ (with vortex number N)} \qquad (25)$$
$$\Delta H - H = -2\pi \sum_{i=1}^{N} \delta(x - x_i) \quad \text{where } H\to 0 \text{ at } \infty$$

so that in this case, H=curlA is the sum of the Green's function with singularities at $x=x_i$. The mathematical problem is what occurs precisely when λ varies between 1 and infinity. The process of finite smooth nonlinear equations degenerating to a sum of delta functions as a parameter goes to infinity while the resulting smooth solutions develop a singularity, I call "Nonlinear Desingularization." Indeed, the effect of the nonlinearity is to smooth out the singularity of the associated Green's function.

In order to study this phenomena, (*) is crucial since it holds for all values of λ. In the simplest nontrivial case N=1, I was able to show:

Theorem (i) For each finite $\lambda\geq 1$, the functional $I_\lambda(\Psi,A)$ has a smooth rotationally symmetric vortex solution $(A_\lambda, \Psi_\lambda)$ with N=1, that is an absolute minimum of (*) among rotationally symmetric smooth finite energy fields $(\Psi_\lambda, A_\lambda)$ with fixed vortex number N. Moreover, as $\lambda\to 1$, these vortex solutions tend to the N=1 vortex solution of Taubes [1].

Theorem (ii) As $\lambda\to\infty$ for the solutions stated in (i) above, the N=1 vortex solutions $h_\lambda = \text{curl}A_\lambda$ tend in the Sobolev space $W_{1,p}(R^2)$ $(1 < p < 2)$ to the Green's function solution of (25).

I am currently studying what happens for N>1 and attempting to prove the existance of a one-parameter family of solutions with topological invariance N that tends as $\lambda\to\infty$ to the Green's function of (25).

Another important aspect of this problem is the confinement problem I mentioned earlier. The problem is to discuss in what sense h=curlA is confined to a bounded domain for topological invariant $N\neq 0$. This is of special interest since the topological invariant equation

$$(2\pi)^{-1}\int_{R^2}\text{curl}A = N \tag{26}$$

indicates that the magnetic field h is restricted in an interesting way. In this case for N=1, it can be shown that confinement occurs on a bounded spherically symmetric set apart from an exponentially decaying remainder. In the next section, we shall show that confinement can occur strictly for a more general action functional.

SECTION 4 CREATION AND BREAKING OF SELF-DUALITY IN YANG-MILLS THEORY IN FOUR DIMENSIONS

4(i) Creation of Self-Duality. Currently a nonlinear generalization of Maxwell's equations in four dimensions embodies the self-duality principle. This was first discovered by the Russian mathematical physicists Polyakov and Schwartz et al. and was carried through by Atiyah, Hitchin and Singer [9]. They observed that if Maxwell's equations are made elliptic by appropriate analytic continuation then the self-duality principle that we have described in the first two examples can be elegantly carried over to the Yang-Mills equations. The argument requires four steps which we now summarize briefly. First the linear Maxwell's equations without sources are written in terms of differential forms with the appropriate ellipticity. Then the resulting linear elliptic Maxwell equations are written in terms of a variational formulation via and appropriate action functional. The resulting fields in the variational principle are then written as the derivative of an appropriate potential A. In this formulation the gauge invariance of the Maxwell equations is particularly clear. The second step in exhibiting the self-duality is to modify the "Gauge Principle" that we mentioned previously. To achieve this we choose a nonabelian Lie group G. This group will act as the symmetry group for the problems we shall discuss and the nonabelian nature of the group will be the main ingredient in the nonlinearity of the resulting partial differential equations. The Gauge Principle comes into play when we rewrite the problems in terms of calculus of variations. Then, in place of the field F and its linear relation to the potential, a nonlinear equation between F and A results :

$$F(A) = dA + [A, A] \tag{27}$$

The resulting variational integral is then extended by writing the associated field in terms of the covariant derivative of the potential with respect to the group G. Differential geometrically, this relates to the notion of curvature with the connection A, [9].

The resulting calculus of variations problem is formulated in

$$I_G(A) = \int_{R^4} |F(A)|^2 \tag{28}$$

and has the effect of an Euler-Lagrange equation for extrema that involves cubic nonlinearities as in our previous examples. The third step in the creation of self-duality involves imposing boundary conditions on the associated calculus of variations problems. The key idea is to restrict attention to finite energy solutions so that F(A) is in $L_2(R^4)$. This imposes an interesting topological restriction on the potentials A. In particular, the differential geometry of connections on S^4 turns out to be an appropriate method for understanding the topological invariance involved. In fact, Atiyah and Singer observed (after earlier important Russian research) that the Pontrajgin index associated with the connection A was involved. Using conformal invariance they transferred the resulting problem to the 4-sphere S^4. In that case the Pontrajgin index can be expressed by an integral invariant as in the equation below

$$(8\pi^2)^{-1} \int_{R^4} F(A) \wedge F(A) = N \tag{29}$$

The final step in tis self-duality argument is particularly simple and elegant. It involves merely minimizing the action functional over potentials A that satisfy the topological invariantive restriction (29) on S^4.

In particular, for an appropriate Sobolev space H_N of connections A satisfying equation (29),

$$\inf_{H_N} (F, F)_{L_2} \tag{30}$$

can be obtained in an extremely simple way by a Hilbert space argument as shown below, as $(*F, *F) = (F, F)$

$$||F - *F||^2_{L_2} = 2(F, F) - 2(F, *F) \tag{31}$$

since the second term on the right side of (31) is constant for fixed N via (29), the action functional is minimized by setting

$$F(A) = *F(A)$$

We call this the self-duality equation. Here * is the Hodge operator on two forms. This particular approach although elegant has not yielded the solution of the deep physical problems for which it was invented. Indeed the deep problems concerning quark confinement do not seem to be understandable from this so called "pure" Yang-Mills Theory. Thus we are concerned with the possible breaking of self-duality.

4(ii) Breaking of Self-Duality. The physical problem that enables us to motivate the breaking of self-duality is the quark confinement problem known more generally as the "bag model" of high energy physics. We wish to show from a minimization of an action functional that a free boundary problem results as an appropriate description of quark confinement. To this end, we make three changes in the pure Yang-Mills functional. First, we add a source term to the Lagrangian consisting of a delta function of magnitude Q at a point x_1 and another delta function of magnitude -Q at a point x_2. Then we alter the Lagrangian to account for the nonlinearity of the associated nonlinear partial differential equation to the form

$$L(F) = F^2(1 + \varepsilon \log F^2) \tag{32}$$

here $\varepsilon > 0$ is an associated physical parameter associated with a nonabelian Lie group that measures the degree of nonlinearity in the problem. This is especially emphasized because we shall not use the previous formula to discuss F as the curvature associated with the connection A. The third step is to consider a purely static case so that the problem is reduced to three dimensions. We shall use the classical abelian formulation obtained by using the equations

$$F^2 = E^2 + B^2 \qquad E = -\nabla\phi \tag{33}$$
$$\tilde{A} = (\phi, A) \qquad B = \mathrm{curl} A$$

The resulting Euler equations for the minimum of the associated action functional

$$I(\phi, A) = \int [L_B(F^2) - \phi \cdot j] \tag{34}$$

are

$$\begin{aligned} &\mathrm{curl} E = 0 \\ &\mathrm{curl}(\varepsilon B) = 0 \qquad \text{with } \varepsilon = c\log(F^2/K^2) \\ &\mathrm{div} B = 0 \qquad D = \varepsilon E \\ &\nabla(\varepsilon E) = j \end{aligned} \tag{35}$$

From these equations and the associated minimization problem we see that a graph of $L(F^2)$ versus F^2

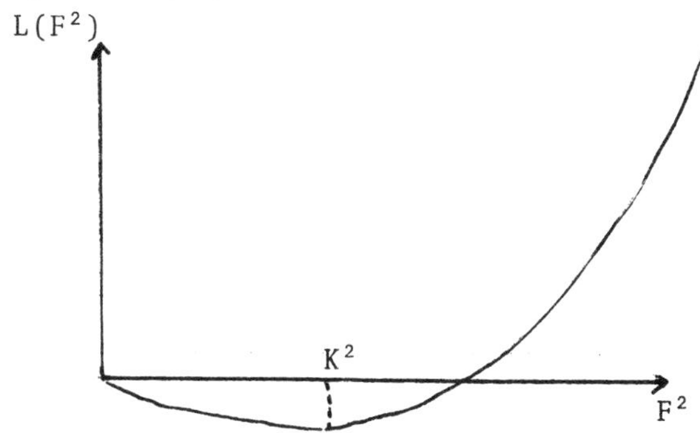

has the following properties. The minimization process automatically provides two regions for the resulting extrema solution. Region (i) is defined by the property $E^2 \leqq K^2$, so we minimize by taking $|B| = (K^2 - E^2)^{\frac{1}{2}}$ and region (ii) is defined by the property $E^2 \geqq K^2$ and minimize by taking $|B| \equiv 0$. Notice in each case, that $E^2 + B^2$ is chosen as close as possible to K^2, where K^2 is the pointwise minimum shown on the graph.

We now derive a nonlinear partial differential equation for the field D. To this end we note that the equations (35) for H imply that

$$\varepsilon(F^2)\underline{B} = 0 \tag{36}$$

Consequently there are two possibilities,

Case I $\quad \underline{B} \equiv 0$ $\qquad\qquad$ Case II $\quad \varepsilon(F^2) = 0$

Near the source charges Case I applies. The relevant nonlinear system is thus

$$\mathrm{div}\, \underset{\sim}{D} = j \tag{37}$$
$$\mathrm{curl}\, \underset{\sim}{E} = 0 \tag{38}$$

with $\underline{D} = \varepsilon(|E|^2)\underline{E}$ and $\varepsilon = \varepsilon\log(|E|^2)$.

To analyze this system we follow the theoretical physicist Stephen Adler [10] and choose cylindrical coordinates in R^3 (r, θ, z) with source charges located symmetrically on the z-axis at $z = \pm a$. We also assume the solution vectors D and E are axi-symmetric, thus reducing the problem to a two dimensional one.

We then treat the system (37) and (38) by noting that off the z-axis, div D = 0. Thus we may reduce the system to a scalar equation for an axi-symmetric "stream" function $\Phi(\rho, z)$ by setting, for the appropriate unit vector $\hat{\theta}$

$$D = \text{curl}\left[\frac{\hat{\theta}}{2\pi\rho}\,\Phi\right] \tag{39}$$

The appropriate boundary conditions for Φ are

$$\begin{aligned} \Phi &= 0 &&\text{for } \rho = 0 \quad |z|>a \\ \Phi &= Q &&\text{for } \rho = 0 \quad |z|<a \\ \Phi &\to 0 &&\text{at } \rho^2 + |z|^2 \to \infty \end{aligned} \tag{40}$$

since these conditions will guarantee that (37) is satisfied.

Moreover following Alder and Piran [11], we rewrite (38) as

$$\text{curl}\left(\frac{D}{\varepsilon(|D|)}\right) = 0 \tag{41}$$

Here we regard ε, the "nonlinear dielectric constant" as a function of $|D|$ since the relation between $|E|$ and $|D|$ is strictly monotone.

Expressing the equation (41) in terms of the stream function Φ, we find after some computation that Φ is determined as a solution to the equation

$$\text{div}[\sigma(\rho, |\nabla\Phi|)\ \text{grad}\,\Phi] = 0 \tag{42}$$

satisfying the boundary conditions (40), where

$$\sigma(\rho, |\nabla\Phi|) = [\rho^2\varepsilon\ (|D|)]^{-1}$$

Now the relationship between the function σ and the variable $|\nabla\Phi|$ is crucial. Indeed for certain σ the equation (42) occurs in subsonic-supersonic potential fluid flow. The present case is quite different. The relationship involved can be graphed as follows

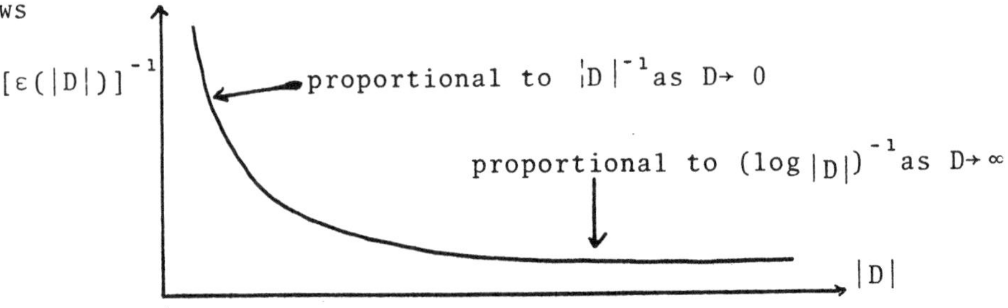

Based on this relationship between $|D|$ and $\varepsilon(|D|)$, it can be shown that this equation (42) has positive semidefinite characteristic form and changes type exactly where $|\nabla\Phi| = 0$. In the linear case, this type of equation was the subject of a recent book by Olenik and Radkevich [12]. Thus the physical problem leads to a new class of nonlinear problems for partial differential equations of this type. The key result to prove is that the vector field D is confined to a connected bounded domain Ω (i.e. $D>0$ in Ω and vanishes outside Ω). Thus the characteristic acts as a free boundary dividing R^3 into two disjoint connected regions. On the bounded domain Ω, physical arguments backed up with computer studies of S. Adler show that the quark-antiquark pair are confined. The rigorous mathematical justification of the fact is under investigation by myself and others. It requires novel ideas in the calculus of variations extending some arguments of [12] to a nonlinear context.

Remark on the Relationship to Other Free Boundary Problems.

This work has an amazing analogy to the mathematical formualtion of global vortex rings in an ideal fluid. In that case, a stream function Φ and axial symmetry are used in the context to isoperimetric variational problems to solve an associated free boundary problem. This free boundary problem is formulated a priori in terms of the type of solution desired. However in this case, the variational problem associated with the action functional (34), dictates the appearence of a bounded sub-domain Ω confining the field D, outside Ω the field D vanishes identically.

BIBLIOGRAPHY

[1]. C. Taubes, "Arbitrary N-Vortex Solutions", Comm. Math. Phy., volume 72 pp. 277-292 (1980). "On the Equivalence of First and Second Order Equations", Comm. Math. Phy., vol. 75 (1980).

[2]. E. Bogomolnyi, "The Stability of Classical Solutions", Sov. Jour. Nucl. Phy., vol.24, pp 449-454 (1976).

[3]. M.S. Berger, Nonlinearity and Functional Analysis, Academic Press, New York 1977.

[4]. M.S. Berger and E. Bombieri, "On Poincare's Isoperimetric Problem", Jour. Funct. Analysis,vol 42, pp. 274-298, 1981.

[5]. M.S. Berger and M. Schechter, "On the Solvability of Semilinear Gradient Operator Equations", Advances in Math, 25, pp 97-132, 1977.

[6]. M.S. Berger and L.E. Fraenkel, "Nonlinear Desingularization in Certain Free Boundary Problems", Comm. Math. Phy., vol 77, pp 149-172, 1981.

[7]. M.S. Berger, Mathematical Aspects of Vorticity, Seminaire College de France (Nonlinear P.D.E.) pp 52-75, Pitman, London 1982.

[8]. A. Jaffe and C. Taubes, Vortices and Monopoles, Birkhauser, Boston 1980.

[9]. M. Atiyah, N. Hitchin and I. Singer, Proc. Nat. Acad. of Sci. vol. 74, pp 2662-5, 1977.

[10]. S. Adler, Nonabelian Statics (to appear in Maxwell Sesquicentennial volume), North-Holland Pub.

[11]. S. Adler and T. Piran, Flux Confinement in the Leading Logarithm Model, (to appear in Physics Letters B, 1982).

[12]. O. Olenik and E. Radkevich, Second Order Equations with Nonnegative Characteristic Form, Plenum Press, New York 1973.

DEPARTMENT OF MATHEMATICS
UNIVERSITY OF MASSACHUSETTS
AMHERST, MA. 01003

Contemporary Mathematics
Volume 17, 1983

ON THE BIFURCATION DIAGRAM FOR A PROBLEM IN BUOYANCY INDUCED FLOW

Stuart P. Hastings and Nicholas D. Kazarinoff*

In this paper we study two related boundary-value problems:

$$(1^{\pm}) \qquad \begin{aligned} f' &= \pm W(\phi,R) \\ \phi'' &= -f\phi' \end{aligned} \qquad (' = d/d\eta)$$

$$(2) \qquad \phi(0) = 1, \quad f(0) = 0, \quad \phi(\infty) = 0,$$

where R is a parameter (called the temperature excess ratio) and the function W has certain physically realistic properties and represents buoyancy force. With the choice

$$(3) \qquad W(\phi,R) = |\phi - R|^q - |R|^q \quad (q > 1),$$

these problems have been proposed to describe plane flows in a porous medium adjacent to a vertical isothermal surface when the saturating fluid (for example, pure or saline water) achieves a density maximum at a temperature T_m above the freezing point. In this case f' and $f - \eta f'$ represent vertical and horizontal components of fluid velocity, respectively, ϕ varies directly with the temperature, and $R = (T_m - T_\infty)/(T_0 - T_\infty)$, where T_∞ is the ambient, constant temperature far from the isothermal surface, which is at temperature T_0. Henceforth, we let P^+ and P^- denote the problems (1^+) - (2) and (1^-) - (2), respectively.

Although it appears that two separate problems are being considered, both are derived from the same set of boundary-layer equations for the flow. An appropriate definition of the "local Grashof number" in terms of the integral of the buoyancy force across the layer, that is

$$I_W = \int_0^\infty W(\phi(\eta),R)d\eta,$$

results initially in a buoyancy force term W/I_W. A change of variables then leads to the problems $(1^{\pm})$ - (2).

*The authors acknowledge support from the National Science Foundation under grants NSF-MCS8101891 (S.P.H.) and NSF-MCS8100727 (N.D.K.).

0271-4132/82/0000-0999/$02.00

A derivation of the problems P^+ and P^- and numerical solutions of them were given by Ramilison and Gebhart (Internat. J. Heat and Mass Transfer, 23(1980), 1521-1530), who found that changes in the force balance in $(1^\pm)$ can produce reversals in the flows. A plus sign is used in (1) when the flow described is largely upward $(I_W > 0)$, while (1^-) represents the opposite situation of predominately downward flow. Using (3) with q at a value (about 1.89) obtained from experimental data on the variation of density with temperature for pure water, Ramilison and Gebhart numerically computed solutions of P^- for values of R decreasing to 0.401. They obtained solutions of P^+ for values of R increasing to 0.194. No solutions were found by them for either P^+ or P^- when R was between 0.194 and 0.401.

More recently these results have been considerably extended by Gebhart, Hassard, Hastings, and Kazarinoff in a paper to appear. For some values of R they found several solutions of P^+, and for other values of R they found two solutions of P^-. It is our purpose here to analyze these problems mathematically. We verify most of the numerical results which have been obtained, and in some respects we are able to go beyond what was found on the computer, for example, see Theorems 3 and 4 in Section II below.

The particular buoyancy function W given in (3) is based upon ad hoc curve fitting of experimental data, and lacks theoretical justification, although it is extremely accurate for a reasonable range of temperatures around T_m. The actual buoyancy force, is, for example, surely not symmetric about $\phi = R$, although the functions in (3) are symmetric about $\phi = R$ on $(0,2R)$. Hence we sought and succeeded in establishing existence of multiple solutions of P^+ and P^- for a whole class of buoyancy functions, including those in (3). This class is defined by the following conditions, all of which are satisfied by the functions in (3):

A) $W(\phi,R)$, $\partial W/\partial\phi$, and $\partial W/\partial R$ are continuous in

$$\{(\phi,R) \mid 0 \le \phi \le 1, \quad R \ge 0\}.$$

B) For $R \ge 0$ there exists a continuous, increasing function $a(R)$ with $a(0) = 0$ and $a(\hat{R}) = 1$ for some $\hat{R} > 0$ and such that:

$$W(0,R) \equiv W(a(R),R) \equiv 0 \quad \text{for } R \ge 0, \quad W(\phi,R) < 0 \quad \text{if } 0 < \phi < a(R),$$

and $W(\phi,R) > 0$ if $1 \ge \phi > a(R)$.

C) $[\partial W(\phi,R)/\partial\phi]|_{\phi=0} < 0$ if $R > 0$ and $\partial W(\phi,R)/\partial R < 0$ if $R > 0$ and $0 < \phi < a(R)$.

These assumptions suffice for several of our results. However, to obtain information, among other things, on the number of solutions for $R = 0$ or R small, our analysis requires further conditions; namely,

D) Let $Q(\phi,R) = W(\phi,R)/\phi$. Then Q can be continuously extended to $[0,1]$, $R \geq 0$, and has the following properties:

(1) $\partial Q(\phi,R)/\partial\phi$ exists and is positive for $R > 0$, $0 \leq \phi < a(R)$, and

(2) $\partial Q(0,R)/\partial R$ exists and is negative if $R > 0$.

E) $(\partial W/\partial R)/W$ is increasing in ϕ for $0 \leq \phi < a(R)$, $R > 0$.

Note that for the functions in (3), $\hat{R} = 1/2$ and $a(R) = 2R$.

II. STATEMENT OF RESULTS

With these assumptions it is clear that for each ζ there is a unique solution of (1^+) or (1^-) such that $\phi(0) = 1$, $\phi'(0) = \zeta$, and $f(0) = 0$. Therefore it is natural to describe results on the existence and multiplicity of solutions of P^+ and P^- with the aid of a "bifurcation diagram", in which are plotted the values of (R,ζ) such that the corresponding solution of (1^+) or (1^-) also satisfies (2). Figure (1-a) gives qualitatively the results of numerical computation, while Figure (1-b) contains a "conjectured" diagram, based on (1-a) and the theorems below.

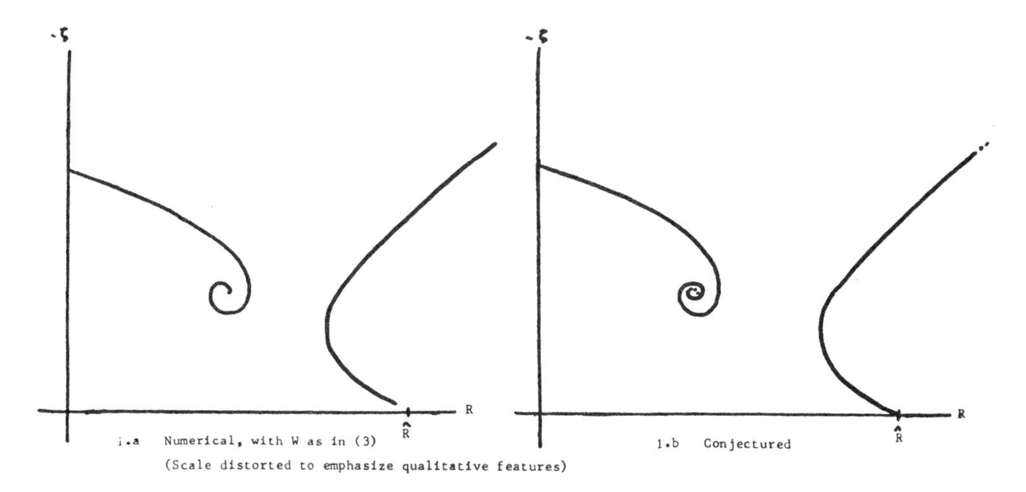

1.a Numerical, with W as in (3)

(Scale distorted to emphasize qualitative features)

1.b Conjectured

Bifurcation diagrams for $(1^{\pm})$-(2) in terms of R and $\zeta = -\phi'(0)$

Figure 1

We conjecture that there is an infinite spiral on the left branch, but have not proved this much. We have shown that there are infinitely many solutions for at least one R, and verified an additional property of the conjectured spiral in (1-b); see Theorem 4 and the subsequent remark.

Other properties of Fig. (1-b) that we have proved include the existence of an interval $R_1 < R < R_2$ in which neither P^+ nor P^- has a solution, and the existence of some interval $\hat{R} - \varepsilon < R < \hat{R}$ in which P^- has exactly two solutions. Several other assertions are also proved.

None of the remarks made so far indicates why the left branch in Figure (1-b) terminates in the center of the spiral, since ordinarily one expects a branch of solutions to continue until a bifurcation point is reached or some variable becomes infinite. (The right branch terminates when $\phi'(0) = 0$, which is not surprising since then $\phi \equiv 1$ and (2) cannot be satisfied.) To explain this it is better to plot $f(\infty)$ against R, with the observation that $f(\infty)$ cannot be negative if $\phi(\infty) = 0$. As a justification for this kind of diagram we state:

THEOREM 1. If conditions (A - D) hold, then for every $\alpha > 0$ and $R \geq 0$ each of the equations (1^+) and (1^-) has a solution such that $\phi(\infty) = 0$, $f(\infty) = \alpha$. Moreover, if $R > 0$, then (1^+) has a solution with $\phi(\infty) = 0$, $f(\infty) = 0$. All of these solutions are unique up to translation of the independent variable.

If (f,ϕ) is a solution of (1^+) or (1^-), then ϕ' does not change sign. The second of equations $(1^\pm)$ can be satisfied only if $\phi' < 0$ on $[0,\infty)$. Since $(1^\pm)$ is autonomous, the boundary conditions can be restated in the following way:

(2') $\quad \phi(\infty) = 0$, and for some η_0, $\phi(\eta_0) = 1$ and $f(\eta_0) = 0$.

The problems $(1^\pm)$ - (2') are translation invariant, and the set of solutions can be described by plotting $\alpha = f(\infty)$ against R. Figure (2) gives the graphs corresponding to Figures (1-a) and (1-b).

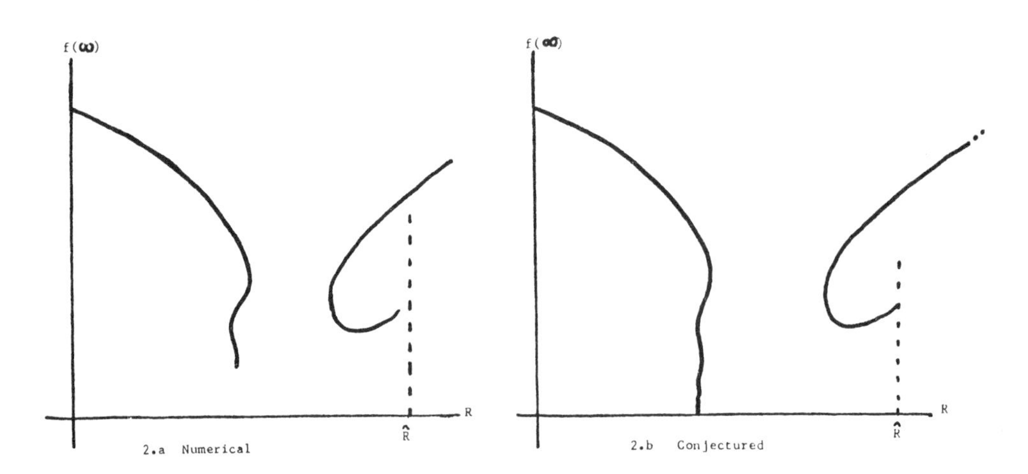

Bifurcation diagrams for $(1^\pm)$-(2) in terms of R and $f(\infty)$.

Figure 2

Conditions A and B imply the following: There is a unique R^*, with $0 < R^* < \hat{R}$, such that

$$\int_0^1 W(\phi, R^*)d\phi = 0,$$

and this integral is positive if $R < R^*$ and negative if $R > R^*$. This leads to our next result.

THEOREM 2. Assume Conditions A and B. Then there exist numbers R_1 and R_2 with $0 < R_1 < R^* < R_2 < \hat{R}$ such that P^+ has no solution if $R > R_1$ and P^- has no solution if $R < R_2$. (In particular, for $R_1 < R < R_2$ neither problem has a solution.

Our next results deal with P^+, so it is assumed that $R \leq R_1$.

THEOREM 3. Suppose conditions (A - E) hold. Then for $R = 0$ and for sufficiently small positive R, P^+ has a unique solution. Let $\Gamma = \{(\alpha, R) \mid$ a solution of (1^+) with $f(\infty) = \alpha$, $\phi(\infty) = 0$ satisfies $(2')\}$. Then Γ is the graph of a smooth function $R = R(\alpha)$, $0 \leq \alpha \leq \alpha_0$, where α_0 is the value of $f(\infty)$ corresponding to the solution of P^+ when $R = 0$.

THEOREM 4. Under assumptions (A - D), let (f_0, ϕ_0) denote the solution of P^+ when $f(\infty) = 0$, $R = R(0)$. If $R = R(0)$, then P^+ has an infinite sequence of distinct solutions $\{(f_j, \phi_j)\}$ such that $f_j(\infty) \to 0$, $\phi_j'(0) - \phi_0'(0) \to 0$, and $\phi_j'(0) - \phi_0'(0)$ alternate in sign.

Remark. This does not exclude the possibility of other solutions when $R = R(0)$. Thus we have not verified the spiral character of the curve seen in Figure (1-b). Our result is consistent with the following bifurcation diagram, for example.

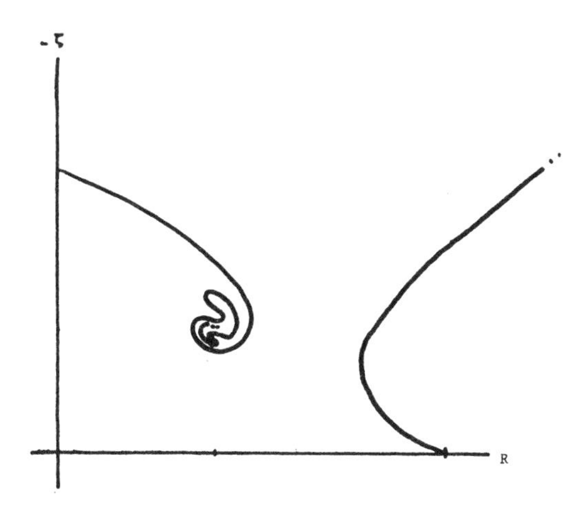

Bifurcation diagram consistent with Theorem 3

Figure 3

In our final results we discuss P^-, with $R \geq R_2$. In particular, we wish to state in what sense P^- has an "even" number of solutions for certain R. This is facilitated by the following preliminary result. For a given R and any $\alpha > 0$, let (f_α, ϕ_α) be a solution of (1^-) such that $\phi(\infty) = 0$, $f(\infty) = \alpha$.

THEOREM 5. Under the assumptions (A - C) and with $0 < R < \hat{R}$, there exists an $\alpha_R > 0$ such that if $0 < \alpha \leq \alpha_R$, then $\phi_\alpha(\eta) \leq a(R) < 1$ for all η such that $(f_\alpha(\eta), \phi_\alpha(\eta))$ is defined, while if $\alpha > \alpha_R$, then $\phi_\alpha(\eta) = 1$ for some η.

We can now specify (f_α, ϕ_α) uniquely for $\alpha > \alpha_R$ by requiring that $\phi_\alpha(0) = 1$. If $\alpha > \alpha_R$ and $f_\alpha(0) = 0$, then (f_α, ϕ_α) solves P^-.

THEOREM 6. Suppose conditions (A - C) are satisfied. Then for $R \geq \hat{R}$, the problem P^- has a unique solution. Also, there is an interval $\hat{R} - \varepsilon < R < \hat{R}$ in which P^- has exactly two solutions.

The essence of our final results is the assertion that generically if $R < \hat{R}$, there exist an even number of solutions of P^-.

THEOREM 7. If, under conditions (A - C), $\alpha_1 - \alpha_R$ is positive and sufficiently small and α_2 is sufficiently large, then $f_{\alpha_1}(0) > 0$ and $f_{\alpha_2}(0) > 0$. Also, $f_\alpha(0)$ is continuous in α for $\alpha > \alpha_R$.

Each zero of $f_\alpha(0)$ corresponds to a solution of P^-. If the degree of a mapping is defined in a standard way, e.g. and if $M(\alpha) = f_\alpha(0)$ for $\alpha > \alpha_R$, then Theorem 7 implies that the number of solutions of P^- is even in the sense of degree.

COROLLARY TO THEOREM 7. $\deg[0;M;[\alpha_1,\alpha_2]]$ is even (in fact, zero) if $\alpha_1 - \alpha_R$ is sufficiently small and α_2 is sufficiently large.

Proofs of these results will appear elsewhere.

DEPARTMENT OF MATHEMATICS
S.U.N.Y. AT BUFFALO
BUFFALO, NEW YORK 14214

Contemporary Mathematics
Volume 17, 1983

FREQUENCY PLATEAUS IN A CHAIN OF WEAKLY COUPLED OSCILLATORS

Nancy Kopell

It is well known that coupled oscillators having nearby frequencies may phase-lock with one another, that is, run at a common frequency and such that phase differences between pairs of oscillators are independent of time. It is also well known, if less understood, that limit cycle oscillators with sufficiently different frequencies may fail to phase lock, and the resulting dynamics may be chaotic. I shall discuss some joint work with G. Bard Ermentrout concerning an intermediate case: a chain of oscillators, uniformly close to one another in dynamics, but with a linear gradient in natural (uncoupled) frequencies. Such chains can display "frequency plateaus," i.e., behavior in which the frequency is constant over portions of the chain. Thus, synchrony is maintained locally, but broken globally.

The investigation was partially motivated by certain phenomena observed in mammalian small intestine, which consists of layers of smooth muscle fiber. The muscle fibers support traveling waves of electrical activity. (These waves modulate waves of muscular contractions via high frequency electrical waves, but the slower electrical activity also exists in the absence of muscular activity.) If a section of the intestine is sliced into pieces of length 1-3 cms., each piece is capable of supporting spontaneous oscillations at a constant frequency, with a wave form that is close to sinusoidal. Furthermore, over a substantial section of the intestine, there is a linear gradient in the frequency of these oscillations. In vivo, the measured electrical activity along the (intact) intestine displays frequency plateaus.

The slow-wave electrical activity has been modelled by several investigators [1-5] as a chain of loosely coupled oscillators, mostly Van der Pols with almost sinusoidal limit cycles. The exact form of the oscillators, the gradient in frequencies, the form and strength of the coupling and the amount of

0271-4132/83/0000-1391/$02.00

inhomogeneity and/or isotropy in the coupling vary among those authors. For a variety of related equations, they produced simulations (digital or electronic) which yield frequency plateaus. Ultimately, the intestinal phenomena should probably be understood in terms of a continuum model; as a first task, we set out to understand the dynamical behavior of a discretized system (as in the above simulations) in a context as free as possible of the (unknown) details of the oscillators and the coupling. Even here, the mathematical questions turn out to be very rich.

We assume that the kth oscillator satisfies

$$(1)_\varepsilon \qquad X_k' = F(X_k) + \varepsilon R_k(X_k,\varepsilon) \equiv F_k(X_k,\varepsilon), \quad k = 1,\dots,n+1$$

where $X_k \in R^m$, F_k and R_k are smooth, $\varepsilon \ll 1$, and $(1)_0$ has a stable limit cycle of period $2\pi/\omega$. The full equations (for isotropic coupling) are

$$(2) \qquad X_k' = F_k(X_k,\varepsilon) + \varepsilon D(X_{k+1}-2X_k+X_{k-1}), \quad X_0 \equiv 0 \equiv X_{n+2}$$

where D is an $m\times m$ matrix.

We first showed that such a system can be reduced to a phase model, i.e., one in which each oscillator is represented by only one dimension, its phase angle. More precisely, we showed that there is an n+1 dimensional attracting invariant submanifold of phase space, parameterized by the phase angles θ_i of the oscillators. To lowest order in ε, this n+1 dimensional system depends on the phase differences $\phi_k \equiv \theta_{k+1} - \theta_k$, and the equations take the form

$$(3) \qquad \frac{d\phi_k}{d\tau} = \Delta_k + H(\phi_{k+1}) + H(-\phi_k) - H(\phi_k) - H(-\phi_{k-1});$$
$$H(-\phi_0) = 0 = H(\phi_{n+1})$$

Here H is a smooth 2π-periodic function depending on F and D, τ is the stretched time, and Δ_k is the (scaled) difference in frequencies between the $(k+1)^{st}$ and k^{th} oscillators.

In [6], we dealt with the case $\Delta_k = -\beta$ (independent of k, corresponding to a decreasing linear gradient in frequency), and H an odd function qualitatively like $\sin\phi$. (For some simple examples, H has the form $H(\phi) = A\sin\phi + B[\cos\phi-1]$, where $A \neq 0$ and B may or may not vanish.) For β small enough, there is a unique phase-locked state, which corresponds to a stable critical point of (3). As β increases to some β_0, all critical points of (3) disappear, and a large amplitude stable limit cycle appears.

For $\beta < \beta_0$, there is an invariant circle which contains the sink and a saddle, i.e., what is sometimes known as a "circle in resonance"; at β_0, these critical points coalesce and the circle in resonance turns into the limit cycle. This limit cycle turns out to correspond to a pair of frequency plateaus with a jump in frequency between them; the homotopy type of the limit cycle as a subset of the phase space of (3) turns out to indicate the position of the jump. To establish the existence of the circle in resonance, we showed the existence of a large invariant region on which a set of inequalities hold, inequalities which are reminiscent of those used by Hirsch [7] in his study of cooperative systems.

The behavior of the solutions in the above case lacks at least one important qualitative feature of the intestinal phenomena (and numerical simulations of them). For a linear gradient in natural frequency, and $H = \sin\phi$ (or any "nearby" odd function) of ϕ), the frequency plateaus are constrained by symmetry considerations to be symmetric with respect to the midpoint of the gradient. In the intestinal data, however, the frequency of each coupled oscillator lies at or above the natural frequency. In current work, we find that this can be accounted for by a phase model, but the oscillators must be more complicated (e.g. H not an odd function). We expect to be able to derive some rigorous results by decomposing H into a sum of its odd and even parts $H_0 + H_e$. If $\Delta_k = -\gamma/n$, then for large n, (3) becomes approximately

$$\phi_t(x) = \frac{1}{n}\{-\gamma+[H_e(\phi)]_x + \frac{1}{n}[H_0(\phi)]_{xx}\}; \quad x = \frac{k}{n}, \quad 0 \leq x \leq 1. \tag{4}$$

One such result is as follows: The time independent states of (4) correspond heuristically to time independent states of (3) for n large. Now for γ not too large and $H_e(\phi) = B\cos\phi$ there is a time independent solution to (4) with the appropriate boundary conditions ($H(-\phi) = 0$ at $x = 0$, $H(\phi) = 0$ at $x = 1$), and the total change in frequency from one end of the chain to the other is γ. This suggests that for $H_e \not\equiv 0$, the chain of oscillators can remain phase-locked with a fixed ($O(\varepsilon)$) total change in frequency as $n \to \infty$. By contrast, it can be explicitly computed that for $H = \sin\phi$, the total change in frequency at the point where phase-locking is lost is $O(\varepsilon/n)$.

BIBLIOGRAPHY

1. Diamont, N.E., Rose, P.K. and Davison, E.J., "Computer simulation of intestinal slow-wave frequency gradient," Amer. J. of Physiology 219, 1684-1690 (1970).

2. Sarna, S.K., Daniel, E.E. and Kingman, Y.J., "Simulation of slow-wave electrical activity of small intestine," Amer. J. of Physiology 221, 166-175 (1971).

3. Robertson-Dunn, B. and Linkens, D.A., "A mathematical model of the slow-wave electrical activity of the human small intestine," Medical and Biological Engineering, 750-757 (Nov. 1974).

4. Brown, B.H., Duthie, H.L., Horn, A.R. and Smallwood, R.H., "A linked oscillator model of electrical activity of human small intestine," Amer. J. Physiology 229, 384-388 (1975).

5. Patton, R.J. and Linkens, D.A., "Hodgkin-Huxley type electronic modelling of gastrointestinal electrical activity," Med. & Biol. Eng. & Computing, 16, 195-202 (1978).

6. Ermentrout, G.B. and Kopell, N., "Frequency plateaus in a chain of weakly coupled oscillators, I," submitted.

7. Hirsch, M., "Systems of differential equations which are competitive or cooperative, I: Limit Sets," SIAM J. Math. Anal. 13, 167-179 (1982).

DEPARTMENT OF MATHEMATICS
NORTHEASTERN UNIVERSITY
BOSTON, MASSACHUSETTS 02115

Contemporary Mathematics
Volume 17, 1983

A SYSTEM OF NONLINEAR PARTIAL DIFFERENTIAL EQUATIONS ARISING IN THE OPTIMAL CONTROL OF STOCHASTIC SYSTEMS WITH SWITCHING COSTS

SUZANNE M. LENHART and STAVROS A. BELBAS

SUMMARY

This note concerns the nonlinear PDE system associated with the problem of optimal switching control of a diffusion process with costly switchings. See [7] for complete proofs of the results stated here.

Let Ω be an open bounded set in $\mathbb{R}^n$ and $A = \{1,2,\ldots,m\}$. For each $a \in A$, let f^a be a function from Ω into $\mathbb{R}$, and L^a an elliptic operator,

$$L^a u = -a^a_{ij} u_{x_i x_j} + b^a_i u_{x_i} + c^a u \quad \text{(summation notation)}.$$

Let $k(a,b)$ be a family of non-negative constants parameterized by $a,b \in A$. We consider the following system:

$$u = (u^1, u^2, \ldots, u^m) : \Omega \to \mathbb{R}^m$$

$$\max\{L^a u^a - f^a,\ u^a - M^a u\} = 0 \quad \text{in } \Omega,\ u^a = 0 \quad \text{on } \partial\Omega \tag{1}$$

where

$$M^a u(x) = \min_{\substack{b \in A \\ b \neq a}} \{u^b(x) + k(a,b)\}$$

The solution of (1) has the stochastic interpretation as the minimal cost function with initial regime a, associated with the control of a diffusion process satisfying

$$dy^a_x(t) = -b^a(y^a_x(t))dt + \sigma^a(y^a_x(t))dw(t)$$
$$y^a_x(0) = x.$$

The functions f^a are the running costs and $k(a,b)$ is the cost of switching from regime a to regime b. See [7] for details of the stochastic background. If $L^a u^a - f^a = 0$, it is optimal to let the system evolve in regime a, otherwise switch to regime b where

$$u^a = u^b + k(a,b).$$

0271-4132/82/0000-1050/$01.50

We make the following assumptions:

$a_{ij}^a,\ b_i^a,\ c^a,\ f^a \in W^{2,\infty}(\Omega)$

$a_{ij}^a(x)\zeta_i\zeta_j \geq \theta|\zeta|^2$ for $\theta > 0$, $\zeta \in \mathbb{R}^n$, $x \in \Omega$

$c^a \geq 0$ on Ω

$\partial\Omega$ is smooth,

and

the switching costs are subadditive, $k(a,b) \leq k(a,c) + k(c,b)$.

A set of regimes $\{a, a_1, a_2, \ldots, a_j, a\}$ constitutes a loop of zero cost if

$$k(a,a_1) = k(a_1,a_2) = \ldots = k(a_j,a) = 0.$$

If we assume there are no loops of zero cost, the solution of (1) is unique in $W^{2,\infty}(\Omega)$. In the presence of loops of zero cost, equation (1) is not sufficient to characterize the minimal cost. See [7] for this case.

The main result is:

THEOREM. Under the above assumptions, there exists a unique solution of (1) such that

$$u^a \in W^{2,\infty}(\Omega).$$

We briefly mention the techniques used in the proof of the theorem. We approximate the solution u^a by solutions u_ε^a of the penalized problem

$$L^a u_\varepsilon^a + \sum_{b;b\neq a} \beta_\varepsilon(u_\varepsilon^a - u_\varepsilon^b - k(a,b)) = f^a$$

where

$$\beta_\varepsilon(t),\ \beta_\varepsilon'(t),\ \beta_\varepsilon''(t) \geq 0$$

$$\beta_\varepsilon(t) \to +\infty \quad \text{for} \quad t > 0, \text{ as } \varepsilon \to 0,$$

and

$$\beta_\varepsilon(t) = 0 \quad \text{for} \quad t < 0.$$

We make pointwise estimates using comparison functions and the maximum principle. The Nirenburg-Kohn device is used for second derivative estimates on the boundary. The techniques are similar to those in P. L. Lions [8], Evans-Friedman [2], Evans [1], Evans-Lions [3], and Lenhart [6].

We show that the solutions u_ε^a converge in $C^1(\Omega)$ to a solution u^a of (1). A uniform bound on the β_ε terms gives

$$u^a \leq M^a u.$$

In the case $k(a,b) > 0$, $W^{2,p}$ regularity results for (1) have been proven by Evans-Friedman [2], Hanouzet-Joly [5], and Garroni-Hanouzet-Joly [4].

REFERENCES

1. Evans, L. C., "A convergence theorem for solutions of nonlinear second order elliptic equations", Indiana U. Math. J., 27 (5) (1978), 875-887.

2. Evans, L. C. and A. Friedman, "Optimal stochastic switching and the Dirichlet problem for the Bellman equation", Trans. AMS, 253, (1979), 365-389.

3. Evans, L. C. and P. L. Lions, "Résolution des équations de Hamilton-Jacobi-Bellman pour des opérateurs uniformément elliptiques", C.R.A.S. Paris, 260, (1980), 1049-1052.

4. Garroni, M. G., B. Hanouzet, and J. L. Joly, "Regularite pour la solution d'un systeme I.Q.V.", Report, Universite de Bordeaux I, Talence, 1978.

5. Hanouzet, B. and J. L. Joly, "Convergence uniforme des iteres definissant la solution d' inequations quasi-variationnelles et applications a la regularite", Report, Universite de Bordeaux I, Talence, 1978.

6. Lenhart, S. M., "Bellman equation for optimal stopping time problem", to appear in Indiana U. Math. Journal.

7. Lenhart, S. M. and S. A. Belbas, "A system of nonlinear partial differential equations arising in the optimal control of stochastic systems with switching costs", to appear in SIAM J. of Applied Math.

8. Lions, P. L., "Resolution de problems de Bellman-Dirichlet, C.R.A.S. Paris, 287, 1978, 474-750; detailed article in Acta Mathematica.

MATHEMATICS DEPARTMENT
UNIVERSITY OF TENNESSEE
KNOXVILLE, TENNESSEE 37996

DEPARTMENT OF ELECTRICAL ENGINEERING
UNIVERSITY OF MARYLAND
COLLEGE PARK, MARYLAND 20742

Contemporary Mathematics
Volume 17, 1983

L^1 STABILITY OF TRAVELLING WAVES WITH APPLICATIONS TO CONVECTIVE POROUS MEDIA FLOW

Stanley Osher and James Ralston

INTRODUCTION

This talk is to be divided into two parts. In Part I, we obtain a stability result, as $t \to \infty$, for fixed points of a general nonlinear semigroup, $T(t)$, acting on the space L^∞. We assume that $T(t)$ is a conservative semigroup of contractions in the L^1 norm, and that $T(t)$ commutes with translations. In addition, we assume that $T(t)$ is eventually strictly contractive on pairs of data u,v such that $u - v$ is of two signs. This hypothesis holds when $T(t)$ is the generator of a strictly monotone difference approximation to a single conservation law in one space dimension. In this case, Theorem A of Part I gives the stability announced by Jennings [6]. Some monotone difference approximations are not strictly monotone. A version of Theorem A, without hypothesis A, was obtained for such a case in [3].

In Part II we apply the ideas from Part I to prove stability of travelling waves for the quasi-linear parabolic equation

$$Lu = u_t + f(u)_x - A(u)_{xx} = 0 \tag{0.1}$$

for $-\infty < x < \infty$, $t > 0$. Here $A'(u) = a(u) \geq 0$. The functions f and A are assumed to be smooth, except perhaps at the zeros of $a(u)$, and the possibility of such zeros means the equation may be degenerate.

The study of this equation is motivated by the case $a(u) = m|u|^{m-1}$, with $m > 1$. This is the convective (if $f(u) \not\equiv 0$) porous media equation. See [8], for a survey of the applications and the known analytic results for the porous media equation. Some stability results for the case $f(u) \equiv 0$ are also mentioned there. For $A(u) = u$ (linear viscosity), and general $f(u)$, one has the stability result of Il'in and Oleinjk [5]. On the other hand, existence, uniqueness, and regularity results for (0.1) were obtained by Vol'pert and Hudjaev, [9], and these results are crucial for our stability theorem.

We do not prove this stability as a direct application of the abstract result of Part I. Instead, we weaken the strictly contractive hypothesis by restricting it to a special case. We then show that this restriction together with the other easily verificable hypotheses of Part I, suffices to prove this result (Theorem B). More precisely, the above hypothesis is replaced by a lemma which uses the fundamental results of Nash, DeGiorgi et. al. on the Hölder continuity of solutions to parabolic equations.

This work was motivated by the study of discrete shock solutions to numerical approximations of nonlinear conservation laws, [6, 10]. Recently, Daniel Michelson [11] obtained a very general existence theorem for such solutions to approximations for systems of conservation laws. Their stability remains to be investigated.

REFERENCES

[1] M. G. Crandall and L. Tartar, "Some relations between nonexpansive and order preserving mappings", Proc. A. M. S. 78 (1980), 385-390.

[2] N. Dunford and J. T. Schwartz, Linear Operators, Part I, General Theory, Interscience, New York, 1958.

[3] B. Engquist and S. Osher, "One sided difference approximations for nonlinear conservation laws", Math. Comp. 36 (1981), 321-351.

[4] A. Friedman, Partial Differential Equations of Parabolic Type, Prentice-Hall, Englewood Cliffs, N. J., 1964.

[5] A. M. Il'in and O. A. Oleinik, "Behavior of the solutions of the Cauchy problem for certain quasilinear equations for unbounded increase of the time", A. M. S. Trans. (2) 42 (1964), 19-23.

[6] G. Jennings, "Discrete Shocks", Comm. Pure Appl. Math. 27 (1974), 25-37.

[7] O. A. Ladyzhenskaia, V. A. Solonikov, and N. N. Ural'tseva, Linear and Quasi-linear Equations of Parabolic Type, A. M. S., Providence, Rhode Island, 1968.

[8] L. A. Peletier, "The porous media equation", Proc. Cong. on Bifurcation Theory; Application of Nonlinear Analysis in the Physical Sciences, Bielefeld, 1979.

[9] A. I. Vol'pert and S. I. Hudjaev, "Cauchy's problem for degenerate second order quasilinear parabolic equations", Math. USSR Sb. 7 (1969), 365-387.

[10] A. Majda and J. Ralston, "Discrete shocks for systems of conservation laws", Comm. Pure Appl. Math. 32 (1979), 445-482.

[11] D. Michelson, "Discrete shocks for difference approximations to systems of conservation laws", Comm. Pure. Appl. Math., to be submitted.

DEPARTMENT OF MATHEMATICS
UNIVERSITY OF CALIFORNIA
LOS ANGELES, CALIFORNIA 90024

Contemporary Mathematics
Volume 17, 1983

A ONE-VARIABLE MAP ANALYSIS OF BURSTING IN THE BELOUSOV-ZHABOTINSKII REACTION

John Rinzel and William C. Troy[1]

ABSTRACT. We consider a three-variable model for the Belousov-Zhabotinskii reaction run in a continuous flow stirred tank reactor (CSTR). We focus on the complex oscillation (burst) patterns for relatively low flow rates. From the chemical model we derive a one-variable discontinuous mapping for the concentration of a critical species from one spike to the next. We consider a reasonable piecewise linear approximation to the map whose solution behavior is described analytically. Various features (e.g. bistability) of the map solution predict burst patterns for the continuous model.

1. INTRODUCTION

The Belousov-Zhabotinskii [1, 17, 18] (BZ) reaction is one of the best characterized chemical systems which exhibits both self-sustained oscillations and chemical excitation as well. The reaction consists of the metal ion oxidation by bromate ion of easily brominated organic materials. In a batch reactor, when the reagent is in the oscillatory state, the oscillations are most apparent in the ratio of the oxidized and reduced forms of the metal ion catalyst. For other recipes, in which the oscillations are suppressed, the system may exhibit excitability. Recent experiments reveal a wide range of possible behavior when the BZ reaction is run in a CSTR (continuously stirred tank reactor). In addition to regular periodic oscillations at some flow rates [2, 4, 7, 10, 11, 16], the observed phenomena include: bursts of oscillations in which one or more nearly identical pulses are separated by more or less regular intervals of quiescence [2, 7, 13], multiple steady states [2, 7], and irregular patterns such as quasi-periodic and chaotic oscillations [5, 10, 11, 16]. Different approaches have been followed to model the CSTR system [6, 8, 11, 12, 14, 15]. Here, we focus our attention on the five variable model of Janz, Vanecek and Field [6]. Their model is based on an irreversible, batch reactor Oregonator

[1]Research supported in part by: NIHRCDA K04 NS00306-05 and NSF Res. Grant No. NCS8002948

0271-4132/82/0000-1164/$03.50

[3] with flow terms, and with instantaneous dependence of the stoichiometric parameter f on the brominated organic substance. Recently, Rinzel and Troy [9] have simplified the five variable model into a system of three equations, namely

$$\dot{y} = \frac{1}{s}(-y - g(y,r)y + f(p)z) + \frac{r}{s}(y^0-y) \tag{1.1}$$
$$\dot{z} = w(g(y,r) - z) + \frac{r}{s}(z^0-z)$$
$$\dot{p} = \frac{1}{s}(y + 2g(y,r)y + \frac{q}{2}[g(y,r)]^2 - f(p)z) + \frac{r}{s}(p^0-p)$$

where

$$g(y,r) = (1-y-r/s^2 + [(1-y-r/s^2)^2 + 4q(y + rx^0/s^2)]^{1/2})/2q, \tag{1.2}$$
$$f(p) \equiv Fp^2/(K\bar{p}^2 + p^2).$$

Here $"\cdot" = \frac{d}{dt}$, $y \propto Y = [Br^-]$, $z \propto Z = [M^{+(n+1)}]$ (the higher oxidation state of the metal ion catalyst), $p \propto P$ = the concentration of the brominated derivative of the organic substrate. Also, y^0, z^0 and p^0 are proportional to the concentrations of Y, Z and P in the feed stream. Other parameter values are $s = 77.27$, $w = .161$, $q = 8.375 \times 10^{-6}$, $F = 4.0$, $K = .0005$ and $\bar{p} = (1/3) \times 10^7$. The parameter r is proportional to the flow rate and it is the easiest physical parameter to tune in the CSTR system.

Our computations show that Eqns. (1.1)-(1.2) exhibit qualitatively different oscillatory responses for high, low or intermediate flow rates. Relatively small values of r lead to bursts of several oxidation pulses separated by an interval of quiescence (IQ) during which the system is near a steady state of reduction. Such a pattern (see Fig. 1) corresponds to bursts computed in [6], also [15, (Fig. 5)], and observed experimentally in [2, 7, 13]. Bursting at relatively larger r (and K), is characterized by single reduction pulses separated by quiet periods of high oxidation. These solutions share qualitative features with calculated ones of Showalter et al. [12], and experimental ones of Schmitz et al. [11], except that the IQ's in these cases exhibit noticeable small oscillations. For intermediate r (and K) we find regularly repetitive pulses of large amplitude with no identifiable IQ between pulses. The system tends to a steady state for extremely high or low r. Experimental responses exhibit a similar qualitative dependence on flow rate [4, 10]: regular oscillations for intermediate r with complex oscillation patterns between this range and very high or very low flows where there is steady state behavior; in some cases, small amplitude oscillations are found between the steady state and complex oscillation regimes.

Our goal here is to understand low flow rate bursting patterns of the model (1.1)-(1.2) similar to that shown in Figure 1. For this it is convenient to

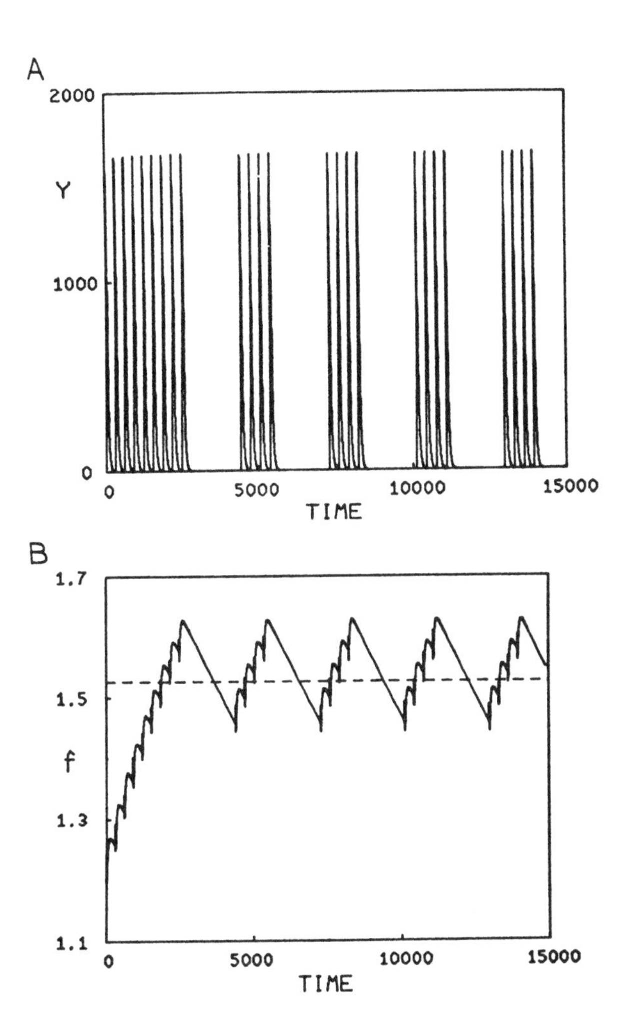

FIGURE 1. Bursts of oxidation pulses (from [9]). Solution, y vs. t and f vs. t, of simplified CSTR model, Eqns. (1.1)-(1.2) for $r = 0.00414$. After initial transient stage the response tends to a periodic pattern of four-pulse bursts separated by regular intervals of quiescence (IQ). Initial conditions are $y(0) = 0.5$, $z(0) = 11.0$, $f(p(0)) = 1.2$. The dashed horizontal line in B is for $f = f^H \approx 1.52$ (see Section 2).

view the CSTR as a "batch reactor" subsystem (i.e. the y and z equations) coupled with the dynamics for the stoichiometric parameter f. As we see in Fig. 1B there is a small net change in f from one spike to the next. Thus we are led naturally to a one dimensional mapping equation for f which describes the net change in f per spike. We compute the map numerically and derive a piecewise linear approximation. In our analysis of the piecewise linear approximation we examine the behavior of solutions as a function of the flow rate. In agreement with solutions to (1.1)-(1.2) we find that the number of pulses per burst increases with r. Moreover, we find that there are r-intervals in which the map exhibits two different stable bursting patterns. This predicts, and we have verified, the simultaneous existence of two stable bursting solutions in the continuous system (see Fig. 6). For some parameter ranges, the map exhibits aperiodic behavior as the flow rate increases. Further studies are now underway to determine the behavior of the continuous system at these higher values.

In Section 2 we analyze the batch reactor subsystem and develop the one-dimensional (1-D) mapping equation. Following that, in Section 3, we describe and analyze the bursting patterns predicted from the 1-D map. Section 4 is discussion.

2. THE BATCH REACTOR AND A ONE DIMENSIONAL MAPPING EQUATION

As was noted in the Introduction, we find it convenient to view the CSTR as a batch reactor subsystem coupled with the dynamics for the stoichiometric parameter f. The batch subsystem consists of the equations

$$\begin{aligned} \dot{y} &= \frac{1}{s}(-y - g(y,0)y + fz) \\ \dot{z} &= w(g(y,0) - z) \end{aligned} \tag{2.1}$$

where, from (1.2),

$$g(y,0) = (1-y + [(1-y)^2 + 4qy]^{1/2})/2q.$$

Here f is held constant and the flow terms are ignored ($r = 0$). Rinzel and Troy [9] have numerically shown (Fig. 2) that there is a Hopf bifurcation of unstable periodic orbits in Eqn. (2.1) which occurs as follows: there are values $f_H \approx .5$ and $f^H \approx 1.52$ such that if $f \in (f_H, f^H)$ then Eqn. (2.1) has an unstable steady state solution (denoted by $(y_0,z_0) = (y_0(f),z_0(f))$ surrounded by a large amplitude stable periodic orbit (of relaxation type). As f passes through f_H (from above) or f^H (from below) there occurs a Hopf bifurcation of small amplitude unstable periodic solutions from the steady state. On intervals of the form $(f_H - \mu, f_H)$ and $(f^H, f^H + \eta)$ there coexist a large amplitude stable periodic solution, a smaller unstable periodic orbit and

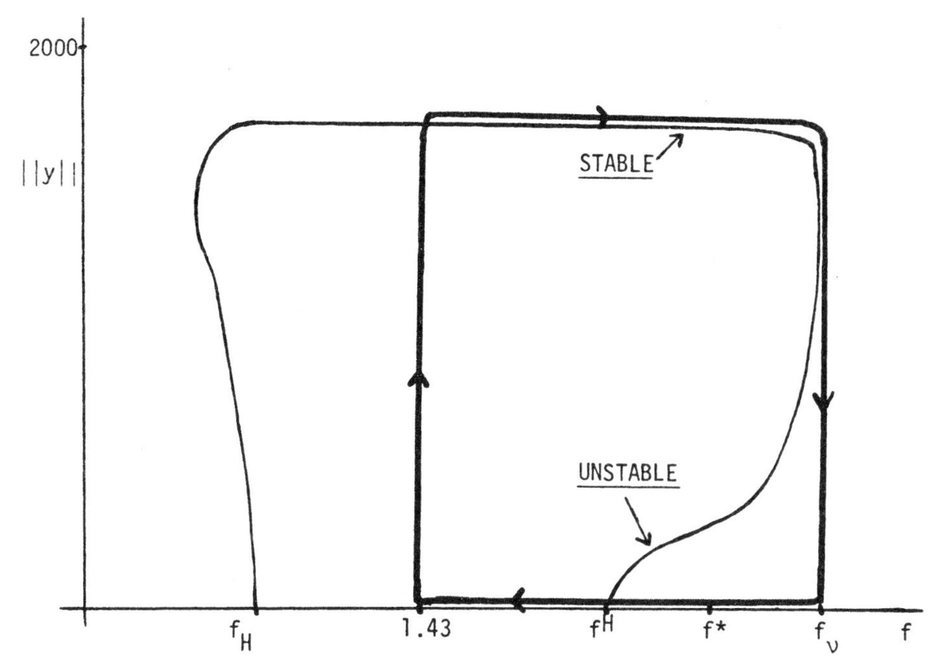

FIGURE 2. The bifurcation of small amplitude unstable periodic orbits of the batch reactor subsystem, Eqns. (2.1), occurs at $f_H \approx 0.5$ and $f^H \approx 1.52$. The knee of the curve occurs at $f_\nu \approx 1.579$. The path with arrows represents the hysteresis loop followed by the bursting solution of Fig. 1 for $r = 0.00414$. While following the upper branch of the curve the solution is in the spiking mode. After falling off the knee the solution follows the f-axis while (y,z) is near the batch steady state during the IQ. The value f^*, 1.544, refers to the discrete map of Fig. 3.

a locally stable steady state. For $f > f^H$ the two families of periodic orbits coalesce at a value $f_\nu \approx 1.58$ called the "knee" of the bifurcation diagram. For $f > f_\nu$ the steady state is globally stable and no periodic solutions exist. The bifurcation diagram has the form shown in Fig. 2.

To obtain a one dimensional simplification of the CSTR equations (1.1)-(1.2) we find it useful to interpret the bursting solution (Fig. 1) in terms of the batch reactor bifurcation diagram (Fig. 2). Initially for Fig. 1, $f(0) = 1.2$. Thus, $f_H < f(0) < f^H$ and the (y,z) subsystem immediately enters the spiking mode. During each large cycle in (y,z) space f undergoes a corresponding small increase Δf. We let f^* denote the critical value of f such that if $f(0) = f^*$ at the start of a spike then $\Delta f = f_\nu - f^*$. After a sufficient number of spikes, f enters the interval (f^*, f_ν). Subsequently, one more large oscillation in (y,z) space causes f to exceed f_ν. At this point (y,z) quickly approaches the stable steady state of the batch system and the large oscillations are extinguished. The p dynamics now cause f to decrease while (y,z) remains close to the steady state. This period is referred to as the interval of quiescence (IQ) between bursts. It persists until f dips below f^H and the steady state of the batch system becomes unstable. Rinzel and Troy [9, pp. 1782-1784] show that in fact f must decrease sufficiently below f^H, to a value approximately equal to 1.43, before the large oscillations in (y,z) space are rekindled (Fig. 1B). Again, each of these spikes causes a net increase in f and the entire loop described above is repeated (see schematic representation of this loop in Fig. 2).

We now derive the one dimensional mapping. We first note that during the spiking mode the trajectories traced out in (y,z) space are nearly identical [9], independent of f. Thus, whenever y is at its peak value during a spike we compute the corresponding value of f. This leads to a discrete sequence $\{f_n\}_{n \in Z}$ of f values. We seek to determine the functional relationship $f_{n+1} = \phi(f_n)$. The periodic bursting pattern of Fig. 1 only gives four points. We determine $\phi(f)$ for other values of f as follows. First, choose f and compute the large amplitude periodic solution of the batch system. Let $\bar{y}$ denote the maximum value of y, and $\bar{z}$ the corresponding value of z. Next, determine from Eqn. (1.2) the value of $\bar{p}$ which corresponds to $\bar{f}$, i.e. such that $f(\bar{p}) = \bar{f}$. Finally, let $(y(0),z(0),p(0)) = (\bar{y},\bar{z},\bar{p})$ in the full system, Eqns. (1.1)-(1.2). At the next spike in y the corresponding value of f is $\phi(f)$. Our numerics show that to a good approximation $\phi(f)$ is linearly increasing for $f < f^*$, and $\phi(f)$ approaches f_ν as f tends to f^* from below (see crosses of Fig. 3). If $f^* < f < f_\nu$ then recall (Fig. 1B) that the next value of f where a spike occurs is $f \approx 1.43 < f^*$. Our numerics show that $\phi(f)$ is discontinuous at $f = f^*$, and monotonically decreasing for

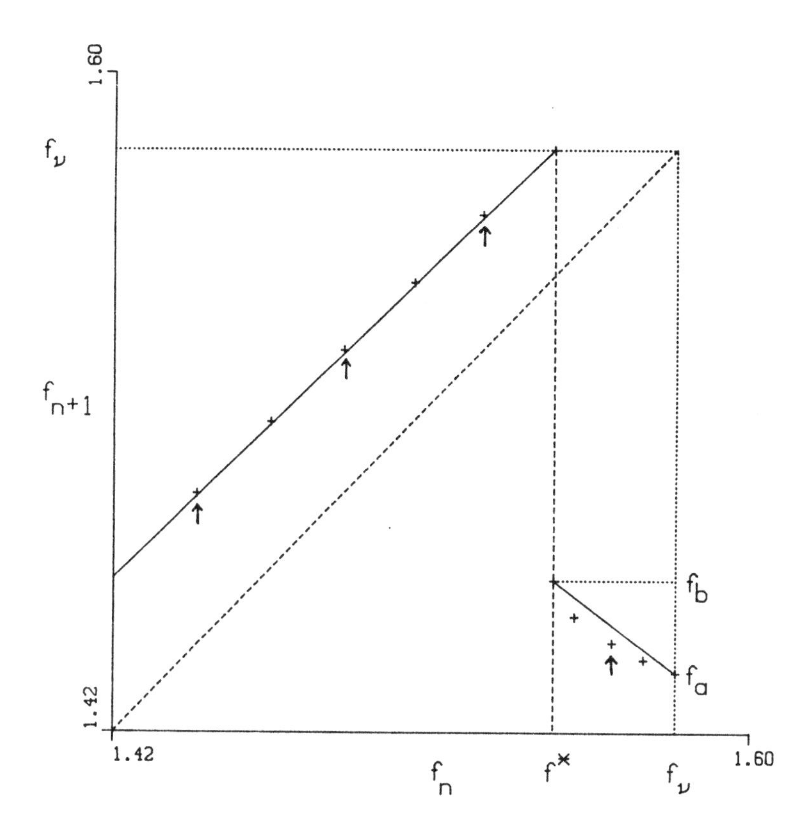

FIGURE 3. The piecewise linear approximation (solid) for the discontinuous map $f_{n+1} = \phi(f_n)$ of f from one spike to the next. The data points (6 above and 5 below the 1:1 line), shown as crosses, were determined by solving the CSTR model, Eqns. (1.1)-(1.2), for different initial conditions (see text). Vertical arrows indicate the periodic, four pulse, burst pattern of Fig. 1. The piecewise linear map also has a four-point periodic solution for these map parameters: $f^* = 1.544$, $f_\nu = 1.579$, $f_a = 1.436$, $f_b = 1.46$ and slope .94 for $1.42 \leq f < f^*$.

$f^* < f < f_\nu$ (again, see the crosses). As a simplification we approximate $\phi(f)$ by the discontinuous piecewise linear function in Fig. 3. This map also has a periodic solution; each of the four points in this discrete periodic solution corresponds to a pulse.

3. BURST PATTERNS FOR THE 1-D MAP

In the preceding section we developed a discrete mapping $f_{n+1} = \phi(f_n)$. Here we restrict attention to the piecewise linear approximation

$$\begin{aligned} \phi(f) &= f_\nu + m(f-f^*), \quad f < f^* \\ &= f_a + \left(\frac{f_b-f_a}{f^*-f_\nu}\right)(f-f_\nu), \quad f^* \le f \le f_\nu. \end{aligned} \tag{3.1}$$

Figure 4 empirically justifies using the discrete map to describe the bursting behavior of the CSTR model, (1.1)-(1.2), for $r = 0.00414$. To apply this description for other flow rates requires that we make explicit the assumed dependence of ϕ upon r. Certain features may be deduced from the parametric dependence of bursts upon r as reported in Section 3.A2 of [9]. There we found that during the spiking phase the net increment of f per spike decreases for larger flow rate. From this it follows that, as r increases, the left segment of ϕ (i.e., for $f < f^*$) moves closer to the 1:1 line and f^* is nearer to f_ν. We make the simplest assumption by supposing that m is independent of r and f^* increases with r. Since we have not yet computed the exact dependence of f^* on r we shall here study the map behavior as a function of f^* to qualitatively mimic dependence of the CSTR responses upon r.

Next we consider the right segment of ϕ (i.e., for $f > f^*$). From our analysis of the IQ [9], this segment must have negative slope for each r. That is, the further f is above f^H at the start of the IQ, the further f falls below f^H before the next spike. Our numerical experiments to date indicate a small relative variation of f_a and f_b with respect to r (e.g., $d(\ln f_a)/dr$). Initially, we suppose f_a and f_b are independent of r but later we shall consider weakening this restriction. We summarize the above assumptions on (3.1) as follows:

i) f^* increases with r,

ii) m is independent of f^*, $m < 1$.

iii) f_a is independent of f^*,

iv) f_b is independent of f^*.

Figure 4 illustrates periodic solutions of the map (3.1) for several values of f^*. An N-point discrete solution corresponds to an N-pulse burst for the continuous model, (1.1)-(1.2). Consistent with our findings [9] for (1.1)-(1.2), the number of pulses per burst increases with flow rate, i.e., with f^*.

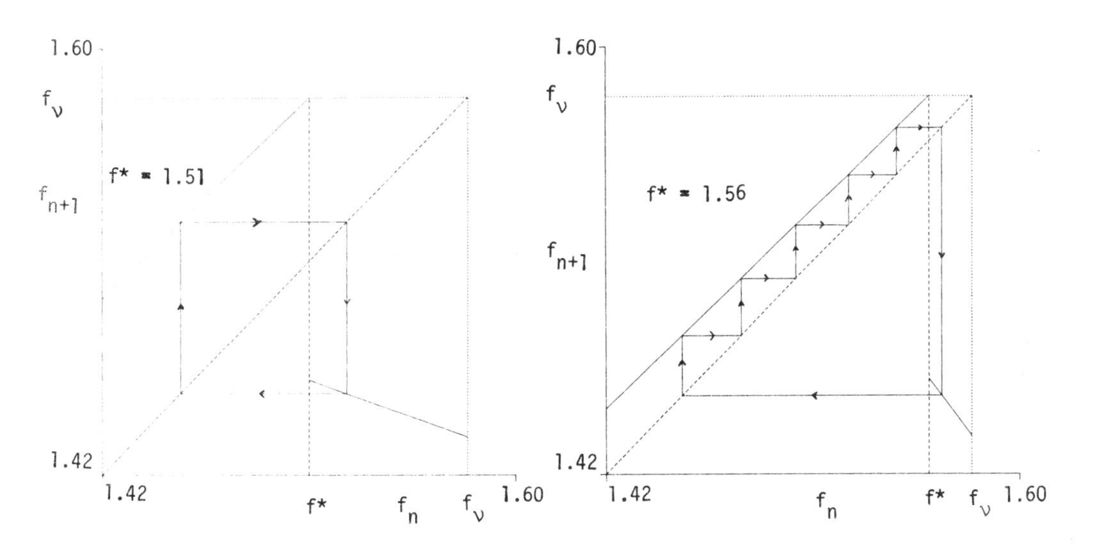

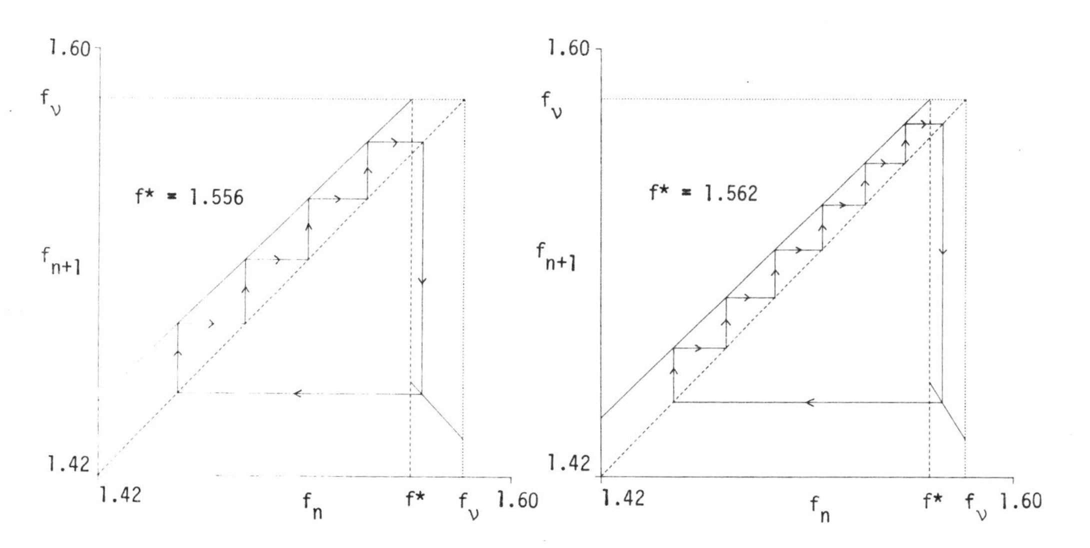

FIGURE 4. Discrete stable periodic solutions to map Eqn. (3.1) for different values of f^*, a parameter corresponding to flow rate of the CSTR. Parameter values f_ν, f_a, f_b, and m are as in Fig. 3. An N-point discrete solution should correspond to an N-pulse periodic solution to the continuous model, Eqns. (1.1)-(1.2). Here observe that N increases with f^*.

Since the approximate map is piecewise linear we may calculate explicitly the discrete periodic orbits shown in Fig. 4. Let a periodic solution of N points be represented by $\{f_1,\ldots,f_N\}$ where $f^* \le f_N \le f_\nu$. Then, from (3.1), we have

$$f_1 = f_a + (f_N - f_\nu)\left(\frac{f_b - f_a}{f^* - f_\nu}\right) \tag{3.2a}$$

and

$$\begin{aligned} f_n &= f_\nu + m(f_{n+1} - f^*) \\ &= \sum_{j=0}^{n-2} (f_\nu - mf^*) + m^{n-1} f_1, \quad 2 \le n \le N. \end{aligned} \tag{3.2b}$$

Now evaluating (3.2b) for $n = N$ and using (3.2a) we obtain a linear equation for f_N in terms of N and f^*:

$$\begin{aligned} f_N = f_a &+ m^{N-1}\left(\frac{f_b - f_a}{f^* - f_\nu}\right)(f_N - f_\nu) \\ &+ \left(\frac{1-m^{N-1}}{1-m}\right)[f_\nu - f_a - m(f^* - f_a)]. \end{aligned} \tag{3.3}$$

This equation has a unique solution f_N, however only for appropriate ranges of N and f^* is the solution f_N in the interval $[f^*, f_\nu]$ so that it corresponds to a periodic burst pattern. It is not difficult to show that in this range f_N is a decreasing function of f^*. Thus, for a given N, the N-point periodic solution first <u>appears</u> as f^* increases through $f^*_{N,a}$ at which value f_N enters the interval $[f^*, f_\nu]$ from above. Thus from (3.3) $f^*_{N,a}$ is defined by the condition $f_N = f_\nu$:

$$f^*_{N,a} = \frac{(m^{N-1} - m^{N-2})f_a + (m^{N-2} - 1)f_\nu}{m^{N-1} - 1} \tag{3.4}$$

The N-point periodic solution then <u>disappears</u> for $f^* = f^*_{N,d}$ $(f^*_{N,d} > f^*_{N,a})$ when f_N disappears into the map's discontinuity at f^*. Thus we have

$$f^*_{N,d} = \frac{(m^N - m^{N-1})f_b + (m^{N-1} - 1)f_\nu}{m^N - 1} \quad . \tag{3.5}$$

For this piecewise linear map we can also determine the stability of the N-point periodic solution. Namely, the solution is stable if the multiplier σ_N, given by

$$\sigma_N = m^{N-1}(f_b - f_a)/(f^* - f_\nu), \tag{3.6}$$

satisfies $|\sigma_N| < 1$; the solution is unstable if $|\sigma_N| > 1$. Since $\sigma_N < 0$ and σ_N decreases with f^*, we find instability for $f^* > f^*_{N,\nu}$ where $\sigma_N = -1$ for $f^* = f^*_{N,\nu}$, i.e., from (3.6)

$$f^*_{N,\nu} = f_\nu - m^{N-1}(f_b - f_a).$$

In Fig. 5 we indicate (for m, f_a, f_b as in Fig. 3) the intervals of f^* over which an N-point periodic solution exists. The vertical bar represents the value of $f^*_{N,\nu}$ if it falls in the interval $(f^*_{N,a}, f^*_{N,d})$; to the right of the bar the N-point solution is unstable. For the assumptions and parameter values of this figure, all N-point periodic solutions for $2 \leq N \leq 5$ are stable while all those for $N \geq 9$ are unstable.

An immediately noticeable feature of Fig. 5 is that more than one periodic solution may exist for certain intervals of f^*. One may see this analytically by comparing (3.4) and (3.5) to find $f^*_{N,a} < f^*_{N-1,d}$. We emphasize that such coexistence is a qualitative consequence of the right branch of ϕ having negative slope. Because periodic solutions for low values of N are stable, these overlapping intervals of existence imply bistability (of periodic solutions) for certain ranges of f^*. For example, for $f^* = 1.5413$ (the vertical arrow in Fig. 5), both a stable 3-point and stable 4-point periodic solution coexist. These two solutions are shown in the upper two panels of Fig. 6. The value of f_N for the $N = 3$ orbit lies to the left of f_N for the $N = 4$ orbit. With respect to initial values in the interval (f^*, f_ν), each orbit's domain of attraction is an interval about its respective f_N value. These two domains of attraction are separated by the intersection of the dotted path (traced backwards from the point (f^*, f^*)) with the right branch of ϕ. For initial values less than f^*, the domains of attraction are alternating bands.

The 1-D map thus makes a definite prediction for coexistence and hysteresis of N-pulse periodic solutions for the continuous model. The lower panels of Fig. 6 verify this prediction. They show that for flow rate, $r = 0.003750$, the Eqns. (1.1)-(1.2) have both a stable 3-pulse pattern and a stable 4-pulse pattern. The initial conditions were chosen consistent with the domains of attraction for the 1-D map solutions. We have also seen coexistence of solutions to the continuous model for some other values of r (e.g., near the r-value at which an N-pulse solution disappears).

Returning to Fig. 5, one sees for increasing f^* how solutions progress from an N-point to an (N+1)-point periodic pattern. One should ask however what the observed discrete pattern is when f^* exceeds the value $f^*_{8,\nu}$, 1.5634..., beyond which no stable N-point periodic solutions exist. In this situation only unstable periodic orbits exist and moreover there are at least two of them for each such f^*. The map response is then aperiodic mixing among patterns which have different numbers of points on the left branch corresponding to the different coexistent unstable, N-point, periodic solutions. For example, from Fig. 5, if f^* satisfies $f^*_{8,d} < f^* < f^*_{11,a}$ $(< f^*_{10,d})$ then the

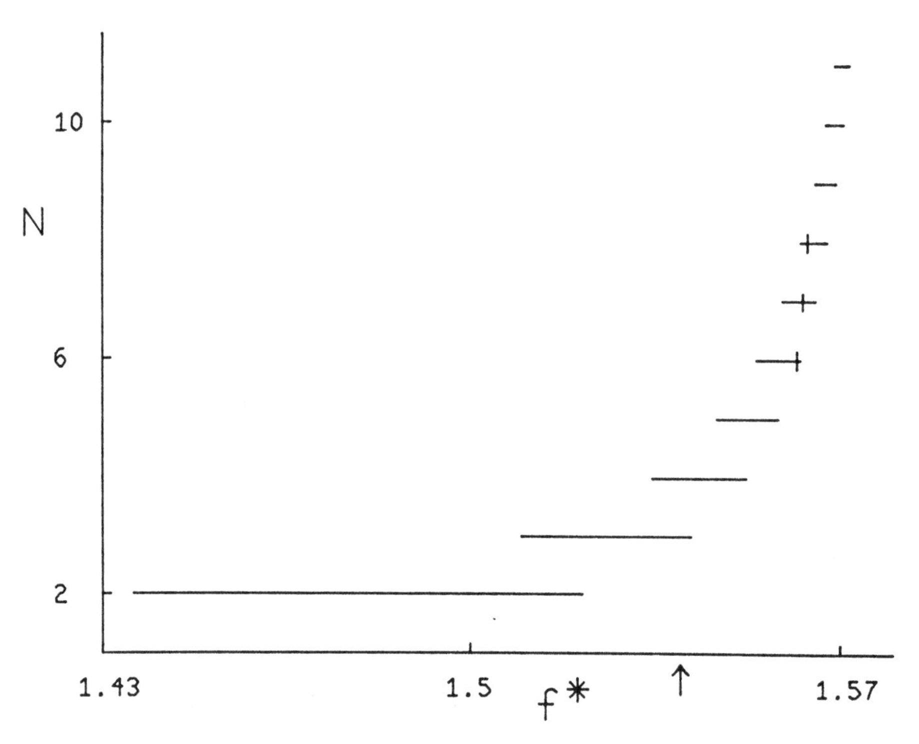

FIGURE 5. Intervals of existence, $f^*_{N,a} \leq f^* \leq f^*_{N,d}$, for N-point periodic solutions to map Eqn. (3.1). Parameter values f_ν, f_a, f_b, and m are as in Fig. 3. Periodic solutions with $N \leq 5$ are stable; those with $N \geq 9$ are unstable. Solutions with $6 \leq N \leq 8$ are stable for f^* to left of vertical bar (at $f^* = f^*_{N,\nu}$) and unstable to right. For $f^* = 1.5413$ (vertical arrow) both three-point and four-point stable orbits exist.

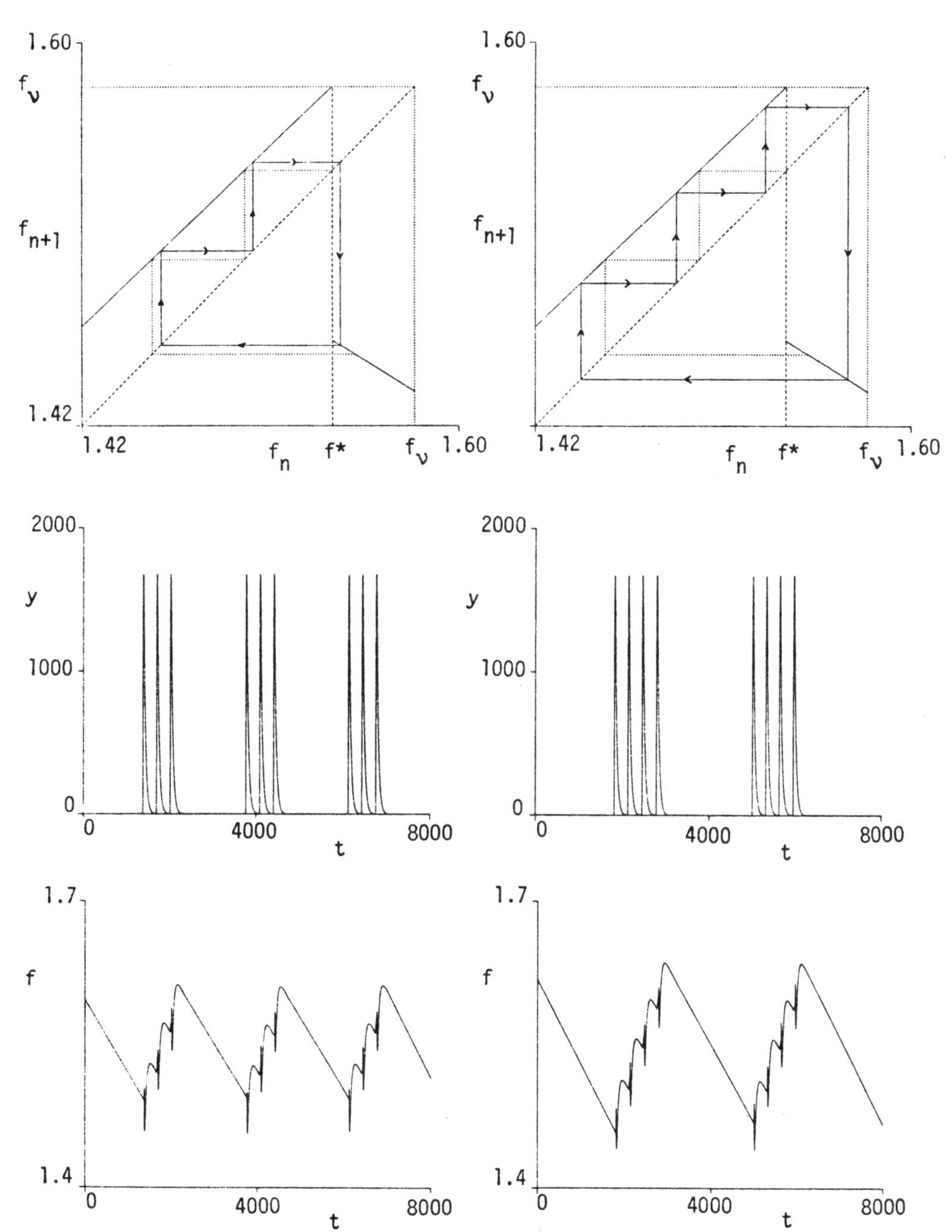

FIGURE 6. For $f^* = 1.5413$ the map Eqn. (3.1) has both a three and four-point stable periodic solution as shown in the upper two panels. The dotted path leading from (f^*,f^*) divides the interval (f^*,f_ν) into two subintervals which define the domains of attraction for the two periodic orbits. For $r = 0.00375$ Eqns. (1.1)-(1.2) exhibit the corresponding coexistence of stable three and four-pulse periodic bursting solutions. The lower four panels show the y and f components of these solutions.

response would be aperiodic mixing between 9- and 10-point patterns. To see this, we follow a suggestion by J. Keener (private communication) and consider the return map for ϕ on the interval (f^*, f_ν). This return map is also a discontinuous, piecewise linear function. Each segment has negative slope, intersects the 1:1 line, and corresponds to a pattern with a distinct number of points on the left branch of ϕ. The aperiodic mixing behavior for f^* large enough corresponds to iterating between the various different segments of this return map.

One wonders if such aperiodic behavior, e.g. mixing between N- and (N+1)-pulse bursts, can be found for the continuous CSTR model. So far we have not observed such aperiodic solutions to Eqns. (1.1)-(1.2). We have computed <u>stable periodic</u> burst patterns with as many as 28 pulses per burst.

On the other hand, it is not requisite that the 1-D map exhibits aperiodic behavior for large N; whether or not it does, depends on the parameters. In particular, if the lower branch is not too steep, i.e. if

$$\frac{f_b - f_a}{f_\nu - f_a} < \frac{1-m}{2-m}, \tag{3.7}$$

then each N-point periodic solution is stable and the map response is always periodic. For a slowly increasing flow rate one would observe an orderly insertion of an additional spike (because $f^*_{N+1,d} > f^*_{N,d}$) into the periodic burst pattern. Note, the inequality (3.7) follows from (3.5) and (3.6) by requiring for all N that $\sigma_N > -1$ for $f^* \leq f^*_{N,d}$. The case of Fig. 4 however falls into the parameter regime

$$\frac{f_b - f_a}{f_\nu - f_a} > \frac{1-m}{m} \tag{3.8}$$

for which all N-point periodic solutions are unstable for large enough N. In this case, the inequality (3.8) follows from (3.4) and (3.6) and the demand that $\sigma_N < -1$ for $f^* = f^*_{N,a}$ when N is large enough.

To correlate further the 1-D map predictions with solutions to (1.1)-(1.2) we should consider more carefully the dependence of the map parameters on the flow rate r (or f^*). For example, we might relax assumptions iii) and iv) and allow f_a and f_b to depend on f^*. If f_b were decreased with f^* appropriately the map would exhibit only stable periodic solutions.

4. DISCUSSION

We have considered a simplified three variable model, Eqns. (1.1)-(1.2), for the BZ chemical system in a CSTR. We focus on the complex oscillation (burst) patterns for relatively low flow rate (Fig. 1). For these bursting responses there is a small positive increment in the species P (and therefore f) during each spike of the burst and then a slow substantial decrease between bursts. By interpreting such patterns and the dynamics of f in terms of the batch reactor subsystem and its bifurcation diagram (Fig. 2) we derive a one variable discontinuous mapping for the value of f from one spike to the next (Fig. 3). We consider a reasonable piecewise linear approximation to the map whose solution behavior is described analytically.

Various features of the map solutions mimic, and moreover predict, burst patterns for the continuous model. For example, the number of spikes per burst increases by one as the flow rate r increases through critical values (Figs. 4 and 5). For r above the low flow rate regime, the discrete and continuous models both exhibit regular (non-bursting) oscillations; for the map this occurs as f^* increases above f_ν and a stable fixed point appears. In addition, the 1-D map predicts hysteresis between the N-pulse and (N+1)-pulse patterns as r is tuned; i.e. for a certain interval of r values these periodic solutions coexist. This prediction is a qualitative one (depending primarily on the negative slope of the map's right branch) and has been verified (in several cases) for the continuous model (Fig. 6). Depending on the map parameters and their assumed dependence on r, the map response may become aperiodic for large enough r. In this case the map predicts aperiodic mixing between bursts with different numbers of pulses. Such behavior has not yet been observed for the continuous model. On the other hand, for different map parameter assumptions all the N-point periodic solutions are stable and this predicts a regular progression from the N-pulse to the (N+1)-pulse periodic burst as flow rate increases. We expect to examine other parameter ranges and assumptions for both models to check for further, and perhaps improved, comparability.

We also seek comparison with experimental data. For example, the models treated here predict hysteresis and bistability which might be investigated experimentally. Experimental work is very active but there is disparity between the different chemical recipes and flow rate regimes which the different groups consider [10]. Our bursts resemble those in [2, 7, 13]. Extensive low flow rate experiments have been done by the Texas group [16] (also, some in Bordeaux: see [10]). Their observed complex oscillation patterns are somewhat different from that in Fig. 1. Their patterns are similar to high flow rate patterns in experiments of [5, 11] but the qualitative dependence on r is reversed. In our terminology, their burst has a single spike and the IQ exhibits small oscillations. As r increases the number of small cycles between spikes decreases until a periodic response of only large spikes is observed (this is the intermediate flow rate regime). The progression from one periodic pattern to the next, e.g. one with N+1 small cycles to one with N small cycles, is not sharp (as in Fig. 5). Rather the experimental data indicate transition regions where aperiodic mixing between the two patterns occurs. There is some evidence that period-doubling cascades lead to these chaotic regimes. Many of these data have been interpreted in terms of 1-D maps obtained from experimental data [10, 14, 16] or differential equation models without chemical interpretation [8].

A discontinuous map of the form we study (but not necessarily piecewise linear) may also exhibit period-doubling behavior and chaos. With curvature in the map one expects a cascade of period-doubling bifurcations subsequent to the destabilization of an N-point periodic solution. However, we view this as fine-structured chaos characterized by irregular timing of pulses in N-pulse bursts. We distinguish this from the chaos of aperiodic mixing between N and (N+1)-point patterns which we have described above. This latter behavior occurs when the return map of the interval $[f^*, f_\nu]$ has multiple branches and more than one branch is expanding, i.e. has slope less than minus one. In this case, the unstable fixed point of each such branch acts as a repeller. A typical trajectory spirals outward on one branch, jumps to a different branch where it spirals out and then is reinjected back to the first branch.

REFERENCES

1. Belousov, B. P., A periodic reaction and its mechanism, Ref. Radiats. Med., 1958 (1959), 145- .

2. DeKepper, P., A. Rossi, and A. Pacault, Etude experimentale d'une reaction chimique periodique. Diagramme d'etat de la reaction de Belousov-Zhabotinskii, C. R. Acad. Sci. Ser., C 283 (1976), 371-375.

3. Field, R. J. and R. M. Noyes, Oscillations in chemical systems. IV. Limit cycle behavior in a model of a real chemical reaction, J. Chem. Phys., 60 (1974), 1877-1884.

4. Graziani, K. R., J. L. Hudson, and R. A. Schmitz, The Belousov-Zhabotinskii reaction in a continuous flow reactor, Chem. Eng. J., 12 (1976), 9-21.

5. Hudson, J. L., M. Hart, and D. Marinko, An experimental study of multiple peak periodic and nonperiodic oscillations in the Belousov-Zhabotinskii reaction, J. Chem. Phys., 71 (1979), 1601-1606.

6. Janz, R. D., D. J. Vanecek, and R. J. Field, Composite double oscillation in a modified version of the Oregonator model of the Belousov-Zhabotinskii reaction, J. Chem. Phys., 73 (1980), 3132-3138.

7. Marek, M. and E. Svoboda, Nonlinear phenomena in oscillatory systems of homogeneous reactions - experimental observations, Biophys. Chem., 3 (1975), 263-273.

8. Pikovsky, A. S., A dynamical model for periodic and chaotic oscillations in the Belousov-Zhabotinskii reaction, Phys. Lett., 85A (1981), 13-16.

9. Rinzel, J. and W. C. Troy, Bursting phenomena in a simplified Oregonator flow system model, J. Chem. Phys., 76 (1982), 1775-1789.

10. Roux, J. C., Experimental studies of bifurcations leading to chaos in the Belousov-Zhabotinskii reaction, Proc. of Los Alamos Conf., Order in Chaos, (1982), to appear in Physica D.

11. Schmitz, R. A., K. R. Graziani, and J. L. Hudson, Experimental evidence of chaotic states in the Belousov-Zhabotinskii reaction, J. Chem. Phys., 67 (1977), 3040-3044.

12. Showalter, K., R. Noyes, and K. Bar-Eli, A modified Oregonator model exhibiting complicated limit cycle behavior in a flow system, J. Chem. Phys., 69 (1978), 2514-2524.

13. Sorensen, P. G., in general discussion, Proc. Faraday Soc. Symp., 9 (1974), 88-89.

14. Tomita, K. and I. Tsuda, Towards the interpretation of Hudson's experiment on the Belousov-Zhabotinskii reaction, Prog. Theoret. Phys., 64 (1980), 1138-1160.

15. Turner, J. S., Periodic and nonperiodic oscillations in the Belousov-Zhabotinskii reaction, Preprint from Discussion Meeting, Kinetics of Physiochemical Oscillations, Aachen, Sept., 1979.

16. Turner, J. S., J.-C. Roux, W. D. McCormick, and H. J. Swinney, Alternating periodic and chaotic regimes in a chemical reaction-experiment and theory, Phys. Lett., 85A (1981), 9-12.

17. Zhabotinskii, A. M., Dokl. Acad. Sci. Nauk SSSR, 157 (1964), 392- .

18. Zhabotinskii, A. M., Periodic course of oxidation of malonic acid in solution (investigation of the kinetics of the reaction of Belousov), Biophysics, 9 (1964), 329-335.

MATHEMATICAL RESEARCH BRANCH
NATIONAL INSTITUTE OF ARTHRITIS, DIABETES
AND DIGESTIVE AND KIDNEY DISEASES
NATIONAL INSTITUTES OF HEALTH
BETHESDA, MARYLAND 20205

DEPARTMENT OF MATHEMATICS
UNIVERSITY OF PITTSBURGH
PITTSBURGH, PENNSYLVANIA 15260

Contemporary Mathematics
Volume 17, 1983

STABLE AND UNSTABLE STATES OF NONLINEAR WAVE EQUATIONS

Walter A. Strauss[1]

ABSTRACT. We present a survey of recent results on stationary states, standing waves and periodic solutions of finite energy of semilinear wave equations and of their stability properties.

For wave equations, in contrast to diffusion equations, there is usually no dissipation and no maximum principle to work with. Instead there is the propagation of waves along characteristics. If a wave does not encounter a boundary or other special phenomenon, it keeps going to infinity, so that a localized initial disturbance spreads out in all directions. (Dimension $n = 1$ is special, there being only two possible directions of spreading, and we exclude it. It is also sometimes convenient to exclude $n = 2$.) This spreading, however, can be counteracted by nonlinearities which can concentrate the wave so that it locally forms peaks.

We mostly limit our discussion to the semilinear wave equation

$$u_{tt} - \Delta u + f(u) = 0 \tag{1}$$

in the whole of space $x \in \mathbb{R}^n$, $n \geq 3$, where $f(0) = 0$ and f is smooth. We consider solutions of <u>finite energy</u>, by which we mean that all the first derivatives (u_t and ∇u = spatial gradient) are continuous functions of time with values in $L^2(\mathbb{R}^n)$. Also the solution itself should be continuous with values in $L^{2n/(n-2)}(\mathbb{R}^n)$, which just means that it is approximable by nice functions in the energy norm.

1. STATIONARY STATES. A <u>solitary wave</u> is a solution of (1) which has finite energy but whose amplitude (norm in $L^\infty(\mathbb{R}^n)$ does not tend to zero as $t \to \infty$. In this survey we discuss several kinds of solitary waves. A <u>stationary wave</u> is a solution of finite energy which does not depend on time:

$$-\Delta\phi + f(\phi) = 0 \tag{2}$$

1980 Mathematics Subject Classification 35L70; 35B35
Supported by National Science Foundation Grant MCS79-01965

THEOREM 1. There exists a solution $\phi \in H^1(\mathbb{R}^n)$, not identically zero of equation (2), provided

(i) $f'(0) > 0$

(ii) $\int_0^s f(u)du < 0$ for some $s > 0$

(iii) $\liminf_{u\to+\infty} f(u)/u^{1+4/(n-2)} \geq 0$.

In fact, there exists such a $\phi(x)$ which is smooth, positive, radially symmetric, decreasing and exponentially decaying.

This solution should be viewed as the ground state of (2), the analogue of the principal eigenfunction of a linear operator such as $-\Delta + V(x)$. It is easily constructed via the variational principle

$$\text{(3)} \qquad \operatorname{Min}\int |\nabla u|^2 dx \quad \text{subject to the constraint}$$

$$\text{(4)} \qquad \int F(u(x))dx = 1 \quad , \quad u \in H_r^1$$

where $F' = f$, $F(0) = 0$ and H_r^1 is the subspace of $H^1(\mathbb{R}^n)$ consisting of radial functions. Note that functions in H_r^1 are uniformly small at infinity:

$$|u(x)| \leq c|x|^{(1-n)/2}\|u\|_{H_r^1} \quad \text{for} \quad |x| > 1 .$$

The solution of (3), (4) satisfies the equation $-\Delta u + \lambda f(u) = 0$. One shows that $\lambda > 0$ and then the change of scale $\phi(x) = u(\sqrt{\lambda}x)$ gives the solution of (2).

From an ordinary differential equation point of view, the existence of a stationary state was considered by Synge [30], Berger [4] and others. The partial differential equation point of view and the consideration of a general f was taken by Strauss [27] and Berestycki and Lions [3].

Condition (i) is almost necessary. If $f'(0) < 0$ we have a positive eigenvalue of the operator $-\Delta + V(x)$ where $V(x) = [f(\phi(x)) - f'(0)\phi(x)]/\phi(x)$ with $V(x)$ smooth and rapidly decaying at infinity. This is impossible.

Condition (ii) is also necessary. Indeed the Laplacian is invariant under dilations $\phi(x) \to \phi(cx)$. The generator of dilations is $x\cdot\nabla = r(\partial/\partial r)$. Multiplying equation (2) by $r\phi_r$ and integrating by parts leads to the identity

$$\text{(5)} \qquad (\frac{1}{2} - \frac{1}{n})\int |\nabla\phi|^2 dx + \int F(\phi)dx = 0 .$$

Since we have assumed $n > 2$ and since $\phi(x) \not\equiv 0$, F must be somewhere negative.

Condition (iii) is not necessary because $f(u)$ can be modified arbitrarily beyond the range of $u = \phi(x)$. However some condition is necessary in addition to (i) and (ii). Indeed if we multiply (2) by $\phi(x)$ and integrate, we get

$$\int |\nabla\phi|^2 dx + \int \phi f(\phi) dx = 0 \tag{6}$$

Identities (5) and (6) together imply

$$\int [F(\phi) - (\tfrac{1}{2} - \tfrac{1}{n})\phi f(\phi)] dx = 0 \tag{7}$$

Therefore

$$H(u) \equiv F(u) - (\tfrac{1}{2} - \tfrac{1}{n}) u f(u) < 0 \quad \text{somewhere} \tag{8}$$

since $H(u)$ is positive near $u = 0$. Condition (8), which is due to Derrick [9] and Pohozaev [19], is probably not sufficient for Theorem 1. A necessary and sufficient condition is not known.

EXAMPLE. $f(u) = u - u^\gamma$ with $\gamma > 1$. Clearly (i) and (ii) are satisfied. Both (iii) and (8) are satisfied if and only if $\gamma < 1 + 4/(n-2)$.

This example is related to a Sobolev inequality. Indeed, the scale change $u(x) \to c^{(n-2)/2} u(cx)$ leaves the Dirichlet integral invariant. It also leaves the integral (4) invariant if and only if $F(u) = \text{const}|u|^{\gamma+1}$ where $\gamma = 1+4/(n-2)$. Thus the variational problem (3), (4) is the same as the problem of finding the best constant k in the Sobolev inequality

$$\|u\|_{\gamma+1} \le k \|\nabla u\|_2$$

where $\gamma+1 = 2n/(n-2)$ is the familiar Sobolev index.

A direct method which gives the value of the ground state energy $\frac{1}{2}\int |\nabla\phi|^2 dx$ without requiring a scale change is to minimize the Dirichlet integral (3) using the identity (5) as the constraint with $u \in H^1_r$, $u \not\equiv 0$. In fact, (5) defines a manifold in H^1_r in which the zero solution is isolated. This method is due to C. Keller [14].

Serrin, McLeod and Peletier [22] have found conditions for uniqueness of the ground state and an example of where it is not unique.

If f is an odd function, there exist an infinite number of stationary states: see [3]. This is a deeper result since it requires a nonlinear minimax procedure. In contrast, in the one dimensional case there is only the ground state.

2. STABILITY THEORY.

In this section we consider the stability of stationary states and of standing waves. First let $\phi(x)$ be as in Theorem 1. Then the spectrum of the linearized operator $A = -\Delta + f'(\phi)$ on $L^2(\mathbb{R}^n)$ looks like this

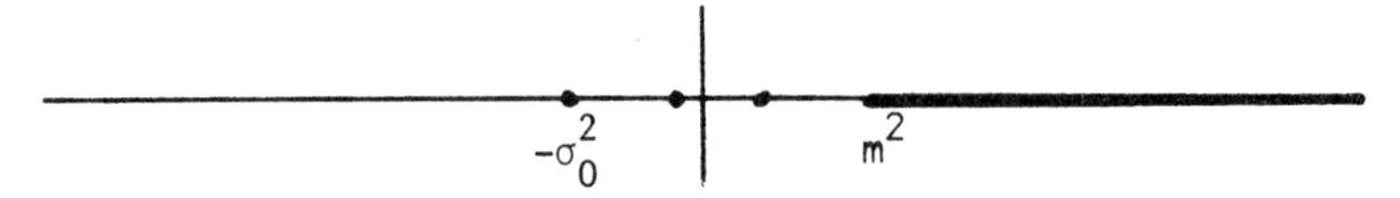

where $m^2 \equiv f'(0) > 0$. It was Derrick [9] who first observed that it has at least one negative eigenvalue. This is a consequence of dilation invariance. Using the generator of dilations $r\partial/\partial r$ and some integration by parts, we have

$$\begin{aligned} &([-\Delta + f'(\phi)][r\phi_r], r\phi_r) \\ &= (r[-\Delta\phi + f(\phi)]_r, r\phi_r) - (n-2)(\nabla\phi, \nabla\phi) \end{aligned}$$

where $(\ ,\ )$ means the L^2 inner product. So a solution of equation (2) satisfies

$$(A[r\phi_r], r\phi_r) = -(n-2)\int |\nabla\phi|^2 dx < 0 . \tag{9}$$

Thus A has some negative spectrum, say at $-\sigma_0^2$.

We are interested, not in the corresponding diffusion equation $v_t + Av = 0$, but in the wave equation

$$v_{tt} + Av = v_{tt} - \Delta v + f'(\phi)v = 0 ,$$

which is equation (1) linearized around the ground state ϕ. Its spectrum is $\sqrt{-1}$ times the square root of the spectrum of A, namely

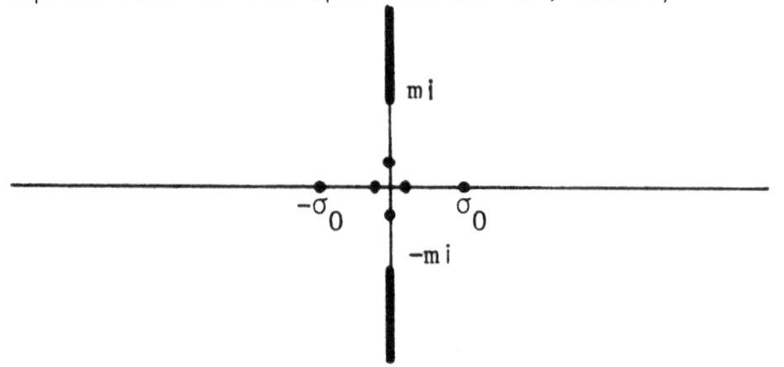

Thus we would expect ϕ to be an unstable state of equation (1). However this is not such an easy matter because none of the general theorems (of the type

that linearized instability implies true instability) are applicable when so much of the spectrum lies on the imaginary axis. Recently Shatah [24] has succeeded in proving it.

THEOREM 2. Under the same conditions on f as in Theorem 1, the ground state $\phi(x)$ is unstable for (1) in the energy norm.

In fact Shatah shows that there are initial data for (1) arbitrary close to $\phi(x)$ (initialy velocity arbitrarily close to 0) for which the energy norm of the solution tends to infinity on some sequence of times. That is, either blow up occurs at a finite time or $\|u(t)\|$ becomes unbounded as $t \to +\infty$. For a particular class of nonlinearities, blow up was previously proved by Berestycki and Cazenave [2].

By way of contrast, in one dimension it is well known that stationary states can be stable. A hint is given by (9). A familiar example is afforded by the Sine-Gordon equation where $f(u) = \sin u$.

Now let us consider <u>standing waves</u>, by which we mean solutions of (1) of the form

$$u = e^{i\omega t}\phi(x) \qquad (\omega \text{ real}). \tag{10}$$

When $\omega = 0$ we get the stationary states. We assume "gauge invariance": $f(\rho e^{i\theta}) = f(\rho)e^{i\theta}$. From (1)

$$-\Delta\phi - \omega^2\phi + f(\phi) = 0 . \tag{11}$$

By Theorem 1 this has a solution provided $\omega^2 < m^2 \equiv f'(0)$ and $-\omega^2u^2/2 + F(u)$ is somewhere negative and (iii) holds. We denote the ground state of (11) by $\phi = \phi_\omega$. Theorem 2 is also valid for these standing waves for certain f and ω. For example, if $f(u) = u - |u|^{\gamma-1}u$ with $\gamma > 1+4/n$, then for any frequency ω for which a standing wave exists, it is unstable.

Stable states are more interesting since they are more likely to be physically observable. Shatah [24] has shown that this is sometimes the case, thereby justifying the numerical evidence in [1]. His theorem applies to a large class of nonlinear terms but here we illustrate it with an example.

THEOREM 3. Let $f(u) = u - |u|^{\gamma-1}u$ where $1 < \gamma < 1+4/n$. Let S_ω be the set of standing waves (10) of lowest energy. By Theorem 1, $S_\omega \neq \emptyset$ for $0 \le \omega < 1$. If ω is sufficiently near 1, then S_ω is a <u>stable</u> set of solu- of (1).

In the stable and in some of the unstable cases, no conclusion can be reached from a linearized stability analysis since the whole spectrum lies on the imaginary axis! Theorem 3 is the analogue of a recent result of Cazenave and Lions [7] for the Schrödinger equation, which is an easier case due to the control of the L^2 norm.

If dissipation is added to the equation, a much more detailed analysis is possible, as the next theorem due to C. Keller [12] shows.

THEOREM 4. For the equation

$$u_{tt} - \Delta u + \alpha u_t + f(u) = 0 \tag{12}$$

with $\alpha > 0$ and f satisfying the conditions of Theorem 1, the phase space near the ground state $\phi(x)$ looks like a saddle point with a finite-dimensional (non-trivial) unstable manifold and an infinite-dimensional stable manifold.

3. STABILITY OF THE ZERO SOLUTION.

By assumption, $u \equiv 0$ is a solution of (1). It is stable if all solutions with sufficiently small initial data (in the energy or some other norm) exist and remain small for all time. This is possible due to the spreading nature of waves in free space. The spreading should imply a pointwise decay of the solutions (as $t \to \infty$ for fixed x) so that nonlinear terms (typically quadratic near $u = 0$) decay faster than linear terms. The higher the dimension, the more spreading and the more likely this reasoning may be justified. In fact, for high enough dimension we have global existence and spreading for practically any smooth nonlinear terms.

THEOREM 5. The zero solution is stable for the equation

$$u_{tt} - \Delta u + u + f(u, u_x, u_t, u_{xx}, u_{xt}) = 0 \tag{13}$$

where the subscript x denotes any spatial derivative and the Taylor expansion of $f(u,p,q,\ldots)$ at $(0,0,0,\ldots)$ has only quadratic and higher degree terms, provided $n \geq 5$. It is stable in the energy norm and asymptotically stable in the L^∞ norm as well as in certain Sobolev norms of higher order.

This remarkable theorem, which is the culmination of a long line of work beginning with that of Segal [21], is due to Klainerman [14], who introduced a Nash-Moser iteration technique and used L^∞ norms. Subsequently a much simpler proof using L^p norms was found by Shatah [23] and by Klainerman and Ponce [15]. The utility of L^p estimates for hyperbolic equations comes as a pleasant surprise.

The dimension condition in Theorem 5 is better understood if we allow f to act like a power $f = au^{\gamma} + bu_x^{\gamma} + \ldots$ near $u = 0$ where now γ can be any real number larger than one. Neglecting the complication caused by the lack of smoothness of f if γ is not an integer, the dimension condition is

$$\frac{n}{2}(\gamma-1) > \frac{\gamma}{\gamma-1} . \tag{14}$$

For $\gamma = 2$, this means $n > 4$ which is the condition in Theorem 5. Here $n/2$ is the L^{∞} decay rate due to the spreading of solutions with smooth, compactly supported initial data (that is, $|u(x,t)| \leq c\, t^{-n/2}$). Heuristically, the support expands with radius $O(t)$ and volume $O(t^n)$ while the L^2 norm is roughly constant.

If we specialize to a semi-linear f which does not depend on derivatives: $f(u) \sim a\, u^{\gamma}$, then the better condition

$$\frac{n}{2}(\gamma-1) > \frac{\gamma+1}{\gamma} \tag{15}$$

is known: see [28]. For $\gamma = 2$ this means $n > 3$. It is not known whether (14) and (15) are optimal. Probably (15) is optimal even when f depends on derivatives.

If the linearized equation is changed to the wave equation without the "mass term", then the L^{∞} decay rate should be only $(n-1)/2$ and conditions (14), (15) should be modified accordingly. In fact, we have

THEOREM 6. For the equation

$$u_{tt} - \Delta u \pm |u|^{\gamma} = 0 \tag{16}$$

the zero solution is stable if

$$\frac{n-1}{2}(\gamma-1) > \frac{\gamma+1}{\gamma} \tag{17}$$

Conversely if the minus sign is chosen in (16) and the inequality (17) is reversed, then essentially all solutions blow up in a finite time.

This theorem is due to F. John [11] in three dimensions. For $n = 3$, (17) becomes $\gamma^2-2\gamma-1 > 0$ or $\gamma > 1+\sqrt{2}$. The blow up of arbitrarily small solutions is a remarkable phenomenon which is possible only because of the absence of the mass term. Subsequent to John's paper and in view of condition (15) for the positive mass case, it was conjectured that (17) should be the correct generalization of John's exponent $1+\sqrt{2}$. Theorem 6 was then proved by Glassey [10] for $n = 2$. For $n > 4$ the blow up statement was proved by Sideris [25]. The stability statement in higher dimensions is not yet completed but is surely true. The three dimensional case is by far the easiest because of the positivity and the simple form of the Riemann function of the wave equation.

4. GLOBAL STABILITY.

Our goal is to find all the initial data which launch a spreading solution. We may recognize it as a solution whose L^p norm tends to zero as $t \to \infty$ for some $p > 2$. More naturally we may recognize it as a scattering state. We say that a solution u of (1) is a <u>scattering state</u> if it exists for all time as a solution of finite energy and there exists a pair $v_\pm$ of solutions of the linearized equation $v_{tt} - \Delta v + f'(0)v = 0$ (the free equation) such that

$$\|u(t) - v_\pm(t)\| \to 0 \quad \text{as} \quad t \to \pm\infty$$

where $\| \ \|$ is the energy norm. We begin with a theorem which is local but states that the set of scattering data fills out a whole neighborhood of zero in the energy norm and so is related to Theorem 5.

THEOREM 7. Consider equation (1) where $f(u) = u + O(|u|^\gamma)$ as $u \to 0$ where

$$1+4/n \le \gamma \le 1+4/(n-1) .$$

If u is a solution with initial data of sufficiently small energy norm, then u is a scattering state.

The lower bound $1+4/n$ is sharp. Indeed, consider a standing wave (10). Its local energy (energy in a fixed spatial ball) obviously does not tend to zero while the local energy of a free solution does. So a standing wave cannot be a scattering state. But if $f = u - |u|^{\gamma-1}u$ and $1 < \gamma < 1+4/n$, then the standing waves tend to zero in energy norm as $\omega \nearrow 1$.

Theorem 7 is proved by Strauss [28] by making use of L^p estimates of Strichartz [29] and Marshall, Strauss and Wainger [16] for the linearized equation.

The strongest conceivable stability theorem would state that <u>every</u> solution is a scattering state.

THEOREM 8. Let $f(u) = u + |u|^{\gamma-1}u$ with

$$1+4/n < \gamma < 1+4/(n-2)$$

a stricter upper bound if $n \geq 11$. Then every solution of (1) with initial data with sufficiently many integrable derivatives, but not necessary small, is a scattering state.

This was first proved by Strauss [26] for $f(u) = |u|^{\gamma-1}u$ and by Morawetz and Strauss [18] for $n = 3$, $3 \leq \gamma < 5$. Extensions were made by Pecher and, in this form, recently by Brenner [5]. Brenner has just announced the ultimate result, allowing arbitrary data of finite energy in Theorem 8 for certain γ and n .

5. PERIODIC SOLUTIONS.

Standing waves are periodic in time. Are there more general periodic solutions? Rabinowitz [20] has proved the following remarkable theorem in the one-dimensional case.

THEOREM 9. There exists a solution, not indentically zero, of the problem

$$u_{tt} - u_{xx} + f(u) = 0 \qquad \text{for} \quad 0 < x < \pi$$

$$u = 0 \quad \text{at} \quad x = 0, \pi$$

$$u(x,t+T) = u(x,t) \qquad \text{for all} \quad x \in (0,\pi) \quad \text{and all} \quad t$$

where T is given as a rational multiple of π , provided that f is a continuous nondecreasing function, $f(0) = 0$ and

$$u\, f(u) \geq (2+\varepsilon)F(u) \qquad \text{for} \quad u \quad \text{large and} \quad \varepsilon > 0 .$$

A simpler proof was subsequently given by Brezis, Coron and Nirenberg [6] using the "mountain pass lemma". The distinctive feature which permits periodic solutions to exist for such f is the boundedness of the domain. In fact we have the following theorem.

THEOREM 10. If f satisfies the condition

$$u\, f(u) > 2F(u) \qquad \text{for} \quad u \neq 0$$

and $\arg f(u) = \arg u$, then there does not exist any periodic solution u of equation (1) in $\mathbb{R}^n$ for $n \geq 1$ such that u has finite energy, $u\, f(u)$ and $F(u) \in L^1((0,T) \times \mathbb{R}^n)$, and $x\, f(u) \in L^1((0,T) ; L^2(\mathbb{R}^n))$.

This theorem was announced in A.M.S. Abstracts 3, August 1982, page 398 and we present a complete proof here. See also Coron [8] for the case $n = 1$.

Periodic solutions other than standing waves do exist for some f in unbounded space. For instance, there is the breather solution of the sine-Gordon equation.

PROOF OF THEOREM 10. First we give a formal explanation. Let

$$M = x\cdot\nabla + \frac{n}{2}I = \frac{1}{2}[x\cdot\nabla + \nabla\cdot x] .$$

Let $B = \nabla$ = gradient so that $B^* = -\nabla\cdot$ = -divergence. Then M is a skew-symmetric operator on $L^2(\mathbf{R}^n)$ and $-B^*B$ is the Laplacian Δ. Let $u = u(x,t)$ be a solution of (1) of period T. Then

$$(u_{tt},Mu) = (u_t,Mu)_t-(u_t,Mu_t) .$$

Upon integration over a period, the first term drops out because u is periodic and the second term because M is skew. Furthermore,

$$(-\Delta u,Mu) = (B^*Bu,Mu) = (Bu,BMu) = (Bu,[B,M]u) = (Bu,Bu)$$

because M is skew and $[B,M] = B$. Denote $\langle v,w\rangle = \int_0^T\int_{\mathbb{R}^n} vw\, dx\, dt$, $\langle v\rangle = \int_0^T\int_{\mathbf{R}^n} v\, dx\, dt$. Then

$$\langle u_{tt} - \Delta u + f(u), Mu\rangle =$$

$$\langle|\nabla u|^2\rangle + \frac{n}{2}\langle u\, f(u) - 2F(u)\rangle = 0 , \tag{18}$$

from which Theorem 10 obviously follows. We now give a precise proof under the stated conditions on u.

LEMMA 1. If u satisfies the $\underline{\text{linear}}$ equation $u_{tt} - \Delta u + f = 0$, has period T in time, has finite energy, and $x\, f \in L^1((0,T)\ ;\ L^2(\mathbb{R}^n))$ and $u\, f \in L^1((0,T)\ ;\ L^1(\mathbf{R}^n))$, then

$$0 = \langle|\nabla u|^2 + \frac{n}{2}\, u\, f + \nabla u\cdot xf\rangle \tag{19}$$

PROOF. If u has smooth initial data of compact support then an elementary computation gives

$$(u_{tt} - \Delta u)(Mu) = \partial_t\{u_t Mu\} + |\nabla u|^2$$
$$+ \nabla\cdot\{-\nabla u Mu + \frac{x}{2}|\nabla u|^2\} .$$

Upon integration we get (19). If we approximate the general u by nice solutions and pass to the limit, the lemma follows.

LEMMA 2. If $\nabla u \in L^2(R^n)$, u is approximable in $\|\nabla u\|_2$ norm by C_c^∞ functions, $x\, f(u) \in L^2(\mathbb{R}^n)$ and $F(u) \in L^1(\mathbb{R}^n)$, then

$$(\nabla u,\ x\, f(u)) = -n\int F(u)dx . \tag{20}$$

PROOF. (20) is true for $u \in C_c^\infty$ because

$$\nabla u\cdot x\, f(u) = \nabla\cdot(xF(u)) - nF(u) .$$

Integrate this identity over a ball $|x| < R$ to get

$$\int_{|x|<R} \nabla u\cdot x\, f(u)dx = R\int_{|x|=R} F(u)dx - n\int_{|x|<R} F(u)dx .$$

Apply this to a mollified u (say $u^\rho \to u$) and to a truncated f (say $f^\nu \to f$). Passing to the limit $u^\rho \to u$, we have $f^\nu(u^\rho) \to f^\nu(u)$ in L^2 so that

$$\int_{|x|<R} [\nabla u\cdot x\, f^\nu(u) + nF^\nu(u)]dx = R\int_{|x|=R} F^\nu(u)dx \tag{21}$$

Then remove the truncation. We have $F^\nu(u) \to F(u)$ a.e. and $x\, f^\nu(u) \to x\, f(u)$ in L^2, so that (21) is valid with the ν's removed. Since $F(u) \in L^1(R^n)$ we have $R_n\int_{|x|=R_n} |F(u)|dx \to 0$ for some $R_n \to \infty$. This implies (20).

Now (19) and (20) imply (18).

REMARK 1. If $n \geq 3$, the assumption on $x\, f(u)$ can be changed to $f(u) \in L^1((0,T)\ ;\ L^2(\mathbb{R}^n))$ because in that case we can use the multiplier

$$M = \frac{1}{2}[\frac{x}{|x|}\cdot\nabla + \nabla\cdot\frac{x}{|x|}]$$

which gives Morawetz's identity [17].

REMARK 2. A similar computation using the multiplier $x\cdot\nabla u - (n/2-1)u = (M-I)u$ gives the identity

$$0 = \langle |u_t|^2 \rangle + n\langle (\frac{1}{2} - \frac{1}{n}) u\, f(u) - F(u) \rangle \; . \tag{22}$$

Thus Theorem 10 is also valid with the alternative condition

$$(\frac{1}{2} - \frac{1}{n}) u\, f(u) > F(u) \qquad \text{for} \quad u \neq 0 \; . \tag{23}$$

Condition (23) generalizes the Derrick-Pohozaev condition (8) to the time-periodic case.

REMARK 3. For the Schrodinger equation

$$i\, u_t - \Delta u + f(u) = 0 \qquad (x \in \mathbb{R}^n) \tag{24}$$

the same identity (18) is valid because

$$\mathrm{Re}(i\, u_t, Mu) = \frac{1}{2}(i\, u, Mu)_t$$

since M is skew symmetric. So Theorem 10 is also valid for (24) where "finite energy" means u is continuous with values in $H^1(\mathbb{R}^n)$.

REFERENCES

1. D. Anderson, J. Math. Phys. 12 (1971), 945-952.
2. H. Berestycki and T. Cazenave, C. R. Acad. Sci. 293 (1981), 489-492.
3. H. Berestycki and P. L. Lions, Arch. Rat. Mech. Anal., to appear.
4. M. Berger, J. Funct. Anal. 9 (1972), 249-261.
5. P. Brenner, to appear.
6. H. Brezis, J. Coron and L. Nirenberg, Comm. Pure Appl. Math. 33 (1980), 667-684.
7. T. Cazenave and P. L. Lions, to appear.
8. J. Coron, C. R. Acad. Sci. 294 (1982), 127-129.
9. G. Derrick, J. Math. Phys. 5 (1964), 1252.
10. R. Glassey, Math. Zeit. 177 (1981), 323-340 and Math. Zeit. 178 (1981), 233-261.
11. F. John, Manuscr. Math. 18 (1979), 235-268.
12. C. Keller, J. Diff. Eqns., to appear.
13. C. Keller, Ph.D dissertation, Brown Univ., 1981.
14. S. Klainerman, Comm. Pure Appl. Math. 33 (1980), 43-101.
15. S. Klainerman and G. Ponce, to appear.
16. B. Marshall, W. Strauss and S. Wainger, J. Maths. Pures et Appl. 59 (1980), 417-440.
17. C. Morawetz, Proc. Roy. Soc. A306 (1968), 291-296.

18. C. Morawetz and W. Strauss, Comm. Pure Appl. Math. 25 (1972), 1-31.

19. S. Pohozaev, Sov. Math. Doklady 6 (1965), 1408-1411.

20. P. Rabinowitz, Comm. Pure Appl. Math. 31 (1978), 31-68.

21. I. Segal, Ann. Sci. Ecole Norm. Sup. (4) 1 (1968), 459-497.

22. J. Serrin, K. McLeod and L. Peletier, to appear.

23. J. Shatah, J. Diff. Eqns., to appear.

24. J. Shatah, to appear.

25. T. Sideris, to appear.

26. W. Strauss, J. Funct. Anal. 2 (1968), 409-457.

27. W. Strauss, Comm. Math. Phys. 55 (1967), 149-162.

28. W. Strauss, in Nonlinear Evolution Equations, ed. M. Crandall, Academic Press, 1978, 85-102; and J. Funct. Anal. 41 (1981), 110-133.

29. R. Strichartz, Duke Math. J. 44 (1977), 705-714.

30. J. L. Synge, Proc. Roy. Irish Acad. 62 (1961), 17-41.

Contemporary Mathematics
Volume 17, 1983

EQUILIBRIUM STATES OF AN ELASTIC CONDUCTOR IN A MAGNETIC FIELD

A PARADIGM OF BIFURCATION THEORY

by Peter Wolfe

SUMMARY

Let a nonlinearly elastic conducting wire have a natural length 1. We identify each material point of the wire by its coordinate $s \in [0,1]$. Let $\underset{\sim}{r}(s)$ represent the deformed position in Euclidean 3-space E^3 of the material point s. Let $\nu(s) = |\underset{\sim}{r}'(s)|$ $(' = \frac{d}{ds})$ and $\underset{\sim}{e}(s) = \frac{\underset{\sim}{r}'(s)}{(s)}$. The wire is assumed stretched between fixed supports located at $\underset{\sim}{0}$ and $b\underset{\sim}{k}$ where $b > 1$. (Here $\{\underset{\sim}{i},\underset{\sim}{j},\underset{\sim}{k}\}$ represents the standard orthonormal basis for E^3). The wire carries a current I. There is a constant magnetic field $\underset{\sim}{B} = B\underset{\sim}{k}$ present. The force on the wire is then given by

$$\underset{\sim}{f}(s) = I\underset{\sim}{r}' \times B\underset{\sim}{k} = I\nu\underset{\sim}{e} \times B\underset{\sim}{k}. \tag{1.1}$$

Let $\underset{\sim}{n}(s)$ denote the resultant contact force exerted by the material of $(s,1]$ on the material of $[0,s]$. Then the equation of equilibrium for the wire has the form

$$\underset{\sim}{n}'(s) + \underset{\sim}{f}(s) = \underset{\sim}{0}. \tag{1.2}$$

We assume the wire is perfectly flexible. This means that it has the property that

$$\underset{\sim}{n}(s) = N(s)\underset{\sim}{e}(s). \tag{1.3}$$

0271-4132/82/0000-1049/$01.50

Recall that $\underset{\sim}{e}(s)$ is the unit tangent to the wire at s. The function $N(s)$ is the tension at s. We assume that the wire is nonlinearly elastic so that the tension at s depends on the stretch $\nu(s)$. Specifically we assume that there is a continuously differentiable function $(0,\infty)\times[0,1] \ni (\nu,s) \to \hat{N}(\nu,s) \in R$ with

$$\hat{N}_\nu(\nu,s) > 0, \qquad \hat{N}(1,s) = 0 \tag{1.4}$$

$$\hat{N}(\nu,s) \to \infty \quad \text{as} \quad \nu \to \infty, \qquad \hat{N}(\nu,s) \to -\infty \quad \text{as} \quad \nu \to 0 \tag{1.5}$$

so that

$$N(s) = \hat{N}(\nu(s),s). \tag{1.6}$$

Our boundary value problem is thus to find functions $\underset{\sim}{r}(s)$ satisfying

$$(N(s)\underset{\sim}{e})' + I\nu\underset{\sim}{e} \;\; B\underset{\sim}{k} = \underset{\sim}{0}, \qquad 0 < s < 1 \tag{1.7}$$

$$\underset{\sim}{r}(0) = \underset{\sim}{0}, \qquad \underset{\sim}{r}(1) = b\underset{\sim}{k}, \qquad b > 1. \tag{1.8}$$

Equation (1.7) with boundary conditions (1.8) can be solved by elementary means. We let

$$\lambda = IB \tag{1.9}$$

and assume $\lambda > 0$. We first consider the case of the homogeneous wire, e.g., we assume the function $\hat{N}$ is independent of s. For all values of $\lambda > 0$ we have the "trivial" solution

$$\underset{\sim}{r}(s) = bs\underset{\sim}{k}. \tag{1.10}$$

In addition we have the following "non trivial" solutions: Suppose $N = \hat{N}(\nu)$. Let $\lambda > 0$ and $\nu > b$ satisfy

$$\frac{\nu\lambda}{N} = 2\pi m \tag{1.11}$$

for some positive integer m. Then the problem (1.7), (1.8) has solutions

$$\begin{aligned}(1.12)\qquad \underset{\sim}{r}_{\theta_0}(s) &= \frac{N}{\lambda}[\sin(\theta_0 + \frac{\nu\lambda}{N})s - \sin\theta_0]\sin\psi\underset{\sim}{i} \\ &+ \frac{N}{\lambda}[\cos\theta_0 - \cos(\theta_0 + \frac{\nu\lambda}{N})s]\sin\psi\underset{\sim}{j} \\ &+ \nu\cos\psi s\underset{\sim}{k}, \qquad 0 \le \theta_0 < 2\pi.\end{aligned}$$

Here

$$(1.13)\qquad \psi = \arccos\frac{b}{\nu}.$$

Thus we see that for each pair (ν,λ) satisfying (1.11) for some m we get a 1-parameter family of solutions parameterized by S^1. The reason for this multiplicity is that the problem (1.7), (1.8) is invariant under rotations about the z axis.

The set of solutions exhibits the classical bifurcation phenomenon as can be seen by linearizing (1.7) about the trivial solution (1.10). The bifurcation points are

$$(1.14)\qquad \lambda_m = 2\pi m\frac{\hat{N}(b)}{b}.$$

For a given λ the number of solution branches present depends on the behavior of the function $\hat{N}(\nu)/\nu$. If we assume

$$(1.15)\qquad \frac{d}{d\nu}\left(\frac{\hat{N}(\nu)}{\nu}\right) > 0$$

then we get a bifurcation picture reminiscent of the buckling of a beam.

The problem (1.7), (1.8) admits the (rotationally invariant) potential energy functional

$$(1.16)\qquad \Phi = \int_0^1 \{\Gamma(\nu,s) + \frac{\lambda}{2}(\underset{\sim}{r}\times\underset{\sim}{k})\cdot\underset{\sim}{r}'\}\,ds$$

with

$$\Gamma(\nu,s) = \int_1^{\nu} \hat{N}(\sigma,s)\, d\sigma. \tag{1.17}$$

On a solution branch (1.12) we have

$$\Phi = \Gamma(\nu) - \frac{1}{2}\nu\hat{N}(\nu) + \frac{1}{2}b^2\frac{\hat{N}(\nu)}{\nu}. \tag{1.18}$$

An easy calculation proves

THEOREM: Assume (1.15) holds. Then for a given $\lambda > 2\pi \frac{\hat{N}(b)}{b}$ among all solutions of (1.7), (1.8) the potential energy Φ is minimized at solutions lying on the first solution branch (i.e., the set of solutions (1.12) with $\frac{\nu}{N} = \frac{2\pi}{\lambda}$).

This theorem indicates that we may expect that the first branch of solutions is stable while the others are not.

We can also treat the case in which the function $\hat{N}$ depends explicitly on s as well as ν. The results are a generalization of those discussed above and the qualitative picture is exactly the same.

ABCDEFGHIJ–AMS–89876543